Digital Techniques for Wideband Receivers

Second Edition

For a complete listing of the *Artech House Radar Library,*
turn to the back of this book.

Digital Techniques for Wideband Receivers

Second Edition

James Tsui

Artech House
Boston • London
www.artechhouse.com

Library of Congress Cataloging-in-Publication Data
Tsui, James Bao-yen.
Digital techniques for wideband receivers / James Tsui. — 2nd ed.
p. cm. — (Artech House radar library)
Includes bibliographical references and index.
ISBN 1-58053-299-3 (alk. paper)
1. Broadband communication systems. 2. Signal processing—Digital techniques.
3. Wireless communication systems. I. Title. II. Series.
TK5103.4.T88 2001
621.382'2—dc21

2001018810

British Library Cataloguing in Publication Data
Tsui, James Bao-yen.
Digital techniques for wideband receivers. — 2nd ed. — (Artech House radar library)
1. Broadband communication systems 2. Signal processing—Digital techniques
I. Title
621.3'822

ISBN 1-58053-299-3

Cover design by Yekaterina Ratner

685 Canton Street
Norwood, MA 02062

International Standard Book Number: 1-58053-299-3
Library of Congress Catalog Card Number: 2001018810

10 9 8 7 6 5 4 3 2 1

To my mother and my wife;
In memory of my father and my parents-in-law

Contents

Preface xvii

Chapter 1 Introduction 1
1.1 Wideband Systems 1
1.2 Digital Approach 1
1.3 Obstacles in the EW Receiver Development 4
1.4 Organization of the Book 4
1.5 Specific Remarks 5
References 6

Chapter 2 Requirements and Characteristics of Electronic Warfare Receivers 7
2.1 Introduction 7
2.2 Introduction to Electronic Warfare 7
2.3 Difference Between Intercept and Communication Receivers 9
2.4 Signal Environment for EW Receivers 10
2.5 Requirements of EW Receivers 12
2.6 Parameters Measured by an EW Receiver 13
2.7 Frequency Information 13
2.8 AOA Information 17
2.9 Outputs of an EW Receiver 18
2.10 Overview of Analog EW Receivers 19
2.11 Instantaneous Frequency Measurement (IFM) Receivers 20
2.12 Channelized Receivers 21
2.13 Bragg Cell Receivers 22
2.14 Compressive (Microscan) Receivers 23
2.15 Digital Receivers 25
2.16 Characteristics and Performance of EW Receivers 26
2.16.1 Single Signal 27
2.16.2 Two Simultaneous Signals 28

2.17 Potential Trend in EW Receiver Development 29
2.17.1 Theoretical Problem Solutions 29
2.17.2 Queuing Receiver 29
2.17.3 Compressive Receiver Used for AOA Measurement 29
2.17.4 Channelized IFM Receiver 30
2.17.5 Digital EW Receiver 30
2.18 Electronic Warfare Processor 31
2.18.1 Deinterleaving 31
2.18.2 PRI Generation 34
2.18.3 Radar Identification 34
2.18.4 Tracking 34
2.18.5 Revisiting 34
2.19 EW Receiver Design Goals 35
2.20 Summary 36
References 36

Chapter 3 Fourier Transform and Convolution 39
3.1 Introduction 39
3.2 Fourier Transform 39
3.3 Impulse Function 42
3.4 Properties of Fourier Transform 43
3.4.1 Linearity 44
3.4.2 Even and Odd Functions 44
3.4.3 Duality 45
3.4.4 Scaling 47
3.4.5 Time Shift 48
3.4.6 Frequency Shift 49
3.4.7 Derivative 49
3.4.8 Integral 49
3.5 Fourier Series 50
3.6 Comb Function 51
3.7 Convolution 55
3.8 Parseval's Theorem 58
3.9 Examples 59
3.10 Summary 71
References 76

Chapter 4 Discrete Fourier Transform 77
4.1 Introduction 77
4.2 Signal Digitization 77
4.3 Graphical Description of Discrete Fourier Transform (DFT) 78
4.4 Analytic Approach to Discrete Fourier Transform 81
4.5 Properties of Discrete Fourier Transform 83
4.5.1 Limited Frequency Bandwidth 83
4.5.2 Unmatched Time Interval 84
4.5.3 Overlapping Aliasing Effect on Real Data 84

4.6	Window Functions	87
	4.6.1 Rectangular Window	89
	4.6.2 Gaussian Window	90
	4.6.3 Cosine Window Raised to the *α*th Order	91
	4.6.4 Generalized Hamming Window	92
4.7	Fast Fourier Transform (FFT)	93
4.8	Possible Advantages of DFT Over FFT in Receiver Applications	101
	4.8.1 Initial Data Accumulation	102
	4.8.2 Sliding DFT	102
4.9	Periodogram	104
4.10	Averaged Periodogram	105
References		108
Chapter 5	Fourier Transform–Related Operations	111
5.1	Introduction	111
5.2	Zero Padding	111
5.3	Periodic and Linear Convolutions	113
5.4	Peak Position Estimation for Rectangular Window	119
5.5	Peak Position Estimation for Hanning Window	123
5.6	Peak Position Estimation Through Iteration	124
5.7	Actual Frequency Determination by Fast Fourier Transform	126
5.8	Real Input Computed by a Complex FFT Operator	128
5.9	Autocorrelation	130
5.10	Autocorrelation (Blackman-Tukey) Spectrum Estimation	132
5.11	Application of FFT to Spectrum Estimation From Autocorrelation Function	134
5.12	Basic Idea of Sub-Nyquist Sampling Scheme	137
5.13	Phase Relation in a Sub-Nyquist Sampling System	140
5.14	Problems and Potential Solutions of Sub-Nyquist Sampling Scheme	143
5.15	Discrete Fourier Transform Through Decimation	146
5.16	Applications of Decimation Method to EW Receivers	148
5.17	Simplified Decimation Method	150
References		152
Chapter 6	Analog-to-Digital Converters	155
6.1	Introduction	155
6.2	ADC Through Folding Technique	155
6.3	ADC Through Sigma-Delta Modulation	157
6.4	Basic Sample and Hold Circuit	160
6.5	Basic ADC Performance and Input Bandwidth	162
6.6	Apparent Maximum and Minimum Signals to an ADC	163
6.7	Quantization Noise of an Ideal ADC	166
6.8	Noise Level Determined by Processing Bandwidth and Dithering Effect	168

6.9 Spurious Responses 169
6.10 Analysis on Spur Amplitudes 172
6.11 Further Discussion on Spur Amplitudes 176
6.12 Noise Effects in ADC 179
6.13 Sampling Window Jittering Effect 182
6.14 ADC Test Through Histogram 186
6.15 ADC Test Through Sine Curve Fitting 188
6.16 ADC Test Through FFT Operation 189
6.17 Requirements on ADC 209
References 212
Appendix 6.A 214
Appendix 6.B 216
Appendix 6.C 216

Chapter 7 Amplifier and Analog-to-Digital Converter Interface 219
7.1 Introduction 219
7.2 Key Component Selection 220
7.3 Notations 220
7.4 Comparison Sensitivity of Analog and Digital Receivers 222
7.5 Noise Figure and Third-Order Intercept Point 223
7.6 Characteristics of the Amplifiers in Cascade 225
7.7 Analog-to-Digital Converter 230
7.8 Noise Figure of Amplifier and ADC Combination 231
7.9 Amplifier and ADC Interface 232
7.10 The Meaning of M and M' 234
7.11 Computer Program and Results 235
7.12 Design Example 237
7.13 Experimental Results 238
7.13.1 Noise Figure Measurement 239
7.13.2 Dynamic Range Test 240
References 244
Appendix 7.A 245
Appendix 7.B 246

Chapter 8 Frequency Downconverters 249
8.1 Introduction 249
8.2 Baseband Receiver Frequency Selection 250
8.3 Frequency Conversion 250
8.4 In-Phase (I) and Quad-Phase (Q) Channel Conversion 253
8.5 Imbalance in I and Q Channels 255
8.6 Analog I and Q Downconverters 259
8.7 Digital Approach to Generate I and Q Channels 262
8.8 Hilbert Transform 262
8.9 Discrete Hilbert Transform 265

8.10	Examples on Discrete Hilbert Transform	268
8.11	Narrowband *I* and *Q* Channels Through a Special Sampling Scheme	271
8.12	Wideband *I* and *Q* Channels Through a Special Sampling Scheme	271
8.13	Hardware Considerations on Filter Design for Wideband Digital *I-Q* Channels	274
8.14	Digital Correction of *I* and *Q* Channel Imbalance	276
	References	279
	Appendix 8.A	280
Chapter 9	Sensitivity and Detection Problems	283
9.1	Introduction	283
9.2	Electronic Warfare Receiver Detection Approach	284
9.3	Potential Detection Advantages in a Digital EW Receiver	285
	9.3.1 Frequency Domain Detection	285
	9.3.2 Time Domain Detection	286
9.4	False Alarm Time and Probability of False Alarm for One Data Sample	287
9.5	Threshold Setting for One Data Sample	288
9.6	Probability of Detection for Single-Sample Detection	289
9.7	Detection Based on Multiple Data Samples	291
9.8	Detection Scheme for Multiple Samples (*L*-Out-of-*N*)	292
9.9	Probability Density Function and Characteristic Function	293
9.10	Probability Density Function of Sum Samples With a Square Law Detector	294
9.11	Detection of Multiple Samples Based on Summation	296
9.12	An Example of Single-Sample Detection	297
9.13	An Example of Multiple-Sample (*L*-Out-of-*N*) Detection	299
9.14	Selection of Threshold Level	302
9.15	Optimizing the Selection of Threshold	304
9.16	An Example of *N* Sample Detection (Summation Method)	306
9.17	Introduction to Frequency Domain Detection	308
9.18	A Suggested Approach to Frequency Domain Detection	309
9.19	Probability of False Alarm in Frequency Domain	311
9.20	Input Signal Conditions in Frequency Domain Detection	312
9.21	Probability of Detection in Frequency Domain	314
9.22	Examples on Frequency Domain Detection	317
9.23	Comments on Frequency Domain Detection	318
	References	319
	Appendix 9.A	320
	Appendix 9.B	321
	Appendix 9.C	322

Appendix 9.D 322
Appendix 9.E 323
Appendix 9.F 324
Appendix 9.G 324
Appendix 9.H 324
Appendix 9.I 325
Appendix 9.J 325
Appendix 9.K 326
Appendix 9.L 327
Appendix 9.M 327

Chapter 10 Phase Measurements and Zero Crossings 329
10.1 Introduction 329
10.2 Digital Phase Measurement 330
10.3 Angle Resolution and Quantization Levels 332
10.4 Comparison of Phase Measurement and FFT Results 333
10.5 Application of the Phase Measurement Scheme 334
10.6 Analysis of Two Simultaneous Signals 335
10.7 Frequency Measurement on Two Signals 338
10.8 Single-Frequency Measurement From Zero Crossing 340
10.9 Ill Condition in Zero Crossing for Single-Signal and Remedy 343
10.10 Simplified Zero Crossing Calculation for Single-Signal 344
10.11 Experimental Results From Single-Frequency Zero Crossing Methods 347
10.12 Application to Coherent Doppler Radar Frequency Measurement 349
10.13 Zero Crossing Used for General Frequency Determination 352
10.14 Basic Definition of the Zero Crossing Spectrum Analysis 353
10.15 Generating Real Zero Crossings 354
10.16 Calculating Coefficients for Zero Crossing Spectrum Analysis 356
10.17 Possible Configuration of Zero Crossing Spectrum Analyzer 359
References 360

Chapter 11 Frequency Channelization 363
11.1 Introduction 363
11.2 Filter Banks 364
11.3 FFT and Convolution Operations 365
11.4 Overlapping Input Data in the FFT Operations 366
11.5 Output Data Rate from FFT Operation 372
11.6 Decimation and Interpolation 374
11.7 Decimation and Interpolation Effects on the Discrete Fourier Transform 377
11.8 Filter Bank Design Methodology 378
11.9 Decimation in the Frequency Domain 380

11.10 Output Filter Shape from a Decimated FFT 384
11.11 Using Weighting Function to Widen the Output Filter 384
11.12 Changing Output Sampling Rate 387
11.13 Channelization through Polyphase Filter 388
11.14 Operation of the Polyphase Filter 390
11.15 Filter Design 390
References 395

Chapter 12 Monobit Receiver 397
12.1 Introduction 397
12.2 Original Concept of the Monobit Receiver 398
12.3 Monobit Receiver Idea 398
12.4 Design Criteria 399
12.5 Receiver Components 403
12.6 RF Chain, ADC, and Demultiplexer 403
12.7 Basic FFT Chip Design 408
12.8 Frequency Encoder Design 410
12.9 Selection of Thresholds 412
12.10 Preliminary Performance of a Monobit Receiver 413
12.11 Possible Improvements 417
12.12 Chip Layout 418
References 419

Chapter 13 Processing Methods After Frequency Channelization 421
13.1 Introduction 421
13.2 Basic Considerations of Channelized Approach 422
13.3 Filter Shape Selection 423
13.4 Analog Filters Followed by Phase Comparators 428
13.5 Monobit Receiver Followed by Phase Comparators 429
13.6 Digital Filters Followed by Phase Comparators 431
13.7 Analog Filters Followed by Monobit Receivers 434
13.8 Considerations of Digital Filters Followed by Monobit Receivers 440
13.9 Increase the Output Sampling Rate by Two 440
13.10 Digital Filters Followed by Monobit Receivers 443
13.11 Digital Filter Bank Followed by Monobit Receivers and Phase Comparators 446
13.12 Digital Filter Bank Followed by Another FFT 446
References 447

Chapter 14 High-Resolution Spectrum Estimation 449
14.1 Introduction 449
14.2 Autoregressive (AR) Method 450
14.3 Yule-Walker Equation 452
14.4 Levinson-Durbin Recursive Algorithm 455

14.5 Input Data Manipulations 457
14.5.1 Covariance Method 459
14.5.2 Autocorrelation Method 461
14.6 Backward Prediction and Modified Covariance Method 462
14.7 Burg Method 465
14.8 Order Selection 467
14.9 Prony's Method 470
14.10 Prony's Method Using the Least Squares Approach 475
14.11 Eigenvectors and Eigenvalues 477
14.12 MUSIC Method 479
14.13 ESPRIT Method 482
14.14 Minimum Norm Method 487
14.15 Minimum Norm Method With Discrete Fourier Transform 489
14.16 Adaptive Spectrum Estimation 491
References 497
Appendix 14.A 499
Appendix 14.B 500
Appendix 14.C 501
Appendix 14.D 501
Appendix 14.E 502
Appendix 14.F 504
Appendix 14.G 505
Appendix 14.H 506
Appendix 14.I 508

Chapter 15 Angle of Arrival Measurements 509
15.1 Introduction 509
15.2 Queuing Concept 510
15.3 Digital Data From a Linear Antenna Array 512
15.4 Outputs From a Circular Antenna Array 514
15.5 Two-Element Phase Array Antenna 517
15.6 AOA Measurement Through Zero Crossing 519
15.7 Phase Measurement in AOA Systems With Multiple Antennas 520
15.8 Fourier Transform Over Space Domain 521
15.9 Two-Dimensional Fourier Transform 524
15.10 Frequency Sorting Followed by AOA Measurements 527
15.11 Minimum Antenna Spacing 529
15.12 Chinese Remainder Theorem 531
15.13 Application of Chinese Remainder Theorem to AOA Measurements 532
15.14 Practical Considerations in Remainder Theorem 535
15.15 Hardware Considerations for Digital AOA Measurements 536
References 538

Chapter 16 Receiver Tests 539
16.1 Introduction 539
16.2 Types of Receiver Tests 540
16.3 Preliminary Considerations in Laboratory Receiver Tests 542
16.4 Receiver Tests Through Software Simulation 544
16.5 Laboratory Test Setup 545
16.6 Anechoic Chamber Test Setup 546
16.7 Preliminary Tests 547
16.8 Single-Signal Frequency Test 549
16.8.1 Frequency Accuracy Test 549
16.8.2 Frequency Precision Test 551
16.9 False Alarm Test 551
16.10 Sensitivity and Single-Signal Dynamic Range 552
16.11 Pulse Amplitude and Pulse Width Measurements 553
16.12 AOA Accuracy Test 554
16.13 TOA Test 555
16.14 Shadow Time, Throughput Rate, and Latency Tests 557
16.15 Two-Signal Frequency Resolution Test 558
16.16 Two-Signal Spurious Free Dynamic Range Test 558
16.17 Instantaneous Dynamic Range Test 559
16.18 Anechoic Chamber Test 560
16.19 AOA Resolution Test 560
16.20 Simulator Test 562
16.21 Field Test 563
References 564
Appendix 16.A 564
Appendix 16.B 565

Glossary 567
About the Author 569

Index 571

Preface

The purpose of this book is to introduce digital signal processing approaches that are potentially applicable to wideband receiver designs. The emphasis is on techniques rather than theoretical discussions. Since the final goal in receiver design is to process the input data in near real time, the calculation speed of these processing methods is of primary concern.

Digital signal processing has been widely applied in many technical areas. In the past, digital signal processing has been used after crystal detectors in electronic warfare (EW) receivers. The technology advances in analog-to-digital converters (ADCs) opens a new era in receiver design. The ADC can replace the crystal detector and keep valuable information that would otherwise be lost after detection.

The requirements of EW and communication used to be different. EW receivers require a very wide bandwidth and communication receivers require a relatively narrow bandwidth. However in recent years, the requirements in these two areas have become more closely aligned. The most significant requirements are wide input bandwidth coverage and high dynamic range in both types of systems. As a result, many techniques that were previously considered unique for EW receivers are now applicable to communication systems also.

This book is divided into 16 chapters. Chapter 1 is an introduction. Chapters 2 and 16 are the only chapters devoted to electronic warfare. Chapter 2 provides a very brief review of EW and Chapter 16 discusses the evaluation of EW receivers. Fourier transforms, discrete Fourier transforms, and some related subjects that are especially interesting for wide bandwidth receiver designs are included in Chapters 3, 4, and 5. Chapters 6 through 8 concern receiver hardware. Chapter 6 discusses ADCs and their impact on receiver performance. Chapter 7 shows the design of receiver front-ends with ADCs at the outputs. Chapter 8 discusses in-phase and quadrature-phase converter designs. Probability of false alarm and probability of detection are discussed in Chapter 9. Phase measurement and zero crossing methods to measuring frequency are discussed in Chapter 10. Chapters 11 through 14 discuss methods closely related to receiver designs. Chapter 11 discusses frequency channelization. Chapter 12 presents a simple design of an EW receiver. Chapter 13 presents some possible methods to process signals after channelization. High frequency resolution is discussed in Chapter 14. Angle measurements are discussed in Chapter 15.

This book is written at a senior or graduate school engineering level. It is written for researchers in the electronic warfare and communication areas. In order to help the readers to understand the subjects, many examples are included. Many computer programs are also included to further illustrate some of the ideas.

The author has very much appreciated valuable discussions with Dr. William McCormick and Dr. Arnab Shaw from Wright State University and Dr. Richard Sanderson from Air Force Research Laboratory. Special thanks to two engineers: Mr. David Sharpin from Air Force Research Laboratory and Mr. Timothy Fields from System Research Laboratory for their technical discussions. I am in debt to Mr. Rudy Shaw, Mr. Anthony White, Mr. Emil Martinsek, Mr. William Moore and Dr. Paul Hadorn from Air Force Research Laboratory for their encouragement. I would also like to thank many of my colleagues: Mr. Robert Davis, Mr. Joseph Caschera, Mrs. Debbie Ables, Mr. Nicholas Pequignot, Mr. James Hedge, Capt. Daniel Zahirniak, Lt. Christine Montgomery, Mr. Scott Rodrique, Mr. Keith Graves, Mr. John McCall, Mr. Joseph Tenbarge, Dr. Steve Schneider, Mr. David Jones, Ms. Darnetta Meeks, Mr. Vasu Chakravarthy, Mr. Keith Jones, Lt. Jamie Marciniec, Lt. George Dossot, Lt. Jason Shtrax, Cap. Daniel Richards, Mr. Ed Culpepper from Air Force Research Laboratory, and Mr. James McCartney, Mr. Jerry Bash, Mr. Mark Thompson, and Mr. Jeff Wagenbach from System Research Laboratory.

Last, but not least, I wish to thank my wife, Susan, for her encouragement and understanding of my spending lots of time on this book.

CHAPTER 1

Introduction

1.1 WIDEBAND SYSTEMS

This book discusses digital signal processing schemes that are potentially applicable to electronic warfare (EW) receivers. These receivers must have very wide instantaneous input bandwidth (about 1 GHz) to fulfill their operational requirements. This means any signal within the input bandwidth will be received all the time without tuning the receiver. On the contrary, a communication receiver has very narrow bandwidth. For example [1], television channels are allotted 6 MHz, frequency modulated (FM) radio channels are allotted about 200 kHz, and amplitude modulation (AM) stations are allotted only 10 kHz. If one turns on ten television sets simultaneously and each one is receiving a different channel, the instantaneous bandwidth of such an arrangement can be considered as 60 MHz (ten 6-MHz channels).

However, communication bandwidth is increasing because the wider the bandwidth, the more information per unit time can be transmitted from one point to another. Some anticipated communication systems would require 1 GHz of instantaneous bandwidth with 100 10-MHz channels [2]. If this trend continues, the difference between an EW receiver and a communication receiver will diminish. Further discussion with communication engineers reveals that many of the hardware considerations and digital signal processing approaches primarily designed for EW receivers are equally applicable to communication receivers. That is the reason for selecting the name of this book. The primary emphasis of this book would still be on EW receivers rather than on communication receivers. Most examples used in this book are from EW receivers.

1.2 DIGITAL APPROACH [3–9]

Many communication and control problems have been solved through digital approaches. Today there is little doubt that many engineering problems (i.e., communication and control) can best be solved through digital signal processing. Digital circuits have long been used in EW receiver applications, such as digital controlling of the receiver operation modes. In addition, once the radio frequency (RF) input signal is converted into video signals through crystal detectors, the video signals will be further processed digitally. However, to date all the EW receivers have crystal video detectors to convert RF into video signals. The detector destroys the carrier frequency and phase information in the signal. If the detector can be replaced by an analog-to-digital converter (ADC), all the information will be maintained. The wideband receiving system is in the developing stage.

The first device needed to convert analog signals into digital data is the ADC. In order to convert signals in a wideband receiver, the ADC must operate at a very high sampling speed. To digitize signals with less quantization errors, the ADC must also have a large number of bits. It is difficult to achieve both goals in an ADC at the same time, but the advancement in the ADC technology is moving at an astonishing speed. It is even difficult to make a meaningful assessment of the ADC technology because of its fast development speed. Figure 1.1 shows a survey on ADCs as of April 1993. This figure is based on information obtained from [3]. The x sign with ARPA indicates that the Advanced Research Projects Agency (ARPA) has a program to build a 100 MHz 12-bit ADC, while the x mark with WL indicates

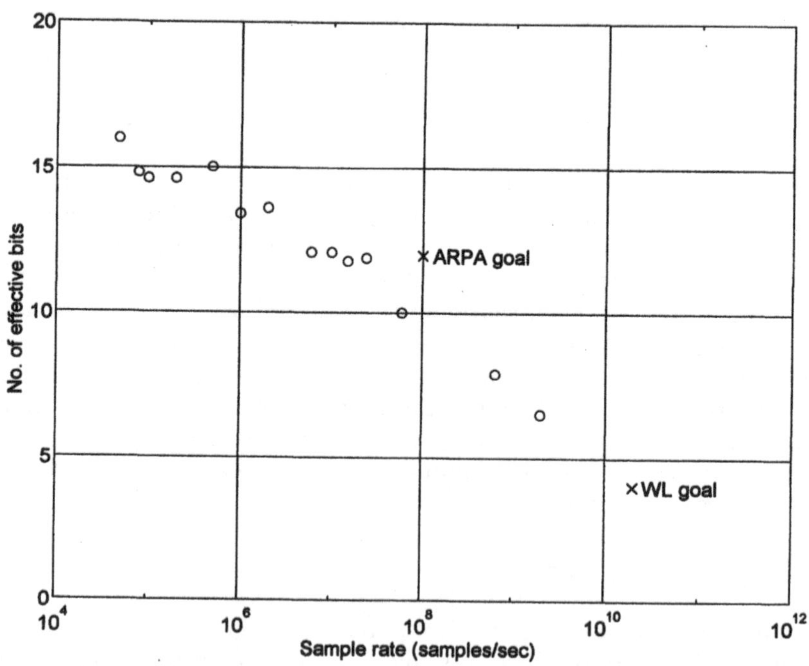

Figure 1.1 Performance of ADCs.

that the Wright Laboratory (WL) has a project for building 20 GHz 4-bit ADC. It is comforting to know that some of the existing ADCs operate near microwave frequencies (i.e., 6 GHz at 6 bits [4]).

High-speed ADC outputs must be processed by high-speed digital circuits. Otherwise, the digitized data will be lost and the system will operate in a nonreal-time mode. Current operating speeds of digital hardware do not yet match the speed of the state-of-art ADC. The speed of digital processors may never catch up with the speed of the ADC. However, in the future, digital processing might be fast enough to be used in a wideband receiver. Some types of system designs (i.e., parallel processing) may solve the speed problem. From present trends, one can realize the speed increasing in digital signal processing. Using the personal computer (PC) as an example, when the first PC was built in the late seventies, it operated at a few MHz, but today 66 MHz is a common clock speed.

Digital circuits applicable to wideband receiving systems must have higher operating speed than in a PC circuit. For example, a high-speed fast Fourier transform (FFT) chip operating at 250 MHz has been demonstrated [5]. Further development is expected to extend the speed to over 500 MHz. Figure 1.2 shows the speed of some common digital hardware up to 1993. It is anticipated that the clock speed will approach 1 GHz for specially designed hardware.

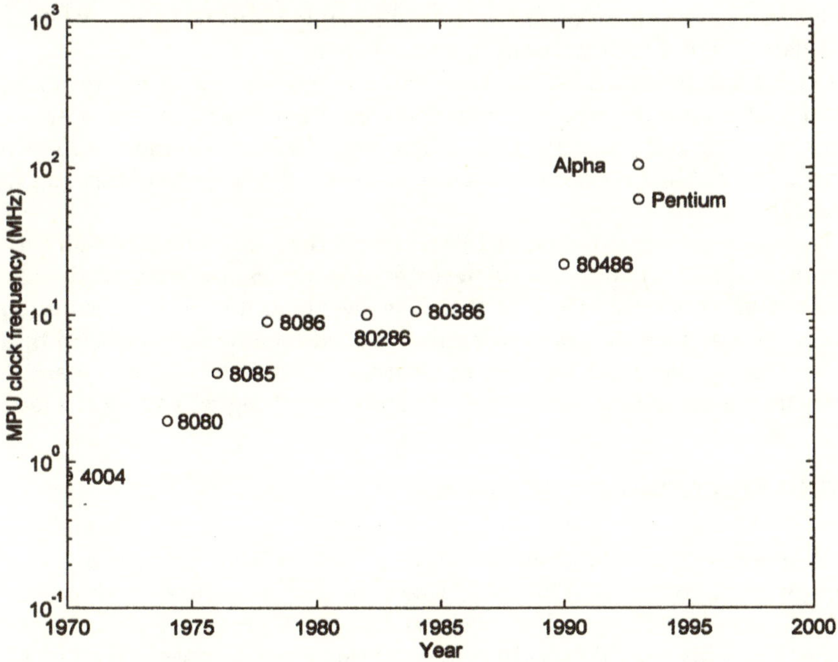

Figure 1.2 Performance of digital processors.

Due to the improvement in ADCs and digital circuits, it is anticipated that ADCs will replace the crystal detectors in an RF receiver in the future, thereby preserving phase information. In some narrowband communication systems, this approach already exists [7]. Furthermore, the ADC might move toward the front end of the receiver; that is, moving from the intermediate frequency (IF) toward the RF end. In the future, it might be possible to design receivers with only RF amplifiers and bandpass filters between the antenna and the ADC [8].

1.3 OBSTACLES IN THE EW RECEIVER DEVELOPMENT

Technically, it appears that there are at least two research and development obstacles in the field of EW receivers. First, scientists and engineers not directly working in the EW field have very little knowledge of it and are not aware of the requirements. The communication field is very much different from EW. In engineering, college curriculum communication is often a required course; therefore, most electrical engineers have some basic concept of communication systems. (It is interesting to notice that the word communication is used both in liberal arts and engineering schools with the same meaning but with different emphases. As a result, this word has a totally different interpretation. In liberal arts, it means improving one's interpersonal communication skills. In engineering, it means how to modulate and demodulate a signal for transmitting and receiving.)

On the other hand, the EW knowledge is only taught in a very few military academies or in short courses. Therefore, it is difficult to communicate with scientists and engineers outside this field to solicit new ideas. To remedy this problem, Chapter 2 of this book is devoted to a discussion on EW with emphasis on intercept systems.

The other problem is that in EW receivers there are no universally accepted evaluation standards. All kinds of performance values are used to describe EW receivers. Unfortunately, many of these receivers cannot even be considered as receivers because they do not provide the desired data format required by an EW receiver. This problem is addressed in Chapter 14. Chapters 2 and 14 are the two chapters in this book that deal with both analog and digital EW receivers.

1.4 ORGANIZATION OF THE BOOK

Many different subjects are discussed in this book. It is intended to arrange these subjects in a coherent way. The subjects are divided into the following groups.

Chapters 2 and 16 cover EW receivers. Chapter 2 provides a general discussion on EW with emphasis on EW receivers as mentioned in the previous section. Chapter 16 presents some measurement methods on EW receivers to obtain the values discussed in Chapter 2.

Fourier transform and related subjects are covered in Chapters 3, 4, and 5. Because Fourier transform is needed in discussing other subjects, this subject is discussed first. Some commonly used Fourier transform examples are collected in Chapter 3 as a quick reference. Readers familiar with Fourier transform may skip this chapter. Chapter 4 discusses discrete Fourier transform (DFT) and fast Fourier transform (FFT). Chapter 5 presents several subjects that are closely related to DFT.

Hardware used in wideband receivers is covered in Chapters 6, 7, and 8. Chapter 6 discusses ADCs and emphasizes the performance of the device rather than the design of it. Because the ADC is a nonlinear device, mathematical analysis is difficult and limited. A large number of simulations through computer runs are used to provide at least a rough idea of its performance. In Chapter 7, the RF front-end design of a receiver is considered. The design only concerns digital receivers that use ADCs to replace crystal detectors. It involves the tradeoff in terms of sensitivity and dynamic range. Chapter 8 discusses frequency downconverters. Both analog and digital approaches are included. Their imbalance impacts the receiver performance and this effect is predicted.

Sensitivity, frequency measurement, and receiver designs are covered in Chapters 9 through 13. Chapter 9 presents several signal detection schemes. The approaches include detection in both the time and frequency domains. Chapter 10 presents two simple schemes to improve frequency measurement accuracy. These methods have the potential to make the frequency resolution dependent on pulse width such that higher frequency resolution can be obtained on longer pulses. This is a very desirable feature in EW receivers. Chapter 11 presents an effective approach to channelize the frequency. The concept of decimation and multirate will be introduced. Chapter 12 presents a simple receiver design with slightly inferior performance. The receiver design has the potential to be fabricated on a single chip. Chapter 13 discusses some potential approaches of improving the frequency measurement after the channelization. These methods might be adopted in the encoder design, which is often considered as the most difficult function of a wideband receiver.

High-frequency resolution and angle of arrival (AOA) measurements are covered in Chapters 14 and 15. Chapter 14 presents several high-resolution parametric spectrum estimation methods developed in the last three decades or so. These methods usually provide finer frequency resolution than the FFT approach, but they are computationally intensive. Chapter 15 discusses AOA measurements. Some practical problems and suggested solutions are also included.

1.5 SPECIFIC REMARKS

In this book many computer programs are included to help readers understand the problems and design of a receiver. All these programs are written in MATLAB. Most of the high resolution schemes in Chapter 14 have a computer program associated with them.

Many of the figures presented in this book are produced using MATLAB. The time scale is often labeled as time sample because it represents samples of the time domain. The corresponding frequency plot is labeled as frequency bin because it is the spectrum calculated from the FFT operation on the time domain samples.

If one is not familiar with this technical area or does not start from the beginning of the book, it may be difficult to recognize some of the acronyms. Acronyms and abbreviations are useful to express technical terms. For convenience, all acronyms are spelled out the first time they appear in a chapter, even if it is a very common one. In addition, a list of all the acronyms is provided in the back of the book.

REFERENCES

[1] *Reference Data for Radio Engineers,* 5th edition, Indianapolis, IN: Howard W. Sams & Co. Inc., 1968.
[2] Budinger, J., et al. NASA Lewis, Cleveland, OH, Private communication.
[3] Walden, R. Hughes Research Laboratories, Malibu, CA, Private communication.
[4] Wang, K. C. Science Center, Rockwell International, Thousand Oaks, CA, Private communication.
[5] Spaanenburg, H. Honeywell Inc., Bloomington, MN, Private communication.
[6] Lemnios, Z. Advanced Research Projects Agency, Arlington, VA, Private communication.
[7] Lennen, G. R., and Daly, P. "A NAVSTAR GPS C/A Code Digital Receiver," *Navigation: Journal of the Institute of Navigation,* Vol. 36, No. 1, Spring 1989, pp. 115–126.
[8] Brown, A., and Wolt, B. "Digital L-BAnd Receiver Architecture With Direct RF Sampling," *IEEE Position Location and Navigation Symposium,* pp. 209–216, Las Vegas, April 1994.
[9] Sharpin, D. Wright Laboratory, Dayton, OH, Private communication.

CHAPTER 2

Requirements and Characteristics of Electronic Warfare Receivers

2.1 INTRODUCTION

One of the most difficult problems in inviting researchers to work on electronic warfare (EW) receivers is that the subject is not well understood, especially by academic scholars (perhaps due to lack of exposure to the problem). The main purpose of this chapter is to introduce the concept of an EW receiver. In order to provide a broader view, the subject of EW will be briefly discussed first. The signal environment and the requirements of the EW receiver will also be discussed. Since EW is basically a responsive action to a hostile electronic environment, the requirements will change with time. If the enemy creates some types of new threats, the EW engineers and systems must respond in a timely manner.

In this chapter, the information contained in a radar pulse will be presented first. Then some of the difficulties encountered in receiver research will be stated. A very simple discussion on analog and digital receivers will then be presented and, finally, the characteristics of EW receivers will be discussed. If several definitions of one term are available, only the ones that have a direct impact on EW receivers will be discussed. All the characteristics presented here are measurable quantities and the measurements will be presented in Chapter 14. Finally, the research trend in EW receivers will be discussed. Readers with EW background can skip this chapter.

2.2 INTRODUCTION TO ELECTRONIC WARFARE [1–6]

An EW system is used to protect military resources from enemy threats. The field of EW is recognized as having three components:

Electronic support measure (ESM), which collects information on the electronic environment.

Electronic countermeasures (ECM), which jam or disturb enemy systems.
Electronic counter-countermeasures (ECCM), which protect equipment against ECM.

Because it does not radiate electromagnetic (EM) energy, the first is often referred to as a passive EW system. The second is referred to as an active EW system, since it radiates EM energy. Because they do not emit EM energy, such techniques as stealth targets (that avoid being detected by enemy radars) and deployment of decoys or chaff (thin metallic wires) to confuse enemy radars are also considered as passive EW. ECCM is usually included in radar designs; hence, it will not be discussed here.

EW intercept systems can be divided into the following five categories:

1. Acoustic detection systems are used to detect enemy sonar and noise generated by ship movement. These systems detect acoustic signals and usually operate at frequencies below 30 kHz.
2. Communication intercept receivers are used to detect enemy communication signals. These systems usually operate below 2 GHz, although a higher operating frequency is required to intercept satellite communication. These receivers are designed to receive communication signals.
3. Radar intercept receivers are used to detect enemy radar signals. These systems usually operate in the range of 2 GHz to 18 GHz. However, some researchers intend to cover the entire 2 to 100 GHz range. These receivers are designed to receive pulsed signals.
4. Infrared intercept receivers are used to detect the plume of an attacking missile. These systems operate at near through far infrared (wavelengths from 3 to 15 μm).
5. Laser intercept receivers are used to detect laser signals, which are used to guide weapon systems (i.e., attack missiles).

The intercept receivers often operate with EW signal processors. The processors are used to process the information intercepted by the receivers to sort and identify enemy threats. After the threats are identified, the information is passed to an ECM system. The ECM system must determine the most effective way to disturb the enemy operation, which may include throwing out chaff. The actions of an ECM system against radars include noise and deceptive jamming. Noise jamming is intended to mask by noise the radar's return signals from targets so that the radar cannot detect any signal and its screen is covered with noise. Deceptive jamming creates false targets on the radar screen such that the radar will lose the true targets.

An EW system with all the different functions is shown in Figure 2.1. Except for the summary material in this chapter, this book is devoted to only a small part of the EW system, namely, the EW radar intercept (or just EW) receiver section that is used to intercept radar signals and convert them to digital pulse descriptor words (PDWs).

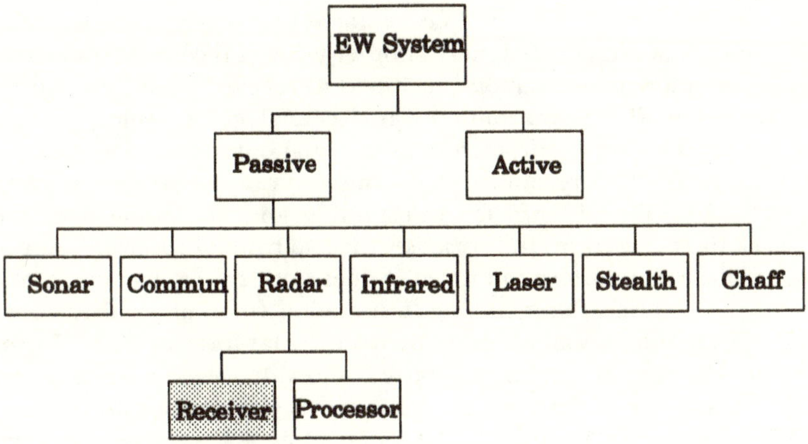

Figure 2.1 Different functions of an EW system.

In the past, the EW receiver has been used to demodulate radar signals, convert them into video pulses, and generate a tone in an earphone. An EW officer would listen to the tone and determine whether it was a threat radar. Under this type of operation, the EW operator works as the processor. This operation cannot satisfy modern requirements. In a modern EW system, in order to cope with the complexity of electronic environment, digital EW processors are used to identify threats. As a result, the EW receivers must generate digital outputs, used as the input of EW processors.

2.3 DIFFERENCE BETWEEN INTERCEPT AND COMMUNICATION RECEIVERS [7–13]

Most people have some exposure to communication receivers (i.e., a television set or a car radio). In designing these kinds of receivers, the frequency, types of modulation, and bandwidth of the incoming signal are known. Thus, the input signal can be considered as a cooperative type and the receiver can be designed very efficiently. A radar receiver that is part of the radar can be considered a communication receiver because the input signal is known. In an intercept (or EW) receiver, not only is the information of the input signal unknown, but the transmitting signal may be specifically designed to avoid detection by an intercept receiver.

The other major difference between an EW receiver and other types of receivers is that the outputs of an EW receiver are digital words describing the characteristic of every individual radar pulse intercepted. The receiver will generate a PDW that includes frequency, incident direction, pulse width, pulse amplitude, and time of arrival on each pulse. This unique characteristic of EW receivers sometimes causes misunderstanding by designers of other types of receivers. For example, the main purpose of a communication receiver is to recover the information emitted by the transmitter. If the transmitted signal is analog (i.e., voice, pictures), the receiver will produce voice and picture as the final output. In most cases, a digital output is not required. A recent trend in communication systems is to convert the analog signal into digital form for transmitting and processing and, finally, to convert back to analog form. In this sense, the two types of receivers are becoming alike, but still the communication receiver does not perform parameter encoding.

In an EW receiver, the incoming information (the radar pulses) are analog; however, the outputs are always presented in digital form as PDW. In most EW receivers, the RF signals are first converted into video signals by diode envelope video detectors. Specially designed encoders are used to convert these video signals into PDW, as shown in Figure 2.2. Although some EW receivers may have very impressive video outputs, matching computer-simulated results, these outputs do not guarantee generating satisfactory PDWs. Past experience indicates that most of the receiver design problems are in converting video signals into PDW. Thus, a good EW receiver must produce satisfactory PDWs as output.

2.4 SIGNAL ENVIRONMENT FOR EW RECEIVERS [1–3, 10, 12–19]

Since an EW receiver is used to intercept radar signals, the signal environment an EW receiver operates in will be presented here. Weapon radars are the primary interest of an EW receiver. In contrast to a communication signal, a weapon radar generates very simple waveforms. Most radars generate pulsed RF signals. Some radars generate frequency modulated (FM) pulsed signals, which are often referred to as chirp signals. The RF ranges roughly from 2 GHz to 100 GHz, but the most popular frequency range is from 2 to 18 GHz. The duration of these pulses may

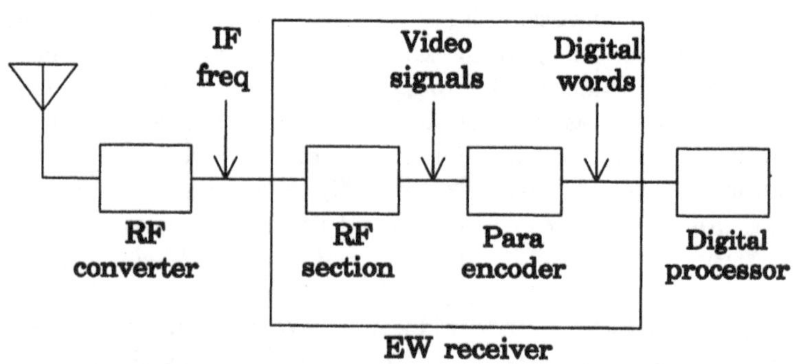

Figure 2.2 A conventional EW receiver.

be between tens of nanoseconds and hundreds of microseconds. Some radars produce continuous wave (CW) signals for low altitude surveillance or weapon guidance. The pulse repetition frequency (PRF), or its reciprocal pulse repetition interval (PRI), is an important parameter of pulsed radar signals. The PRF ranges roughly from a few hundred hertz to about one megahertz. Most of the radars have stable PRF, which means the PRF is a constant. Some radars have staggered PRI; that is, a group of pulses (i.e., a few to tens) repeat themselves at a certain PRF. Some radars even generate agile or random PRI; that is, the PRI varies from pulse to pulse. Agile PRI usually means that the PRI varies in a certain pattern and random PRI means the PRI does not have any predetermined pattern.

There are radars called low probability of intercept (LPI) radars. One of their main design goals is to avoid detection by enemy intercept receivers. These types of radars can either control their radiation power or generate wideband (spread spectrum) or frequency agile signals. A radar with power control capability only radiates enough power for target detection. If a detected target is getting closer to the radar, the radar can reduce its transmitter power. Its main goal is to provide just enough power to keep the detected target in range. This operation reduces the probability of being detected by an intercept receiver.

Some radars generate wideband signals to improve ranging resolution. The radar receiver can use matched filtering or signal processing to produce processing gain. It is difficult to detect a wideband signal with an intercept receiver because if the exact waveform is not known, one cannot produce the matched-filter gain. However, the spread spectrum generated by a radar is relatively simple in comparison with communication signals. Only three types of spread spectra are of concern to EW receiver designers. They are the FM (chirp) pulsed signal, pulsed biphase shift keying (BPSK), and polyphase coded signals. The pulse width of these types of signals can range from several ms to a few hundred ms. Against these spread spectrum signals, detecting the signal becomes the primary task. Once the signal is detected, one can perform identification. In a frequency hopping radar, the RF of the pulse changes from pulse to pulse. This kind of radar is usually not of concern to an intercept receiver because the receiver can intercept these pulses without much difficulty. However, this type of radar may cause problems to the EW signal processor following the receiver because it is difficult to deinterleave them into a pulse train.

A threat radar can obtain the necessary information and take action against an airplane or a ship in a few seconds. If a missile guidance signal is detected by an EW receiver, the impact could be imminent. Thus, an EW system must respond to the input signals as soon as possible. If an EW system cannot respond within some critical time, it is equivalent to no EW system at all because it cannot protect the aircraft or ship as desired.

During conventional EW operation, a jammer works almost continuously. When a jammer is working, it usually disturbs the operation of the EW information collection because the jammer is located close to the receiver. Its power might block the receiver from receiving. In actual operation, the jammer usually stops jamming temporarily for the receiver to collect information, such as to determine whether the signals being jammed are still in operation. This time duration is called the look-through time and has a 5% or less duty cycle as shown in Figure 2.3. The actual data collecting time is a few to tens of milliseconds.

In an electronic order battle (EOB), many different radars are present, even including friendly ones. Although the pulse density depends on the location of the intercept receiver and the scenario, it is usually assumed that a receiver will face a few million pulses per second. This signal environment determines the requirement of EW receivers.

2.5 REQUIREMENTS OF EW RECEIVERS [1–3, 13, 20–28]

From the discussion in the previous section, an EW receiver should fulfill the following requirements.

1. It requires near real response time. In general, after the receiver intercepts a pulse, the measured information (i.e., the PDW) must be passed to an EW processor within a few μs.
2. The input signal range (say, 2 to 18 GHz) is often divided into many subbands. The frequencies of these subbands will be converted to outputs with a common intermediate frequency (IF). An EW receiver will be time-shared among all the outputs. In order to cover the input frequency band in a rapid manner, the IF bandwidth must be wide. This means the instantaneous bandwidth of the EW receiver must be wide. Instantaneous bandwidth means all signals with enough energy in this bandwidth will be detected instantly. An optimum bandwidth of an EW receiver has not been obtained yet, because it depends on the input scenario, the capability of the digital EW processor following the receiver, and so forth. As a result, too many parameters need to be

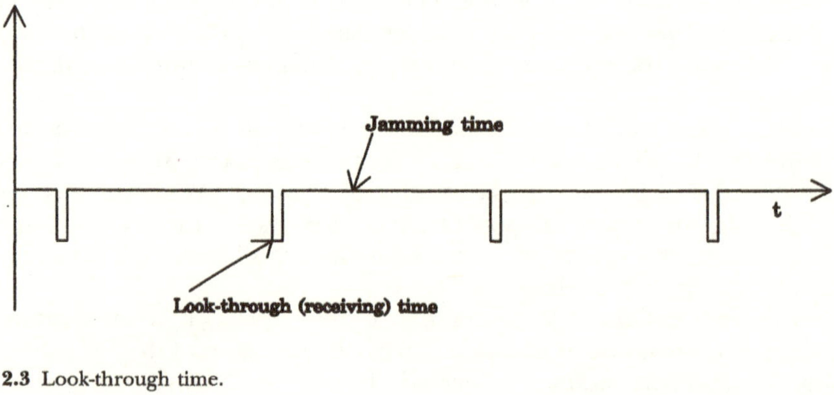

Figure 2.3 Look-through time.

optimized to obtain the shortest response time. The present way to determine the bandwidth is to make it as wide as technology allows. It usually ranges from about 0.5 to 4 GHz, and a receiver with a bandwidth of less than 500 MHz is usually considered unacceptable for EW applications. Of course, if one can make a narrowband (less than 500 MHz) receiver compact and low cost, theoretically one can use many such receivers in parallel to cover a wide bandwidth.

3. The receiver is required to process simultaneous signals. If more than one pulse arrives at the receiver at the same instant of time, the receiver should obtain the information on all the pulses. The maximum number of simultaneous pulses a receiver is required to process is often considered to be four.
4. A proper tradeoff of sensitivity and the dynamic range of an EW receiver must be achieved. Of course, high sensitivity is always desirable because one can detect a radar at a far distance, which provides more time to respond, or it can detect radars from their antenna sidelobes. Receivers with high dynamic range can receive simultaneous signals without generating spurious signals. In receiver design, these two parameters work against each other. Higher sensitivity almost always leads to lower dynamic range. Hence, a compromise between these two quantities must be carefully evaluated.

2.6 PARAMETERS MEASURED BY AN EW RECEIVER [29, 30]

An EW receiver must be able to obtain all the information from a pulse transmitted by a radar. Figure 2.4 illustrates a pulse transmitted by a radar. When the pulse reaches the intercept receiver at the aircraft, the following information can be measured: pulse amplitude (PA), pulse width (PW), time of arrival (TOA), carrier frequency (also referred to as the RF), and angle of arrival (AOA). In very limited cases, the electric polarization of the input signal is also measured. Pulse amplitude and pulse width measurements are self explanatory. In EW application, an input signal may be designated as CW when the pulse width is longer than a certain predetermined value (i.e., tens to a few hundreds μs). Time of arrival measurement assigns a specific time tag from an internal clock in the receiver to the leading edge of a received pulse. The TOA information is used to generate the PRF of a radar. The differences in EW receiver designs are largely based on the techniques to measure carrier frequencies of pulses. The AOA information is of primary importance, and it is also the most difficult one to obtain. In the following sections the discussion will concentrate on the frequency and AOA measurements.

2.7 FREQUENCY INFORMATION [31–42]

An intercept receiver measures only the center of the carrier frequency of a pulse. In general, the distribution of the spectrum is not needed. If the input is a chirp

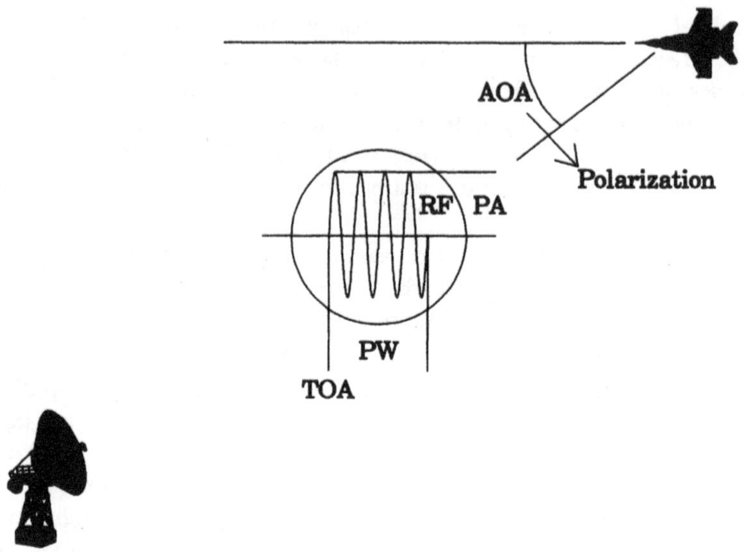

Figure 2.4 Parameters in a radar pulse.

signal, the information of interest is usually the starting and ending frequencies and the PW. The chirp is often assumed linear. The chirp rate R_c can be calculated by dividing the difference in frequency from the leading and trailing edges by the PW. Mathematically this can be expressed as

$$R_c = \frac{f_t - f_l}{PW} \tag{2.1}$$

where f_l and f_t are the frequencies at the leading and trailing edges, respectively.

In order to measure the frequency of a signal with variable frequency, theoretically, the concept of instantaneous frequency should be considered [29, 30]. In an EW receiver, these frequencies are measured as the average frequency over a short period (i.e., 100 ns at the leading and trailing edges). If the input is phase coded, the carrier frequency and the chip rate (the phase shift clock) are of interest. In comparison with a communication receiver, the desired information from a pulse is quite different.

It is desirable to obtain fine frequency resolution on the input signal because the resulting high accuracy permits the jammer to concentrate its power on the victim radar. However, in the conventional approach, a receiver is designed to intercept the signal with the minimum anticipated PW. A typical minimum PW can be considered as 100 ns. In order to design a bandpass filter to process this signal, the required bandwidth is approximately 10 MHz (or 1/(100 ns)). This bandwidth limits the frequency resolution to about 10 MHz. Various designs have been attempted to obtain accuracy of frequency measurement to a small fraction of the frequency resolution (e.g., by comparing outputs of adjacent filters to interpolate within the filter and signal bandwidth). Some of these approaches do provide limited success, but at the cost of causing other problems such as increased spurious signal detections.

In a receiver, the system noise bandwidth is usually set by the narrowest filter bandwidth in the RF chain. The noise floor may be defined as the effective input noise level of a system operating with an input temperature $T_o = 290$K. For example, the noise floor [N(dBm)] of a 10-MHz system is

$$N(\text{dBm}) = 10 \log (kTB) = -104 \text{ dBm} \tag{2.2}$$

where k is the Boltzmann's constant (= 1.38×10^{-20} mjoule/K), $B = 10^7$ Hz, and

$$P(\text{dBm}) = 10 \log [P(\text{mW})] \tag{2.3}$$

If the receiver has a noise figure of 15 dB and a threshold of 15 dB, the sensitivity is about –74 dBm (–104 + 15 + 15).

It is also desirable to build an intercept receiver that has adaptive frequency resolution (i.e., to make the frequency resolution PW dependent). For short pulses, the receiver can only generate coarse frequency resolution, whereas for long pulses, the receiver can generate fine frequency resolution. This concept can be extended to the sensitivity design of a receiver (i.e., the receiver would have moderate sensitivity for short pulses and higher sensitivity for longer pulses). The adaptivity requirement might be difficult to attain in an analog receiver, which is hardware oriented, but it should be easier to accomplish with software in a digital design.

Another problem concerning the frequency measurement is from a theoretical point of view. The question is what frequency accuracy can a receiver measure on two signals given frequency separation, PW, and signal-to-noise ratio (S/N). In most signal processing articles the Cramer-Rao bound is used as the maximum likelihood upper bound. The Cramer-Rao bound determines the minimum variance of the unbiased estimate. Two examples are shown here [38, 39]. In Figure 2.5(a), the $PW = 0.1$ μs, $S/N = 20$ dB, and the desired frequency measurement accuracy is 1 MHz. Under this condition, one should be able to measure two signals to 1-MHz accuracy when they fall on the right side of the curve. For example, if two frequencies are separated by 10 MHz and 18 dB in amplitude, one can measure both of them to 1-MHz accuracy. Figure 2.5(b) shows similar results for two 1-μs signals. Under this condition, if two signals are separated by 2 MHz in frequency and 50 dB in amplitude, one can measure both of them to 1-MHz accuracy.

In reality, there is no known EW receiver that can measure two simultaneous signals of 0.1-μs pulse width with 1-MHz resolution, no matter how far apart they are separated in frequency and how close they are in amplitude. The only receiver

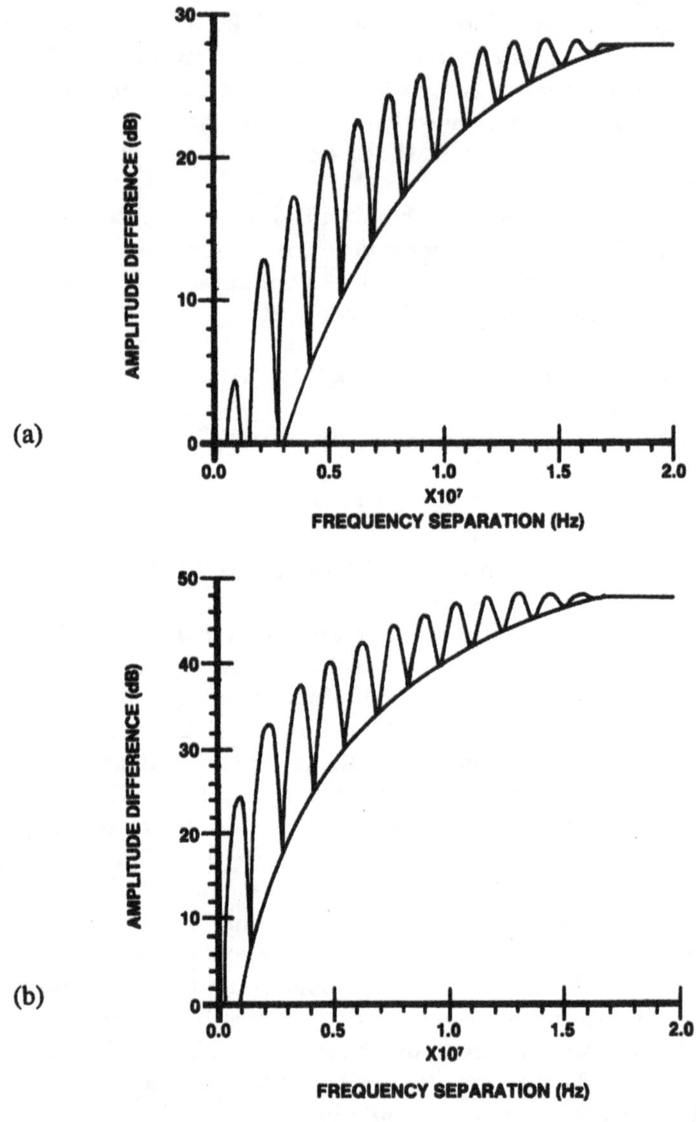

Figure 2.5 Cramer-Rao bound: (a) for 0.1 μs pulse, (b) for 1 μs pulse.

that can measure a 0.1-μs pulse with 1-MHz accuracy is the instantaneous frequency measurement (IFM) receiver, which cannot measure simultaneous signals at all. The IFM receiver will be briefly discussed in Section 2.11. Most of the EW receivers cannot detect two signals separated by more than 40 dB in amplitude, but according to Figure 2.5(b), 55 dB should be achievable. Both figures predict that when two signals are separated by a few megahertz with close amplitudes, their frequencies can be measured. However, most EW receivers are designed to begin measuring signals separated by at least 20 MHz.

From the above discussion, it is obvious that there is a large discrepancy between the theory and a practical EW receiver. Either the receivers are poorly designed or the Cramer-Rao bound is not suitable for EW applications. The latter might be true. Because the input signals of an EW receiver are unknown, one can not design an optimum receiver under this condition. Thus, when an EW receiver design is initialized, the desired performance is obtained on past experience rather than from a theoretical analysis. Definitely, this kind of approach is not very scientific. One open problem is to study whether there are other bounds besides the Cramer-Rao bound that apply to actual intercept receiver design.

2.8 AOA INFORMATION [43–46]

The AOA is a valuable parameter to be used in deinterleaving radar signals because a radar cannot change its position rapidly. Even an airborne radar cannot significantly change its position in the few milliseconds of the PRI time. As a result, the AOA measured by an intercept receiver on the radar is a relatively stable value. Unfortunately, the AOA parameter is also the most difficult one to measure. It requires a large number of antennas and receivers in addition to the necessary AOA measurement circuits. All these antennas and receivers must be matched, either in amplitude or in phase. Thus, the cost of such systems is usually very high.

A narrowband AOA system can be cost effective. For example, one can measure the frequency of an incoming pulse, then tune narrowband receivers connected to different antennas to that frequency to measure the AOA of the next incoming pulse. As mentioned previously, an EW receiver should measure the AOA information on a pulse-by-pulse basis; thus, this approach cannot satisfy this requirement. In addition, the receivers must be able to measure the AOA on simultaneous signals, and this requirement makes the design even more difficult.

Commonly used AOA measurement methods in intercept receivers are amplitude comparison and phase comparison methods. If the angle coverage of the two approaches are the same, they produce similar AOA accuracy. An amplitude comparison system can be easily designed to cover a wide angle, while a phase comparison system is used for narrow angle coverage. An airborne amplitude comparison scheme usually covers 360-deg azimuth and can produce an AOA accuracy of about ±15 deg. In this approach, the amplitude of every receiver must be matched from the antenna to the AOA measurement circuit. If multiple signal capability is required, even this approach can be very complicated.

The phase comparison system usually covers a much narrower angle range with approximately ±1-deg angle accuracy, which is desirable for modern EW applications. A phase measurement system requires all phase measurement channels to be phase matched. If the system must cover a wide instantaneous bandwidth and measure AOA on simultaneous signals, the phase among different channels must be matched, and it is definitely not a trivial task. If the phase cannot be matched among different channels, theoretically a calibration table can be used to remedy this shortcoming. However, if the phase is poorly matched, the calibration table can be very large and this requires lots of memory.

2.9 OUTPUTS OF AN EW RECEIVER

The output of an EW receiver is the PDW. Depending on its design, each receiver will have a unique PDW format. The PDW usually includes all five parameters discussed in Section 2.6, but each parameter may have a different number of bits. For example, a receiver that can only detect the existence of biphase shift keying (BPSK) and chirp signal may report the data shown in Table 2.1.

In this example, the word length is 75 bits. In some receivers the word length can be much longer, but usually less than 128 bits.

The range is an approximate value obtained from the resolution and total number of bits. For example, the 32-GHz frequency range is obtained from 2^{15} times 1 MHz. It does not represent the capability of the receiver. These PDWs are often passed to the digital processor through three 32-bit words. If there are two simultaneous signals, two PDWs will be generated by the receiver and the two PDWs will have the same TOA values. Thus, there is no need to flag simultaneous signals. They can be detected by the TOA readings in the PDWs.

The TOA may be reported in reverse order in some special cases. For example, a long pulse may arrive before a short one, but the trailing edge of the long pulse is after the short one, as shown in Figure 2.6. Under this condition, the measurement on the short pulse will be completed first and the corresponding PDW will be sent out to the EW processor. However, the TOA data of the short pulse is later than

Table 2.1
A Typical PDW Format

Parameters	*Range*	*No. of Bits*
Frequency	Up to 32 GHz	15 (1-MHz resolution)
Pulse amplitude	Up to 128 dB	7 (1-dB resolution)
Pulse width	Up to 204 μs	12 (0.05-μs resolution)
TOA	Up to 50 sec	30 (0.05-μs resolution)
AOA	360 deg	9 (1-deg resolution)
BPSK signal flag		1
Chirp signal flag		1
Total no. of bits		75

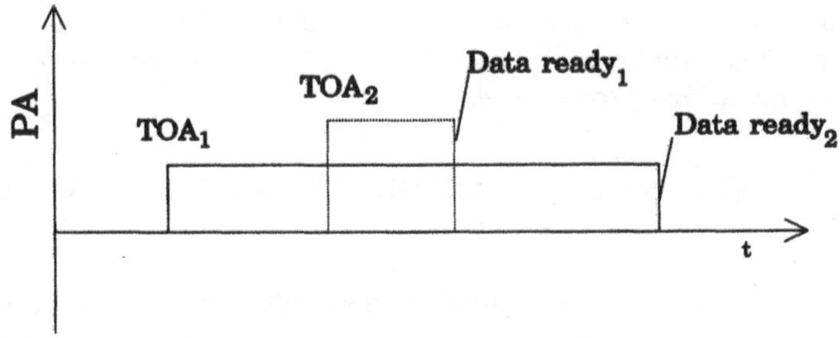

Figure 2.6 Condition TOA reported in reverse order.

the long pulse, which is reported at the end of the trailing edge of the long pulse. Under this condition, the first reported TOA corresponds to the short pulse whereas the second reported TOA is meant for the longer pulse. The EW processor must have the capability to process signals that arrive in this order.

2.10 OVERVIEW OF ANALOG EW RECEIVERS [47, 48]

Traditionally, EW receivers are classified into six categories by their structures. These are, crystal video, superheterodyne, IFM, channelized, compressive (microscan), and Bragg cell receivers. These receivers are referred to as analog receivers. The input signals are converted into video signals through crystal detectors. These video signals are further processed to generate the PDW, which includes all the desired parameters. The classification can be considered somewhat arbitrary. For example, a channelized receiver may use the superheterodyne technique and a digital receiver may use the channelization approach. The discussion of these types of EW receivers can be found in [13]. Crystal and superheterodyne receivers cannot process simultaneous signals; therefore, these two types of receivers will not be discussed. An IFM receiver cannot process simultaneous signal either, but the operation concept will be used in later chapters; thus, it will be included.

Channelized, compressive, and Bragg cell receivers can process simultaneous signals. In all these receivers a critically important topic is the parameter encoder, which is shown in Figure 2.2. Almost all receiver problems occur in the parameter encoder design. The front-end designs (i.e., the RF input to the video outputs) usually produce satisfactory results. Converting these video outputs into digital frequency data sometimes produces deficiencies, such as reporting erroneous frequency. These deficiencies often occur in those receivers that can process simultaneous signals.

In almost all the well-designed receivers, the parameter encoder and the RF section are designed as a single unit. The RF components (i.e., filter shape and delay line weighting) must generate the desired video signals to feed the parameter encoder. Many times an RF front end is constructed first, meaning the video signal is available, but it cannot be made into a functioning receiver because a satisfactory encoder design is difficult to achieve.

2.11 INSTANTANEOUS FREQUENCY MEASUREMENT (IFM) RECEIVERS [49–51]

An IFM receiver cannot process simultaneous signals; however, this receiver is very attractive in terms of instantaneous bandwidth, frequency measurement accuracy, size, weight, and cost. This is the type of receiver that can measure frequency accuracy to 1 MHz on a 0.1-μs pulse. The instantaneous input bandwidth can achieve 16 GHz (from 2 to 18 GHz). Since the IFM receiver has such a good performance, it will be discussed briefly here.

Basically, an IFM receiver uses the nonlinear property of crystal detectors to generate the autocorrelation of the input signal. The correlator (or frequency discriminator) is the heart of an IFM receiver. A basic configuration of a correlator is shown in Figure 2.7. The delay line with delay time τ in combination with the correlator generates the autocorrelation of the input signal with lag τ, which can be used to determine the input frequency. Theoretically, one can solve for multiple signals if the autocorrelations with many lags can be obtained. Therefore, one should be able to solve the simultaneous signal problem in the IFM receiver. Many attempts have been made to improve its capability to process simultaneous signals, though with only very limited success so far, due mainly to the following reason.

In an actual receiver, there are four crystal detectors in the correlator. The detectors have a dynamic range of about 15 dB. In order to increase the single frequency dynamic range of the receiver, a limiting amplifier is used in front of the correlator. The limiting amplifier is a nonlinear device. If there is only one

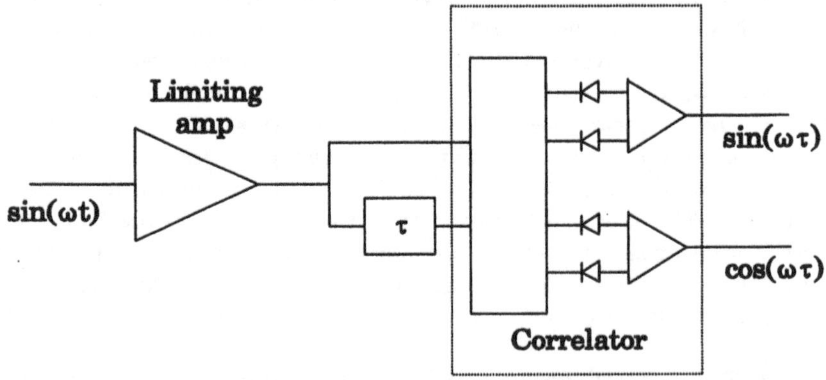

Figure 2.7 A basic IFM receiver.

signal, the strongest output from the nonlinear device is the true signal. This signal is measured by the receiver. If there are multiple inputs at the input of the limiting amplifier, the nonlinear effect cannot be neglected. As a result, the outputs of the correlator are no longer the desired autocorrelation for multiple signals. This is one of the main difficulties in solving the simultaneous signal problem in this type of receiver.

If the autocorrelations can be obtained, some of the high-frequency solution methods discussed in Chapter 14 can be used to solve for multiple signals. Even if these approaches are theoretically viable, they must be implemented in real time.

2.12 CHANNELIZED RECEIVERS [42–55]

The idea of a channelized receiver is very simple, and it uses a filter bank to sort signals with different frequencies. Amplifiers are used after the filter outputs to improve the receiver sensitivity. These amplifiers placed after the filter bank can improve sensitivity without affecting the dynamic range. Since at most one signal appears in one channel after the filter bank, intermodulation (often referred to as intermod) is not a problem. If two signals appear in one channel, this input condition is beyond the capability of the receiver and it may generate erroneous frequency information. Two types of amplifiers are often used: limiting amplifiers (or linear amplifiers used at saturation level) and log video amplifiers. Log video amplifiers can be used to measure the pulse amplitude information at the outputs of the filter bank. When limiting amplifiers are used, the amplitude information is lost, and therefore the pulse amplitude information must be measured somewhere else in the receiver.

To find the center frequency of the input signal, intuitively one will look for the filter with the highest output compared to its adjacent ones. As a result, amplitude comparators between adjacent channels are often used to determine the frequency of the input signal. This approach can successfully provide the correct frequency information if the required instantaneous dynamic range is low. If high instantaneous dynamic range is required, this approach often generates spurious responses. This deficiency can be explained with the help of Figure 2.8. In this figure, the spectrum of a square pulse is displayed. There is one mainlobe and many sidelobes. The energy difference between successive sidelobes that are close to the mainlobe are relatively significant. The energy differences between two successive sidelobes that are far away from the mainlobe are very small.

The amplitude comparison scheme works well when the filters A, B, and C are close to the mainlobe, as shown in Figure 2.8. In this case, the outputs are A < B > C. Because the outputs from these filters are far apart in amplitude, this condition can be easily detected and the correct frequency will be reported. On the other hand, the filters M, N, and O are far away from the mainlobe. The outputs for this case should be M > N > O. Under this condition, no frequency should be

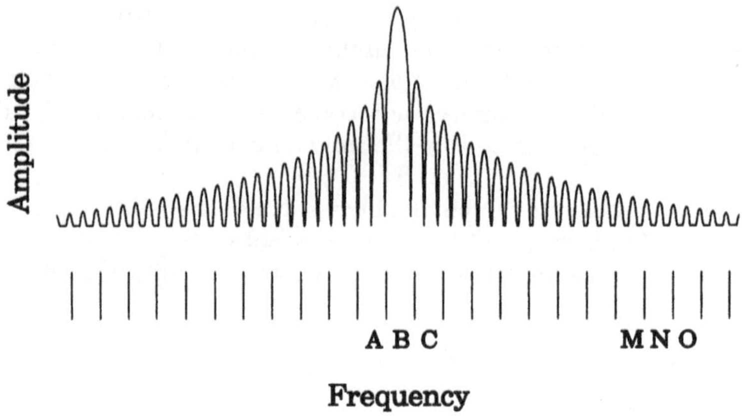

Figure 2.8 Spectrum display and filter bank.

reported. Since the amplitudes of these three filter outputs are very close, any gain imbalance may violate the above conditions. If the outputs are M < N > O, a false frequency report will be generated at output N. Balancing the gains among the channels appears to be an impossible task. When the required instantaneous dynamic range is low, outputs from filters far away from the mainlobe are neglected and spurious responses can be avoided.

In many channelized receiver designs, techniques to determine whether a signal is inside or outside of a certain filter are used. These approaches do not compare outputs from adjacent channels, but use the output from one single filter. The time domain response (the transient effect) of a signal passing through a filter is used to make the decision. Circuits following the filter are designed to measure the shape of the output. If the output shape meets certain criteria, the signal frequency is considered inside the filter; otherwise, it is outside the filter. In this type of design, the detection filter bandwidth is usually 1.5 times wider than the separation between filters to avoid channel boundary alignment problems. As a result, the frequency resolution is half the separation between filters. These types of approaches are among the most successful ones.

2.13 BRAGG CELL RECEIVERS [56–60]

A Bragg cell receiver uses an optical Bragg cell to perform frequency separation. The input RF signal is converted into an acoustic wave traveling in the Bragg cell, which diffracts a collimated laser beam. The position of the diffracted laser beam is a function of the input frequency. A photodetector array is used to convert the laser output into a video signal. In this arrangement, the input is RF and the outputs are channelized video signals. It is equivalent to the front end of a channelized receiver, including video detectors. The major advantage of a Bragg cell receiver ignoring the parameter encoder is its simplicity. A large number of channels (i.e., 100) can be accommodated in a very few components: a laser, a collimator, two optical lenses, a Bragg cell, and a photodetector array. This arrangement can be built very small.

The major disadvantage of a Bragg cell receiver is that the Bragg cell has optical outputs. With today's technology, it is difficult to place optical amplifiers between the Bragg cell and the photodetectors to improve sensitivity and not affect the dynamic range. Amplifiers can be added in front of the Bragg cell to improve the receiver sensitivity. The intermodulation generated in these amplifiers by simultaneous signals limits the instantaneous dynamic range. Because of the lack of low-cost optical amplifiers, the dynamic range of a Bragg cell receiver is usually low. This type of receiver is also referred to as the power Bragg cell receiver because the power of the laser output is measured, in contrast to the interferometric Bragg cell receiver discussed in the following paragraph.

The interferometric approach has been developed to improve the dynamic range of the Bragg receiver. In this approach, two Bragg cells are used: one as a reference cell and a second one as the signal cell. The laser beam is divided into two paths, each containing a Bragg cell. The two output beams beat against each other through a photodetector to produce an IF. This IF output is an electric signal; therefore, IF amplifiers can be added to improve the dynamic range. Theoretically, this approach may improve the dynamic range of the receiver. However, due to the limited power in the source laser and the difficulty in generating a proper reference signal, very limited improvement in dynamic range has been realized. The configuration of an interferometric Bragg cell receiver is very complicated. After the Bragg cells, each channel has a photodetector working as a mixer to generate the desired IF. This IF signal is lowpass (or bandpass) filtered, amplified, and converted into a video signal by a crystal video detector. Taking all the components into consideration, it is probably more complicated than a conventional channelized receiver that uses bandpass filters.

In conclusion, the optical portion of the power Bragg receiver can be considered as a combination of a filter bank followed by crystal detectors. In comparison with an RF channelizer, the Bragg cell approach is simpler, but the performance is also inferior.

The most common encoder design for a Bragg cell receiver is the amplitude comparison scheme against adjacent channels. As mentioned before, this approach usually has limited dynamic range. In order to improve the dynamic range of a Bragg cell receiver, not only the optical arrangement should be improved, but the frequency encoder should be studied as well.

2.14 COMPRESSIVE (MICROSCAN) RECEIVERS [51–63]

In a compressive receiver, the Fourier transform is performed on the input signal to convert signals with different frequencies into short pulses in the time domain.

A simple front end of a compressive receiver is shown in Figure 2.9. The input signal is converted into a chirp signal through a mixer fed by an FM local oscillator. The chirp signal is compressed into short pulses through a compressive (or dispersive delay) line. These short pulses pass through a log video amplifier and are converted into video signals. The video circuit must have very wide bandwidth to process the narrow video pulses. The time position of each short output pulse, relative to the initiation of the LO sweep, represents the frequency of the corresponding input signal.

A parameter encoder is required to convert these video pulses into the desired PDW. Since the video pulses come out in time sequence from one output port, when compared to a channelized receiver, less hardware is required in the parameter encoder. However, this hardware must operate at very high speed, equal to the bandwidth of the receiver. In most receivers, the input bandwidth equals the bandwidth of the dispersive delay line. Under this condition, if a receiver has an instantaneous bandwidth of 2 GHz, the logic circuit must also operate at 2 GHz. Each compressed pulse has a mainlobe with some sidelobes. The parameter encoder must be able to detect the mainlobe and neglect the sidelobes. Detection of sidelobes will produce spurious signal reports. In general, a pulsed signal can be intercepted in many consecutive scans. The information needed by an EW processor is on a pulse-by-pulse basis and not on a scan-by-scan basis. The parameter encoder must combine all the information generated by each scan within the same pulse and produce the PDW at the end of the pulse.

The most attractive feature of a compressive receiver is its potential to simplify AOA measurement. All the information on the input signal containing amplitude and phase is maintained at the compressed pulse before the log video amplifier.

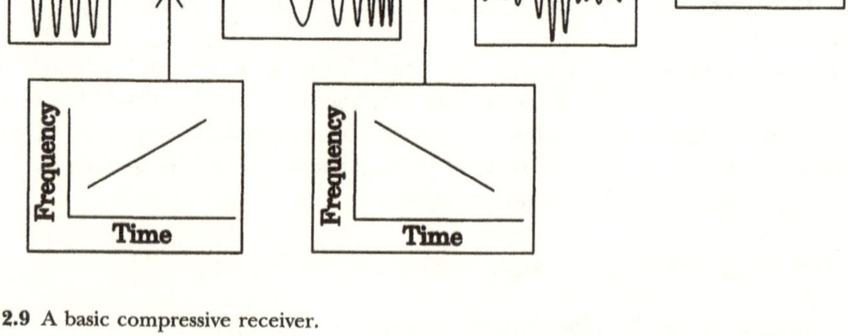

Figure 2.9 A basic compressive receiver.

Both amplitude comparison and phase interferometric approaches can be used to measure AOA. Since the information coming out of the receiver is in series, less hardware is required to measure the AOA. For example, assume that four antennas/receivers are required to measure the AOA through phase comparison and each receiver generates 100 frequency resolution cells. Four microscan receivers with four outputs and four phase comparison circuits are needed. If channelized receivers are used to achieve the same results, 400 channels and comparators would be required, which is impractical to build.

2.15 DIGITAL RECEIVERS

Digital receivers are the main subject of this book. Because of the advancements in analog-to-digital converters (ADC) and the increase in digital signal processing speed, present research has concentrated on digital EW receivers. In this type of receiver, the input is downconverted into an IF, which is then digitized with high-speed ADCs with large number of quantization levels. Digital signal processing is then used to produce the desired PDW.

A digital receiver does not have a crystal video detector. The output from the ADC is digital. Some of the major advantages are related to digital signal processing. Once a signal is digitized, the following processing will all be digital. Digital signal processing is more robust because there is no temperature drifting, gain variation, or dc level shifting as in analog circuits. Therefore, less calibration is required. The frequency resolution can be very fine if high-resolution spectrum estimation techniques can be applied. In many spectrum estimation schemes, the results are comparable with the Cramer-Rao bound at high signal-to-noise ratios, which analog receiver cannot achieve.

The two areas in a digital EW receiver that need to be investigated are increasing the input instantaneous bandwidth and real-time processing to produce the desired PDW. These requirements can be solved by increasing the ADC and digital signal processing speeds. The Nyquist sampling criterion limits the input bandwidth. In order to cover 1-GHz bandwidth for real data (contrast to complex data), the ADC must operate at least at 2 GHz. Due to vigorous research in ADC, the operating speed and number of bits are now increasing at a surprising rate. The allowable receiver bandwidth is directly proportional to the ADC sampling rate and the number of bits is directly related to the dynamic range.

The main problem in a digital EW receiver is to process the ADC output at a rate as high as 1 GHz at 8 bits. One possible approach is to multiplex the ADC output. If an ADC operates at 1,000 MHz and the fast Fourier transform (FFT) chip can only operate at 250 MHz, one can divide the output of the ADC into four parallel outputs feeding an FFT chip placed at each of the outputs. Another approach is to use conventional multirate digital filter designs. In this approach, the output of the ADC is also multiplexed and many parallel filters are used to sort the signals.

A brute force approach is to build many narrowband digital receivers. A number of these receivers are combined together to cover a wide instantaneous bandwidth. All the receiver outputs must be properly combined to determine the number of input signals and their center frequencies. In essence, this approach may have similar design criteria as the analog channelized receiver.

A digital EW receiver can be represented in functional blocks, as shown in Figure 2.10. This figure is similar to the analog receiver as shown in Figure 2.2. The output from the ADC is digital. These data are in the time domain and must be converted to frequency domain. In the frequency domain, the information is available as spectral lines or spectrum density. However, these outputs do not satisfy the EW requirements. The spectral lines must be converted into the carrier frequencies of the input signals. In order to emphasize this process, a parameter encoder is identified separately from the spectrum estimator. The parameter encoder converts the frequency information into the desired PDW.

Research on the digital receiver is at the beginning stage. It is necessary to provide solutions to some critical problems. Research should be concentrated on many areas, including the sensitivity and dynamic range of receivers as well as the nonlinear effect of the ADC.

2.16 CHARACTERISTICS AND PERFORMANCE OF EW RECEIVERS [64–66]

The most prominent problem in EW receiver development is the lack of a universally accepted performance standard. Making this matter even worse is the intentional and unintentional reporting of misleading results, which tend to confuse researchers in the field. The researchers may not be aware of where the deficiencies in the receivers lie, and as a result may not know where they should put their research resources. Sometimes the performance of an uncompleted receiver is reported.

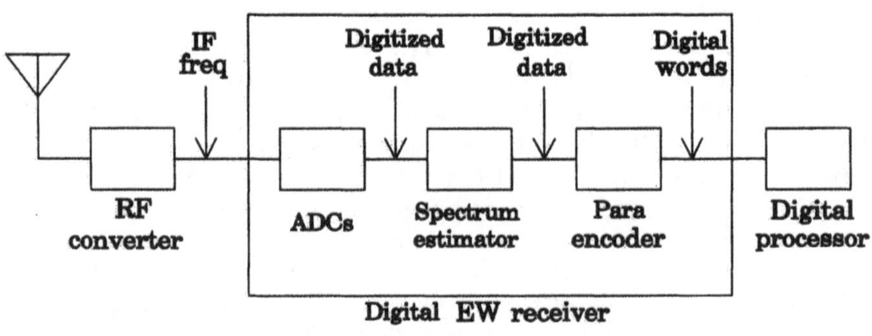

Figure 2.10 Function of a digital EW receiver.

For example, the receiver may not even produce a PDW in real time. In other words, the reported data may not be of a complete receiver, but from some video outputs or some types of displays. It is not even uncommon to have different performances reported on the same receiver. For example, the sensitivity of a wideband receiver can change from –55 to –65 dBM across the band. The optimistic engineer may report the best result, and the pessimistic engineer may report the worst one. Worse yet, some person may report –75 dBM by observing the video output. The correct way is to report the sensitivity as the minimum power level at which the receiver can repetitively generate the correct PDW. In addition, the sensitivity versus frequency or the maximum and minimum values should be reported.

For example, there are three different types of dynamic ranges, and they are the single signal, third-order intermod, and instantaneous dynamic ranges. All these dynamic ranges are important to EW receiver performance. A receiver may have 70-dB single-signal dynamic range, but only 20-dB instantaneous dynamic range. If one person reports the best value and another one reports the worst without clarification, one can imagine the potential confusion. Of course, the correct way is to report all three values.

In order to keep the discussion simple and exact, only the definitions that can be measured on receivers with PDW output will be presented here. These definitions can be applied to analog as well as digital receivers. The inputs are limited to one signal and two simultaneous signals, although some receivers can process more than two simultaneous signals. The measurements to obtain these defined values will be presented in Chapter 16. These characteristics are as follows.

2.16.1 Single Signal

1. *Frequency data resolution:* This is the finest increment in measured frequency data.
2. *Accuracy of frequency measurement:* This is the error between the measured frequency and the input frequency.
3. *Precision of frequency measurement:* This is the repeatability of the frequency measurement.
4. *False alarm rate:* This is the number of false alarms per unit time when there is no signal applied to the input of the receiver.
5. *Sensitivity:* This is the lowest signal power that can be properly detected and encoded by the receiver. Properly encoded means the measured parameter must be within a predetermined tolerance.
6. *Dynamic range (single signal):* This is the ratio of power of the strongest signal that the receiver can properly detect without generating spurious responses to the signal at sensitivity level.

7. *Pulse amplitude data resolution:* This is the finest increment in measured amplitude data. It is usually measured in decibels.
8. *Pulse width data resolution:* This is the finest increment in measured pulse width data. Pulse width is often measured in nonuniform scale. High pulse width data resolution is used to measure short pulse and low pulse width data resolution to measure long pulse.
9. *Angle of arrival data resolution:* This is the finest increment in measured AOA data.
10. *Time of arrival data resolution:* This is the finest increment in measured TOA data. Because the TOA is referenced to an internal clock in the receiver, it is impractical to compare the measured TOA against the incoming pulses. The common approach is to measure the TOA difference (ΔTOA).
11. *Throughput rate:* The throughput rate is the maximum number of pulses that can be processed by the receiver per unit time.
12. *Shadow time:* This is the minimum time between the trailing edge of one pulse and the leading edge of the next that permits the receiver to properly encode both of them. This quantity is usually PW-dependent and it is defined here at the minimum PW.
13. *Latency time:* This is the delay between the arrival time of the pulse at the receiver and the output of the digital word from the receiver.

2.16.2 Two Simultaneous Signals

In order to keep the discussion simple, the following definitions are applicable only for two input signals of same pulse width and coincident in time.

1. *Frequency Resolution:* This is the minimum frequency separation of two simultaneous signals with the same incident angle that permits the receiver to properly encode both of them.
2. *Spurious free dynamic range:* This is the power ratio of the strongest signal (one of two equal amplitude signals) that the receiver can properly encode without generating detectable third-order intermodulation to the power at the sensitivity level. When two strong signals at frequencies f_1 and f_2 arrive at the receiver, third-order intermod (short for intermodulation) will be generated. The third-order intermod is often measured with the two input signals kept at the same amplitude. The third-order intermod occurs at frequencies $2f_1 - f_2$ and $2f_2 - f_1$.
3. *Instantaneous dynamic range:* This is the power ratio of the maximum and minimum simultaneously received pulses that can be properly encoded by the receiver.

4. *Angle of arrival resolution:* This is the minimum angular separation between two sources received simultaneously at the same frequency that permits the receiver to properly encode both of them.

2.17 POTENTIAL TREND IN EW RECEIVER DEVELOPMENT [13]

It is really difficult to correctly assess the future trend in EW receiver development. For example, the Bragg cell receiver was first built as an electronic intelligent (ELINT) receiver around 1974. One unit was used in an airborne system to collect data and the results were very impressive. However, after many years of research, some key problems still do not have satisfactory solutions. The discussion in this section is based on the present needs and anticipated technology trend.

2.17.1 Theoretical Problem Solutions

The solutions of two theoretical problems are needed in EW receivers. They are the determination of the optimum bandwidth of an EW receiver and a theoretical bound for a receiver that can process two simultaneous signals. The first problem is very much system-oriented and a solution may not be reached easily from a purely theoretical point of view.

The second problem is to find the instantaneous dynamic range and the frequency resolution if the pulse width and signal-to-noise ratio are given. This question should be answered with the real-time processing in mind. This is a problem similar to the Cramer-Rao bound. The analog receiver performance is far from the Cramer-Rao bound because the receiver has not so far been designed as a maximum likelihood receiver. It is useful to find a different bound. It is possible that the bound is receiver design dependent. If this is the case, it is difficult to specify the requirements of a receiver without designing it first.

It is anticipated that future developments of EW receivers will be concentrated in four areas: queuing receiver, compressive, channelized IFM, and digital receivers.

2.17.2 Queuing Receiver

This receiver actually consists of two or more types of receivers: at least one coarse and one fine measurement receiver. The basic idea is to measure one parameter coarsely (i.e., frequency or AOA) and use this information to direct other receivers to further measure the information. There can be many different types of designs. This subject will be further discussed in Chapter 15.

2.17.3 Compressive Receiver Used for AOA Measurement

As mentioned before, less hardware is needed in a compressive receiver when it is used to measure AOA. The research is expected to concentrate on the phase comparison system because it can produce better AOA accuracy with narrower angle coverage in comparison with an amplitude comparison system. For this approach, the receivers must be phase-matched. One of the anticipated problems is to measure the phase difference in the compressive pulse. The compressive pulse is very short (e.g., a 1-GHz input bandwidth generates a compressive pulse of 1 ns). It is not an easy task to measure the phase difference in such a short time. One possible approach is to stretch the compressed pulse artificially to provide more time for the measurement. Because the width of the compressed pulse is related to frequency resolution, increasing the compressed pulse width artificially reduces the receiver's capability to separate signals close in frequency. For example, it is possible that a compressive receiver can measure the frequencies of two signals 20-MHz apart, but the AOA measurement circuit may not be able to measure them because the extended compressed pulse width degrades the frequency separation capability.

2.17.4 Channelized IFM Receiver

The concept of channelized IFM receiver is not new. An IFM receiver can be very small and measure frequency accurately on a short pulse, but it cannot process simultaneous signals. Placing a narrowband filter in front of the receiver to limit the probability of simultaneous signal occurrence seems to be an obvious solution. However, the narrowband filter will cause transients on the pulsed signal. This effect must be carefully studied. The bandwidth of the filters needs to be determined based on the total number of channels and the minimum pulse width anticipated. The major problem in this concept is still in the parameter encoders. One strong signal may be detected in several channels, but the receiver must report the one frequency correctly. On the other hand, two simultaneous signals reaching two adjacent channels must be reported as two signals with the correct frequency information. Usually, when two simultaneous signals reach the same channel, one may expect some probability of erroneous frequency data.

2.17.5 Digital EW Receiver

Because of the advancement in ADC technology, it is likely that ADCs can be used to build a receiver with wide instantaneous band (1 GHz) and reasonable dynamic range (50 dB) in the near future. Although some spectrum estimation schemes can generate very high frequency resolution, they are usually computation intensive and may not suitable for real-time application. In the near future, the fast Fourier transform (FFT) might be the most promising method because high-speed FFT chips are available. Another promising approach is to use decimation in multirate signal processing to build a digital channelized receiver. There is one possible advantage in digital channelization in comparison with analog approach in that all the channels may be better balanced. It may be noted that because the ADC response is frequency-dependent, even the digital channelizer may not have perfectly balanced outputs.

Even in a digital receiver, the parameter encoder will remain one of the most important components. This work has seldom been discussed in literature. After the frequency analysis (i.e., FFT), the carrier frequency of the input signal and its amplitude must be obtained. The frequency encoding scheme should avoid the sidelobes and recognize the mainlobes. It is inevitable in an EW receiver that a strong signal will saturate the ADC; this saturation problem and many other problems not found in a communication receiver must be investigated in the future.

2.18 ELECTRONIC WARFARE PROCESSOR

In this section, it is intended to introduce the very basic concept of an EW processor. An EW processor is expected to perform the following functions: deinterleaving, generating PRI data, identifying individual radar, tracking, and revisiting. Each of the functions will be discussed separately.

2.18.1 Deinterleaving

Let us use an example to demonstrate this operation. If there are three simple radars with constant PRI, the radars will emit three stable pulse trains as shown in Figure 2.11(a–c). When an EW receiver intercepts these pulse trains, the result can be represented as in Figure 2.11(d). One can imagine that in this figure the receiver can measure the total number of simultaneous signals. This result is from the interleaving of the three radars. From this figure, it is difficult to determine which pulse comes from which radar. If the EW receiver cannot determine the number of simultaneous signals, the result is shown in Figure 2.11(e). Under this condition, it is even more difficult to identify the pulse train because one does not know how many pulses are received at any instant.

An EW processor must deinterleave the intercepted pulse train into individual radar pulse train. In order to solve this problem each intercepted pulse will be compared to see whether they originate from the same radar. The common parameters used to perform this comparison are the center frequency (or RF), TOA difference (ΔTOA), and the AOA of the received pulses. If two pulses are very close in RF, they can be considered as from the same radar. In cases where a radar changes its RF on a pulse by pulse basis, it is difficult to compare the RF to sort out the pulse train. A similar argument can be applied to the TOA difference. For radars with frequency hopping and agile PRI capability, AOA is the most effective parameter to compare. From parameter comparison, the intercepted pulses can be deinterleaved into different radar trains.

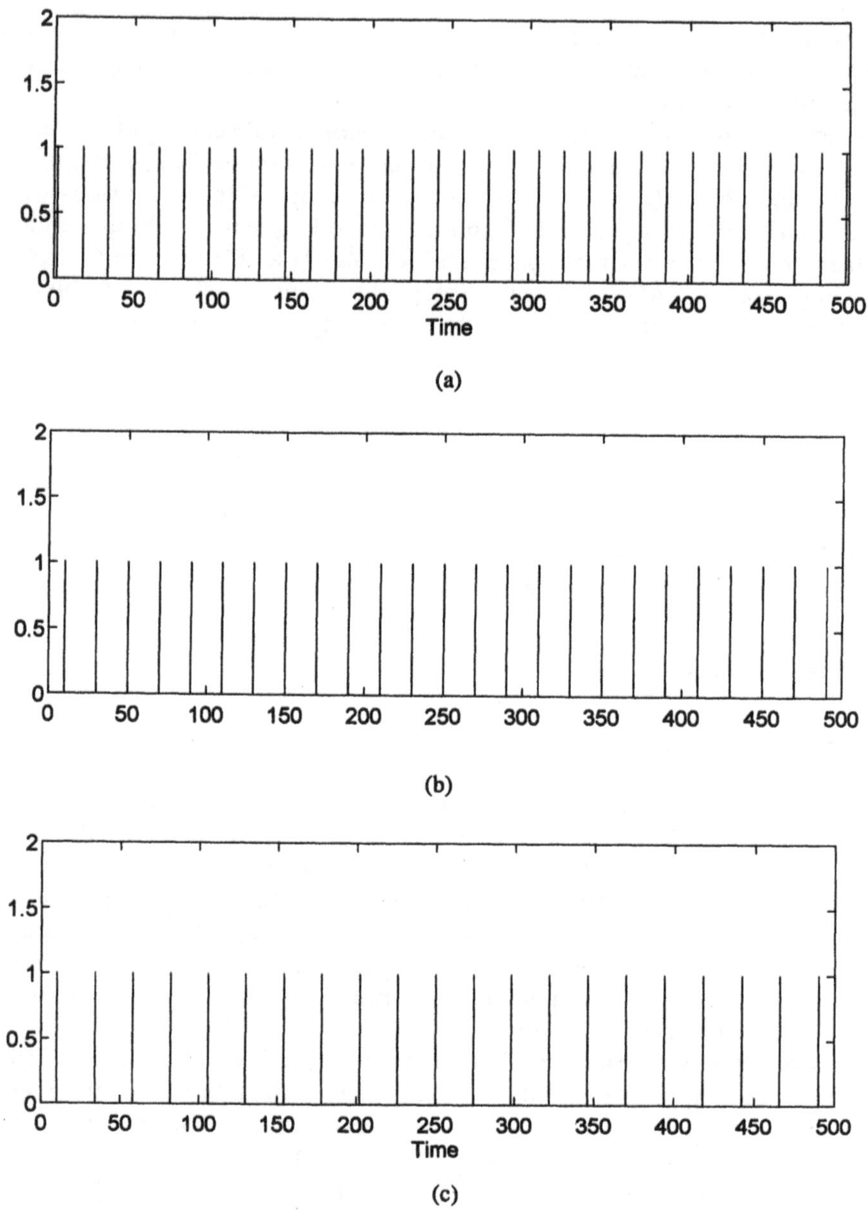

Figure 2.11 Pulse interleaving: (a) radar 1, (b) radar 2, (c) radar 3, (d) intercepted pulses with simultaneous signal identified, (e) intercepted pulses without simultaneous signal identified.

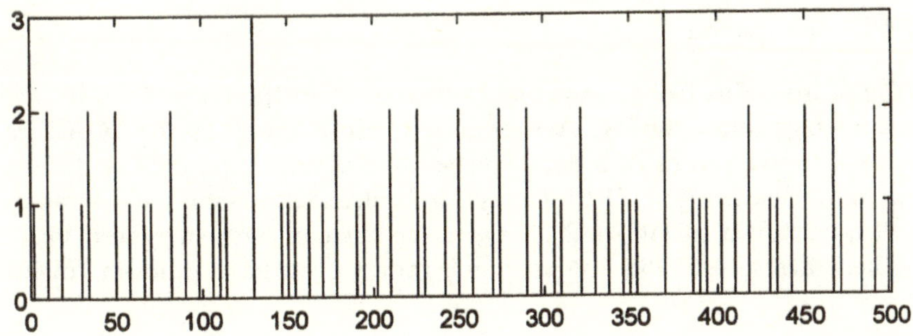

(d)

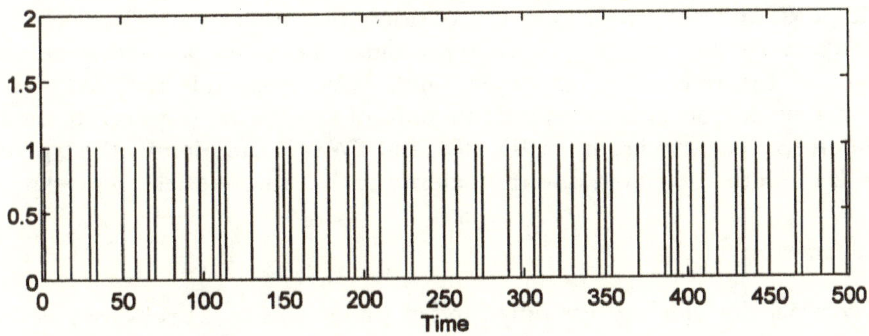

(e)

Figure 2.11 (continued).

Pulse amplitude measured by an EW receiver depends on the directions of the transmitting and receiving antenna; thus, it is not a dependable parameter. Multipath disturbs the PW measurement accuracy. Multipath means that one signal arrives at the receiver through different paths (e.g., reflection from a building). Thus, PA and PW are usually not used in pulse deinterleaving.

One can consider the deinterleaving as a two-dimensional pattern recognition problem. The RF and AOA are the parameters that are used to identify patterns. It is anticipated that new parameters may be generated (i.e., components from certain types of transforms). These new parameters might be easier to obtain than the AOA. Although research in this area has begun, no real-time system that can operate with EW receivers has been reported yet.

The deinterleaving is the major effort of an EW processor. It should be designed to have the maximum efficiency to perform this operation. In other words, the processor should be able to perform deinterleaving with a minimum number of received pulses.

2.18.2 PRI Generation

Once the pulse trains from individual radars are identified, the TOA information can be used to generate the PRI. As mentioned before, the PRI is the TOA difference between successive pulses of a deinterleaved pulse train. If the TOA difference is used in deinterleaving, the PRI information is already available. Some radars have stable PRI, as shown in Figure 2.11. Other radars can have staggered PRI, which means there are several PRI values. Some radars even have random (or jittered) PRI.

2.18.3 Radar Identification

The RF, PRI, and pulse width can be considered as the intrinsic characteristics of a radar because they are generated by the radar. They can be used to determine the type of the radar. On the other hand, pulse amplitude and AOA are not generated by the radar, but they are functions of the relative positions of the radar and intercept receiver. From the RF, PRI, and PW one can identify the type of the radar. If it is a thread radar, one can determine the jamming technique against it.

2.18.4 Tracking

An EW processor can process only limited pulse density. This density is usually lower than the pulse density an EW receiver can intercept. If an EW processor can process 100K pulses/sec and a receiver can intercept 1M pulses/sec, the received pulse density will choke the processor. However, once a radar pulse train is identified, it is no longer necessary to deinterleave on these pulses. Trackers are built to prevent pulses that are a continuation of an identified train from reaching the deinterleaving portion of the processor.

Trackers can be considered as two-dimensional filters that stop pulses from getting to the deinterleaving portion of the processor. One dimension is the PRI in the time domain and the other one is the RF in the frequency domain. It gates pulses in certain periods of time and within certain RF ranges. If the receiver cannot produce frequency information (i.e., a crystal video receiver), the tracker can only work in the time domain on the PRI of the pulse train. A tracker usually tracks one signal, but a processor can have many trackers. In some cases, trackers are considered as part of the active electronic countermeasure (ECM) because it is part of the technique generator. A technique generator provides the desired video pulses modulated by RF signals for jamming.

2.18.5 Revisiting

Once a pulse train is identified as a threat, a jammer may be turned on to jam the signal. At the same time the pulse train is tracked, the information will not reach the deinterleaving circuit. As a result, one does not know whether the signal is still transmitting (or being intercepted by the receiver). It is important to know whether the signal being jammed is still in operation. Otherwise, one may waste the energy on jamming a signal that is no longer in operation. In order to find out whether the signal is still being intercepted, the tracker will temporarily stop tracking and pass the information to the processor. This processing is often called revisiting. If the signal is still being intercepted, the parameters measured on the pulse will be presented in the processor and the tracker can continue to track the signal. If the signal is no longer being intercepted, the jammer and the tracker actions can be stopped.

2.19 EW RECEIVER DESIGN GOALS [67]

This section provides EW receiver design goals. Theoretical limits such as the Cramer-Rao bound discussed in Section 2.7 on the performance of a wideband receiver that can process simultaneous signals should be very useful. Such bounds, however, are not available for EW receivers. Therefore, the performance discussed in this section can be considered as a goal. Whether these goals can be achieved simultaneously is uncertain. Table 2.2 lists the goals.

The instantaneous bandwidth, the spatial coverage, and the number of simultaneous signals usually can be achieved if the receiver is so designed.

The data length used for signal processing usually matches the minimum pulse width. This choice can provide the best signal-to-noise ratio [67]. It also

Table 2.2
EW Receiver Performance Goals

Instantaneous bandwidth		>1GHz
Spatial field of view		4π solid angle
Simultaneous signal capability		up to 4
Minimum pulse width		100 ns[a]
Sensitivity		–65 ~ –90 dBm[b]
Dynamic range	Single signal spur free	75 dB[a]
	Instantaneous	50 dB[a]
	Two signal spur free	55 dB[a]
Two signal frequency resolution		20 MHz[a]
Parameter measurement precision	Radio frequency (RF)	1 MHz[a]
	Angle of arrival	1 deg[a]
	Time resolution	1 ns ~ 1 μs[a]
	Pulse amplitude	1 dB[a]
Real-time operation (latency time)		<2 μs[a]

[a]This can be considered a design goal.
[b]It is desirable to have higher sensitivity on longer pulse.

simplifies the parameter encoder design. Short data length processing is equivalent to a wideband filter, which may have simpler parameter encoder design but it will have low signal-to-noise ratio and low frequency data resolution. Long data processing is equivalent to narrowband filters. A narrowband filter will disturb a pulsed signal both in the carrier frequency and the pulse shape. As a result, the measurements on the carrier frequency, the pulse amplitude, and pulse width may not be accurate. In order to cover the same input bandwidth, many parallel narrowband channels are needed; therefore, it is difficult to build the parameter encoder. A short pulse can produce outputs from many filters. Under this condition, not only the carrier frequency is difficult to determine, even the number of signals is difficult to determine.

2.20 SUMMARY

Although research in EW receivers is an old topic, there are still many problems that need to be solved. In the past, most research has been concentrated on analog receivers. The main effort was to improve the performance of receivers that could process simultaneous signals such as the channelized, compressive, and Bragg cell receivers. Due to technical difficulties and funding limitations, very few receivers of the mentioned types have actually been built. As far as research is concerned, the EW area is still an open field. It is anticipated that many interesting problems will be investigated. Research and development in digital EW receivers will accelerate.

REFERENCES

[1] Boyd, J. A., Harris, D. B., King, D. D., and Welch, H. W., Jr., Editors. *Electronic Countermeasures*, Los Altos, CA: Peninsula Publishing, 1978.

[2] Fitts, R. E., Editor. *The Strategy of Electromagnetic Conflict*, Los Altos, CA: Peninsula Publishing, 1980.

[3] Price, A. "The History of U.S. Electronic Warfare," *The Association of Old Crows*, Alexandria, VA, 1984.

[4] Davies, C. L., and Hollands, P. "Automatic Processing for ESM," *IEEE Proc.*, Vol. 129, June 1982, pp. 164–171.

[5] Hovanessian, S. A. "Noise Jammers as Electronic Countermeasures," *Microwave Journal*, Sept. 1985, p. 113.

[6] Mardia, H. K. "New Techniques for the Deinterleaving of Repetitive Sequence," *IEE Proc.-F*, Vol. 136, pp. 149–154, Aug. 1989.

[7] Skolnik, M. I. *Introduction to Radar Systems*, New York, NY: McGraw Hill Book Co., 1962.

[8] Skolnik, M. I., Editor. *Radar Handbook*, New York, NY: McGraw Hill Book Co., 1970.

[9] Stremler, F. G. *Introduction to Communication Systems*, 2nd Edition, Reading, MA: Addison-Wesley Publishing Co., 1982.

[10] Wiley, R. G. *Electronic Intelligence: the Analysis of Radar Signals*, Norwood, MA: Artech House, 1982.

[11] Rohde, U. L., and Bucher, T. T. N. *Communications Receivers Principles and Design*, New York, NY: McGraw Hill Book Co., 1988.

[12] Ziemer, R. E., and Tranter, W. H. *Principles of Communications, Systems, Modulation and Noise*, 2nd Edition, Boston, MA: Houghton Mifflin Co., 1985.

[13] Tsui, J. B. Y. *Microwave Receivers With Electronic Warfare Applications*, New York, NY: John Wiley & Sons, 1986.

[14] Cook, C. E., and Bernfeld, M. *Radar Signals: an Introduction to Theory and Application*, New York, NY: Academic Press, 1967.

[15] Dixon, R. C. *Spread Spectrum Systems*, New York, NY: John Wiley & Sons, 1976.

[16] Barton, D. K. *Radar System Analysis*, Norwood, MA: Artech House, 1976.

[17] Stimson, G. W. *Introduction to Airborne Radar*, El Segundo, CA: Hughes Aircraft Co., 1983.

[18] Rogers, J. A. V. "ESM Processor System for High Pulse Density Radar Environments," *IEEE Proc.*, Vol. 132, Dec. 1985, pp. 621–625.

[19] Eaves, J. L., and Reedy, E. K., Editors. *Principles of Modern Radars*, New York, NY: Van Nostrand Reinhold, 1987.

[20] Rappolt, F., and Stone, N. "Receivers for Signal Acquisition," *Microwave Journal*, Jan. 1977, p. 29.

[21] Hoffmann, C. B., and Baron, A. R. "Wideband ESM Receiving Systems," *Microwave Journal*, Sept. 1980, p. 24.

[22] Cochrane, J. B., and Markl, F. A. "Broadband Building Blocks Shape Tomorrow's Warning Receivers," *Microwaves*, Nov. 1980, pp. 92–95.

[23] Hinshaw J., and Carpenter, J. A. "A Primer on Digital Output Detector Video Receiver Systems," *Microwave Journal*, Sept. 1984, p. 151.

[24] Moore, R. A., and Marinaccio, R. E. "Advancing EW System Strategies and Supporting Technologies," *Microwave Journal*, Feb. 1986, p. 26.

[25] Mardia, H. K. "New Techniques for the Deinterleaving of Repetitive Sequences," *IEEE Proc.*, Vol. 136, Dec. 1986, pp. 149–154.

[26] Watson, R. "Guidelines for Receiver Analysis," *Microwave & RF*, Dec. 1986, p. 113.

[27] Lochhead, D. L. "Receivers and Receiver Technology for EW Systems," *Microwave Journal*, Feb. 1986, p. 139.

[28] Steinbrecher, D. H. "Achieving Maximum Dynamic Range in a Modern Receiver," *Microwave Journal*, Sept. 1985, p. 129.

[29] Engelson, M. "Sharpen Pulse Signal Measurement Accuracy," *Microwave & RF*, Aug. 1986, p. 81.

[30] Ball, F. "Measure the Real Impulse Bandwidth," *Microwave & RF*, Jan. 1987, p. 93.

[31] Hovanessian, S. A. "Noise Jammers as Electronic Countermeasures," *Microwave Journal*, Sept. 1985, p. 113.

[32] Ackroyd, M. H. "Instantaneous and Time Varying Spectra-an Introduction," *J. Radio and Electron.*, Vol. 39, No. 3, 1970, pp. 145–152.

[33] Boashash, B. "Estimating and Interpreting the Instantaneous Frequency of a Signal-Part I: Fundamentals," *IEEE Proc.*, Vol. 80, April 1992, pp. 519–538.

[34] Boashash, B. "Estimating and Interpreting the Instantaneous Frequency of a Signal-Part 2: Algorithms and Applications," *IEEE Proc.*, Vol. 80, April 1992, pp. 539–568.

[35] Slepian, D. "Estimation of Signal Parameters in the Presence of Noise," *IRE Trans. Information Theory*, PGIT-3, March 1957, pp. 68–89.

[36] Rife, D. C., and Vincent, G. A. "Use of the Discrete Fourier Transform in the Measurement of Frequencies and Levels of Tones," *Bell System Technical Journal*, Vol. 49, Feb. 1970, pp. 197–228.

[37] Rife, D. C. "Digital Tone Parameter Estimation in the Presence of Gaussian Noise," Doctoral dissertation, Electrical Eng. Dept., Polytechnic Institute of Brooklyn, 1973.

[38] Rife, D. C., and Boorstyn, R. R. "Single-Tone Parameter Estimation From Discrete-Time Observations," *IEEE Trans. Information Theory*, Vol. IT-20, Sept. 1974, pp. 591–598.

[39] Rife, D. C., and Boorstyn, R. R. "Multiple Tone Parameter Estimation From Discrete-Time Observations," *Bell System Technical Journal*, Vol. 55, May 1976, pp. 1389–1410.

[40] Friedlander, B., and Porat, B. "A General Lower Bound for Parametric Spectrum Estimation," *IEEE Trans. Acoustic, Speech, Signal Processing*, Vol. ASSP-32, Aug. 1984, pp. 728–732.

[41] Tsui, J. B. Y., Thompson, M. H., and McCormick, W. "Theoretical Limit on Instantaneous Dynamic Range of EW Receivers," *Microwave Journal*, Vol. 30, Jan. 1987, p. 147.

[42] Tsui, J. B. Y. *Digital Microwave Receivers: Theory and Concepts*, Norwood, MA: Artech House, 1989.

[43] Earp, C. W., and Godfrey, R. M. "Radio Direction-Finding by the Cyclical Differential Measurement of Phase," *IEEE Proc.*, Vol. 94, March & April 1947, pp. 705–721.

[44] Chubb, E., Grindon, J. R., and Venters, D. C. "Omnidirectional Instantaneous Direction Finding System," *IEEE Trans. Aerospace Electronic Systems*, Vol. AES-3, March 1967, pp. 250–256.

[45] Baron, A. R., Davis, K. P., and Hofmann, C. P. "Passive Direction Finding and Signal Location," *Microwave Journal*, Sept. 1982, p. 59.

[46] Mosko, J. A. "An Introduction to Wideband, Two-Channel Direction Finding Systems," *Microwave Journal*, Feb. 1984, p. 91.

[47] Harper, T. "Hybridization of Competitive Receivers," *Watkins-Johnson Co. Tech-notes*, Vol. 7, No. 1, Jan./Feb. 1980.

[48] Tsui, J. B. Y. "An Introduction to EW Microwave Receivers," *Journal of Electronic Defense*, Vol. 12, Dec. 1989, p. 39.

[49] Cumming, R. C., and Myers, G. A. "Performance of Receivers and Signal Analyzers Using Broadband Frequency Sensitive Devices," *Technical Report*, No. 1905-1, Stanford Electronic Laboratory, SU-SEL-66-125 Stanford University, March 1976.

[50] East, P. W. "Design Techniques and Performance of Digital IFM," *IEEE Proc.*, Vol. 129, 1982, pp. 154–163.

[51] Bowler, B. L. "Tradeoffs in Digital IFM Receiver Design," Published M1820-38, Anaren Microwave, Inc., presented at Joint DADC/Dmpire AOC Technical Seminar, Griffiss AFB., Nov. 4, 1982.

[52] Harper, T., "New Trends in EW Receivers," *Countermeasures*, Dec./Jan. 1976, p. 34.

[53] "The Channelized Receiving Systems," Staff Report, *Microwave System News*, Vol. 6, Dec. 1975/Jan. 1976, p. 63.

[54] Hennessy, P., and Quick, J. D. "The Channelized Receiver Comes of Age," *Microwave System News*, Vol. 9, July 1979, p. 36.

[55] Anderson, G. W., Webb, D. C., Spezio, A. E., and Lee, J. N. "Advanced Channelization Technology for RF Microwave and Millimeterwave Applications," *IEEE Proc.*, Vol. 79, March 1991, pp. 355–388.

[56] Goodman, J. W. *Introduction to Fourier Optics*, New York, NY: McGraw Hill Book Co., 1968.

[57] Chang, I. C. "Acousto-Optic Devices and Applications," *IEEE Trans. Sonics Ultrasonics*, Vol. SU-23, 1976, pp. 2–22.

[58] Lugt, A. V. "Interferometric Spectrum Analyzer," *Applied Optics*, Vol. 20, Aug. 15, 1981, pp. 2770–2779.

[59] Wilby, W. A., and Gatenby, P. V. "Theoretical Study of the Interferometric Bragg-Cell Spectrum Analyser," *IEEE Proc.*, Vol. 133, Feb. 1986, pp. 47–59.

[60] Goutzoulis, A. P., and Abramovitz, I. J. "Digital Electronics Meets its Match," *IEEE Spectrum*, Aug. 1988, pp. 21–25.

[61] White, W. D. "Signal Translation Apparatus Utilizing Dispersive Network and the Like, for Panoramic Reception, Amplitude-Controlling Frequency Response, Signal Frequency Gating, Frequency Time Domain Conversion, etc.," U.S. Patent 2,954,465, Sept. 27, 1960.

[62] Kincheloe, W. R. "The Measurement of Frequency With Scanning Spectrum Analyzers," Technical Report 557-2, Systems Techniques Laboratory, Stanford Electronic Labs. Stanford University, Oct. 1962.

[63] Daniels, W. D., Churchman, M., Kyle, R., and Skudera, W. "Compressive Receiver Technology," *Microwave Journal*, April 1986, p. 175.

[64] Watson, R. "Guidelines for Receiver Analysis," *Microwave & RF*, Dec. 1986, p. 113.

[65] Watson, R. "Use One Figure of Merit to Compare All Receivers," *Microwave & RF*, Jan. 1987, p. 99.

[66] Tsui, J. B. Y., Shaw, R. L., and Davis, R. L. "Performance Standards for Wideband EW Receivers," *Microwave Journal*, Feb. 1989, p. 46.

[67] Lawson, J. L., and Uhlenbeck, G. E., "Threshold Signals," New York, NY: McGraw-Hill Book Co., 1950, pp. 199–210.

CHAPTER 3

Fourier Transform and Convolution

3.1 INTRODUCTION

In this chapter, both the Fourier transform and Fourier series will be discussed. The properties of the Fourier transform will be presented and the concept of impulse function will be introduced. The definition of convolution and its relation with Fourier transform will be presented. Examples of some commonly used Fourier transforms are given and the results are presented in a table for quick reference. There are many good books on this subject. A few of the books are listed at the end of this chapter. Readers with a background in this area can skip this chapter. However, Examples 3.10 and 3.11 may be of interest, because the former one is related to the radar pulse train and the latter will be used in the Hilbert transform.

3.2 FOURIER TRANSFORM [1–4]

Fourier series was introduced in 1807 by a French engineer, Jean Baptiste Joseph de Fourier (1768–1830). He suggested that any arbitrary function defined over a finite interval by any piecewise graph, continuous or discontinuous, could be represented as an infinite sum of continuous functions such as sine and cosine. Although almost all the members of the French Academy questioned its validity, it turned out to be one of the most powerful tools in signal processing.

The basic concept of the Fourier transform is that any function in the time domain can be represented by an infinite number of sinusoidal functions. The Fourier transform is defined as follows. A function $x(t)$ in the time domain t has a corresponding function $X(f)$ in the frequency domain related by

$$\mathcal{F}[x(t)] \equiv X(f) = \int_{-\infty}^{\infty} x(t)\, e^{-j2\pi ft} dt \tag{3.1}$$

where $j = \sqrt{(-1)}$ is the imaginary number and f represents frequency. The inverse transform is

$$\mathcal{F}^{-1}[X(f)] \equiv x(t) = \int_{-\infty}^{\infty} X(f)\, e^{j2\pi ft} df \tag{3.2}$$

Since an exponential term can be decomposed into sine and cosine functions as

$$e^{-j\theta} = \cos\theta - j\sin\theta \tag{3.3}$$

(3.1) can be written as

$$X(f) = \int_{-\infty}^{\infty} x(t)\cos(2\pi ft)\, dt - j\int_{-\infty}^{\infty} x(t)\sin(2\pi ft)\, dt \tag{3.4}$$

This equation can be used to find the properties of even and odd functions under Fourier transform. A simple example is given below to illustrate the applications of the above equations.

Example 3.1

A simple example of Fourier transform is to find the Fourier transform of

$$\begin{aligned} x(t) &= A \qquad \text{for } -\frac{T}{2} < t < \frac{T}{2} \\ &= 0 \qquad \text{elsewhere} \end{aligned} \tag{3.5}$$

The function $x(t)$ is often referred to as the rectangular function. This function is an even function of t (symmetric with respect to t). However, its product with a sine function is antisymmetric with respect to t, because the sine function is antisymmetric with respect to t. The integral of an antisymmetric function of t from $t = -\infty$ to ∞ is zero, because the positive and negative parts cancel each other. Thus, the Fourier transform of $x(t)$ is

$$\begin{aligned} X(f) &= A\int_{-T/2}^{T/2} e^{-j2\pi ft} dt = A\int_{-T/2}^{T/2} \cos(2\pi ft)\, dt = \frac{A\sin(\pi fT)}{\pi f} \\ &= \frac{AT\sin(\pi fT)}{T\pi f} = AT\,\mathrm{sinc}(\pi fT) \end{aligned} \tag{3.6}$$

This formula is known as the sinc function, which is defined as

$$\text{Sinc}(x) = \frac{\sin(x)}{x} \tag{3.7}$$

The functions $x(t)$ and $X(f)$ are shown in Figure 3.1.

In this figure $A = 2$ and $T = 1$. From (3.6) and (3.7), the first zero of $X(f)$ occurs at $f = 1/T$, (or at $\sin(\pi) = 0$). Thus the width of the mainlobe is $2/T$.

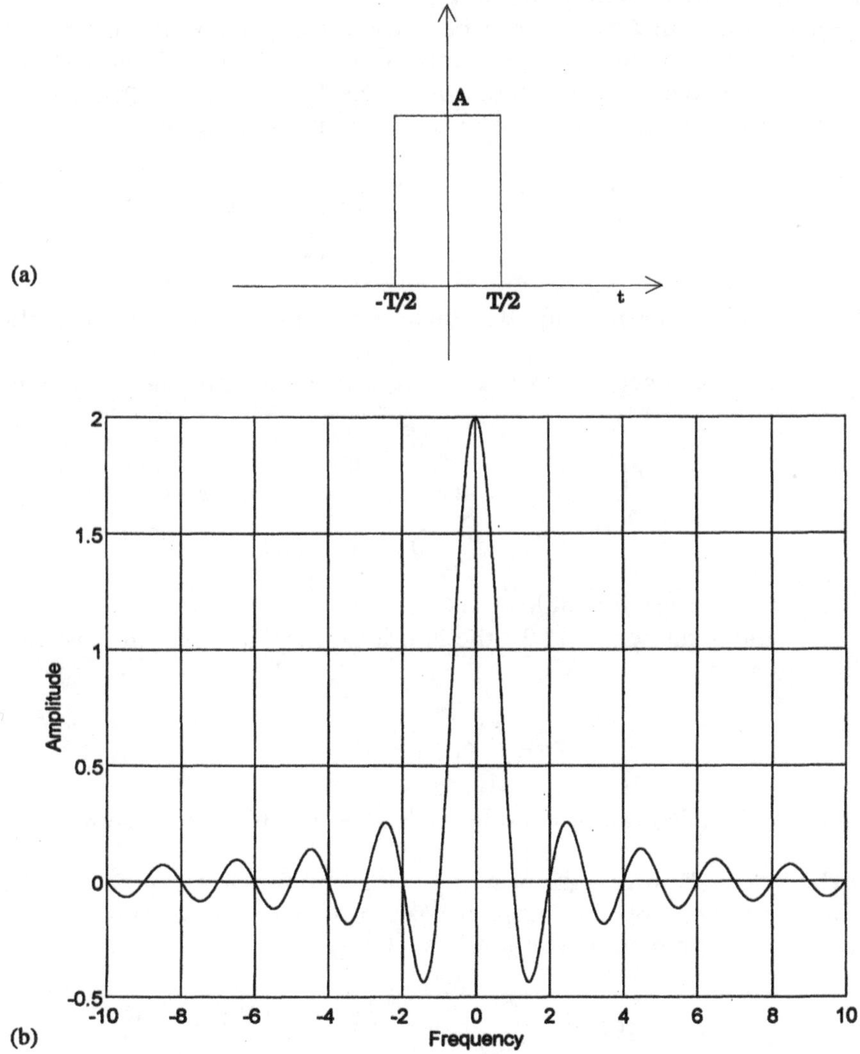

Figure 3.1 A rectangular function and its Fourier transform: (a) time domain, (b) frequency domain.

3.3 IMPULSE FUNCTION [5–11]

An impulse function is represented by the symbol $\delta(t)$. Sometimes, it is also referred to as the Dirac delta function. The output of a system with an impulse function as the input is called the impulse response of the system. The impulse function is only conceptual, $\delta(t)$ cannot be generated in practice. However, it is very useful in digitizing input signals; one can think of "sampling" as multiplication of the input function by a periodic train of impulses.

An impulse function $\delta(t)$ can be considered as a very narrow rectangular function at $t = 0$, as shown in Figure 3.1(a) with $A = 1/T$ and T approaching zero. The area of this rectangular function is 1. The function $\delta(t - t_o)$ is the impulse function at $t = t_o$. Integrating the impulse function, the result is

$$\begin{aligned} \int_a^b \delta(t - t_o)\, dt &= 1 \qquad \text{for } a \le t_o \le b \\ &= 0 \qquad \text{otherwise} \end{aligned} \tag{3.8}$$

If the integral includes the impulse function, the integral is unity; otherwise, the integral is zero.

The impulse function can also be defined more generally by its sampling property as

$$\begin{aligned} \int_a^b x(t)\delta(t - t_o)\, dt &= x(t_o) \qquad a \le t_o \le b \\ &= 0 \qquad \text{otherwise} \end{aligned} \tag{3.9}$$

which represents a specific value of $x(t)$ at $t = t_o$.

From the definition in (3.9), the Fourier transform of the impulse function can be written as

$$X(f) = \int_{-\infty}^{\infty} \delta(t)\, e^{-j2\pi ft} dt = e^0 = 1 \tag{3.10}$$

In this derivation, $\delta(t)$ can be considered as $\delta(t - 0)$. This result is shown in Figure 3.2.

The inverse Fourier transform of the impulse function can be found either from (3.2) or (3.4), and the imaginary part of (3.4) is zero because it is an odd function (or antisymmetry). The result is

$$\begin{aligned} \delta(t) &= \int_{-\infty}^{\infty} e^{j2\pi ft} df = \frac{e^{j2\pi ft}}{j2\pi t}\Big|_{-\infty}^{\infty} = \frac{\cos(2\pi ft) + j\sin(2\pi ft)}{j2\pi t}\Big|_{-\infty}^{\infty} \\ &= \frac{\sin(2\pi ft)}{2\pi t}\Big|_{-f}^{f} = \lim_{f\to\infty} \frac{2\sin(2\pi ft)}{2\pi t} \end{aligned} \tag{3.11}$$

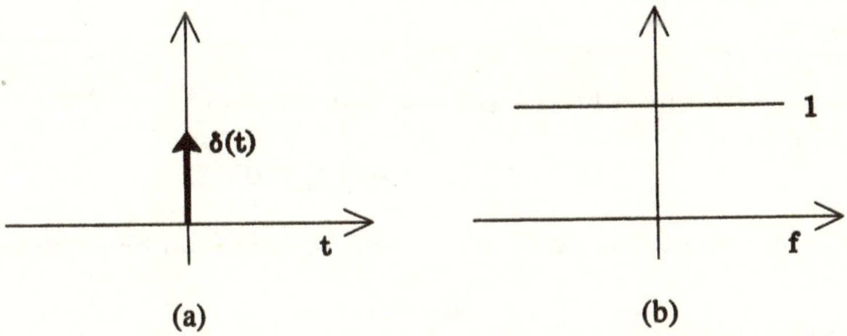

Figure 3.2 An impulse function and its Fourier transform: (a) time domain, (b) frequency domain.

One can consider this as another definition of the impulse function; that is,

$$\lim_{f\to\infty}\frac{\sin(2\pi ft)}{\pi t} = \delta(t) \tag{3.12}$$

A similar result can be obtained for a time-shifted impulse function as

$$\mathscr{F}[\delta(t - t_o)] = \int_{-\infty}^{\infty}\delta(t - t_o)e^{-j2\pi ft}dt = e^{-j2\pi ft_o}$$

or

$$\begin{aligned}\delta(t - t_o) &= \int_{-\infty}^{\infty} e^{j2\pi f(t-t_o)}df = \frac{e^{j2\pi f(t-t_o)}}{j2\pi(t - t_o)}\Bigg|_{-\infty}^{\infty} \\ &= \frac{\sin[2\pi f(t - t_o)]}{2\pi(t - t_o)}\Bigg|_{-f}^{f} \\ &= \lim_{f\to\infty}\frac{\sin[2\pi f(t - t_o)]}{\pi(t - t_o)}\end{aligned} \tag{3.13}$$

In this equation, the impulse function $\delta(t - t_0)$ can be considered as defined by

$$\delta(t - t_o) = \lim_{f\to\infty}\frac{\sin[2\pi f(t - t_o)]}{\pi(t - t_o)} \tag{3.14}$$

More examples of the impulse function will be illustrated in later sections.

3.4 PROPERTIES OF FOURIER TRANSFORM [5–12]

Some of the properties of Fourier transform will be discussed in this section. If the relation is obvious, no proof will be provided. These properties are:

3.4.1 Linearity

The Fourier transform is a linear operator, which means

$$\mathcal{F}[\alpha x(t) + \beta y(t)] = \alpha X(f) + \beta Y(f) \tag{3.15}$$

Using the definition of Fourier transform, this property can be proven easily.

3.4.2 Even and Odd Functions

This property can be obtained by observing (3.4). If $x(t)$ is an even function, the second term on the right (the imaginary part) equals zero and the output is real as

$$X(f) = \int_{-\infty}^{\infty} x(t) \cos(2\pi ft)\, dt \tag{3.16}$$

Since $\cos(\theta) = \cos(-\theta)$, $X(f)$ is symmetrical with respect to $f = 0$.

If $x(t)$ is an odd function in (3.4), the first term on the right (the real part) equals zero and the output is imaginary as

$$X(f) = -j\int_{-\infty}^{\infty} x(t) \sin(2\pi ft)\, dt \tag{3.17}$$

Since $\sin(\theta) = -\sin(-\theta)$, $X(f)$ is antisymmetric with respect to $f = 0$. Let us use $\cos(2\pi f_o t)$ and $\sin(2\pi f_o t)$ to demonstrate these properties.

Example 3.2

Find the Fourier transforms of $\cos(2\pi f_o t)$ and $\sin(2\pi f_o t)$. The Fourier transform of $\cos(2\pi f_o t)$ can be written by the definition of (3.1) as

$$\begin{aligned} X(f) &= \int_{-\infty}^{\infty} \cos(2\pi f_o t)\, e^{-j2\pi ft} dt \\ &= \frac{1}{2}\int_{-\infty}^{\infty} [e^{-j2\pi(f-f_o)t} + e^{-j2\pi(f+f_o)t}]\, dt \\ &= \frac{1}{2}[\delta(f - f_o) + \delta(f + f_o)] \end{aligned} \tag{3.18}$$

In obtaining the final result, the relation in (3.14) is used. The Fourier transform of the cosine function is two impulse functions at $f = \pm f_o$; thus, they are real and symmetrical with respect to $f = 0$. These results are shown in Figure 3.3.

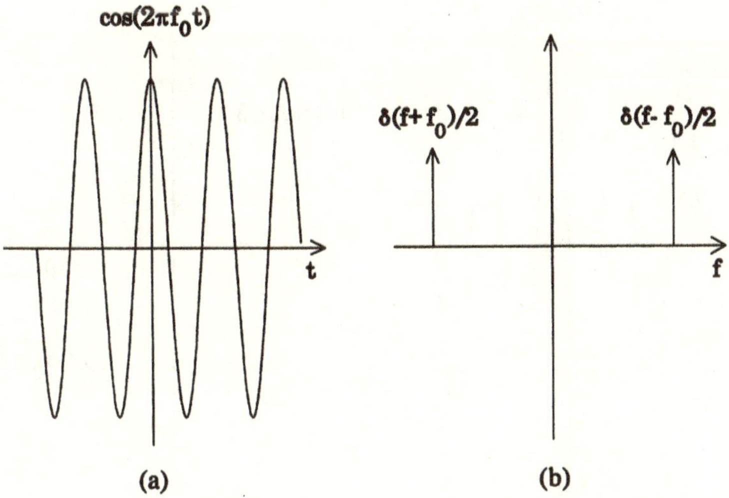

Figure 3.3 Fourier transform of cosine wave: (a) time domain, (b) frequency domain.

To perform the Fourier transform on the sine function, (3.1) is used and the result is

$$\begin{aligned} X(f) &= \int_{-\infty}^{\infty} \sin(2\pi f_0 t)\, e^{-j2\pi ft} dt \\ &= \frac{1}{2j}\int_{-\infty}^{\infty} [e^{-j2\pi(f-f_0)t} - e^{-j2\pi(f+f_0)t}]\, dt \\ &= \frac{-j}{2}[\delta(f-f_0) - \delta(f+f_0)] \end{aligned} \tag{3.19}$$

The outputs are two imaginary impulse functions and they are antisymmetrical with respect to $f = 0$. These results are shown in Figure 3.4.

These results agree with the conclusions drawn from (3.16) and (3.17).

3.4.3 Duality

Duality means that if the function in time domain is $x(t)$ and the Fourier transform is $X(f)$, then $X(t)$ has a Fourier transform of $x(-f)$. The proof of this property can be accomplished from the inverse Fourier equation (3.2). The proof is very simple, it only requires three steps and the change of notations of two variables. In (3.2), one can treat both variables t and f as dummy variables. The first step is to replace f by t'; the result is

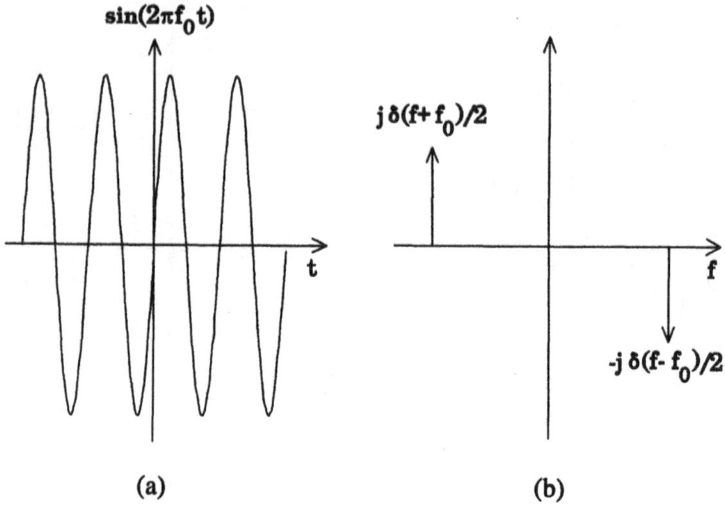

Figure 3.4 Fourier transform of sine wave: (a) time domain, (b) frequency domain.

$$x(t) = \int_{-\infty}^{\infty} X(t') \; e^{j2\pi t' t} dt' \tag{3.20}$$

The second step is to replace t with $-f$; the result is

$$x(-f) = \int_{-\infty}^{\infty} X(t') \; e^{-j2\pi f t'} dt' \tag{3.21}$$

Since t' is a dummy variable it can be written as t; thus, this equation can be written as

$$x(-f) = \int_{-\infty}^{\infty} X(t) e^{-j2\pi f t} dt \tag{3.22}$$

which is the desired result. Let us use an example to demonstrate this property.

Example 3.3

Find the Fourier transform of $x(t)$ where

$$x(t) = A\frac{\sin(\pi F t)}{\pi t} \tag{3.23}$$

This is the result obtained in (3.6) with t replacing f and F replacing T. The direct approach to finding the Fourier transform is quite tedious. This procedure is shown as follows. Since this $x(t)$ is an even function, (3.16) will be used and the imaginary part of $X(f)$ is equal to zero. The Fourier transform can be calculated as

$$X(f) = \int_{-\infty}^{\infty} \frac{A \sin(\pi F t)}{\pi t} \cos(2\pi f t)\, dt$$
$$= \frac{A}{2\pi} \int_{-\infty}^{\infty} \left[\frac{\sin(\pi F + 2\pi f)t}{t} + \frac{\sin(\pi F - 2\pi f)t}{t} \right] dt \tag{3.24}$$

This integral can be found in the integral table as [12]

$$\int_{-\infty}^{\infty} \frac{\sin mx}{x}\, dx = \pi \qquad m > 0$$
$$= -\pi \qquad m < 0 \tag{3.25}$$

Equation (3.24) can be evaluated in three regions in terms of f. If $-F/2 < f < F/2$, then $\pi F + 2\pi f$ and $\pi F - 2\pi f$ are positive. Both integrals result in π, thus $X(f) = A$. If $f < -F/2$, then $\pi F + 2\pi f$ is negative and $\pi F - 2\pi f$ is positive. Thus, the first integral is $-\pi$ and the second integral is π, which results in $X(f) = 0$. If $f > F/2$, then $\pi F + 2\pi f > 0$, and $\pi F - 2\pi f < 0$. The first integral is π and the second is $-\pi$; thus, $X(f) = 0$. If all the three regions of f are considered, the result is the desired result.

$$X(f) = A \qquad \text{for} \quad -\frac{F}{2} < f < \frac{F}{2}$$
$$= 0 \qquad \text{elsewhere} \tag{3.26}$$

The same result can be obtained from the duality property of Fourier transform in a much simpler way. Since $x(t)$ matches the output $X(f)$ of Example 3.1, the result can be directly written as

$$X(-f) = A = X(f) \qquad \text{for} \quad \frac{-F}{2} < f < \frac{F}{2}$$
$$= 0 \qquad \text{elsewhere} \tag{3.27}$$

In this equation, the result of $X(-f) = A$ is not a function of f; thus, the Fourier transform $X(f)$ has the same form.

3.4.4 Scaling

If in a function $x(t)$, the time t is replaced by t/a, the Fourier transform can be found by replacing t/a with t', and the result is

$$\begin{aligned}\mathscr{F}\left[x\left(\frac{t}{a}\right)\right] &= \int_{-\infty}^{\infty} x\left(\frac{t}{a}\right) e^{-j2\pi ft} dt \\ &= a\int_{-\infty}^{\infty} x(t') e^{-j2\pi aft'} dt' \\ &= aX(af)\end{aligned} \tag{3.28}$$

This expression means that when time is divided by a factor a, the amplitude and the frequency of the Fourier are expanded by the factor a.

Similarly, if the frequency is divided by a factor a, the inverse Fourier transform can be written as

$$\mathscr{F}^{-1}\left[X\left(\frac{f}{a}\right)\right] = ax(at) \tag{3.29}$$

The proof of this equation is similar to the proof of (3.28).

3.4.5 Time Shift

When the time of the input signal $x(t)$ is shifted by t_0 as $x(t - t_0)$, the Fourier transform can be obtained by replacing $t - t_0$ with t'. The result is

$$\begin{aligned}\mathscr{F}[x(t - t_0)] &= \int_{-\infty}^{\infty} x(t - t_0)\ e^{-j2\pi ft} dt \\ &= \int_{-\infty}^{\infty} x(t')\ e^{-j2\pi f(t'+t_0)} dt' = e^{-j2\pi ft_0} X(f)\end{aligned} \tag{3.30}$$

This means that a shift in the time domain causes a phase shift in the frequency domain and the amplitude of the output does not change.

Example 3.4

Let us find the Fourier transform of $x(t) = \cos[2\pi f_0(t - t_o)]$. From (3.30), the result can be written directly as

$$\begin{aligned}X(f) &= \int_{-\infty}^{\infty} \cos[2\pi f_0(t - t_0)]\ e^{-j2\pi ft} dt = \frac{1}{2}\ e^{-j2\pi ft_0}[\delta(f - f_0) + \delta\ (f + f_0)] \\ &= \frac{1}{2}\{\cos(2\pi ft_0)[\delta(f - f_0) + \delta(f + f_0)] - j\sin(2\pi ft_0)[\delta(f - f_0) + \delta\ (f + f_0)]\}\end{aligned} \tag{3.31}$$

The output is no longer real, but becomes complex because the function $x(t)$ is not symmetric with respect to $t = 0$. However, the real part is still symmetric and the imaginary part is antisymmetric. The amplitudes of the impulse functions are still 1/2.

3.4.6 Frequency Shift

When there is a frequency shift f_0 in the Fourier output $X(f)$ as $X(f-f_0)$, the effect on the input signal can be found by substituting $f' = f - f_0$ as

$$\begin{aligned}\mathcal{F}^{-1}[X(f-f_0)] &= \int_{-\infty}^{\infty} X(f-f_0)\ e^{j2\pi ft}df \\ &= \int_{-\infty}^{\infty} X(f')\ e^{j2\pi(f'+f_0)t}df' = e^{j2\pi f_0 t}x(t)\end{aligned} \tag{3.32}$$

The corresponding effect is a frequency shift in the time domain, as expected.

3.4.7 Derivative

The derivative of the input $x(t)$ with respect to t can be represented by $x'(t)$. Its Fourier output can be found through integration by parts as

$$\begin{aligned}\mathcal{F}[x'(t)] &= \mathcal{F}\left[\frac{dx(t)}{dt}\right] = \int_{-\infty}^{\infty}\frac{dx(t)}{dt}e^{-j2\pi ft}dt \\ &= x(t)\ e^{-j2\pi ft}\Big|_{-\infty}^{\infty} - (-j2\pi f)\int_{-\infty}^{\infty} x(t)\ e^{-j2\pi ft}dt \\ &= j2\pi f X(f)\end{aligned} \tag{3.33}$$

In the above expression, it is assumed that $x(t)$ is zero at $t = \pm\infty$. This assumption makes the first term equal to zero.

3.4.8 Integral

If the time domain function $\theta(t)$ is

$$\theta(t) = \int_{-\infty}^{t} x(\tau)\,d\tau \tag{3.34}$$

and $\theta(\tau)|_{\tau=\infty} = 0$, the Fourier transform can be found as follows. This equation can be written as

$$\frac{d\theta(t)}{dt} = x(t) \tag{3.35}$$

From the relation in (3.33), the Fourier transform of $d\theta(t)/dt$ is

$$\mathcal{F}\left[\frac{d\theta(t)}{dt}\right] = j2\pi f\Theta(f) \tag{3.36}$$

This equation can be written as

$$\Theta(f) = \mathcal{F}[\theta(t)] = \frac{\mathcal{F}\left[\frac{d\theta(t)}{dt}\right]}{j2\pi f} \tag{3.37}$$

Substituting the results of (3.34) and (3.35) into this equation, the result is

$$\mathcal{F}\left[\int_{-\infty}^{t} x(\tau)\, d\tau\right] = \frac{\mathcal{F}[x(t)]}{j2\pi f} \tag{3.38}$$

This is the desired result.

3.5 FOURIER SERIES [5–11]

A periodic function $x(t)$ can be represented by the sum of an infinite number of sinusoidal waves with discrete frequencies. The Fourier series can be written in several forms. One of these is

$$x(t) = A_0 + \sum_{n=1}^{\infty} A_n \cos\frac{2\pi nt}{T} + \sum_{n=1}^{\infty} B_n \sin\frac{2\pi nt}{T} \tag{3.39}$$

where T is the period of $x(t)$. A_0 is the average value of $x(t)$, which can be found as

$$A_0 = \frac{1}{T}\int_{-\infty}^{\infty} x(t)\, dt \tag{3.40}$$

The constants A_n and B_n can be found as

$$\begin{aligned} A_n &= \frac{2}{T}\int_{-T/2}^{T/2} x(t) \cos\frac{2\pi nt}{T} dt \\ B_n &= \frac{2}{T}\int_{-T/2}^{T/2} x(t) \sin\frac{2\pi nt}{T} dt \end{aligned} \tag{3.41}$$

Another form of the Fourier series is combining the sine and cosine terms into exponential terms as

$$x(t) = \sum_{-\infty}^{\infty} C_n e^{\frac{j2\pi nt}{T}} \tag{3.42}$$

where C_n can be found from

$$C_n = \frac{1}{T}\int_{-T/2}^{T/2} x(t)\, e^{\frac{-j2\pi nt}{T}}\, dt \tag{3.43}$$

In the above equation, when T approaches infinity, one can think that $f = n/T$ represents a continuous frequency and the amplitude of the integral is normalized to one. Then, the equation is actually the Fourier transform.

In order to have a physical picture, one can imagine that a periodic function in the time domain can be represented by an infinite number of frequencies in the frequency domain and these frequencies are discrete. This is the Fourier series. If the function in the time domain is no longer a periodic function (or, in other words, the period is infinite), it can be represented by an infinite number of frequencies and these frequencies are continuous.

3.6 COMB FUNCTION [5–11]

A comb function is defined as the sum of an infinite number of uniformly distributed impulse functions, and they are shaped like a comb. A comb function is shown in Figure 3.5. The function can be defined in either the time or frequency domain. This function can be used to represent digitized data. Mathematically, the comb function in the time domain can be written as

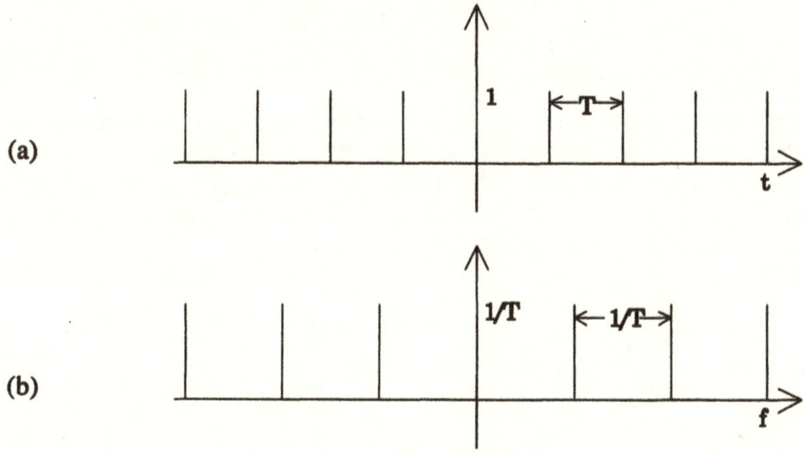

Figure 3.5 Fourier transform of a comb function: (a) time domain, (b) frequency domain.

$$x(t) = \text{comb}(t) = \sum_{n=-\infty}^{\infty} \delta(t - nT) \tag{3.44}$$

where T is the separation between the delta functions.

The Fourier transform of the comb function can be obtained from directly integrating the above equation. The following manipulation demonstrates the approach [5]:

$$\begin{aligned} X(f) &= \int_{-\infty}^{\infty} \sum_{n=-\infty}^{\infty} \delta(t - nT)\, e^{-j2\pi ft} dt \\ &= \sum_{n=-\infty}^{\infty} \int_{-\infty}^{\infty} \delta(t - nT)\, e^{-j2\pi ft} dt = \sum_{n=-N}^{N} e^{-j2\pi fnT} \Big|_{N\to\infty} \end{aligned} \tag{3.45}$$

Using the relation

$$\sum_{n=-N}^{N} z^n = \frac{z^{N+1} - z^{-N}}{z - 1} \tag{3.46}$$

(3.45) can be written as

$$\begin{aligned} X(f) &= \frac{e^{-j2\pi f(N+1)T} - e^{j2\pi fNT}}{e^{-j2\pi fT} - 1} \\ &= \frac{e^{-j2\pi fT/2}\left[e^{-j2\pi f(N+1/2)T} - e^{j2\pi f(N+1/2)T}\right]}{e^{-j2\pi fT/2}\left[e^{-j2\pi f1/2T} - e^{j2\pi f1/2T}\right]} \\ &= \frac{\sin\left[2\pi\left(N + \frac{1}{2}\right) fT\right]}{\sin(\pi fT)} \end{aligned} \tag{3.47}$$

In this function, f is the variable. The result of this equation is shown in Figure 3.6. In this figure, $N = 7$ and f ranges from -3.9 to 3.9. Where $f = n/T$, whenever n is an integer, there is a peak. This figure has the rough shape of a comb. The last step is to prove that when $N \to \infty$, the result of (3.47) approaches a delta function. The result of the above function can be written as

$$\frac{\sin[2\pi(N+1) fT]}{\sin(\pi fT)}\Bigg|_{N\to\infty} = \frac{\sin\left[2\pi(N+1)\left(f - \frac{n}{T}\right)T\right]}{f - \frac{n}{T}} \; \frac{f - \frac{n}{T}}{\sin(\pi fT)}\Bigg|_{N\to\infty} \tag{3.48}$$

Using the result of (3.13), one of the definitions of delta function, the first part of the right side of the above equation is

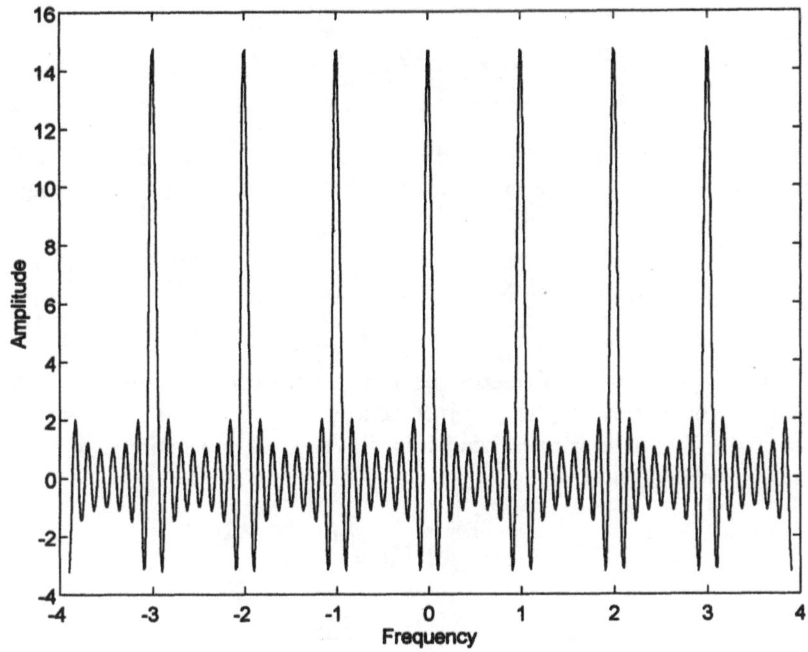

Figure 3.6 Plot of $\sin[2\pi(N+1)fT]/\sin(\pi fT)$.

$$\lim_{N\to\infty} \frac{\sin\left[2\pi(N+1)\left(f-\frac{n}{T}\right)T\right]}{f-\frac{n}{T}} = \pi\delta\left(f-\frac{n}{T}\right) \tag{3.49}$$

The second part of the right-side equation can be calculated from the L'Hospital's rule as

$$\left.\frac{f-\frac{n}{T}}{\sin(\pi fT)}\right|_{f=\frac{n}{T}} = \left.\frac{\frac{d(f-n/T)}{df}}{\frac{d[\sin(\pi fT)]}{df}}\right|_{f=\frac{n}{T}} \tag{3.50}$$

$$= \left.\frac{1}{\pi T\cos(\pi fT)}\right|_{f=\frac{n}{T}} = \frac{1}{\pi T}$$

Thus, any peak can be written as

$$\frac{\sin[(N+1)2\pi fT]}{\sin(\pi fT)} = \frac{\delta\left(f - \frac{n}{T}\right)}{T} \tag{3.51}$$

Thus, the Fourier transform of the comb function is another comb function with amplitude of $1/T$ and separation of $1/T$. This result can be mathematically written as

$$X(f) = \mathscr{F}\left[\sum_{n=-\infty}^{\infty} \delta(t - nT)\right] = \frac{1}{T}\sum_{n=-\infty}^{\infty} \delta\left(f - \frac{n}{T}\right) \tag{3.52}$$

A different approach to obtain the Fourier transform of the comb function is through a Fourier series. The comb function is a period function with a period of T; therefore, it can be represented by a Fourier series as shown in (3.39). The constants can be obtained as

$$\begin{aligned} A_0 &= \frac{1}{T}\int_{-T/2}^{T/2} \delta(t)\ dt = \frac{1}{T} \\ A_n &= \frac{2}{T}\int_{-T/2}^{T/2} \delta(t)\ \cos\frac{2\pi nt}{T} dt = \frac{2}{T} \\ B_n &= \frac{2}{T}\int_{-T/2}^{T/2} \delta(t)\ \sin\frac{2\pi nt}{T} dt = 0 \end{aligned} \tag{3.53}$$

It should be noted that $\delta(t)$ exists only inside the range of one period ($T/2 \leq t \geq T/2$) and all the other delta functions are outside this period. Substituting these constants into (3.44), the result is

$$x(t) = \frac{1}{T} + \frac{2}{T}\sum_{n=1}^{\infty} \cos\frac{2\pi nt}{T} \tag{3.54}$$

This equation is still in the time domain, but in a different form. Taking the Fourier transform of this function, the first term can be obtained from the symmetry property of the Fourier transform. The result is $\delta(f)/T$, which represents a delta function at $f = 0$. The Fourier transform of the rest of the terms can be obtained by using the result of (3.18) with $f_o = n/T$. Thus, the result is

$$X(f) = \mathscr{F}[\text{comb}(t)] = \frac{1}{T}\sum_{n=-\infty}^{\infty} \delta\left(f - \frac{n}{T}\right) \tag{3.55}$$

This is the same result as obtained from (3.52). This result is shown in Figure 3.5. This result is very important for processing radar signals because a radar signal with stable pulse repetition frequency (PRF) can be represented by a comb function.

3.7 CONVOLUTION [5–11]

If an input signal is applied to a linear and time-invariant system, the output can be found through two approaches. The time domain approach is through convolution and the frequency domain approach is through Fourier transform. The results obtained from these two approaches are identical.

Let us assume that the input signal to the system is a delta function $\delta(t)$ and the corresponding output is $h(t)$, where $h(t)$ is called the impulse response of the system. Because the system is time-invariant, an input $\delta(t - \tau)$ will produce an output of $h(t - \tau)$. This means that when the input is delayed time τ, the output is also delayed time τ, and the system response does not change with time. An input function $x(t)$ can be imagined as consisting of many rectangular functions, each with the same infinitesimal width but different amplitude and delay time. Thus, the input signal $x(t)$ and the output $y(t)$ can be written as

$$
\begin{aligned}
x(t) &= \lim_{\Delta\tau_i \to 0} \sum_i \Delta\tau_i x(t - \tau_i) \\
y(t) &= \lim_{\Delta\tau_i \to 0} \sum_i \Delta\tau_i x(t - \tau_i) h(\tau_i) = \lim_{\Delta\tau_i \to 0} \sum_i \Delta\tau_i x(\tau_i) h(t - \tau_i) \\
&= \int_{-\infty}^{\infty} x(t - \tau) h(\tau)\, d\tau = \int_{-\infty}^{\infty} x(\tau) h(t - \tau)\, d\tau \equiv x(t) \bigstar h(t)
\end{aligned}
\tag{3.56}
$$

where $\Delta\tau_i$ is the width of the rectangular function, t is time, and τ can be considered as a dummy variable. After the integration, the variable τ disappears and the output is a function of t as expected. The $\bigstar$ notation in the equation is the conventional way to represent convolution. The term $h(t - \tau)$ means inverting the $h(t)$ in the time domain and delay by time τ. From the above equation, one can see that either the input signal or the impulse response can be inverted in time.

The convolution is graphically illustrated in Figure 3.7. Figures 3.7(a) and 3.7(b) show the function of $x(\tau)$ and $h(\tau)$, respectively. Figure 3.7(c) shows $h(\tau)$ reversed in time and represented by $h(-\tau)$. Figure 3.7(d) shows $h(-\tau)$ is shifted by time t and represented by $h(t - \tau)$. Figure 3.7(e) represents the result of $x(\tau)h(t - \tau)$ and Figure 3.7(f) represents the result obtained from the convolution. The point a on the curve (or $y(t) = a$) of Figure 3.7(f) represents the total area shown in Figure 3.7(e). If $x(\tau)$ is reversed instead of $h(\tau)$, the same result will be obtained.

Another important example is to find the convolution of an impulse function with function $x(t)$. If system response is an impulse function (i.e., $h(t) = \delta(t - t_o)$), the output is

$$
y(t) = \int_{-\infty}^{\infty} h(\tau) x(t - \tau)\, d\tau = \int_{-\infty}^{\infty} \delta(\tau - t_o) x(t - \tau)\, d\tau = x(t - t_o)
\tag{3.57}
$$

(a)

(b)

(c)

(d)

(e)

(f)

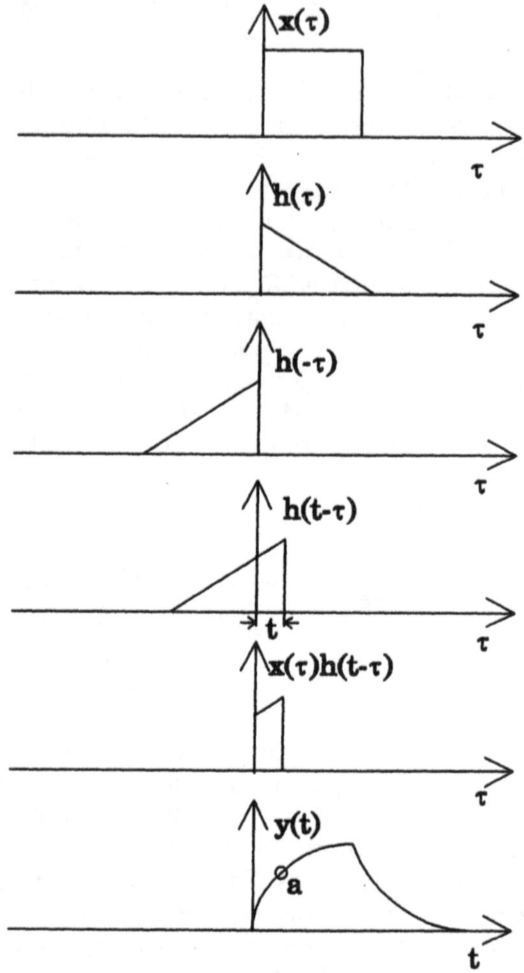

Figure 3.7 A graphic display of convolution of $x(t)$ and $h(\tau)$: (a) $x(\tau)$, (b) $h(\tau)$, (c) $h(-\tau)$, (d) $h(t-\tau)$, (e) $x(\tau)h(t-\tau)$, (f) $y(t)$.

The operation shifts the function $x(t)$ to $x(t - t_o)$. Figure 3.8 shows the result graphically.

Now we try to solve the output $y(t)$ from the frequency domain approach. The input and impulse response of the system are still $x(t)$ and $h(t)$, respectively. This operation can be found from taking the Fourier transform of $y(t)$. The result is

$$Y(f) = \int_{-\infty}^{\infty}\int_{-\infty}^{\infty} x(\tau)\,h(t-\tau)\; d\tau e^{-j2\pi ft}dt \tag{3.58}$$

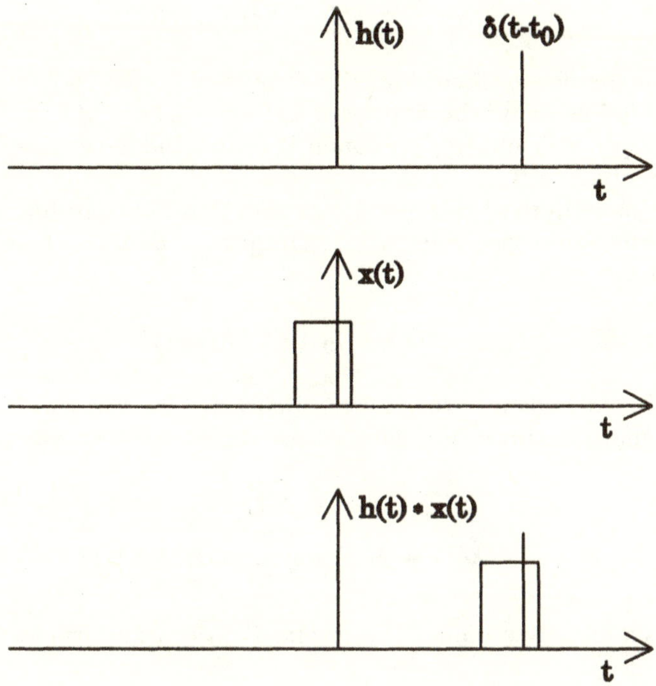

Figure 3.8 Convolution of $\delta(t-t_o)$ with $x(t)$: (a) $\delta(t-t_o)$, (b) $x(t)$, (c) $y(t)$.

Changing the order of integration, this equation can be written as

$$Y(f) = \int_{-\infty}^{\infty} x(\tau)\left(\int_{-\infty}^{\infty} h(t-\tau)\, e^{-j2\pi ft}dt\right)d\tau \tag{3.59}$$

Changing the variable by letting $t-\tau = u$, then

$$\begin{aligned} Y(f) &= \int_{-\infty}^{\infty} x(\tau)\left(\int_{-\infty}^{\infty} h(u)\, e^{-j2\pi fu}du\right)e^{-j2\pi f\tau}d\tau \\ &= \int_{-\infty}^{\infty} x(\tau) H(f)\, e^{-j2\pi f\tau}d\tau \\ &= H(f)\int_{-\infty}^{\infty} x(\tau)\, e^{-j2\pi f\tau}d\tau = H(f)X(f) \end{aligned} \tag{3.60}$$

In order to find $y(t)$, an inverse Fourier transform on $Y(f)$ is required. This relation can also be written as

$$x(t) \star h(t) = \mathcal{F}^{-1}[X(f)H(f)] \tag{3.61}$$

From the above discussion, the frequency domain solution can be accomplished as follows. Find the Fourier transforms of both $x(t)$ and $h(t)$ as $X(f)$ and $H(f)$. Multiply $X(f)$ and $H(f)$ to obtain $Y(f)$ and find the inverse Fourier transform of $Y(f)$ to obtain $y(t)$.

The above derived relation also implies that the convolution in time domain is equivalent to multiplication in the frequency domain. This relation is often expressed as

$$x(t) \star h(t) \leftrightarrow X(f)H(f) \tag{3.62}$$

and is referred to as the convolution theorem. One can take the convolution in the frequency domain in the same manner as in the time domain. This is shown as

$$X(f) \star H(f) = \int_{-\infty}^{\infty} X(\lambda)H(f-\lambda)\,d\lambda \tag{3.63}$$

where λ is a dummy variable. The inverse Fourier transform of this result is

$$\begin{aligned} \mathcal{F}^{-1}[X(f) \star H(f)] &= \int_{-\infty}^{\infty}\int_{-\infty}^{\infty} X(\lambda)H(f-\lambda)\,d\lambda\, e^{j2\pi ft}df \\ &= \int_{-\infty}^{\infty} X(\lambda)\left(\int_{-\infty}^{\infty} H(u)\, e^{j2\pi ut}du\right) e^{j2\pi\lambda t}d\lambda = h(t)\,x(t) \end{aligned} \tag{3.64}$$

The proof is identical to (3.60). The above relation can be expressed as

$$X(f) \star H(f) \leftrightarrow x(t)\,h(t) \tag{3.65}$$

Equations (3.62) and (3.65) are referred to as the duality of convolution and Fourier transform.

3.8 PARSEVAL'S THEOREM [5–8]

The Parseval's theorem states that for a given signal, the total energy in the time domain equals the total energy in the frequency domain. This relation can be written as

$$\int_{-\infty}^{\infty} |x(t)|^2 dt = \int_{-\infty}^{\infty} |X(f)|^2 df \tag{3.66}$$

This relation can be proven as follows. From (3.64) or (3.65), the result can be written as

$$\mathscr{F}[h(t)x(t)] = X(f) \star H(f)$$

$$\int_{-\infty}^{\infty} h(t)\, x(t)\, e^{-j2\pi\lambda t} dt = \int_{-\infty}^{\infty} H(\lambda - f) X(f)\, df \tag{3.67}$$

In this equation both sides are functions of λ, a dummy variable. If $\lambda = 0$, this equation becomes

$$\int_{-\infty}^{\infty} h(t)\, x(t)\, dt = \int_{-\infty}^{\infty} H(-f) X(f)\, df \tag{3.68}$$

Let us assume that $h(t) = x^*(t)$ where * represents a complex conjugate; then

$$H(-f) = \int_{-\infty}^{\infty} h(t)\, e^{j2\pi ft} dt = \int_{-\infty}^{\infty} x^*(t)\, e^{j2\pi ft} dt = \left[\int_{-\infty}^{\infty} x(t)\, e^{-j2\pi ft} dt\right]^* = X^*(f) \tag{3.69}$$

Substituting this relation into (3.68), the result of (3.66) can be obtained. This completes the proof of the Parseval's identity.

3.9 EXAMPLES [3, 5, 8, 12, 13]

This section will present some examples that are often encountered in signal processing and wideband receivers. Most of the problems are not solved directly, but through the duality of convolution and Fourier transform.

Example 3.5

Find the Fourier of $e^{j2\pi f_o t}$. This result can be obtained directly from the Fourier transform or from the following relation since the Fourier transform of sine and cosine are already known.

$$e^{j2\pi f_o t} = \cos(2\pi f_o t) + j\sin(2\pi f_o t) \tag{3.70}$$

Using the linearity property of (3.15) and combining with the results of (3.18) and (3.19), the result is

$$\mathscr{F}(e^{j2\pi f_o t}) = \frac{1}{2}[\delta(f - f_o) + \delta(f + f_o) + \delta(f - f_o) - \delta(f + f_o)] = \delta(f - f_o) \tag{3.71}$$

which is a single-sided response. This result is important in discussing I and Q channels of frequency downconverters. This result is shown in Figure 3.9. It should be noted that Figure 3.9(a) is not in the time domain but is in the complex plane, and the time is represented by the rotating vector.

Example 3.6

Find the Fourier transform of a rectangular windowed cosine wave. This waveform is often used to represent the output of pulsed radar. If the window is from $-T/2$ to $T/2$ with amplitude A, the windowed cosine wave can be generated by multiplying the window function and a cosine wave as

$$\begin{aligned} x_1(t) &= \cos(2\pi f_o t) \\ w(t) &= A \qquad \text{for } -\frac{T}{2} < t < \frac{T}{2} \\ &= 0 \qquad \text{elsewhere} \\ x(t) &= w(t)\,x_1(t) \end{aligned} \tag{3.72}$$

where f_0 is the frequency of the input pulse. Since multiplication in time domain is equivalent to convolution in the frequency domain, the Fourier transforms of the rectangular window and cosine wave are needed. The Fourier transform of the window, from Example 3.1, is $W(f) = A\sin(\pi fT)/\pi f$. The Fourier transform of a cosine wave, from Example 3.2, is $X_1(f) = [\delta(f - f_o) + \delta(f - f_o)]/2$. The Fourier transform of $x(t)$ in the above equation is

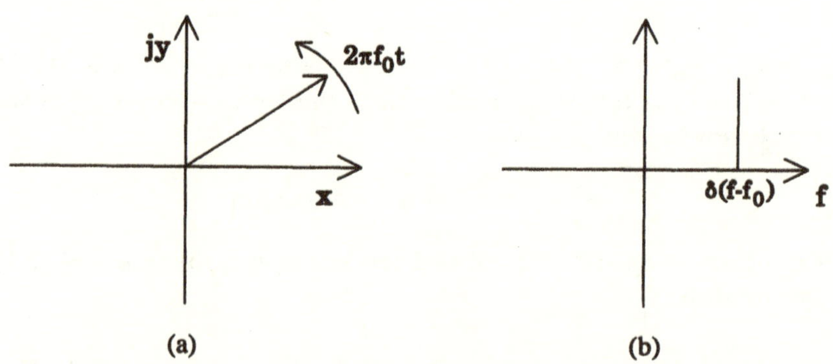

Figure 3.9 Fourier transform of $e^{j2\pi f_0 t}$: (a) complex plane, (b) frequency domain.

$$X(f) = W(f) \star X_1(f) = \int_{-\infty}^{\infty} \frac{A \sin(\pi\lambda T)}{\pi\lambda} \frac{[\delta(f - f_o - \lambda) + \lambda(f + f_o - \lambda)]}{2} d\lambda$$

$$= \frac{A}{2}\left\{\frac{\sin[2\pi(f - f_o)T]}{\pi(f - f_o)} + \frac{\sin[2\pi(f + f_o)T]}{\pi(f + f_o)}\right\} \quad (3.73)$$

The results are shown in Figure 3.10. In this figure, $A = 1$, $T = 1$, and $f_0 = 10$. Since the Fourier transform of a cosine function has two delta functions, the convolution with them will generate two outputs. One shifts by f_o, the other one shifts by $-f_o$.

Example 3.7

Find the Fourier transform of an isosceles triangle. Instead of finding the transform directly, the following approach is used. An isosceles triangle with height $(AT)^2$ and width $2T$ (from $-T$ to T) can be generated by convolving two identical rectangular functions of height A and width T (from $-T/2$ to $T/2$) as shown in Figure 3.11. The Fourier transform of such a rectangular function of unit height is given by (3.6) as $X_1(f) = A\sin(\pi fT)/\pi f$. Since convolving in the time domain is equivalent

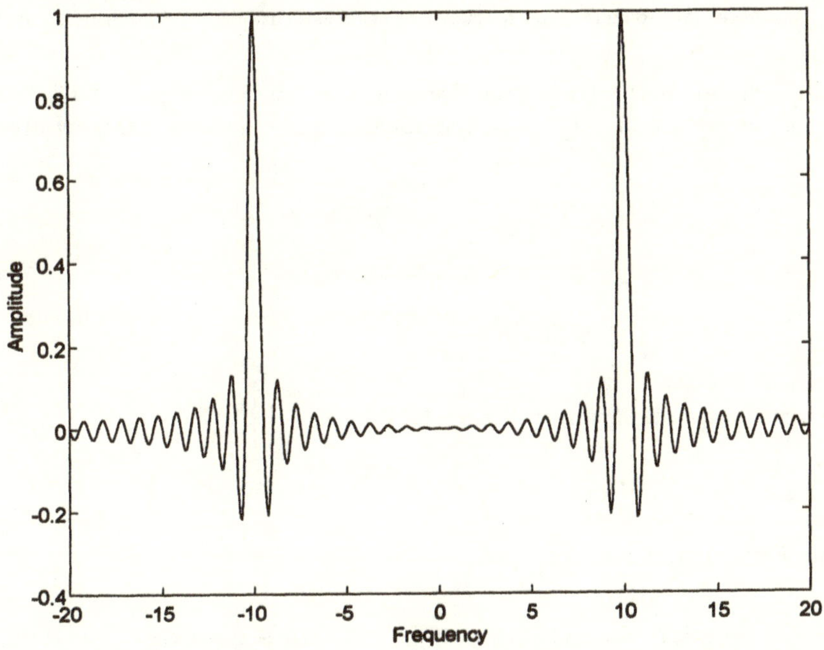

Figure 3.10 Fourier transform of a rectangular windowed cosine wave.

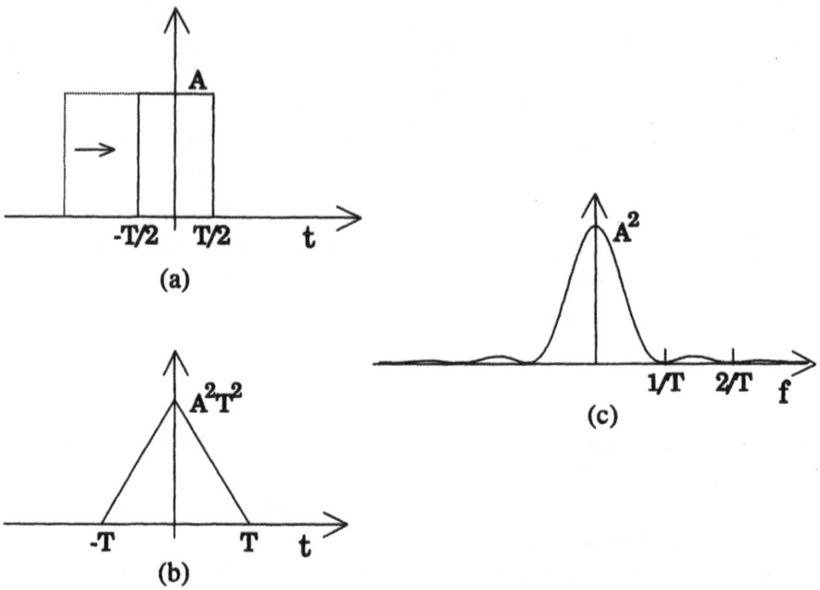

Figure 3.11 Convolving two rectangular functions to generate an isosceles triangle and its Fourier output: (a) two rectangular functions, (b) isosceles triangle, (c) Fourier transform output.

to multiplication in the frequency domain, the Fourier output of the isosceles triangle is the square of $X_1(f)$. Mathematically, the equations can be written as

$$
\begin{aligned}
x(t) &= A^2T^2\left(1 - \frac{|t|}{T}\right) \text{ for } |t| < T \\
&= 0 \qquad \text{elsewhere} \\
X(f) &= \frac{A^2 \sin^2(\pi f T)}{(\pi f)^2}
\end{aligned}
\tag{3.74}
$$

The Fourier output is shown in Figure 3.11(c).

Example 3.8

Find the Fourier transform of

$$
\begin{aligned}
x(t) &= a + (1 - a)\cos\left(\frac{2\pi t}{T}\right) \qquad \text{for } -\frac{T}{2} < t < \frac{T}{2} \\
&= 0 \qquad \text{elsewhere}
\end{aligned}
\tag{3.75}
$$

where $a < 1$. The $x(t)$ is often referred to as a generalized cosine window function. Different window functions are often used in signal processing. When $a = 0.54$, $x(t)$ is called the Hamming window. The limited response in time domain ($-T/2 < t < T/2$) can be achieved by multiplying $a + (1-a)\cos(2\pi t/T)$ by a rectangular function. The duality of the convolution theorem can be used to obtain the desired transform. Let $x_1(t) = a + (1 - a)\cos(2\pi t/T)$ and $x_2(t)$ represent the rectangular function; their Fourier transforms are

$$X_1(f) = a\delta(t) + \frac{1-a}{2}[\delta(f-f_o) + \delta(f+f_o)]$$

$$\text{where} \quad f_o = \frac{1}{T} \tag{3.76}$$

$$X_2(f) = \frac{\sin(\pi f T)}{\pi f}$$

The Fourier transform of $x(t)$ equals the convolution of $X_1(f)$ and $X_2(f)$ as

$$\begin{aligned} X(f) = X_2(f) \bigstar X_1(f) &= \frac{\sin(\pi f T)}{\pi f} \bigstar \{a\delta(f) + \frac{1-a}{2}[\delta(f-f_o) + \delta(f+f_o)]\} \\ &= \frac{a\sin(\pi f T)}{\pi f} + \frac{1-a}{2}\left\{\frac{\sin[\pi(f-f_o)T]}{\pi(f-f_o)} + \frac{\sin[\pi(f+f_o)T]}{\pi(f+f_o)}\right\} \end{aligned} \tag{3.77}$$

The time and frequency domains are shown in Figure 3.12. In this figure, $a = 0.5$ and $T = 5$; thus, $-2.5 < t < 2.5$ and f is from $-10/T < f < 10T$.

Example 3.9

Find the Fourier transform of the following function:

$$x(t) = \frac{1}{\sqrt{2\pi}\sigma} e^{-\frac{t^2}{2\sigma^2}} \tag{3.78}$$

This is a Gaussian (or normal) function of t. The solution of this equation can be obtained from direct integration. The result is

$$\begin{aligned} X(f) &= \frac{1}{\sqrt{2\pi}\sigma}\int_{-\infty}^{\infty} e^{-\frac{t^2}{2\sigma^2}} e^{j2\pi f t} dt \\ &= \frac{2}{\sqrt{2\pi}\sigma}\int_{0}^{\infty} e^{-\frac{t^2}{2\sigma^2}} \cos(2\pi f t)\, dt = e^{-2\pi^2\sigma^2 f^2} \end{aligned} \tag{3.79}$$

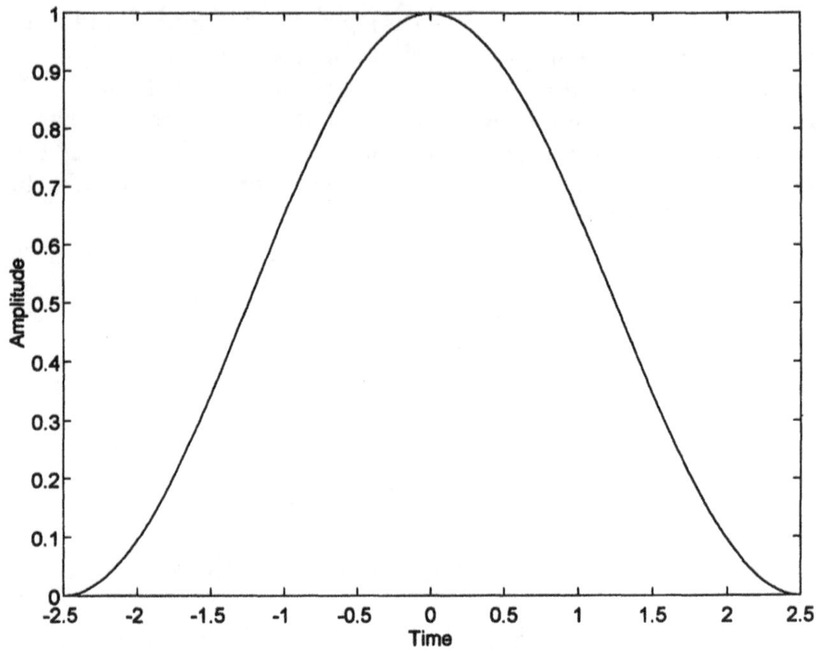

(a)

Figure 3.12 Fourier of a general cosine window with $a = 0.5$: (a) time domain, (b) frequency domain.

This integral can be found from [3] (861.20). This result shows that the Fourier transform of a Gaussian function in time domain is a Gaussian function in the frequency domain also. These results are shown in Figure 3.13. In this figure $\sigma = .8$.

Example 3.10

Find the Fourier transform of a radio frequency (RF) pulse train. This problem can be solved directly, as shown in [13]. The approach provided here is a step-by-step combination of desired signals to provide the pulse train. At each step the duality theorem of convolution is used. There are four steps to generate the pulse train in the time domain. In each step the time domain operations and the corresponding frequency domain operations will be represented. The last step produces the desired spectrum distribution. To keep this discussion simple, let us assume the pulse train has unit amplitude. In order to keep this approach clear, the

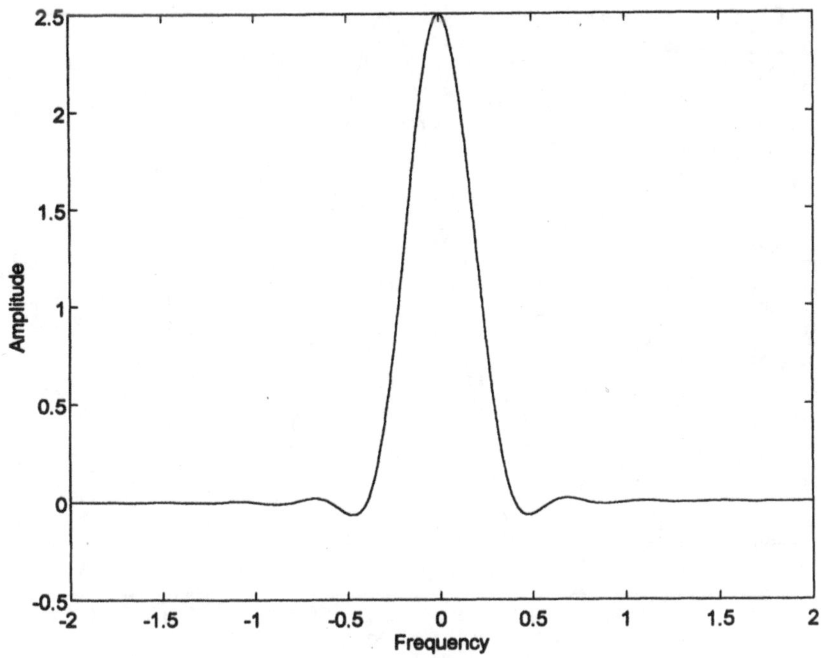

(b)

Figure 3.12 (continued).

selected functions are represented by $c(t)$, $s_i(t)$, and $r(t)$; the resulting functions are represented by $x_i(t)$ where i is an integer.

1. Choose a comb function $c(t)$ with period T as

$$
\begin{aligned}
c(t) &= \sum_{n=-\infty}^{\infty} \delta(t - nT) \\
C(f) &= \frac{1}{T} \sum_{n=-\infty}^{\infty} \delta\left(f - \frac{n}{T}\right)
\end{aligned}
\tag{3.80}
$$

where T represents the pulse repetition interval (PRI) of the pulse train and n is an integer.

2. Choose a rectangular function $s_1(t)$ with width τ and unit amplitude to represent an individual pulse. The time and frequency domains are

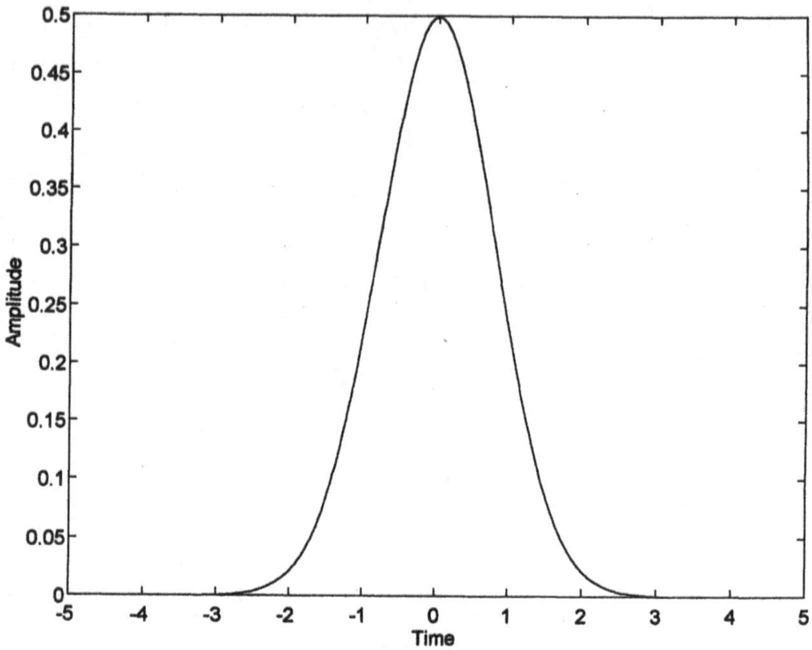

(a)

Figure 3.13 A Gaussian function and its Fourier transform: (a) time domain, (b) frequency domain.

$$
\begin{aligned}
s_1(t) &= 1 \qquad \text{for } |t| < \frac{\tau}{2} \\
&= 0 \qquad \text{elsewhere} \\
S_1(f) &= \frac{\sin(\pi f \tau)}{\pi f}
\end{aligned}
\tag{3.81}
$$

In order to create a pulse train, $x_1(t)$, $s_1(t)$ can be used to convolve with the comb function $c(t)$; thus

$$
\begin{aligned}
x_1(t) &= c(t) \star s_1(t) \\
X_1(f) &= C(f)\ S_1(f) = \frac{\sin(\pi f \tau)}{\pi f} \sum_{n=-\infty}^{\infty} \delta\left(f - \frac{n}{T}\right)
\end{aligned}
\tag{3.82}
$$

In the above operation, $f = n/T$ is substituted into $S_1(f)$ because the delta functions in $C(f)$ only exist at these values.

3. Choose a second rectangular function $s_2(t)$ with width NT and unit amplitude to represent the total length of the pulse train; thus

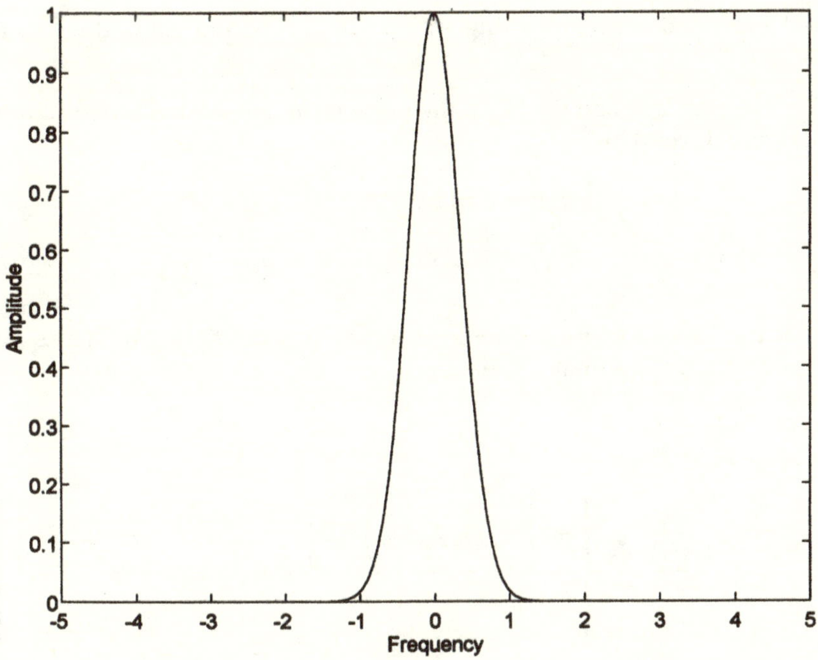

(b)

Figure 3.13 (continued).

$$\begin{aligned} s_2(t) &= 1 \qquad \text{for } |t| < \frac{NT}{2} \\ &= 0 \qquad \text{elsewhere} \\ S_2(f) &= \frac{\sin(\pi f NT)}{\pi f} \end{aligned} \tag{3.83}$$

The function $s_2(t)$ is multiplied by $x_1(t)$ to create the pulse train, which is represented by $x_2(t)$. The frequency domain result is obtained by convolving $X_1(f)$ with $S_2(f)$ as

$$\begin{aligned} x_2(t) &= x_1(t)s_2(t) \\ x_2(f) &= X_1(f) \star S_2(f) \\ &= \frac{\sin(\pi f \tau)}{\pi^2 f} \sum_{n=-\infty}^{\infty} \frac{\sin\left[\pi\left(f - \frac{n}{T}\right)NT\right]}{f - \frac{n}{T}} \end{aligned} \tag{3.84}$$

It should be noted that convolving with a delta function shifts the function by n/T.

4. The last step is to add the RF in the pulse train. Let us choose $r(t)$ to represent a cosine wave; then

$$\begin{aligned} r(t) &= \cos(2\pi f_o t) \\ R(f) &= \frac{1}{2}[\delta(f - f_o) + \delta(f + f_o)] \end{aligned} \tag{3.85}$$

where f_0 is the RF of the pulse. The function $x_2(t)$ is multiplied by $r(t)$ to produce the RF pulse train. The final results are

$$\begin{aligned} x(t) &= x_2(t)r(t) \\ X(f) &= X_2(f) \star R(f) \\ &= \underbrace{\frac{\sin(\pi f \tau)}{\pi^2 f}}_{1} \sum_{n=-\infty}^{\infty} \left\{ \underbrace{\frac{\sin\left[\pi\left(f - f_o - \frac{n}{T}\right)NT\right]}{f - f_o - \frac{n}{T}}}_{2} + \underbrace{\frac{\sin\left[\pi\left(f + f_o - \frac{n}{T}\right)NT\right]}{f + f_o - \frac{n}{T}}}_{3} \right\} \end{aligned} \tag{3.86}$$

The time domain signal generated through the four steps of operations is shown in Figure 3.14(a). It is a series of RF pulses. This signal in radar is called coherent because there is a certain phase relation between the individual pulses. In other words, a continuous sinusoidal wave that is gated by a pulse train generates a coherent pulse train because every pulsed RF signal is part of the same sinusoidal wave.

Now let us discuss the result of (3.86). The first term represents the envelope of the spectrum. The first zero of the envelope can be found from

$$f = \frac{1}{\tau} \tag{3.87}$$

The width of the mainlobe from null to null is $2/\tau$, and this result is shown in Figure 3.14(b). The second and third terms represent upper and lower band spectra. They are located at $\pm f_o$ and are symmetrical with respect to zero frequency. The upper group of spectrum lines occur only at

$$f = f_o + \frac{n}{T} \tag{3.88}$$

The spectrum lines are separated by $1/T$. The fine structure of the individual line can be found from the second term in (3.86) as

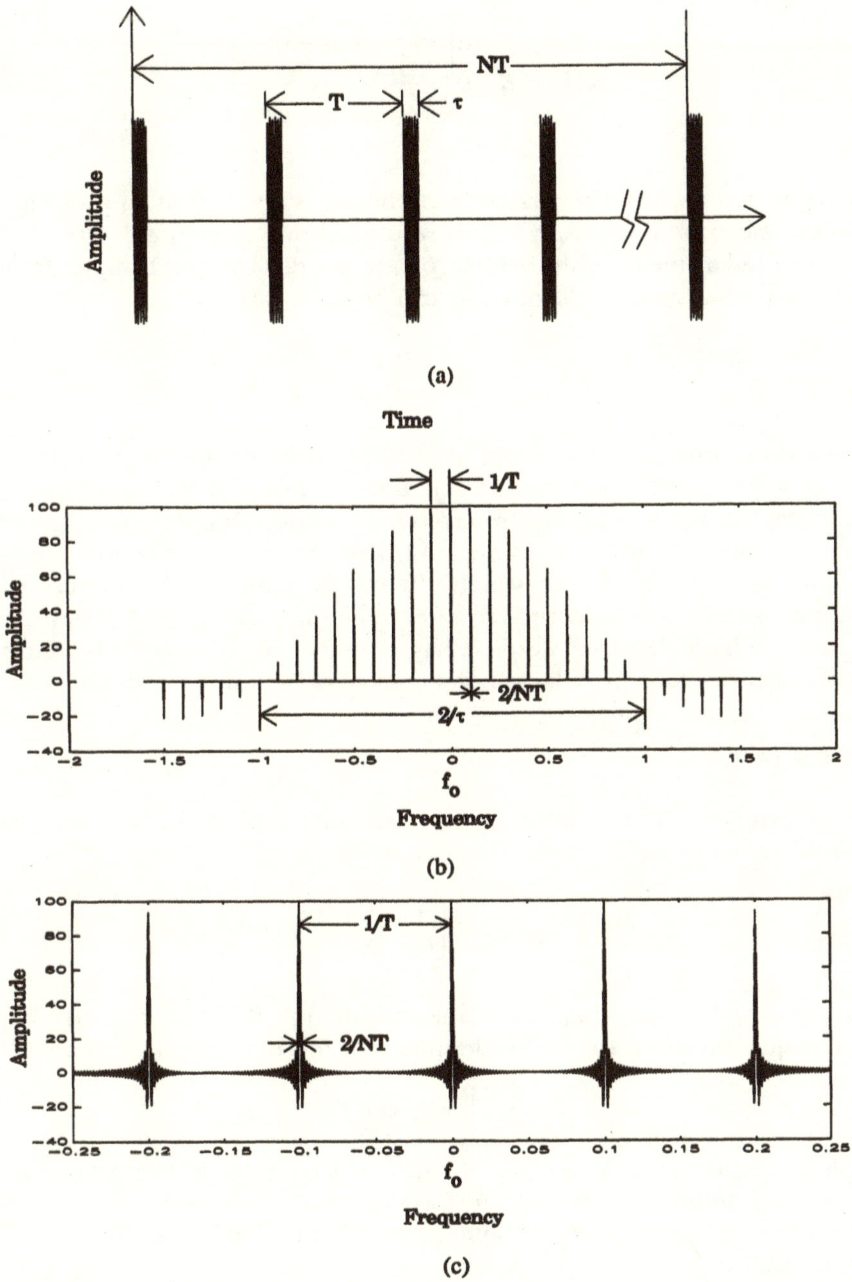

Figure 3.14 Fourier transform of an RF pulse train: (a) RF pulse train, (b) spectrum lines, (c) details of individual lines.

$$X(f) = K\frac{\sin\left[\pi\left(f - f_o - \frac{n}{T}\right)NT\right]}{\pi\left(f - f_0 - \frac{n}{T}\right)NT} \tag{3.89}$$

where K is a constant. The first zero occurs at $NT = 1$; thus, the width of the individual spectrum line is $2/NT$. This result is shown in Figure 3.14(c).

Some radars measure the velocity of a target through the Doppler frequency shift f_d. In a radar system, this quantity can be written as [13]

$$f_d = \frac{2f_0V}{C} \tag{3.90}$$

where f_0 is the operating frequency, V is the velocity of the target, and C is the speed of light. The factor 2 in this equation is caused by the round trip of the radar signal. This type of radar is referred to as a pulse Doppler radar. In order to reduce the ambiguity in frequency measurement, the separation between frequency lines must be wide. This requires high PRF or low pulse repetition interval (PRI), which means short T. In order to have fine spectrum line width, NT must be large or N must be large. In other words, a large number of pulses must be integrated. Therefore, pulse Doppler radar can generate very high pulse density.

Example 3.11

Find the inverse Fourier transform for a quadrature phase shift. Before starting this problem, let us define a sign function as

$$sgn(f) = \begin{cases} 1 & f > 0 \\ -1 & f < 0 \end{cases} \tag{3.91}$$

When $f > 0$, the function is a positive one and when $f < 0$, the function is a negative one. The quadrature phase shift is defined in terms of the sign function as

$$X(f) = -j\, sgn(f) \tag{3.92}$$

The phase relation is shown in Figure 3.15(b), which is referred to as a step function. The approach to find the inverse transform of this function is to use the derivative of $X(f)$. The derivative can be obtained through a similar way as shown in (3.33), and the result is

$$\mathscr{F}^{-1}\left[\frac{dX(f)}{df}\right] = -j2\pi t x(t) \tag{3.93}$$

The next step is to obtain the derivative of $X(f)$. The derivative of a step function is a delta function at the transition position multiplied by a constant. The constant depends on the amplitude change of the step function. In this special case, the amplitude change is $-2j$ (from $+j$ to $-j$). Thus, the derivative is $-j2\delta(f)$. Substituting this relation in the above equation, the result is

$$\begin{aligned} x(t) &= \frac{1}{-j2\pi t}\mathcal{F}^{-1}\left[\frac{dX(f)}{df}\right] = \frac{j}{2\pi t}\int_{-\infty}^{\infty}[-j2\delta(f)]\,e^{j2\pi ft}df \\ &= \frac{j}{2\pi t}(-j2) = \frac{1}{\pi t} \end{aligned} \tag{3.94}$$

This equation has a singularity at $t = 0$. This result will be used in the Hilbert transform in Chapter 8. The time and frequency domain plots are shown in Figure 3.15.

3.10 SUMMARY

This section will list some of the results obtained previously as a quick reference. The following are some of the properties of the Fourier transform.

1. Linearity

$$\mathcal{F}[\alpha x(t) + \beta y(t)] = \alpha X(f) + \beta Y(f) \tag{3.15}$$

2. For even functions $x(t)$

$$X(f) = \int_{-\infty}^{\infty} x(t)\cos(2\pi ft)\,dt \tag{3.16}$$

3. For odd function $x(t)$

$$X(f) = -j\int_{-\infty}^{\infty} x(t)\sin(2\pi ft)\,dt \tag{3.17}$$

4. Duality

$$\text{If} \quad X(f) = \mathcal{F}[x(t)] \qquad \text{then} \qquad x(-f) = \mathcal{F}[X(t)] \tag{3.22}$$

5. Scaling

$$\mathcal{F}\left[x\left(\frac{t}{a}\right)\right] = aX(af) \tag{3.28}$$

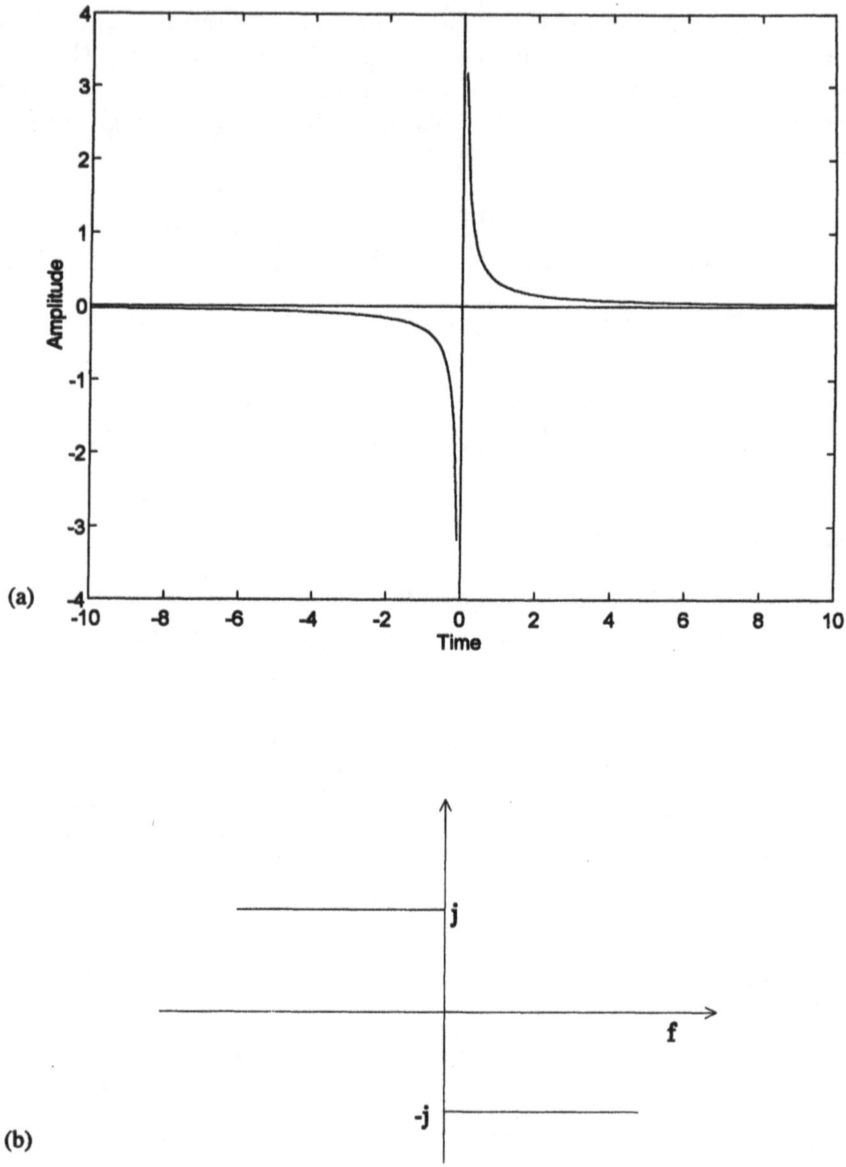

Figure 3.15 A quadrature phase shift: (a) time domain, (b) frequency domain.

$$\mathscr{F}^{-1}\left[X\left(\frac{f}{a}\right)\right] = ax(at) \tag{3.29}$$

6. Time shifting

$$\mathscr{F}[x(t - t_o)] = e^{-j2\pi f t_o}X(f) \tag{3.30}$$

7. Frequency shifting

$$\mathscr{F}^{-1}[X(f - f_o)] = e^{j2\pi f_o t}x(t) \tag{3.32}$$

8. Derivative

$$\mathscr{F}[x'(t)] = j2\pi f X(f) \qquad x(t)\big|_{t=\pm\infty} = 0 \tag{3.33}$$

9. Integral

$$\mathscr{F}\left[\int_{-\infty}^{t} x(\tau)\, d\tau\right] = \frac{\mathscr{F}[x(t)]}{j2\pi f} \qquad \int_{-\infty}^{t} x(\tau)\, d\tau\big|_{t=\infty} = 0 \tag{3.38}$$

10. Convolution

$$x(t) \star y(t) \leftrightarrow X(f)\ Y(f) \tag{3.62}$$

$$x(t)y(t) \leftrightarrow X(f) \star Y(f) \tag{3.65}$$

The properties of impulse function are

$$\int_{-\infty}^{\infty} \delta(t)\ dt = 1 \tag{3.8}$$

$$\int_{-\infty}^{\infty} x(t)\,\delta(t - t_o)\, dt = x(t_o) \tag{3.9}$$

$$\delta(t) = \lim_{f\to\infty} \frac{\sin(2\pi f t)}{\pi t} \tag{3.11}$$

$$\delta(t - t_o) = \lim_{f\to\infty} \frac{\sin[2\pi f(t - t_o)]}{\pi(t - t_o)} \tag{3.13}$$

Next, some of the commonly used Fourier transforms are listed below.

1. Rectangular function

$$X(t) = A \quad \text{for } -\frac{T}{2} < t < \frac{T}{2} \qquad X(f) = A \sin\frac{(\pi f T)}{\pi f} \tag{3.5}$$

$$= 0 \quad \text{elsewhere} \qquad = AT\,\text{sinc}(\pi f T) \tag{3.6}$$

(See Figure 3.1.)

2. Impulse function

$$x(t) = \delta(t) \qquad X(f) = 1 \tag{3.10}$$

(See Figure 3.2.)

3. Cosine function

$$x(t) = \cos(2\pi f_o t) \qquad X(f) = \frac{1}{2}[\delta(f - f_o) + \delta(f + f_o)] \tag{3.18}$$

(See Figure 3.3.)

4. Sine function

$$x(t) = \sin(2\pi f_o t) \qquad X(f) = -\frac{j}{2}[\delta(f - f_o) - \delta(f + f_o)] \tag{3.19}$$

(See Figure 3.4.)

5. Comb function

$$x(t) = \sum_{-\infty}^{\infty} \delta(t - nT) \tag{3.44}$$

$$X(f) = \frac{1}{T}\sum_{-\infty}^{\infty} \delta\left(f - \frac{n}{T}\right) \tag{3.55}$$

(See Figure 3.5.)

6. $x(t) = e^{j2\pi f_0 t}$

$$x(t) = e^{j2\pi f_0 t} \tag{3.70}$$

$$x(f) = \delta(f - f_0) \tag{3.71}$$

(See Figure 3.9.)

7. Windowed cosine function

$$x(t) = A\cos(2\pi f_0 t) \quad \text{for } -\frac{T}{2} < t < \frac{T}{2}; \quad = 0 \quad \text{elsewhere}$$

$$X(f) = \frac{A}{2}\left\{\frac{\sin[2\pi(f - f_0)T]}{\pi(f - f_0)} + \frac{\sin[2\pi(f + f_0)T]}{\pi(f + f_0)}\right\} \tag{3.73}$$

(See Figure 3.10.)

8. For an isosceles triangle

$$x(t) = A^2T^2\left(1 - \frac{|t|}{T}\right) \quad \text{for } |t| < T; \quad = 0 \quad \text{elsewhere}$$

$$X(f) = \frac{A^2\sin^2(\pi f T)}{(\pi f)^2} = [AT\,\text{sinc}(\pi f T)]^2 \tag{3.74}$$

(See Figure 3.11(b,c).)

9. Generalized cosine window function

$$x(t) = a + (1 - a)\cos\left(\frac{2\pi t}{T}\right) \quad \text{for } -\frac{T}{2} < t < \frac{T}{2}; \quad = 0 \quad \text{elsewhere}$$

$$X(f) = \frac{a\sin(\pi f T)}{\pi f} + \frac{1 - a}{2}\left[\frac{\sin[\pi(f - f_0)T]}{\pi(f - f_0)} \right. \tag{3.75}$$

$$\left. + \frac{\sin[\pi(f + f_0)T]}{\pi(f + f_0)}\right] \tag{3.77}$$

(See Figure 3.12.)

10. Gaussian function

$$x(t) = \frac{1}{\sqrt{2\pi\sigma}}e^{\frac{-t^2}{2\sigma^2}} \tag{3.78}$$

$$X(f) = e^{-2\pi^2\sigma^2 f^2} \tag{3.79}$$

(See Figure 3.13.)

11. Coherent RF pulse train

$$x(t) = \left\{\left[\sum_{n=-\infty}^{\infty} \delta(t - nT)\right] \star s_1(t)\right\} s_2(t) \cos(2\pi f_o t)$$

$$\text{where} \quad s_1(t) = 1 \quad \text{for } |t| < \frac{\tau}{2}$$

$$s_2(t) = 1 \quad \text{for } |t| < \frac{NT}{2}$$

$$X(f) = \frac{\sin(\pi f \tau)}{\pi^2 f} \sum_{n=-\infty}^{\infty} \left\{ \frac{\sin\left[\pi\left(f - f_0 - \frac{n}{T}\right)NT\right]}{f - f_0 - \frac{n}{T}} + \frac{\sin\left[\pi\left(f + f_o - \frac{n}{T}\right)NT\right]}{f + f_0 - \frac{n}{T}} \right\} \tag{3.86}$$

(See Figure 3.14(a,b).)

12. Quadrature phase shift

$$x(t) = \frac{1}{\pi t} \qquad x(f) = -jsgn(f) \tag{3.92}$$

(See Figure 3.15.)

REFERENCES

[1] Campbell, G. A., and Foster, R. M. *Fourier Integrals for Practical Applications,* New York, NY: Van Nostrand Reinhold, 1948.

[2] Erdilyi, A. *Table of Integral Transforms,* Vol. 1, New York, NY: McGraw Hill Book Co., 1954.

[3] Dwight H. B. *Tables of Integrals and Other Mathematical Data,* 4th Edition, New York, NY: MacMillan Co., 1961.

[4] Robinson, E. A. "A Historical Perspective of Spectrum Estimation," *IEEE Proc.,* Vol. 70, Sept. 1982, pp. 885–907.

[5] Papoulis, A. *The Fourier Integral and its Applications,* New York, NY: McGraw Hill Book Co., 1962.

[6] Papoulis, A. *Probability, Random Variables, and Stochastic Processes,* New York, NY: McGraw Hill Book Co., 1965.

[7] Bracewell, R. *The Fourier Transform and its Applications,* New York, NY: McGraw Hill Book Co., 1965.

[8] Brigham, E. O. *The Fast Fourier Transform,* Englewood Cliffs, NJ: Prentice-Hall, 1974.

[9] Carlson, A. B. *Communication Systems,* New York, NY: McGraw Hill Book Co., 1975.

[10] Stremler, F. G. *Introduction to Communication Systems,* 3rd Edition, Reading, MA: Addison-Wesley Publishing Co. 1990.

[11] Ziemer, R. E., and Tranter, W. H. *Principles of Communications: System, Modulation, and Noise,* 3rd Edition, Boston, MA: Houghton Mifflin Publishing Co., 1990.

[12] Stimson, G. W. *Introduction to Airborne Radar,* El Segundo, CA: Hughes Aircraft Co., 1983.

[13] Skolnik, M. I. *Introduction to Radar Systems,* New York, NY: McGraw Hill Book Co., 1962.

CHAPTER 4

Discrete Fourier Transform

4.1 INTRODUCTION

In this chapter, the discrete Fourier transform (DFT) will be discussed. The Fourier transform discussed in the previous chapter is quite useful, but the applications will be limited for two reasons. First, the function in the time domain must be representable in closed form so that the Fourier integral can be performed. Thus, unless the input function can be written in closed form, it is impossible to evaluate the integral. Second, even if the time function can be written in closed form, it might also be difficult to find a closed-form solution to the integral.

In a digital receiver, the input data are obtained from digitizing the input signal. In an electronic warfare (EW) environment, the input signal represented by a function in the time domain is usually unknown. Even if the input is known (such as simulated data generated from a sine or cosine function to test a signal processing algorithm), the signal is digitized before it is processed. When the input signal is in digitized form, the DFT is used to implement the Fourier transform discussed in the previous chapter. Unlike the Fourier transform, the DFT can be performed on any kind of digitized input data; therefore, its usage is not limited.

It is important to know that DFT does not provide the same result as the Fourier transform. It only provides an approximate solution. Sometimes these results can be very close to the desired result, but it rarely happens for short data lengths. At other times, the results generated from the DFT could be misleading. In order to better understand the DFT, the basic operation of the DFT will be discussed first. Then, some of the important properties will be presented. Finally, the fast Fourier transform (FFT) will be presented.

4.2 SIGNAL DIGITIZATION

A sine wave in the time domain t can be represented by

$$x(t) = A \sin(2\pi f_0 t) \tag{4.1}$$

where A is the amplitude and f_0 is the frequency. A set of digitized data can be generated from this signal by taking some discrete points on the sine wave. The actual device to perform this operation is called the analog-to-digital converter (ADC), which will be discussed in Chapter 6. The digitizing operation can be represented by a switch, as shown in Figure 4.1. It samples the input signal at a uniform rate. Of course, the input signal can be sampled at a nonuniform rate. However, in this book only the uniform digitized signal will be discussed because it is necessary for the DFT operation.

In this figure, the input is continuous but the output is a pulse train with the envelope following the input. It is obvious that not all the information in the signal is transferred to the output. The information between the sampling points is lost. Thus, the output is no longer exactly equal to the input signal. This loss of information will be covered in Section 4.5, where the bandwidth limitation of the signal is discussed.

Mathematically, the digitizing operation can be represented by multiplying the input signal with a comb function in the time domain, which can be written as

$$x(nt_s) = x(t)\sum_{-\infty}^{\infty}\delta(t - nt_s) \tag{4.2}$$

where n is an integer, t_s is the sampling interval, and $x(nt_s)$ represents the digitized signal. The digitized signal has output only at integer multiples of t_s in time, as shown in Figure 4.1.

4.3 GRAPHICAL DESCRIPTION OF DISCRETE FOURIER TRANSFORM (DFT)

In this section, the DFT procedure will be described through graphical illustration. Two aspects will be discussed: 1) the mathematical formula and 2) the time and frequency domain responses. A step-by-step approach will be taken starting from

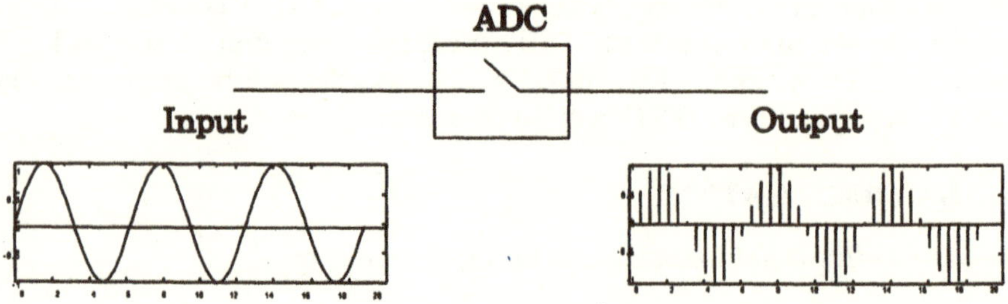

Figure 4.1 A digitized waveform.

signal digitization. The discussion is somewhat similar to Example 3.10. In Example 3.10, the data in the frequency domain are still continuous. In a DFT, the data in the frequency domain are also discrete, and this approach can avoid analytic integrals. Let us use a simple example to demonstrate the operation.

1. The input signal $x(t)$ is a $\text{sinc}^2 x$, which is continuous in the time domain, and its Fourier transform $X(f)$ is an isosceles triangle. The results are shown in Figure 4.2(a).
2. The sampling process can be represented by multiplying $x(t)$ by a comb function $b(t)$ with interval t_s. The Fourier transform of this function, $B(f)$, is also a comb function with amplitude $1/t_s$ and period of $1/t_s$. The results are shown in Figure 4.2(b).
3. The input signal $x(t)$ multiplied by $b(t)$ can be written as $y_1(t) = x(t)b(t)$. The corresponding frequency domain operation is $Y_1(f) = X(f) \bigstar B(f)$. The results are shown in Figure 4.2(c).
4. The input signal must be limited in length. This operation can be achieved by using a rectangular function $w(t)$. This function will cover N points from 0 to $N-1$. The Fourier transform of $w(t)$ is a sinc function $W(f)$, as shown in Figure 4.2(d).
5. The output $y_1(t)$ is multiplied by the rectangular function $w(t)$ to limit data length. This operation can be written as $y_2(t) = y_1(t)w(t)$. The corresponding frequency domain operation is $Y_2(f) = Y_1(f) \bigstar W(f)$. The results are shown in Figure 4.2(e). It should be noted that the output is periodic in the frequency domain, but is still continuous. One may consider that these are all the operations required. However, in order to obtain the frequency information, a Fourier transform of $y_2(t)$ is required. As a result, the frequency information must still be obtained through an analytical operation; hence, this is not the desired final result of the DFT. This analytic operation must be avoided. If one can obtain the frequency information in discrete values, the analytic operation can be avoided. The next two steps are required to eliminate the analytic operation and at the same time generate discrete components in the frequency domain. However, these operations also change the shape of the input signal.
6. In order to obtain discrete frequency components, a second comb function in the frequency domain $C(f)$ is needed. This comb function cannot be created with an arbitrary period. It has a period of $1/(Nt_s)$, which is determined by the data length Nt_s. The inverse Fourier transform is $c(t)$ in the time domain with a period of Nt_s, which matches the length of the rectangular function $w(t)$. The results are shown in Figure 4.2(f).
7. The last step is to obtain the frequency information in discrete form. This operation is achieved by multiplying $Y_2(f)$ by $C(f)$, which can be written as $X(k) = Y_2(k)C(k)$. In this operation, the frequency can be written as

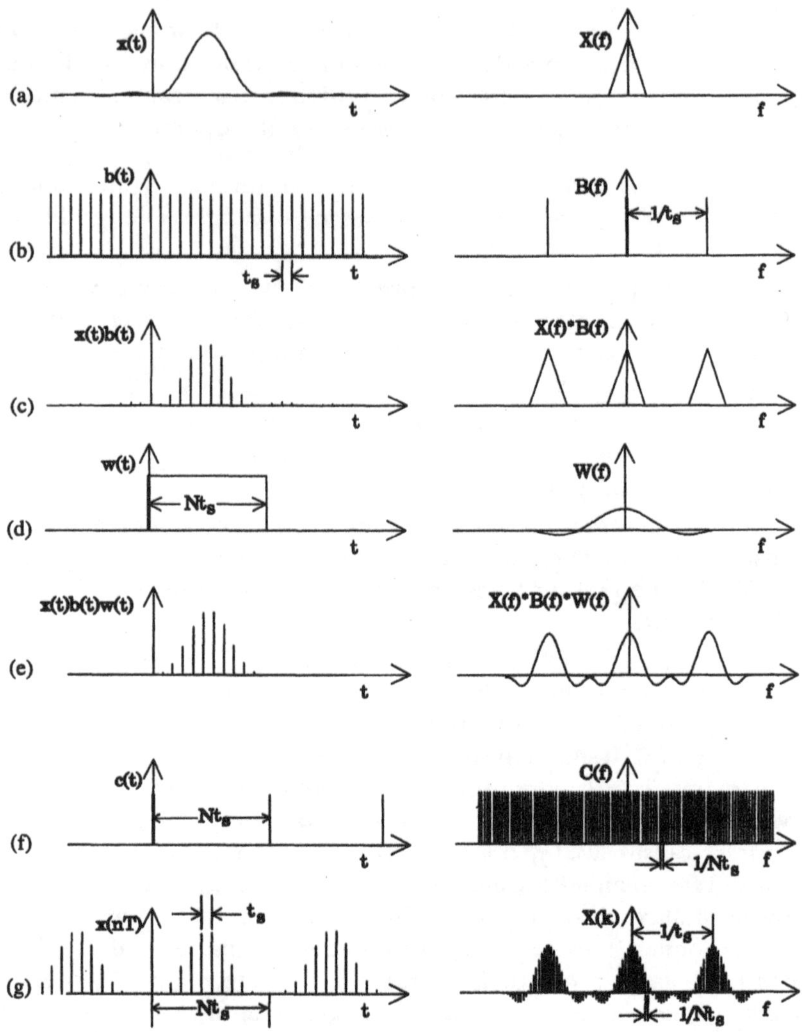

Figure 4.2 Graphic demonstration of discrete Fourier transform: (a) input $x(t)$ and $X(f)$, (b) comb function $b(t)$ and $B(f)$, (c) $x(t)b(t)$ and $X(f) \star B(f)$, (d) rectangular function $w(t)$ and $W(f)$, (e) $x(t)b(t)w(t)$ and $X(f) \star B(f) \star W(f)$, (f) comb function $c(t)$ and $C(f)$, (g) output $x(nt_s)$ and $X(k)$.

$f = k/(Nt_s)$ where k is an integer and $1/(Nt_s)$ is the frequency data resolution. In the time domain, the corresponding operation is $x(nt_s) = y_2(t) \star c(t)$ where $x(nt_s)$ represents discrete time domain data. This operation generates an infinite replica of the digitized input signal. The results are shown in Figure 4.2(g). This is a very important step because after this step, the input signal in the time domain is no longer the digitized signal shown in Figure 4.2(e). The input signal becomes periodic with period Nt_s. If one performs all of the above operations, the approximate frequency components of the input signal can be obtained.

4.4 ANALYTIC APPROACH TO DISCRETE FOURIER TRANSFORM [1–7]

The analytic approach will follow the graphical demonstration discussed in the previous section. The input signal is $x(t)$ and the comb function $b(t)$ is

$$b(t) = \sum_{-\infty}^{\infty} \delta(t - nt_s) \tag{4.3}$$

where t_s is the sampling interval. The function $y_1(t)$ is

$$y_1(t) = x(t)b(t) = x(t)\sum_{-\infty}^{\infty}\delta(t - nt_s) = \sum_{-\infty}^{\infty} x(t)\delta(t - nt_s) \tag{4.4}$$

The rectangular function $w(t)$ covers the delta function from $t = 0$ to $N - 1$, which limits the summation of $y_1(t)$; thus,

$$y_2(t) = y_1(t)w(t) = \sum_{n=0}^{N-1} x(nt_s) \tag{4.5}$$

The next step is to multiply $Y_2(f)$ by a comb function $C(f)$, which has a period of $1/(Nt_s)$. This operation is equivalent to convolving $y_2(t)$ with $c(t)$, and the result is

$$\begin{aligned} y(t) &= y_2(t) \star c(t) = \sum_{n=0}^{N-1} x(nt_s)\delta(t - nt_s) \star \sum_{r=-\infty}^{\infty} \delta(t - rNt_s) \\ &= \sum_{r=-\infty}^{\infty}\sum_{n=0}^{N-1} x(nt_s)\, \delta(t - nt_s - rNt_s) \end{aligned} \tag{4.6}$$

The last step is to find the Fourier transform of $y(t)$. This step can be written as

$$\begin{aligned} X(f) &= \int_{-\infty}^{\infty} \sum_{r=-\infty}^{\infty}\sum_{n=0}^{N-1} x(nt_s)\, \delta(t - nt_s - rNt_s)\, e^{-j2\pi ft} dt \\ &= \sum_{r=-\infty}^{\infty}\sum_{n=0}^{N-1} x(nt_s) \int_{-\infty}^{\infty} \delta(t - nt_s - rNt_s)\, e^{-j2\pi \frac{k}{Nt_s} t} dt \\ &= \sum_{r=-\infty}^{\infty}\sum_{n=0}^{N-1} x(nt_s)\, e^{-j2\pi kn/N - j2\pi kr} \\ &\equiv X(k) \end{aligned} \tag{4.7}$$

In the above equation, the relation $f = k/(Nt_s)$ is used. In the frequency domain, the output only has values at $f = k/(Nt_s)$ because of the sampling property of the delta function. The notation $X(k)$ is used to represent discrete components of the Fourier transform.

From the above result, the DFT has an infinite number of frequency components. However, because of the periodical property of $e^{-j2\pi kr}$ where k and r are integers, the output spectrum repeats the data from $k = 0$ to $N - 1$. Usually only one such period is used to represent the output. If one considers $r = 0$ in the above equation, the result is

$$X(k) = \sum_{n=0}^{N-1} x(n)\, e^{\frac{-j2\pi kn}{N}} \tag{4.8}$$

This is the well-known relation of the Fourier transform in discrete form. In this equation, $t_s = 1$ is used.

The inverse DFT can be written as

$$x(n) = \frac{1}{N}\sum_{k=0}^{N-1} X(k)\, e^{j2\pi kn/N} \tag{4.9}$$

The inverse transform can be proven as follows. By substituting the result in (4.9) into (4.8), the result of (4.8) can be obtained. This can be shown as

$$\begin{aligned} X(k) &= \frac{1}{N}\sum_{n=0}^{N-1} x(n)\, e^{\frac{-j2\pi kn}{N}} \\ &= \sum_{n=0}^{N-1}\left[\frac{1}{N}\sum_{k'=0}^{N-1} X(k')\, e^{\frac{j2\pi k'n}{N}}\right] e^{\frac{-j2\pi kn}{N}} \\ &= \sum_{k'=0}^{N-1} X(k')\left[\frac{1}{N}\sum_{n=0}^{N-1} e^{\frac{j2\pi(k'-k)n}{N}}\right] \\ &= X(k) \end{aligned} \tag{4.10}$$

In the above equation, the relation

$$\begin{aligned} \sum_{n=0}^{N-1} e^{\frac{j2\pi(k'-k)n}{N}} &= N \qquad k' = k \\ &= 0 \qquad k' \neq k \end{aligned} \tag{4.11}$$

is used. Equations (4.8) and (4.9) are the well-known relations of the DFT and the inverse discrete Fourier transform (IDFT).

4.5 PROPERTIES OF DISCRETE FOURIER TRANSFORM

Discrete Fourier transform is different from continuous Fourier transform. Fourier transform of a continuous signal needs not be limited in length in the time domain. The output in the frequency domain is not band-limited either. However, the input signal in a DFT is limited in length in the time domain and becomes periodic due to the sampling effect in the frequency domain. The output in the frequency domain is also periodic. Thus, the DFT is only an approximation of the continuous Fourier transform. One must carefully evaluate the output of the DFT because the result could be erroneous if the input signal does not fulfill certain requirements. The following properties are important for utilization of the DFT.

4.5.1 Limited Frequency Bandwidth

Figures 4.2(b) and 4.2(g) show that the output in the frequency domain has a period of $f_s = 1/t_s$ where f_s is the sampling frequency and t_s is the sampling interval. Since the frequency output from a real signal has two sidebands, the signal frequency must be less than $f_s/2$. If the signal frequency is higher than $f_s/2$, the periodic extensions of the output will overlap as shown in Figure 4.3. Figure 4.3(a) shows the signal $x(t)$ and its Fourier transform. Figure 4.3(b) shows the digitizing comb functions $c(t)$ and $C(f)$. Figure 4.3(c) shows the product of $x(t)$ and $c(t)$ and the corresponding frequency domain output $X(f) \star C(f)$. The true spectrum is shown

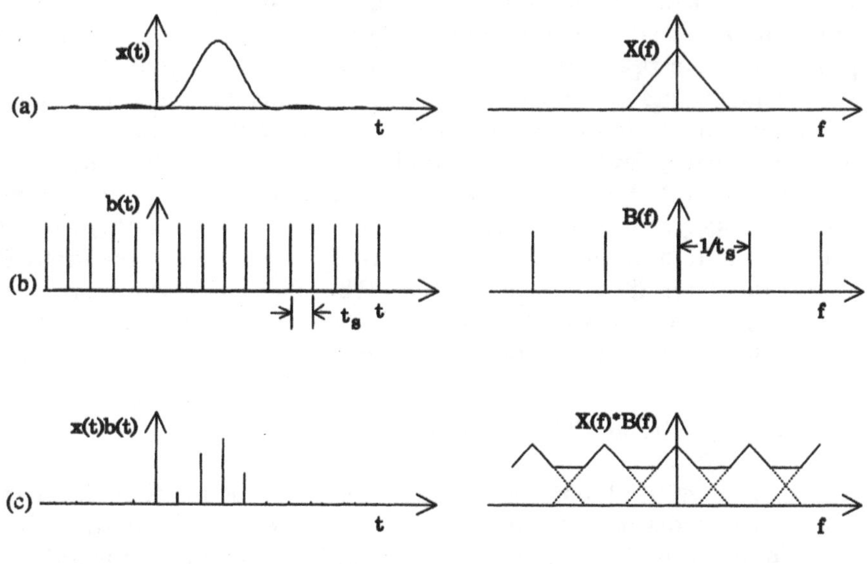

Figure 4.3 Aliasing effect: (a) $x(t)$ and $X(f)$, (b) $c(t)$ and $C(f)$, (c) $x(t)c(t)$ and $X(f)*C(f)$.

by the dotted lines. It is obvious that the output spectrum does not match the input spectrum because of the overlap. Therefore, in general, the input signal bandwidth is kept less than $f_s/2$. This is referred to as the Nyquist sampling theorem. If the input signal has a bandwidth higher than $f_s/2$, the spectrum will alias into the baseband. Sometimes, DFT can be used to solve the problem of several narrowband signals spreading over a bandwidth greater than $f_s/2$, such as in an electronic warfare (EW) receiver. In order to use the DFT for this purpose, special procedures are required in the processing and this problem will be further discussed in Chapter 5.

If the sampling interval t_s is decreased to a very small value, the unambiguous bandwidth will widen. When all the continuous time data in the input signal are used, which is equivalent to making $t_s \rightarrow 0$, the bandwidth of equivalent sampled responses becomes infinitely wide. Therefore, one can consider that the limitation on the bandwidth is due to information being lost between the sampled points.

4.5.2 Unmatched Time Interval

The input signal in the time domain is windowed by a rectangular function, which contains N data points from $n = 0$ to $N - 1$ as discussed in the previous section. If the input is a periodic function, such as $\sin(2\pi f_o t)$, the window may not match the period of the sine function. Since the input signal is always periodic with respect to N, the input signal will not be a continuous sine wave when the window does not match the period of the signal. Figure 4.4 shows the results of such an operation. Figure 4.4(a) shows that the window matches the period of the input signal. Under this condition, the signal appears continuous and its discrete Fourier has two peaks. The positions of the peaks represent the frequency of the signal and the input frequency matches one of the DFT output components. Figure 4.4(b) shows the case when the time window does not match the period of the input signal. Under this condition, the sine wave is no longer continuous, but breaks into many discontinuous sections. As a result, there are many spectrum lines in the frequency domain. The additional spectrum lines are caused by the sharp discontinuity in the input signal. The truncation in time domain is equivalent to convolving the input signal with a sinc function in the frequency domain. This phenomenon is often referred to as the leakage effect because the frequency is leaked from one main component to many sidelobes due to the windowing in the time domain.

4.5.3 Overlapping Aliasing Effect on Real Data

First let us consider the Fourier transform in continuous form. If the data are complex, the frequency domain output will be single sideband as shown in Section 3.9. For example, if the input is $e^{\pm j2\pi f_o t}$, the output is one impulse function on one side of the y-axis depending on the sign of f_o. If the data are real, the frequency

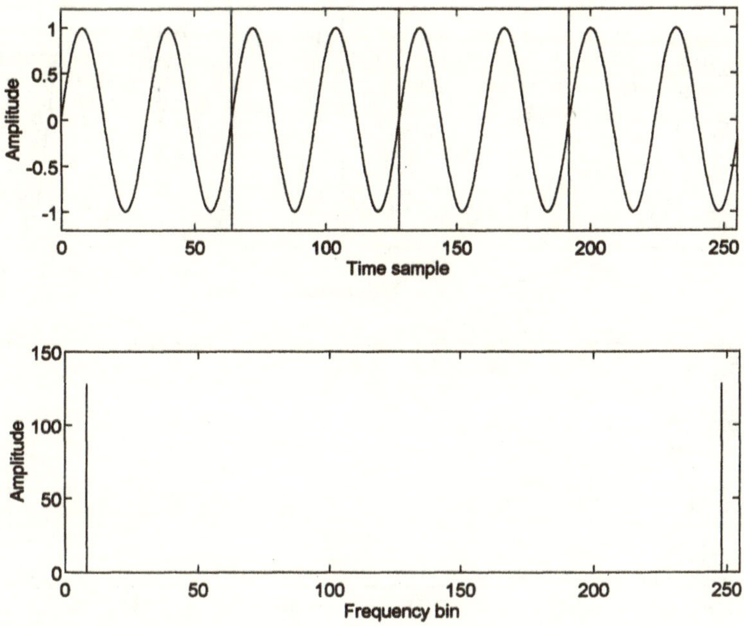

(a)

Figure 4.4 Leakage effect of a sine wave: (a) matching window, (b) mismatching window.

domain output will have positive and negative components. For example, if the input is $\sin(2\pi f_0 t)$, the output will be at $\pm f_0$. When the input frequency is low, the two spectra are close to the y-axis. When the input frequency increases, the two spectra will separate far apart. Let us assume that real data are used as input. In DFT, the output spectra are only shown from 0 to $N-1$. The negative frequency does not show up on the negative side of the y-axis. Both the spectra are shown in the 0 to $N-1$ range. They are symmetrical with respect to $N/2$. One should note that because of the periodic nature of the output, the output at $X(0)$ (the lowest frequency bin) and $X(N-1)$ (the highest frequency bin) should be next to each other. Figure 4.5 shows the power spectra of a sine function. In Figure 4.5(a), the input frequency is very low, the two spectra are slightly overlapped at the low end (close to $X(0)$) and high end (close to $X(N-1)$). This result still resembles the true frequency response.

When the input frequency increases (in the DFT case), the positive spectrum moves away from $X(0)$ toward the right. The negative spectrum should move away from $X(0)$ toward the left, but there is no negative index of X. Since $X(0)$ is next to $X(N-1)$, moving away from $X(0)$ toward the left is equivalent to moving away from $X(N-1)$ toward the left. Thus, the negative spectrum moves away from

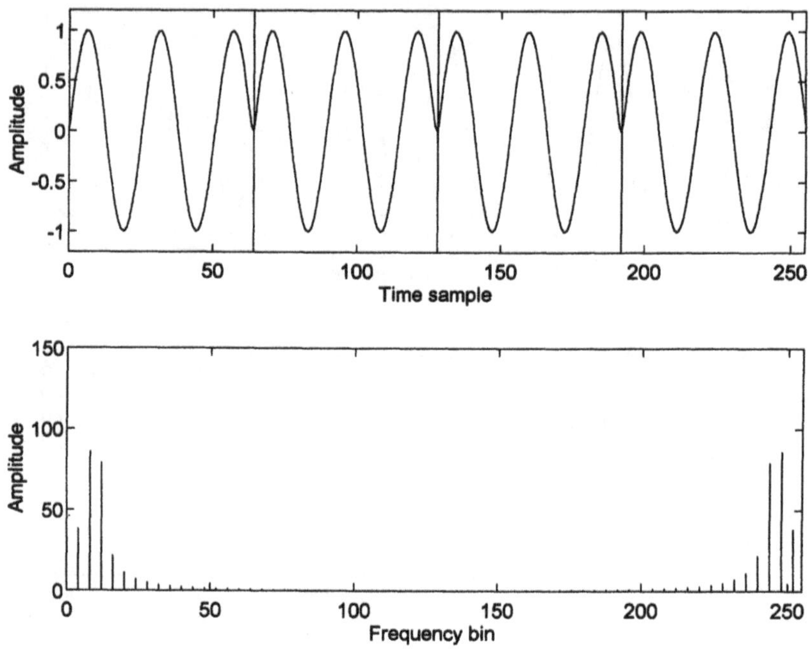

(b)

Figure 4.4 (continued).

$X(N-1)$ toward $X(0)$. When the input frequency increases approaching $f_s/2$, the two spectra are close together at near $X(N/2)$ as shown in Figure 4.5(b). In general, since each spectrum has sidelobes, these two spectra will interfere with each other. When the input frequency equals half the sampling frequency ($f_s/2$), the two responses will overlap and highly interfere with each other.

In general, when the input signal frequency is close to

$$f_i \approx \frac{nf_s}{2} \tag{4.12}$$

where n is an integer including zero, the positive and negative spectra will interfere with each other. The amount of interference depends on the leakage effect and the bandwidth of the input signal and the sidelobes. Even if the input frequency bandwidth is less than $f_s/2$ (i.e., the Nyquist sampling rate is fulfilled), the positive and negative spectra may still interfere with each other. If the input data are complex, this overlapping effect can be avoided.

In summary, if the bandwidth of input signals is less than $f_s/2$ and the input frequency range is selected not near $nf_s/2$, the interference will not be a problem.

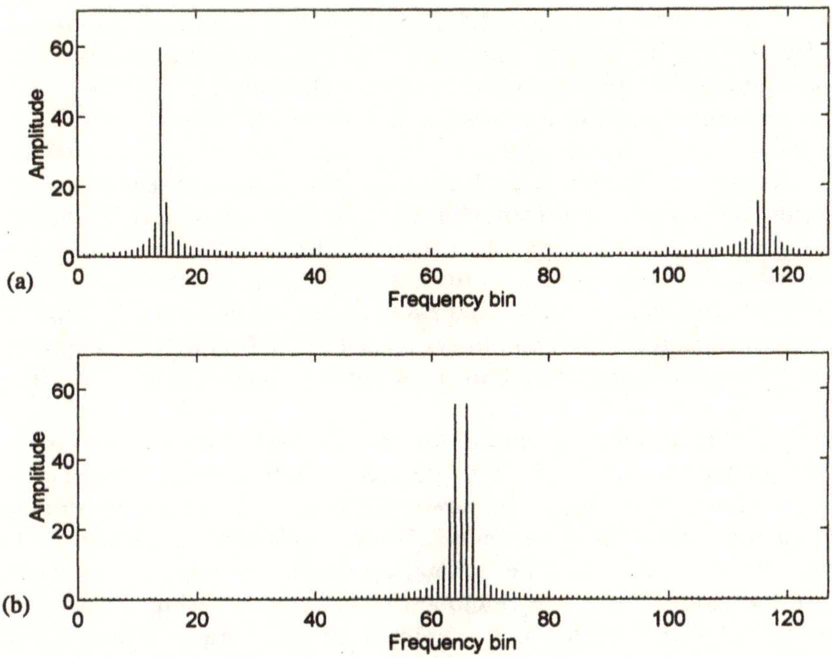

Figure 4.5 Positive and negative spectra: (a) low frequency, (b) high frequency.

However, the leakage phenomenon is not easy to avoid. Special window functions discussed in the next section will reduce the discontinuity caused by rectangular windows.

4.6 WINDOW FUNCTIONS [8]

In the last section, it was shown that when the rectangular window in the time domain does not match an integer number of cycles in the input signal, discontinuity will occur at the window edges. This discontinuity will generate spurious spectrum lines that do not exist in the true Fourier transform. The window size in the time domain is often prefixed, especially when the FFT algorithm (which will be discussed in Section 4.7) is used. The number of sample points used in the FFT is usually 2^m, where m is an integer. Therefore, it is highly unlikely that the window length will match an integer number of periods of the input signal.

It is clearly seen from Figure 4.4(b) that the mismatch occurs at the ends of the rectangular window. One obvious way to reduce the discontinuity is to reduce the amplitudes at the end of the time window to minimize or eliminate the discontinuity. The basic idea is to create some special windows other than the rectangular one discussed in Sections 4.3 and 4.4. Figure 4.6(a) shows the discontinuity at the ends of a rectangular window and Figure 4.6(b) shows the results of a cosine window. Although in both cases the signals are discontinuous at the edges of the windows, the cosine window reduces the severity of the discontinuity.

A more commonly used way to explain the leakage effect is through convolution. As discussed in Section 4.3, selecting a certain length of data is equivalent to multiplying the data by a rectangular window. This operation is equivalent to convolving the signal with a sinc function in the frequency domain. Since the sinc function contains many sidelobes, these sidelobes will appear in the frequency domain. When the rectangular window size matches the period of the input sinusoidal signal, the period of the comb function used in the frequency domain sampling matches the zeros in the sinc function. Thus, there are no sidelobes in the frequency domain.

Under general sampling conditions, in order to reduce the output sidelobes in the frequency domain, a window with low sidelobes in the frequency domain will be used. The cosine window has lower sidelobes in the frequency domain than a rectangular window. There are many different kinds of windows that can be used to reduce the discontinuity. The general approach to designing a window is to minimize the amplitude of the sidelobes. In general, lowering the sidelobes will increase the width of the mainlobe, which is an undesirable effect. One way to evaluate the performance of the window is to measure the width of the main spectrum lobe and the amplitudes of the sidelobes. Low sidelobes result in a high dynamic range for EW receivers, while a wide mainlobe width reflects poor frequency resolution. Therefore, selecting a window is normally based on a tradeoff between frequency resolution and dynamic range.

Usually the frequency domain response determines the performance of the window. The time domain is also very important since it determines the transient

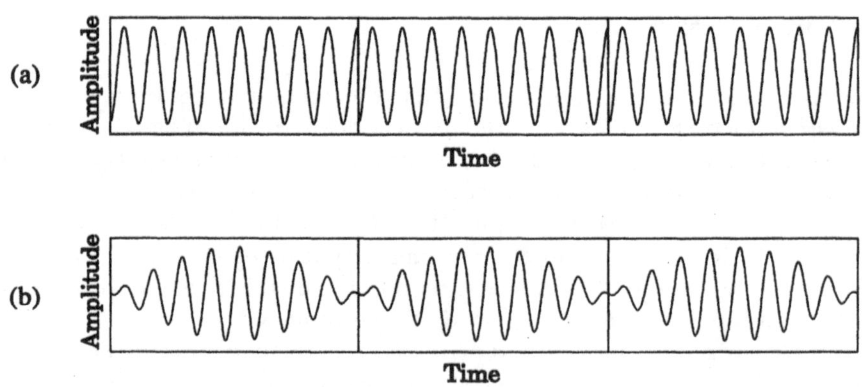

Figure 4.6 Discontinuity at ends of windows: (a) rectangular window, (b) cosine window.

effect. The time and frequency responses of the window functions are related by the Fourier transform. Some common window functions will be discussed as follows.

4.6.1 Rectangular Window

This is often used as a reference to evaluate other types of window functions. The frequency domain response can be obtained from the Fourier transform of a rectangular function. Width of the mainlobe (i.e., from two nodes of the mainlobe) equals to $2/N$. In a comparison of windows, this width is often considered as unity. Figure 4.7 shows the frequency response of a rectangular window. In order to obtain the fine structure of the window (i.e., sidelobes), the figure is obtained through zero padding, which will be discussed in the next section. In this plot, let us define $N = 128$ as the window width rather than the total FFT points, and $31N$ of zeros are added after the data (i.e., 128 1's followed by 3,968 0's) as shown in Figure 4.7(a). The peak of the rectangular window in the frequency domain is at zero (i.e., around $X(0)$ and $X(32N - 1)$), as shown in Figure 4.7(b). Since the peak appears at the two ends, it is difficult to see the window shape. The peak can be

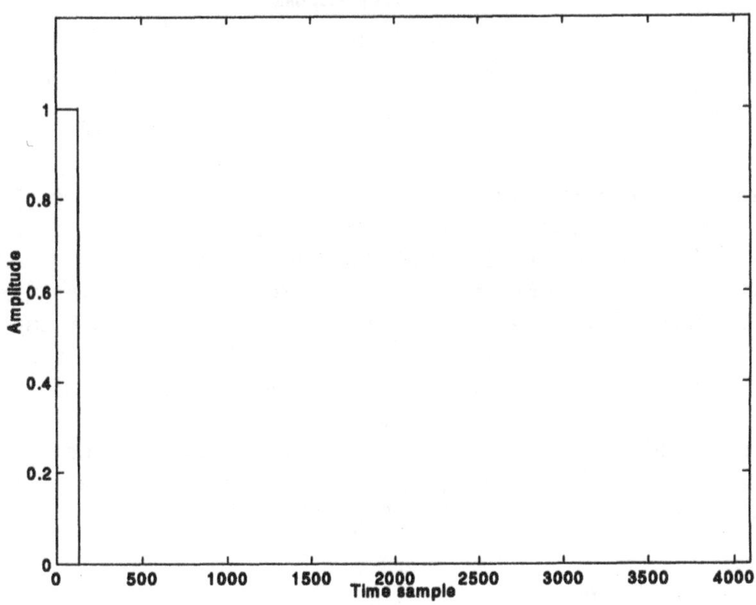

(a)

Figure 4.7 Response of a rectangular window: (a) time domain response, (b) frequency domain response, (c) frequency response after shift, (d) frequency response near the peak.

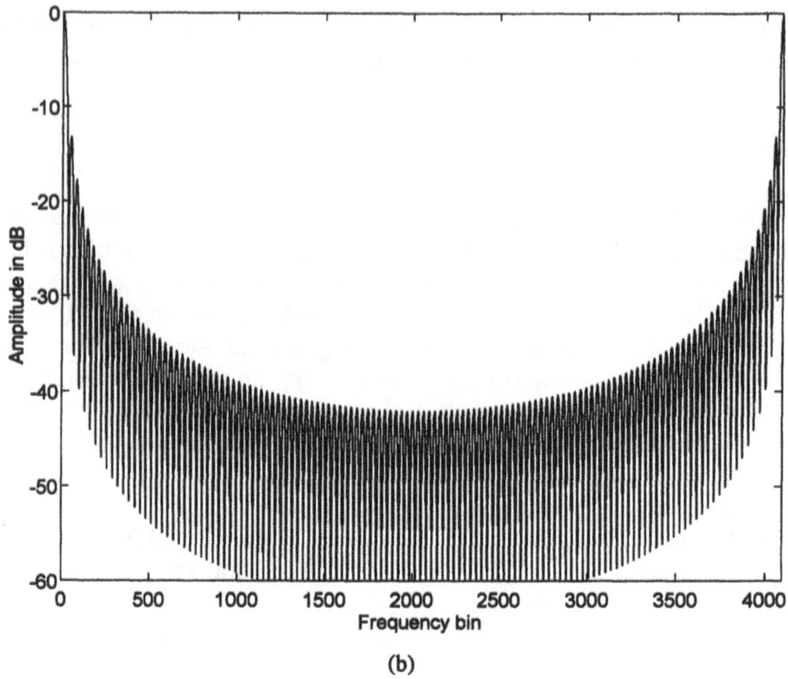

Figure 4.7 (continued).

shifted toward the center of the figure so that the entire peak is displayed undivided. In order to shift the peak to the center of the figure at $X(16N)$, first the points that form $X(16N)$ to $X(32N-1)$ are plotted, then followed by $X(0)$ to $X(16N-1)$. The result is shown in Figure 4.7(c). Figure 4.7(d) only plots the center portion around the peak. The width of the mainlobe is $2/N$ and this is the narrowest frequency response in comparison with other windows. The highest sidelobe is about 13-dB down from the mainlobe. Other windows are designed to reduce the sidelobes.

A zero padding scheme and a shifting of frequency domain data are applied to all the windows in the following discussions in order to produce a better visual display of the shapes of the windows. Many different windows are listed in [8].

4.6.2 Gaussian Window

A Gaussian function is written as

$$w(n) = e^{-\frac{1}{2}\left[\frac{2a(n-N/2)}{N}\right]^2} \tag{4.13}$$

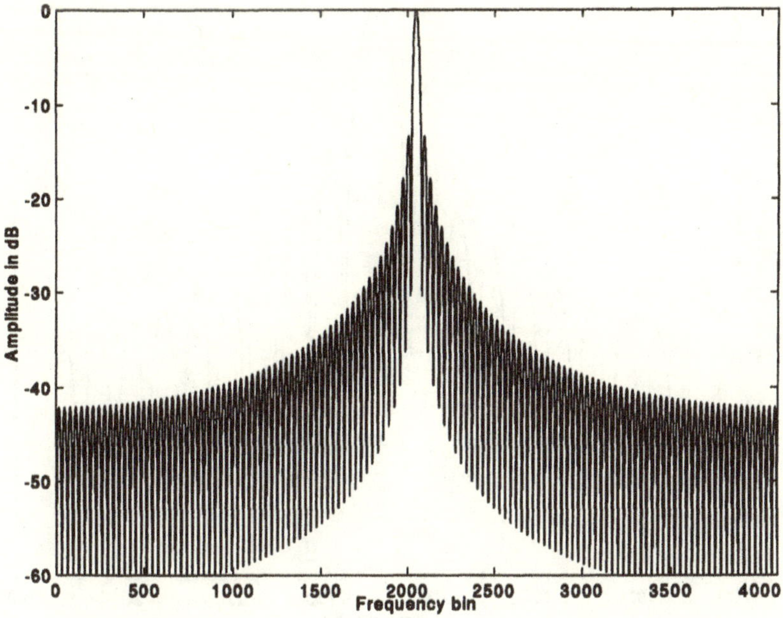

(c)

Figure 4.7 (continued).

where a is a constant and N is the total number of points in the window. The Fourier transform of a Gaussian window in time domain is a Gaussian function in the frequency domain, as shown in (3.79). Theoretically, a Gaussian window should not produce any sidelobes. However, in actual applications the Gaussian window must be truncated to a certain length for the DFT operation. This truncation will generate sidelobes in the frequency domain. When $a = 2.5$, the width of the mainlobe is about 1.33 times that of the rectangular window. The highest sidelobe is 42-dB down from the mainlobe. This result is shown in Figure 4.8. It should be noted that although the data in the time domain are zero padded, the zeros are not shown in Figure 4.8(a).

4.6.3 Cosine Window Raised to the *a*th Order

This window function is written as

$$w(n) = \cos^a\left[\frac{\pi(n - N/2)}{N}\right] \tag{4.14}$$

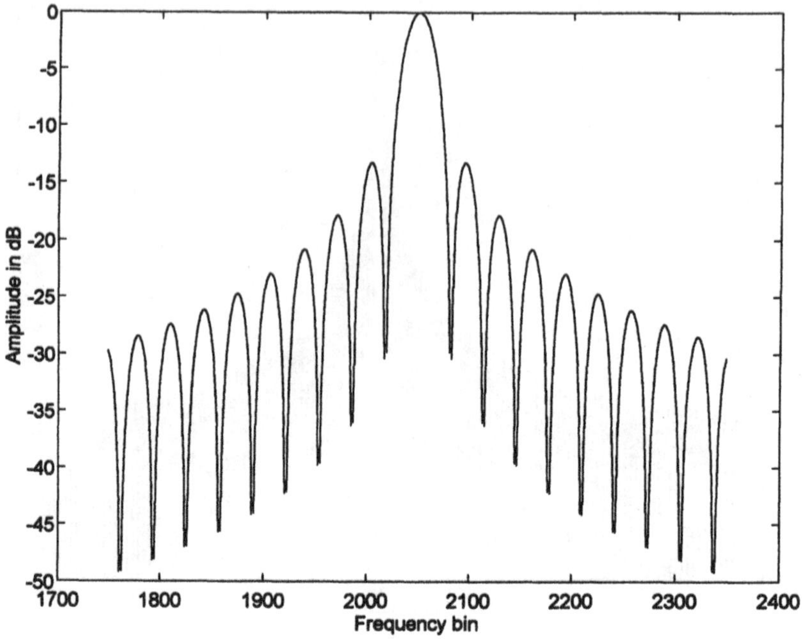

(d)

Figure 4.7 (continued).

When $a = 2$, the window function is called the Hanning window (due to Von Hann). The width of the mainlobe is 1.5 and the highest sidelobe generated by the Hanning window is –32 dB. Figure 4.9 shows the time and frequency domain responses.

4.6.4 Generalized Hamming Window

This window function is written as

$$w(n) = a + (1 - a)\cos\left[\frac{2\pi(n - N/2)}{N}\right] \tag{4.15}$$

where $a < 1$. When $a = .54$, the window is referred to as the Hamming window (due to R. W. Hamming). The mainlobe width is 1.36 and the highest sidelobe is –43 dB. The time and frequency domain responses of the Hamming window are illustrated in Figure 4.10. It is interesting to notice that all the sidelobes shown have approximately the same amplitude. This result is also shown in Chapter 3. However, the result in Chapter 3 is generated analytically. This frequency response is calculated from the DFT.

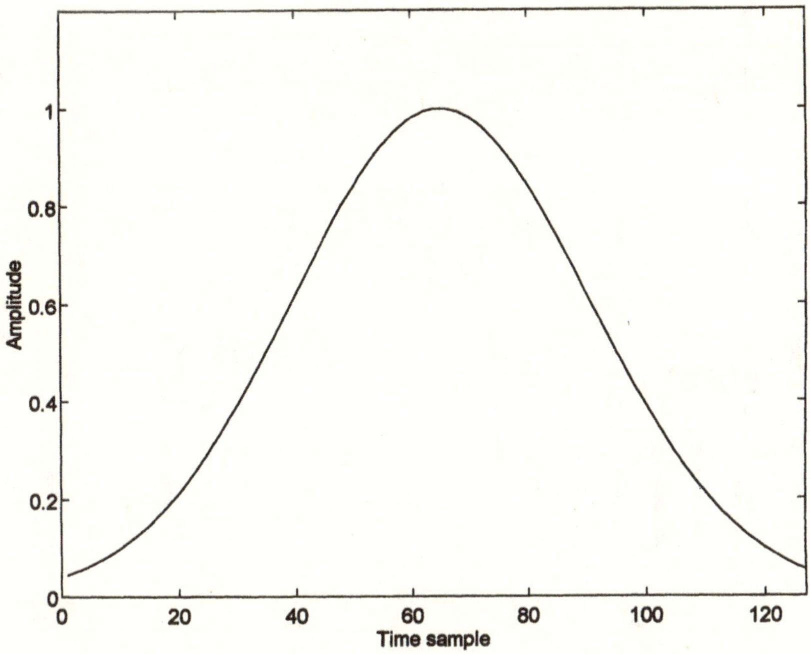

(a)

Figure 4.8 Response of a Gaussian window: (a) time domain, (b) frequency domain.

When $a = .5$, the generalized Hamming window becomes a Hanning window since

$$\cos^2\left[\frac{\pi(n - N/2)}{N}\right] = \frac{1}{2}\left\{1 + \cos\left[\frac{2\pi(n - N/2)}{N}\right]\right\} \tag{4.16}$$

Note that not only the names of the windows (Hamming and Hanning) are close, but the functions can also be related under special conditions.

The Hamming window has lower sidelobes than the Hanning window. Assume that both the Hamming and Hanning windows have N points. Since the endpoints of the Hanning window are zero, in reality the Hanning window has only $N - 2$ points. One can consider that this is the reason that the Hanning window has higher sidelobes, because it is shorter than the Hamming window. In fact, this is the reason why the Hamming window is also called the raised cosine window.

4.7 FAST FOURIER TRANSFORM (FFT) [1–7, 9–27]

The DFT is, in general, computation-intensive. For example, to compute the DFT, (4.7) will require N data points, N multiplications and summations to calculate one

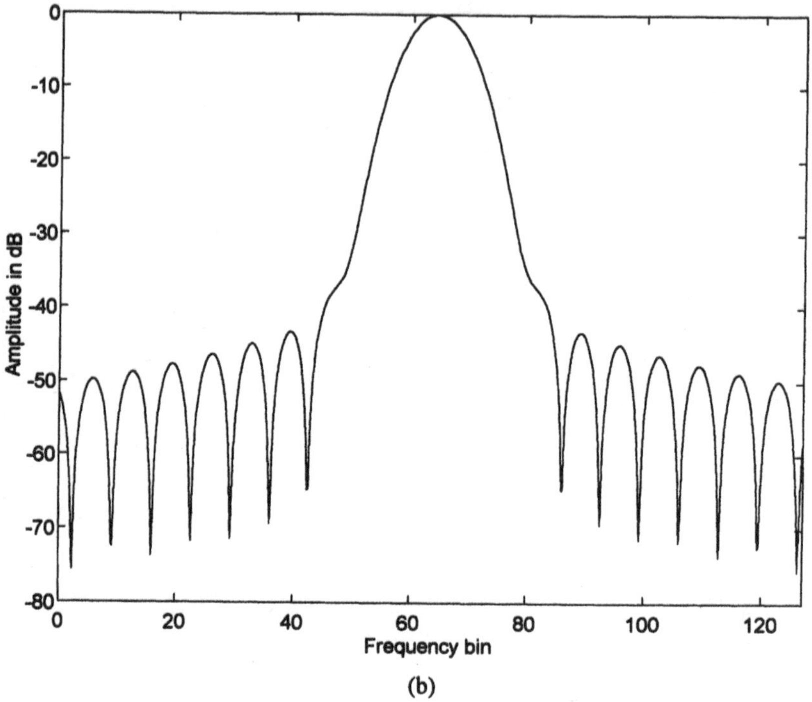

Figure 4.8 (continued).

component in the frequency domain. This implies that N^2 operations are needed in the complete DFT.

In 1965, Cooley and Tukey discovered a very efficient way to calculate the DFT based on the symmetry of the $e^{-j2\pi nk}$ kernel. This discovery has allowed the DFT to become the preferred method of analyzing the spectrum of signals. This method is later referred to as the FFT, because of the improvement in calculation speed. The FFT can simply be considered as an efficient way to calculate the DFT. The results obtained from the FFT are identical to those obtained from the DFT for 2^m data points. All the characteristics and properties of the DFT are also identical in the FFT.

The derivation of the FFT can be obtained from the DFT. The procedure follows the discussion in [17]. The DFT in (4.8) can be rewritten here as

$$X(k) = \sum_{n=0}^{N-1} x(n)\, e^{-j2\pi kn/N} \tag{4.8}$$

where $N = 2^m$ and m is an integer. The notation of the above equation can be simplified by letting

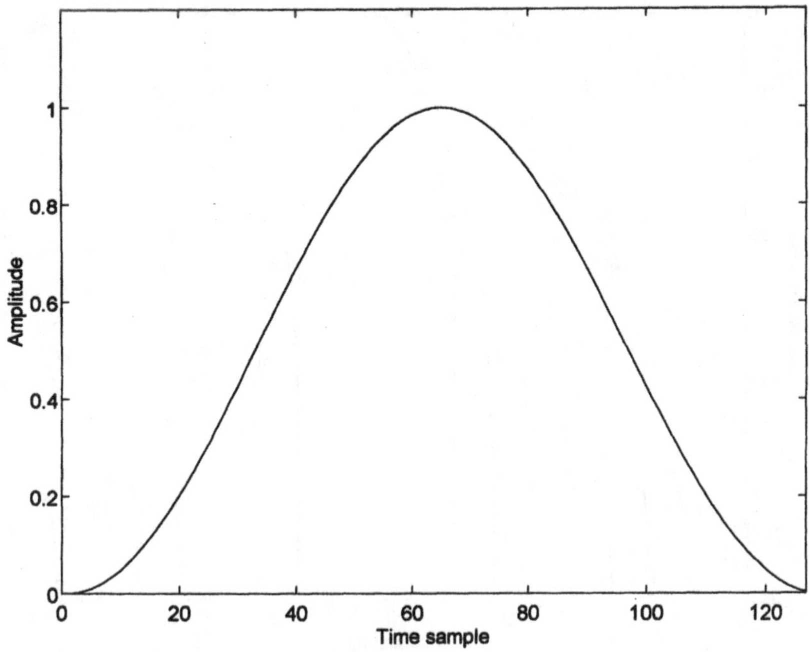

(a)

Figure 4.9 Cosine window raised to *a*th order: (a) time domain, (b) frequency domain.

$$W = e^{\frac{-j2\pi}{N}} \tag{4.17}$$

Equation (4.8) can be written as

$$X(k) = \sum_{n=0}^{N-1} x(n)\, W^{kn} \tag{4.18}$$

Let us use $N = 8$ as an example to illustrate the FFT operation. Under this condition, the k and n values in the above equation are

$$\begin{aligned} k &= 0, 1, \ldots . 7 \\ n &= 0, 1, \ldots . 7 \end{aligned} \tag{4.19}$$

Both k and n can be expressed in binary form as

$$\begin{aligned} k &= k_2 4 + k_1 2 + k_0 \\ n &= n_2 4 + n_1 2 + n_0 \end{aligned} \tag{4.20}$$

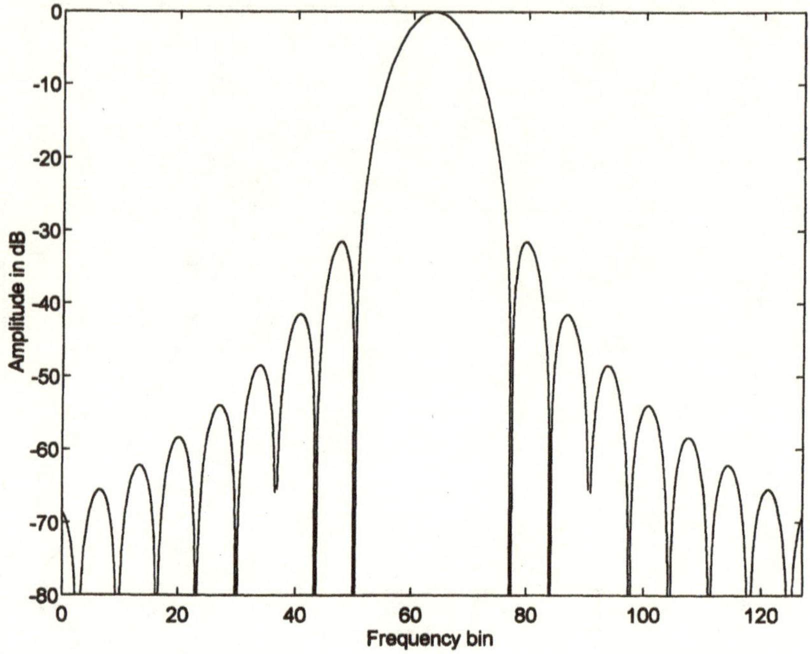

(b)

Figure 4.9 (continued).

where k_i and n_i ($i = 0, 1, 2$) both can take only one of two values: 0 or 1. Using these expressions of k and n, (4.18) can be written as

$$X(k_2, k_1, k_0) = \sum_{n_0=0}^{1}\sum_{n_1=0}^{1}\sum_{n_2=0}^{1} x(n_2, n_1, n_0)\, W^{(k_2 4+k_1 2+k_0)(n_2 4+n_1 2+n_0)} \tag{4.21}$$

It should be noted that

$$W^{m+n} = W^m W^n \tag{4.22}$$

Substituting the k and n relations from (4.20) into this equation, the result is

$$W^{(k_2 4+k_1 2+k_0)(n_2 4+n_1 2+n_0)} = \left[W^{(k_2 4+k_1 2+k_0)\, n_2 4}\right]\left[W^{(k_2 4+k_1 2+k_0)\, n_1 2}\right]\left[W^{(k_2 4+k_1 2+k_0)\, n_0}\right] \tag{4.23}$$

Each individual term can be written as

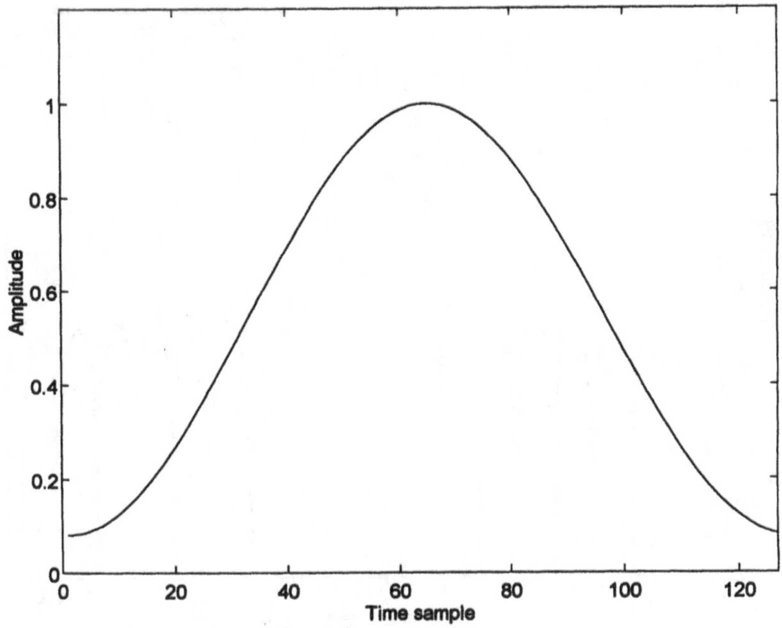

(a)

Figure 4.10 Response of a Hamming filter: (a) time domain, (b) frequency domain.

$$
\begin{aligned}
W^{(k_2 4+k_1 2+k_0)\, n_2 4} &= \left[W^{8(k_2 2+k_1)\, n_2}\right] W^{k_0 n_2 4} = W^{k_0 n_2 4} \\
W^{(k_2 4+k_1 2+k_0)\, n_1 2} &= \left[W^{8 k_2 n_1}\right] W^{(k_1 2+k_0)\, n_1 2} = W^{(k_1 2+k_0)\, n_1 2} \\
W^{(k_2 4+k_1 2+k_0)\, n_0} &= W^{(k_2 4+k_1 2+k_0)\, n_0}
\end{aligned}
\tag{4.24}
$$

In the above equation, the relation

$$W^8 = e^{\frac{-j2\pi 8}{8}} = e^{-j2\pi} = 1 \tag{4.25}$$

is used.

Substituting the results from (3.24) into (4.21), the result is

$$X(k_2, k_1, k_0) = \underbrace{\sum_{n_0=0}^{1} \underbrace{\sum_{n_1=0}^{1} \underbrace{\sum_{n_2=0}^{1} x(n_2, n_1, n_0)\, W^{k_0 n_2 4}}_{x_1(k_0,\, n_1,\, n_0)} W^{(k_1 2+k_0)\, n_1 2}}_{x_2(k_0,\, k_1,\, n_0)} W^{(k_2 4+k_1 2+k_0)\, n_0}}_{x_3(k_1,\, k_2,\, k_3)} \tag{4.26}$$

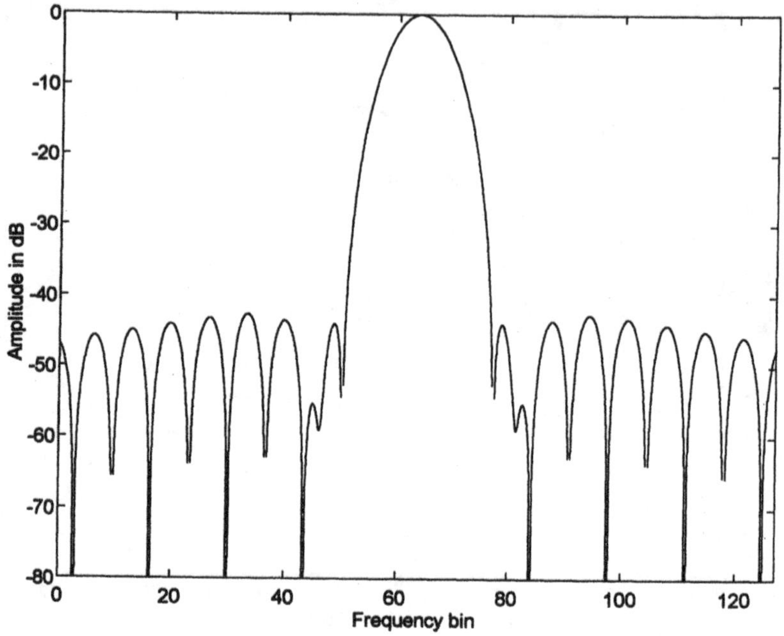

(b)

Figure 4.10 (continued).

It is convenient to perform each of the summations separately and label the intermediate results. Note that each set consists of only eight terms and that only the latest set needs to be saved. Thus, the above equations can be written in three steps as

$$
\begin{aligned}
x_1(k_0, n_1, n_0) &= \sum_{n_2=0}^{1} x(n_2, n_1, n_0) W^{k_0 n_2 4} \\
x_2(k_0, k_1, n_0) &= \sum_{n_1=0}^{1} x_1(k_0, n_1, n_0) W^{(k_1 2+k_0) n_1 2} \\
x_3(k_0, k_1, k_2) &= \sum_{n_0=0}^{1} x_2(k_0, k_1, n_0) W^{(k_2 4+k-12+k_0) n_0}
\end{aligned}
\tag{4.27}
$$

This accomplishes the computation of the FFT, except the output must be properly arranged. The final step in the Fourier transform is

$$X(k_2, k_1, k_0) = x_3(k_0, k_1, k_2) \tag{4.28}$$

because the order of k_is in (4.21) is in a different order than that of (4.27). In order to obtain the result of the FFT, the output has to be put in the correct order. This is referred to as bit reversal.

The result of (4.27) can be expressed graphically. Let us use the first equation in (4.27) to illustrate this operation. In order to simplify the notation, the commas between the numbers are omitted (i.e., $x(100)$ is used to denote $x(1, 0, 0)$). The same two input data points can generate two outputs. For example, $x_1(000)$ and $x_1(100)$ can be generated from $x(000)$ and $x(100)$. It is interesting to notice that from one layer to the next, the indices in the parentheses do not change. These results can be written as

$$x_1(100) = x(100)\,W^0 + x(100)\,W^4 = x(000) - x(100) \tag{4.29}$$
$$x_1(000) = x(000)\,W^0 + x(100)\,W^0 = x(000) + x(100)$$

It should noted that $W^0 = 1$ and $W^4 = -1$, and these relations are used in the above equation. This relation can be graphically expressed as in Figure 4.11. Thus Figure 4.11(a, b) show the same results with only the outputs alternated. Figure 4.11(c)

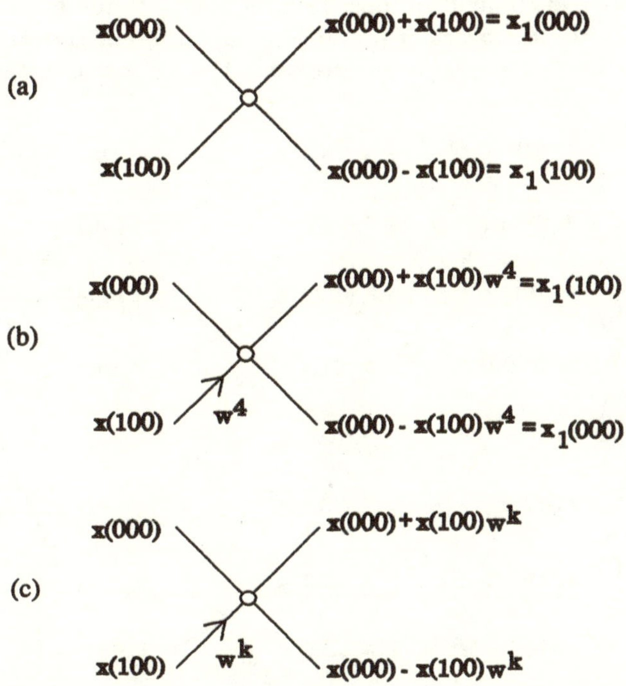

Figure 4.11 FFT butterfly: (a) simple notation, (b) equivalent to, (c) general notation.

shows a general case and this symbol is often referred to as the radix 2 FFT or the FFT butterfly.

The result of (4.26) is shown in Figure 4.12. For this 8-point FFT, there are three layers of FFT butterflies. The last operation shown is the reordering. This is due to bit reversal in the arguments of (4.28).

For $N = 8$, a direct evaluation of the DFT through (4.8) will require nearly 64 complex multiply-and-add operations. The FFT equations need 48 operations. Notice also that the first multiplication in each summation is actually a multiplication by +1, because when n_0, n_1, $n_2 = 0$, $W^0 = 1$. Thus, the number of operations reduces to 24. By further noting that $W^0 = -W^{-4}$, $W^1 = -W^5$, $W^2 = -W^6$, and $W^3 = -W^7$, the number of multiplications can be reduced to 12.

These reductions in operations can be extended to the more general case of $N = 2^m$, where m is an integer. The computations can be reduced from N^2 operations to $(N/2)\log_2(N)$ complex multiplications, $(N/2)\log_2(N)$ additions, and $(N/2)\log_2(N)$ subtractions. This reduction can be found in Figure 4.13. From this figure, one can see that the savings in the number of operations is quite dramatic. The number of multiplications and additions for FFT operations can be found in [27].

This same discussion can be applied to the inverse FFT. The only difference is to change $-j$ in (4.17) to $+j$, and the rest of the discussion will stay the same.

The number of input data must be a power of 2 number from the above discussion. Since the discovery of the FFT, many different approaches similar to the original Cooley-Tukey have been proposed. Fast Fourier transforms with input

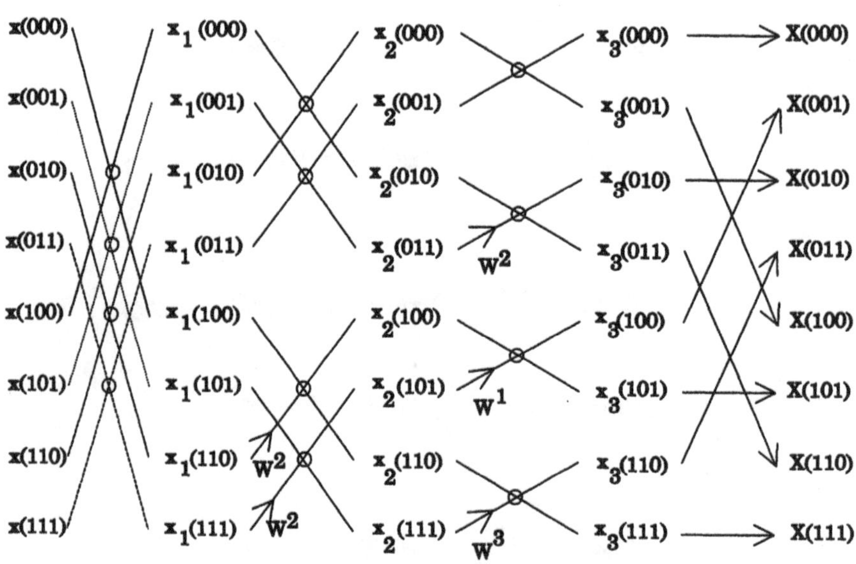

Figure 4.12 A flow diagram of the Cooley-Tukey FFT algorithm for performing an 8-point transform.

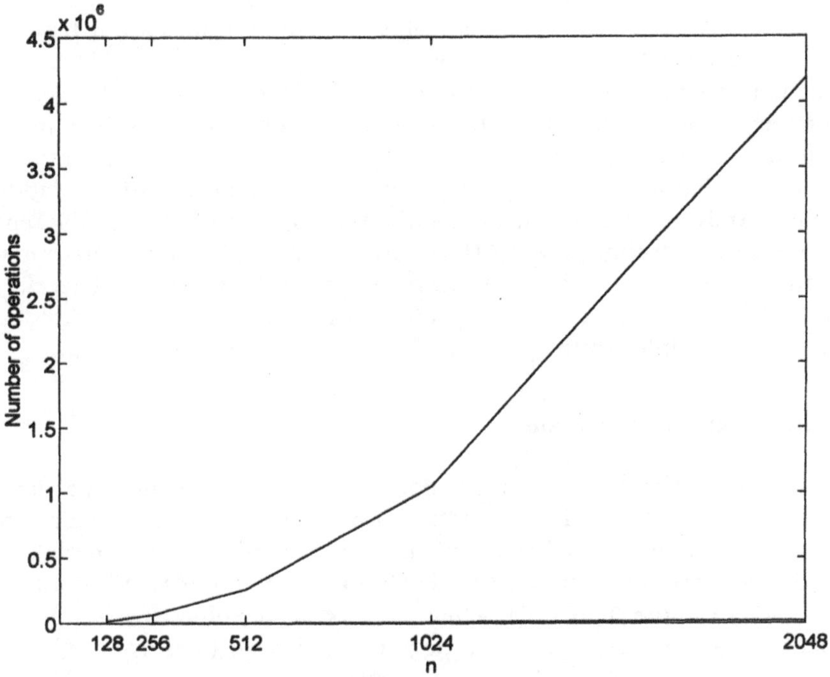

Figure 4.13 The number of operations required for computing FFT and DFT.

data points other than a base 2 number have been developed. In some designs, the output is in the correct order and no bit reversal is required; however, in those approaches the input data points must be rearranged [1]. Thus, there is no significant advantage. Today, it appears that the most popular FFT algorithm is still the original Cooley-Tukey approach.

The most promising approach for microwave receiver applications might be FFT chips built-in hardware. If their operating speeds can match the analog-to-digital converter (ADC), one ADC can be followed by one FFT chip. If the FFT speed is less than the ADC, a number of FFT chips can be connected and operated in parallel or some other approach, such as decimation technique, can be used.

4.8 POSSIBLE ADVANTAGES OF DFT OVER FFT IN RECEIVER APPLICATIONS [27, 28]

Although FFT requires less operations than DFT, for wideband microwave receiver applications it is possible that DFT still can have some advantages over FFT under certain conditions. One should note that in Figure 4.13, when $N = 2{,}048$ there is a large advantage for FFT over DFT. The saving in computation can be expressed as the ratio of $N/\log_2 N$. At $N = 32$, 64, and 128, these savings are 6.4, 10.67, and 18.29, respectively.

If the microwave receivers are for EW application, at least in the near future, the number of input data will probably be 32, 64, or 128 points. The input data length is limited by two factors: the operating speed and the minimum pulse width. If the collected data length is much longer than the input pulse width, the additional data contain only noise. The data points without a signal will degrade the sensitivity of the receiver. With the present ADC operating speed, it is likely that on a short pulse only tens of or a few hundred data points can be collected. Therefore, the DFT and FFT operations are compared for these short data points.

The two possible advantages of DFT over FFT are discussed below.

4.8.1 Initial Data Accumulation

It is worthy to notice that the initial data accumulation in the time domain is different from DFT and FFT. From (4.26), when an N-point FFT is calculated, all N points of data must be available. On the contrary, from (4.8), the data can be collected and processed as they arrive. Each input data point $x(n)$ is multiplied by N terms. For example, if $N = 64$, when the first data point $x(0)$ is collected, 64 intermediate data points $X_i(n, k)$ can be generated from (4.8) as

$$X_i(n, k) = x(0)e^{\frac{-j2\pi k0}{N}} = x(0) \tag{4.30}$$

where $n = 0$ and $k = 0$ to 63. Similar operations can be applied to all the input data. For the nth data $x(n)$, the intermediate data $X_i(n, k)$ is

$$X_i(n, k) = x(n)e^{\frac{-j2\pi kn}{N}} \tag{4.31}$$

where $k = 0$ to 63. When the last data point is collected, there is a total of N^2 intermediate data points. To find the kth components in the frequency domain, all the intermediate data with the same k value summed together will produce the required value.

Of course, there are N^2 operations in the above approach. However, as soon as the last input data are collected, a summation will produce the desired result. For the FFT approach, it must wait until all the data are collected before the operation can be started. Thus, for small N values, the DFT approach might be easier to build as a real-time processor.

4.8.2 Sliding DFT

In an EW receiver, the input data may or may not contain a signal. One common way to divide the input data are from 0 to $N - 1$, then from N to $2N - 1$, and so on. This approach saves processing time, but a signal may be divided into two data groups; thus, the sensitivity and frequency resolution of the receiver are low. One method of improving this approach is to process the data with some degree of overlap. For a 50% overlap, the first group will contain points 0 to $N-1$ and the second group $N/2$ to $(3N/2)-1$. This approach will double the required processing, but improve the receiver performance. The overlapping of data will be further discussed in Chapter 11.

The most sophisticated approach is to slide the DFT by one point at a time. For example, the first group contains 0 to $N-1$, the second group contains 1 to N, and the third group 2 to $N+1$, as shown in Figure 4.14. Although this approach may provide the best receiver performance, it is processing-intensive. If FFT is used for this approach, each set of input data will be calculated independently. If the DFT is used for this type of processing, the new information can be obtained from the previous one. To demonstrate this operation, let us rewrite (4.8) as

$$X_0(k) = x(0) + x(1)e^{\frac{-j2\pi k}{N}} + x(2)e^{\frac{-j2\pi k2}{N}} + \dots + x(N-1)e^{\frac{-j2\pi k(N-1)}{N}} \tag{4.32}$$

where $X_0(k)$ is used to represent the DFT of the first set of data. This is exactly the same equation except in slightly different form. The second set of data will generate a DFT as

$$X_1(k) = x(1) + x(2)e^{\frac{-j2\pi k}{N}} + x(3)e^{\frac{-j2\pi k2}{N}} + \dots + x(N)e^{\frac{-j2\pi k(N-1)}{N}} \tag{4.33}$$

The only difference between this equation and the previous one is the set of data points. The exponential terms of the two equations are identical. One can easily see that (4.32) can be modified slightly to obtain (4.33). This relation can be written as

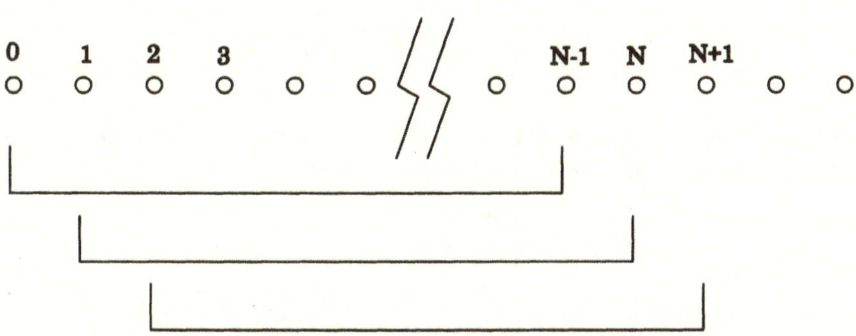

Figure 4.14 Sliding DFT.

$$X_1(k) = [X_o(k) - x(0)]e^{\frac{j2\pi k}{N}} + x(N)e^{\frac{j2\pi k(N-1)}{N}}$$
$$= [X_o(k) - x(0) + x(N)]e^{\frac{j2\pi k}{N}} \tag{4.34}$$

This approach can be generalized as

$$X_{m+1}(k) = [X_m(k) - x(m) + x(N+m)]e^{\frac{j2\pi k}{N}} \tag{4.35}$$

It appears that this approach might be more efficient than the FFT approach. If this operation can be performed efficiently, a real-time sliding FFT is possible.

4.9 PERIODOGRAM [1–7]

Once the FFT or DFT of the digitized input signal is completed, the next step is to find the spectrum or the frequency components. The straightforward way is to find the power spectrum. Since the FFT output is complex, the power spectrum can be written as

$$P(k) = \frac{1}{N}|X(k)|^2 \tag{4.36}$$

where $X(k)$ is the FFT of $x(n)$. This is referred to as the periodogram. Most of the figures in this chapter are generated using the periodogram. The value $P(k)$ can be plotted either directly or in decibels, $(10\log(P(k)))$ as a function of k.

If the input data are real, the power spectrum generated from the above equation has an even symmetry. This can be shown as follows. Let us find the frequency component $X(k)$ and $X(N-k)$. The results are

$$X(k) = \sum_{n=0}^{N-1} x(n)e^{\frac{-j2\pi nk}{N}} \tag{4.37}$$
$$X(N-k) = \sum_{n=0}^{N-1} x(n)e^{\frac{-j2\pi n(N-k)}{N}} = \sum_{n=0}^{N-1} x(n)e^{\frac{j2\pi nk}{N}}$$

The only difference between $X(k)$ and $X(N-k)$ is the phase angle. The power spectrum generated from these two components is

$$|X(k)|^2 = \left(\sum_{n=0}^{N-1} x(n)e^{\frac{-j2\pi nk}{N}}\right)\left(\sum_{n=0}^{N-1} x(n)e^{\frac{j2\pi nk}{N}}\right) \tag{4.38}$$
$$|X(N-k)|^2 = \left(\sum_{n=0}^{N-1} x(n)e^{\frac{j2\pi nk}{N}}\right)\left(\sum_{n=0}^{N-1} x(n)e^{\frac{-j2\pi nk}{N}}\right)$$

It is obvious that these two components are equal.

When $k = N/2$, $X(k)$ and $X(N - k)$ will represent the same component $X(N/2)$. This implies that the output is symmetrical with respect to $X(N/2)$. Thus, if there are N points of real data after the FFT, the power spectrum has N frequency components. However, among the N frequency components, only half of them provide useful information. The other half-power spectrum components are equal to the first half. As a result, if there are N points of real input data, even though after the FFT $X(k)$ has N components, it is common to plot only $N/2$ components.

4.10 AVERAGED PERIODOGRAM [4, 5, 29, 30]

If the input signal contains only noise with a variance of σ^2, one might expect that when the input data length increases, the variance of power spectrum may approach 0. However, this is not true. When the data length increases, the variability in the power spectrum will not decrease. This phenomenon is illustrated in Figure 4.15. In this figure, 1,024 points of data are collected and the data contain only noise. The data are divided into eight subgroups and each contains 128 points. Figure 4.15(a) shows the square root of the periodogram of all 1,024 data points. Figure 4.15(b) shows the square root of the periodogram of the first 128 data points. All the other 7 sets of data points will show similar results. It is clearly shown that more data points used in the calculation do not decrease the noise variability in the frequency domain.

In order to reduce the variance of noise in the output in the above example, one can calculate the periodograms of all eight subgroups of data and take the average on all eight results. The operation of this special case can be written as

$$P_{av}(k) = \frac{1}{8}\sum_{i=0}^{7} X_i(k)$$

where

$$X_i(k) = \sum_{n=0}^{127} x(128i + n)e^{\frac{-j2\pi nk}{N}} \qquad i = 0, 1, \ldots 7 \tag{4.39}$$

Figure 4.15(c) shows the square root of the averaged periodogram. Comparing with Figures 4.15(b, c), the variance of the noise spectrum is reduced through this averaging process. Because the eight sets of data can be considered independent, the sum will increase the noise by $\sqrt{8}$. When this result is divided by 8, the variance of noise will decrease.

Now let us change the input data to include a sine wave as the input signal. This signal is contaminated with noise. A total of 1,024 data points are collected. The same three approaches are used to process the input data. First, the periodogram is obtained by using all the data points. Secondly, the periodogram of the first 128 data points is obtained. Finally, the input data are divided into eight subgroups and their periodograms are obtained and averaged. The square roots of the periodograms are shown in Figure 4.16. In Figure 4.16(a), the signal frequency is clearly

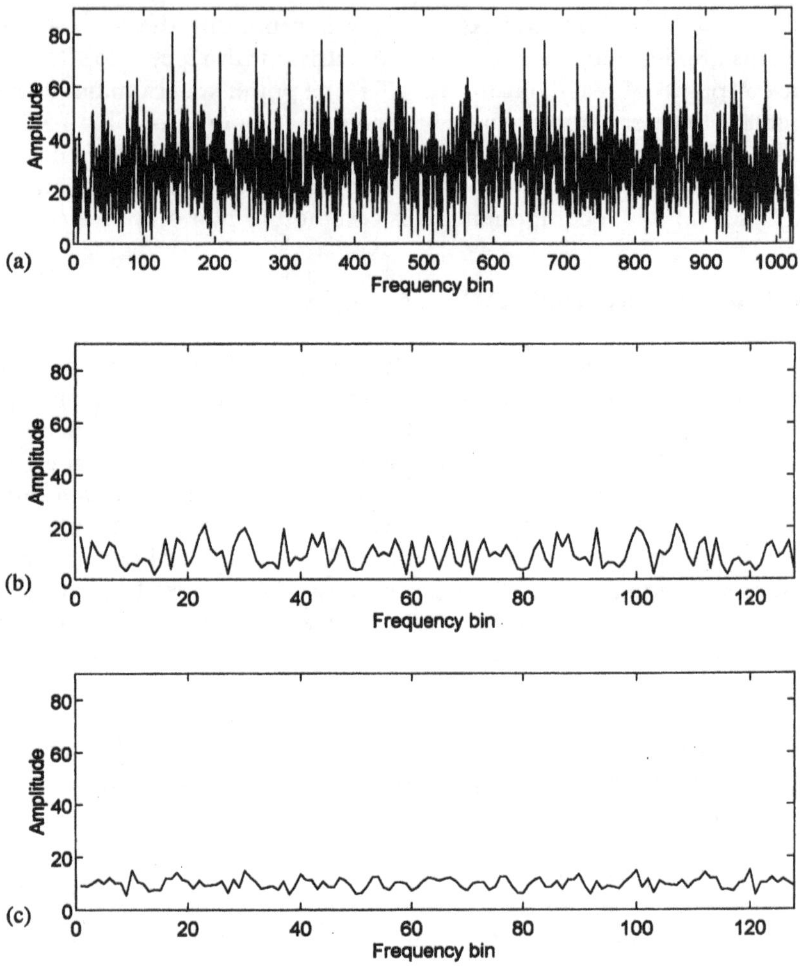

Figure 4.15 Periodogram of noise only input: (a) 1,024 data points, (b) 128 data points, (c) 1,024 data points divided into eight subgroups.

visible for the 1,024-point FFT. In Figure 4.16(b), the signal frequency is difficult to recognize for the 128-point FFT. In Figure 4.16(c), the signal frequency is clearly shown for the averaged case, but the frequency resolution is not as good as that in Figure 4.16(a).

Although averaging the periodogram can decrease the noise contribution, it does lose frequency resolution. In general, if the desired frequency resolution determined by the FFT length is reached, additional sets of data can be collected and, through averaging the signal-to-noise ratio, can be improved.

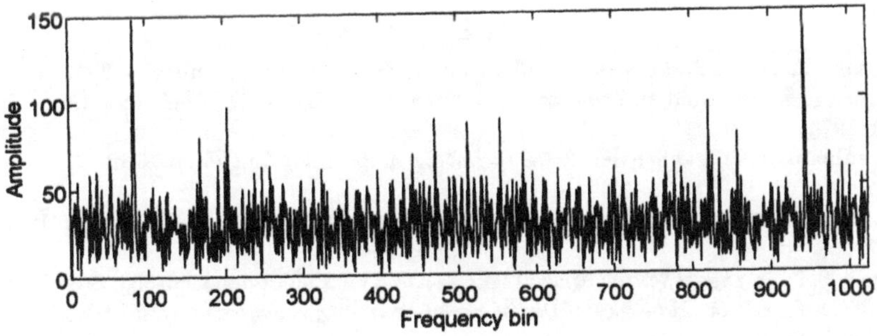

(a)

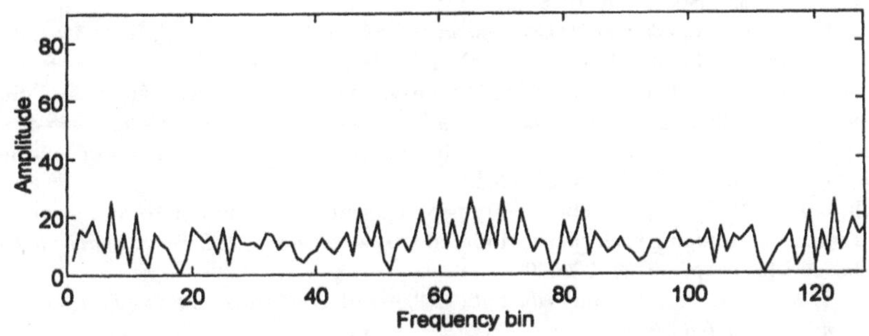

(b)

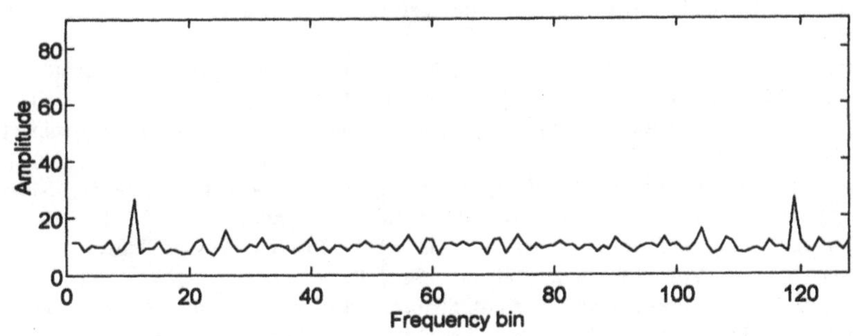

(c)

Figure 4.16 Periodogram of noise only input: (a) 1,024 data points, (b) 128 data points, (c) 1,024 data points divided into eight subgroups.

REFERENCES

[1] Brigham, E. O. *The Fast Fourier Transform*, Englewood Cliffs, NJ: Prentice Hall, 1974.

[2] Rabiner, L. R., and Gold, B. *Theory and Application of Digital Processing*, Englewood Cliffs, NJ: Prentice Hall, 1975.

[3] Oppenheim. A. V., and Schafer, R. W. *Digital Signal Processing*, Englewood Cliffs, NJ: Prentice Hall, 1975.

[4] Kay, S. M. *Modern Spectral Estimation Theory and Application*, Englewood Cliffs, NJ: Prentice Hall, 1987.

[5] Marple, S. L., Jr. *Digital Spectral Analysis With Applications*, Englewood Cliffs, NJ: Prentice Hall, 1987.

[6] Elliott, D. F., Editor. *Handbook of Digital Signal Processing, Engineering Applications*, New York, NY: Academic Press, Inc., 1987.

[7] Tsui, J. B. Y. *Digital Microwave Receivers: Theory and Concepts*, Norwood, MA: Artech House, 1989.

[8] Harris, F. J. "On the Use of Windows for Harmonic Analysis With the Discrete Fourier Transform," *IEEE Proc.*, Vol. 66, No. 1, Jan. 1978, pp. 51–83.

[9] Tukey, J. W. *An Introduction to the Calculations of Numerical Spectrum Analysis*, in *Spectral Analysis of Time* series, ed. by Haaris, B., New York, NY: John Wiley & Sons, 1967.

[10] Blackman, R. B., and Tukey, J. W. "The Measurement of Power Spectra From the Point of View of Communications Engineering," *Bell System Tech. Journal*, 1958, pp. 185–282, 485–569.

[11] Cooley, J. W., and Tukey, J. W. "An Algorithm for the Machine Calculation of Fourier Series," *Math. Comput.*, Vol. 19, 1965, pp. 297–301.

[12] Welch, P. D. "The Use of Fast Fourier Transform for the Estimation of Power Spectra: a Method Based on Time Averaging Over Short, Modified Periodograms," *IEEE Trans. Audio and Electroacoustics*, Vol. AU-15, June 1967, pp. 70–73.

[13] Cooley, J. W., Lewis, P. A. W., and Welch, P. L. "Historical Notes on the Fast Fourier Transform," *IEEE Trans. Audio and Electroacoustics*, Vol. AU-15, June 1967, pp. 76–79.

[14] Cochran, W. T., et al. "What is Fast Fourier Transform?" *IEEE Trans. Audio and Electroacoustics*, Vol. AU-15, June 1967, pp. 45–55.

[15] Brigham, E. O., and Morrow, R. E. " The Fast Fourier Transform," *IEEE Spectrum*, Dec. 1967, pp. 63–70.

[16] Cooley, J. W., Lewis, P. A. W., and Welch, P. D. "The Finite Fourier Transform," *IEEE Trans. Audio and Electroacoustics*, Vol. AU-17, June 1969, pp. 251–259.

[17] Bergland, G. D. "A Guided Tour of the Fast Fourier Transform," *IEEE Spectrum*, July 1969, pp. 41–52.

[18] Sanderson, R. B. "Instrumental techniques," Chapter 7 of *Molecular Spectroscopy: Modern Research*, ed. by Rao, R. N., and Matthews, C. W., New York, NY: Academic Press, 1972.

[19] Palmer, L. C. "Coarse Frequency Estimation Using the Discrete Fourier Transform," *IEEE Trans. Information Theory*, Vol. IT-20, Jan. 1974, pp. 104–109.

[20] Allen, J. B. "Short Term Spectral Analysis, Synthesis, and Modification by Discrete Fourier Transform," *IEEE Trans. Acoustics, Speech, and Signal Processing*, Vol. ASSP-25, June 1977, pp. 235–238.

[21] Sorensen, H. V., Jones, D. L., Heideman, M. T., and Burrus, C. S., "Real-Valued Fast Fourier Transform Algorithm," *IEEE Trans. Acoustics, Speech, and Signal Processing*, Vol. ASSP-35, June 1987, pp. 849–863.

[22] Thong, T. "Practical Consideration for a Continuous Time Digital Spectrum Analyzer," *1989 IEEE International Symposium on Circuits and Systems*, Vol. 2, May 9–11, 1989, pp. 1047–1050.

[23] Jenq, Y. C. "Digital Spectra of Nonuniformly Samples Signals: A Robust Sampling Time Offset Estimation Algorithm for Ultra High-Speed Waveform Digitizers Using Interleaving," *IEEE Trans. on Instrumentation and Measurement*, Vol. 39, Feb. 1990, pp. 71–75.

[24] Agoston, M., and Henricksen, R. "Using Digitizing Signal Analyzers for Frequency Domain Analysis," *Microwave Journal*, Sept. 1990, p. 181.

[25] Fine, B. "DSPs Address Real-World Problems," *Microprocessor*, 1990, p. 72–74.

[26] Sayegh, S. I. "A Pipeline Processor for Mixed-Size FFT's," *IEEE Trans. Signal Processing,* Vol. 40, Aug. 1992, pp. 1892–1900.

[27] Duhamel, P. "Implementation of Split-Radix FFT Algorithms for Complex, Real, Real, and Real-Symmetric Data," *IEEE Trans. Acoustics, Speech and Signal Processing,* Vol. ASSP-34, April 1986, pp. 285–295.

[28] Springer, T. "Sliding FFT Computes Frequency Spectra in Real Time," *Electronic Design News,* Sept. 29, 1988, pp. 161–170.

[29] Shanmugan, K. S., and Breigohl, A. M. *Random Signals Detection, Estimation and Data Analysis,* New York, NY: John Wiley & Sons, 1988.

[30] Papoulis, A. *Probability, Random Variables, and Stochastic Processes,* New York, NY: McGraw-Hill Book Co., 1965.

CHAPTER 5

Fourier Transform–Related Operations

5.1 INTRODUCTION

In this chapter, the operations discussed will be related to the Fourier transform and especially to the discrete Fourier transform (DFT) or fast Fourier transform (FFT). Some of these operations are useful to receiver design and some of these techniques can improve the results of the FFT. The discussion includes zero padding, better localization of FFT peaks, digital convolution, parallel FFT operations to increase overall speed, performing real FFT by using complex FFT operations, and a phase sampling scheme to increase the bandwidth of the receiver.

5.2 ZERO PADDING [1, 2]

Zero padding is the addition of zeros at the end of a digital data string before the FFT is performed. For example, if 64 points of data are collected, usually a 64-point FFT is performed to obtain the frequency. However, one can add 64 zeros at the end of the data string and perform a 128-point FFT. Since zero padding does not add any new information, the resultant FFT form (i.e., the frequency resolution) does not change. For the 64 data points with 64 zeros, the number of output frequency components is doubled.

In general, suppose there are N data points $x(0)$ to $x(N-1)$, and we add N zero points from $x(N)$ to $x(2N-1)$ and perform a $2N$-point DFT. The result will be

$$X(k) = \sum_{n=0}^{2N-1} x(n)e^{\frac{-j2\pi kn}{2N}} \tag{5.1}$$

Note that the summation is from 0 to $2N-1$ and the kernel is $e^{-j2\pi kn/2N}$ rather than $e^{-j2\pi kn/N}$. Since the $x(N) = x(N+1) = \ldots = x(2N-1) = 0$, this equation can be written as

$$X(k) = \sum_{n=0}^{N-1} x(n)\, e^{\frac{-j2\pi kn}{2N}} \tag{5.2}$$

The only change from (5.1) is in the limit of the summation. It should be noted that the value of k in $X(k)$ is from $k = 0$ to $2N - 1$. Thus, the number of spectrum lines is $2N$, which doubles the result of an N point FFT.

In the above equation, if we consider only the even number of spectrum lines, one can write $k = 2k'$. Substituting this relation in (5.2), the result is

$$X(k') = \sum_{n=0}^{N-1} x(n)\, e^{\frac{-j2\pi k'n}{N}} \tag{5.3}$$

which is identical to an N-point FFT. Thus, in this example, zero padding does not change the amplitude and phase of the even-numbered components of the spectrum, but provides interpolated values at the odd number of components.

The above discussion can be generalized to pad the input data with LN of zeros with L as a positive integer. In the above discussion $L = 1$. In general, in order to keep the total number of data points a power of 2 number, the L value is

$$L = 2^i - 1 \qquad \text{where} \qquad i = 0, 1, 2, \ldots \tag{5.4}$$

An example is shown in Figure 5.1. A total of 32 points of data are collected and Figure 5.1(a) shows the 32-point FFT. Figure 5.1(b) shows the result of a 64-point FFT padded with 32 zeros. One can see that the even components of this figure have the same amplitude as in Figure 5.1(a). Figure 5.1(c) shows the result of a 128-point FFT padded with 96 zeros. One can see that every fourth component is identical to the result in Figure 5.1(a). Figure 5.1(d) shows the result of a 1,024 point FFT padded with 992 zeros. Therefore, it is obvious that the zero padding does not change the spectrum shape, but only interpolates between the original N-point FFT. In these figures, the frequency range is from 0 to $f_s/2$, where f_s is the sampling frequency. If the input data are complex, the frequency is from 0 to f_s. The frequency index is based on the number of the FFT points.

From this example, one can see that in the 32-point FFT, it might be difficult to choose the peak of the power spectrum. The fine structure of the sidelobes is even difficult to observe. With zero padding, the location of the peak of the power spectrum appears more clearly and the sidelobes are clearly shown. Thus, the additional processing provided by zero padding can improve the capability of interpolating the peak of the power spectrum. The price paid is greater processing time. If zero padding is not used, the fine structure in the frequency domain cannot be obtained. Many of the plots generated in this book use zero padding.

A very important point is that zero padding does not increase the fundamental resolution of the FFT. In other words, the width of the main lobe of the FFT does not change due to zero padding. The frequency resolution depends only on the

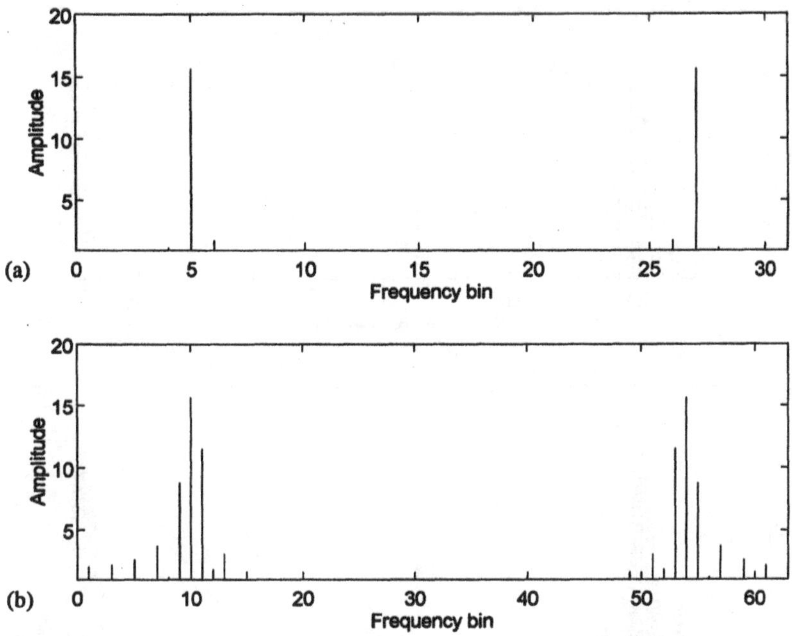

Figure 5.1 Effect of zero padding: (a) 32-point FFT, (b) 64-point FFT, (c) 128-point FFT, (d) 1,024-point FFT.

true data length. Zero padding can only improve the capability of selecting the peak frequency component in the main lobe.

5.3 PERIODIC AND LINEAR CONVOLUTIONS [3, 4]

It has been shown in Section 3.7 that convolution in the time domain is equivalent to multiplication in the frequency domain. Equation (3.62) is rewritten here as

$$x(t) \bigstar y(t) \leftrightarrow X(f)Y(f) \tag{5.5}$$

The same relation can be obtained in discrete form. The relation can be written as

$$\begin{aligned} z(n) &= \sum_{m=0}^{N-1} x(m)y(n-m) \\ \text{or} \quad Z(k) &= X(k)Y(k) \end{aligned} \tag{5.6}$$

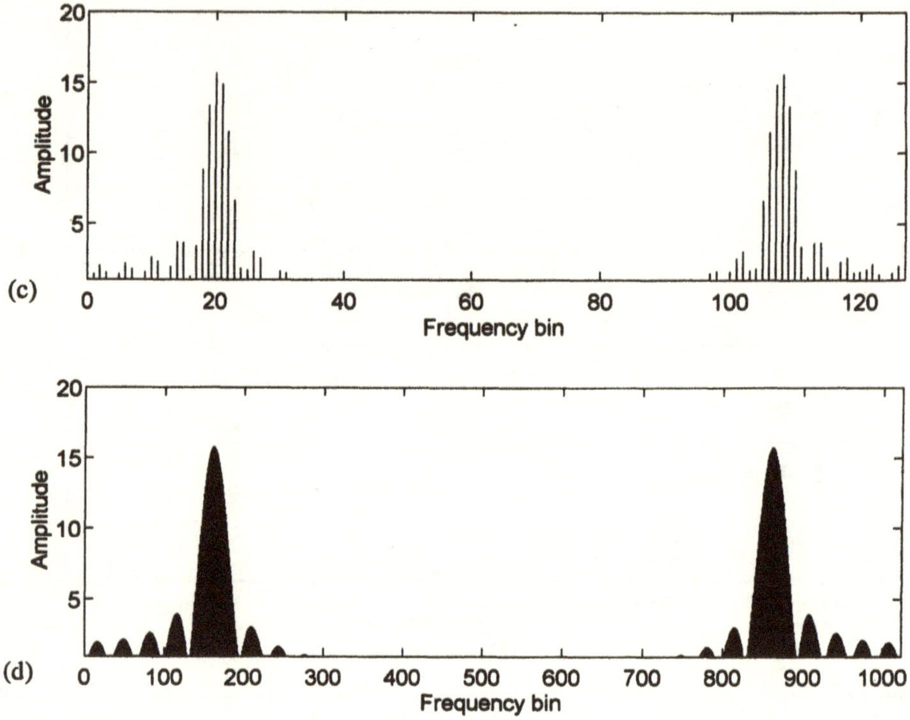

Figure 5.1 (continued).

It should be noted that the time shift in $y(n-m)$ is circular. This relation can be proven as follows. By taking the discrete Fourier transform of (5.6), one obtains $Z(k)$ as

$$\begin{aligned} Z(k) &= \sum_{n=0}^{N-1}\sum_{m=0}^{N-1} x(m)y(n-m)\, e^{\frac{-j2\pi kn}{N}} \\ &= \sum_{m=0}^{N-1} x(m)\left[\sum_{n=0}^{N-1} y(n-m)\, e^{\frac{-j2\pi k(n-m)}{N}}\right] e^{\frac{-j2\pi km}{N}} \\ &= Y(k)\sum_{m=0}^{N-1} x(m)\; e^{\frac{-j2\pi km}{N}} = X(k)\,Y(k) \end{aligned} \tag{5.7}$$

Equation (5.6) is referred to as the periodic convolution (or circular convolution). It does not provide the expected result of a linear convolution. The reason is due to the periodic nature of the discrete Fourier transform.

Before we explain this effect, let us use a simple example to demonstrate this effect. Let $x(n)$ = 0, 1, 2, 3 and $y(n)$ = 1, 1, 1, 1 for n = 0, 1, 2, 3, respectively. The correct linear convolution is $z(n) = x(n) \bigstar y(n)$ = 0, 1, 3, 6, 6, 5, 3 for n = 0 to 6. The result is shown in Figure 5.2. Figure 5.2(a) and 5.2(b) show $x(n)$ and $y(n)$,

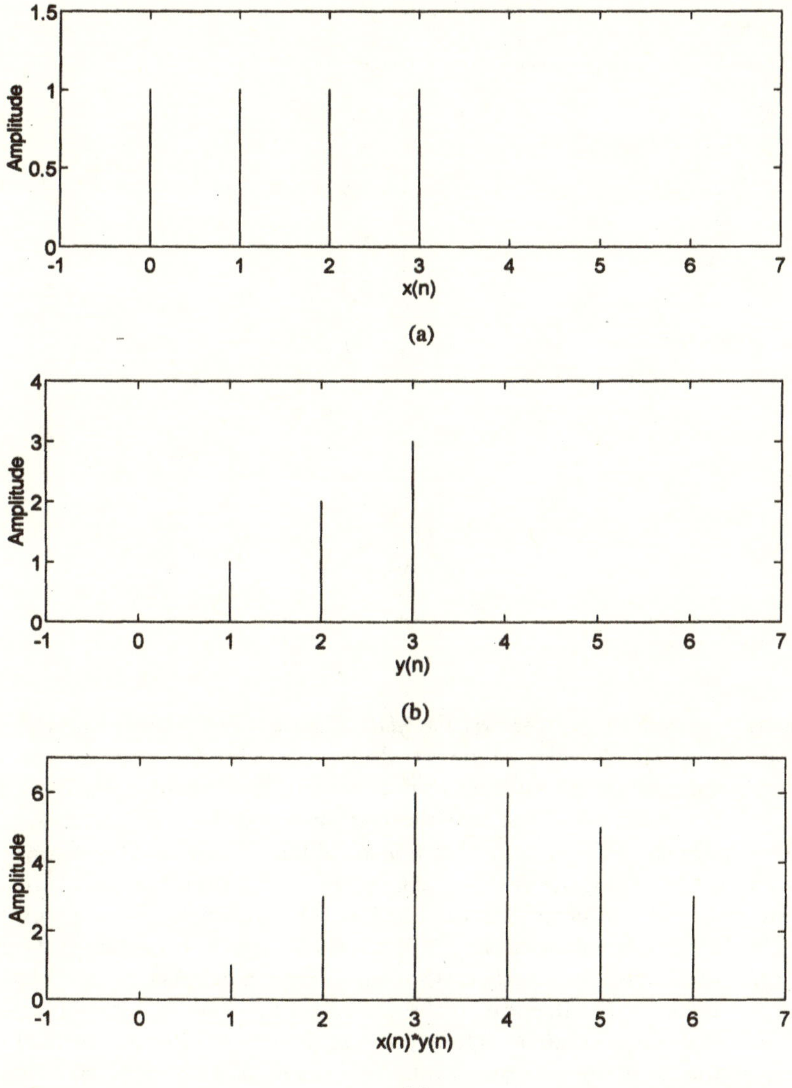

Figure 5.2 Result of linear convolution: (a) $x(n)$, (b) $y(n)$, (c) $z(n) = x(n) \bigstar y(n)$, (d) sliding of $x(n-m)$.

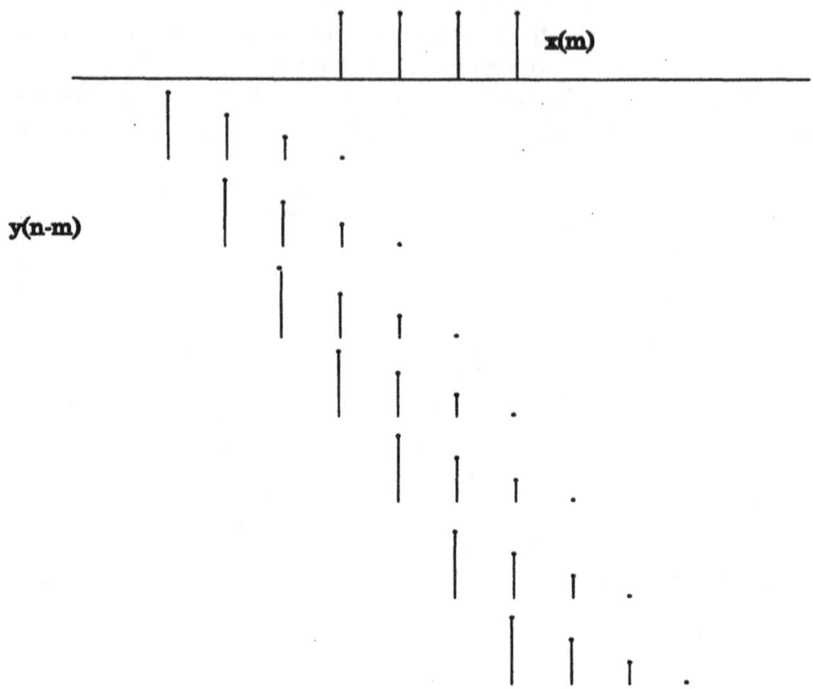

(d)

Figure 5.2 (continued).

respectively. Figure 5.2(c) shows $x(n) \star y(n)$. Figure 5.2(d) shows that the $y(n-m)$ shifts seven times (or $N + N - 1$ times where N is 4 in this specific case) to the right. If one times $y(n - m)$ and $x(n)$ at each shift, the result in Figure 5.2(c) will be obtained. It should be noted that the output has seven components.

If one takes the FFT of $x(n)$ and $y(n)$ to obtain $X(k)$ and $Y(k)$, then takes the inverse FFT of $X(f)Y(f)$, the result is $z(n) = 6, 6, 6, 6$, and it has only four outputs. This is the result of the periodic convolution. This result is wrong if one expects it to be the linear convolution, which is useful in signal processing. The reason that the undesired result is obtained is shown in Figure 5.3. As discussed in Chapter 4, due to the periodicity inherent in the DFT and inverse DFT (IDFT), both $x(n)$ and $y(n)$ are periodic in nature. When $y(n - m)$ is sliding one unit toward to right, the entire sequence is moved one unit. As a result, the pattern will repeat itself after four shifts. For this example, the output is the same at each step and the result is always 6 (3 + 2 + 1), which is the same as that obtained from the FFT operation.

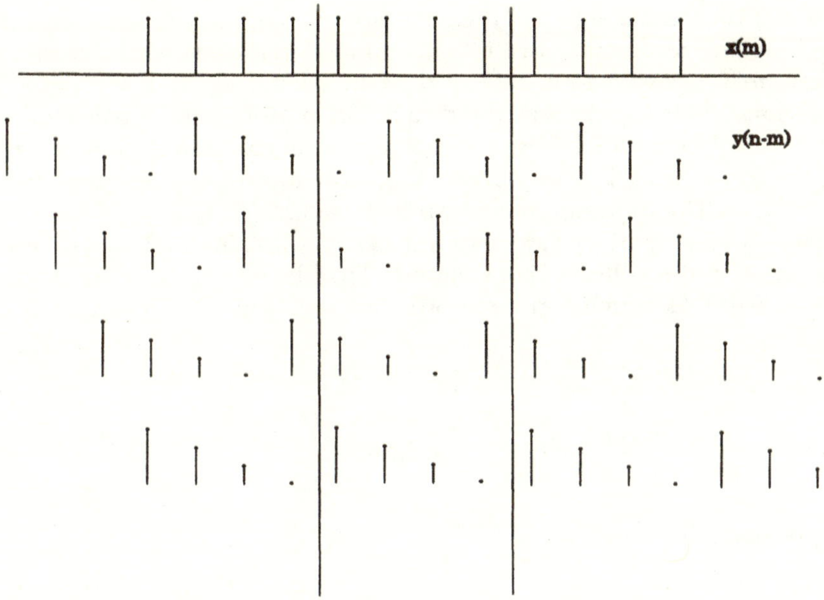

Figure 5.3 Periodic convolution.

In signal processing, one would like to avoid periodic or circular convolution and instead obtain linear convolution. One way to accomplish this goal is through zero padding. The question is how many zeros are needed. Let us discuss this problem from a more general approach. It should be noted that to perform convolution, the data length of x and y need not be the same. From Figure 5.2(d), it is easy to see that the total length of the convolution is $N + N - 1$. If $x(n)$ has M data points and $y(n)$ has N points, the linear convolution output has $M + N - 1$ points and one of the inputs must shifted $M + N - 1$ times during convolution. Therefore, both the length of $x(n)$ and $y(n)$ must be padded with a sufficient number of zeros so that the minimum length is $M + N - 1$ points. This creates enough room to perform the linear convolution and avoid circular convolution.

The example in the beginning of this section will be solved with zero padding as a demonstration. Let us pad $x(n)$ and $y(n)$ each with a minimum of 3 zeros. Thus, $x(n)$ = 1, 1, 1, 1, 0, 0, 0 and $y(n)$ = 0, 1, 2, 3, 0, 0, 0. If one now takes the FFT of $x(n)$ and $y(n)$ to obtain $X(k)$ and $Y(k)$, then takes the inverse FFT of $X(k)\,Y(k)$, the correct answer will be obtained. However, in this example, the length of the new data is 7 bits long and it is not a power of 2 number. The FFT program usually processes faster with power of 2 numbers. If one desires, the $x(n)$ and $y(n)$ can be padded with 4 zeros to make the length a power of 2 number.

In practical applications, $x(n)$ can be a long signal in the time domain and $y(n)$ may be the filter response with a relatively short time response. Under this condition, it is impractical to pad the data with zeros because first, the signal could be very long and the required FFT too computationally intensive. Second, in this operation, it is necessary to collect all the data on the signal. If the signal is very long, it is impossible to collect all the data before performing the convolution. The periodic convolution can be used to obtain the linear convolution in this case. There are two approaches to achieve this; only one of the approaches will be discussed here. The other approach can be found in [3, 4].

Let us assume the filter has a response of $h(n)$ with N points in the time domain and the signal $x(n)$ is collected continuously. First the signal will be segmented into $x_k(n)$ with a size M, which can be approximately equal to N. Then $x_k(n)$ can be written as

$$x_k(n) = \begin{cases} x(n) & kM \le n \le (k+1)M - 1 \\ 0 & \text{otherwise} \end{cases} \tag{5.8}$$

The input signal is

$$x(n) = \sum_{k=0}^{\infty} x_k(n) \qquad k = 0, 1, 2 \ldots \tag{5.9}$$

If $x_k(n)$ and $h(n)$ are properly padded with enough zeros, the linear convolution of each section $z_k(n)$ can be found as

$$z_k(n) = h(n) \bigstar x_k(n) \tag{5.10}$$

It should be noted that $Z_k(n)$ has $M + N - 1$ points. The overall linear convolution is the summation of $z_k(n)$ as

$$z(n) = \sum_{k=0}^{\infty} z_k(n) \tag{5.11}$$

In this operation, the summation is performed every M points. Since $z_k(n)$ is $M + N - 1$ long and summed every M points, there is data overlapping.

The phenomenon discussed above is illustrated graphically in Figure 5.4. In this figure, let us assume $N = 4$, $M = 5$, and the output of $z_k(n)$ is 8 (5 + 4 − 1) points long, but the summations are made every 5 points. Therefore, when $z_0(n)$ + $z_1(n)$ is performed there are three points overlapping. This figure shows the results of $z_0(n)$, $z_1(n)$, $z_2(n)$, and their summation $z(n)$. This approach is referred to as the overlap-add method.

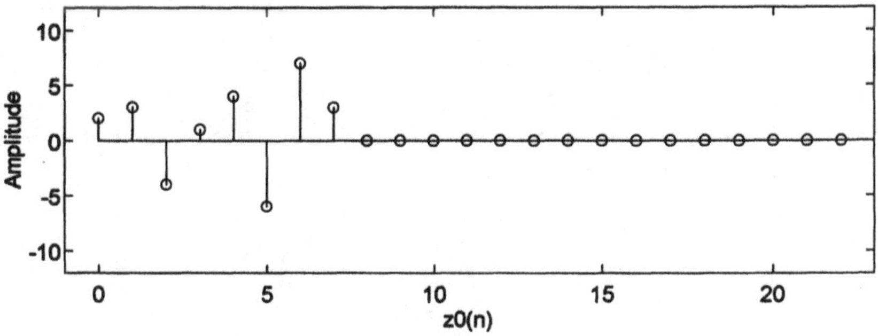

(a)

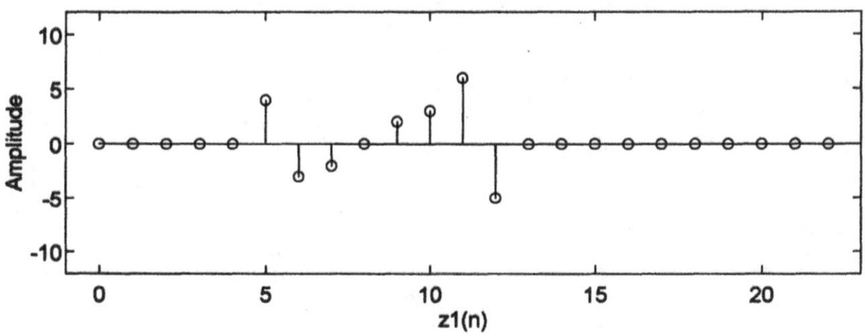

(b)

Figure 5.4 Linear convolution through overlap-add: (a) $z_0(\mathrm{n})$, (b) $z_1(n)$, (c) $z_2(n)$, (d) $z(n)$.

5.4 PEAK POSITION ESTIMATION FOR RECTANGULAR WINDOW [5–11]

This section discusses a simple way to extrapolate the peak of the frequency component from FFT outputs. The zero padding operation discussed in Section 5.2 can be used to better locate the peak, but the length of the FFT must be increased. In this method, a minimum of two points from the FFT output will be enough.

First, let us consider the continuous case and assume a rectangular window. This case is discussed in Section 3.2 and the results will be presented here again. If the input is a rectangular window in the time domain with unity amplitude and width T as shown in Figure 5.5(a), the Fourier transform in the frequency domain is a sinc function (from Section 3.2) as

$$X(f) = \frac{\sin(\pi f T)}{\pi f} \tag{5.12}$$

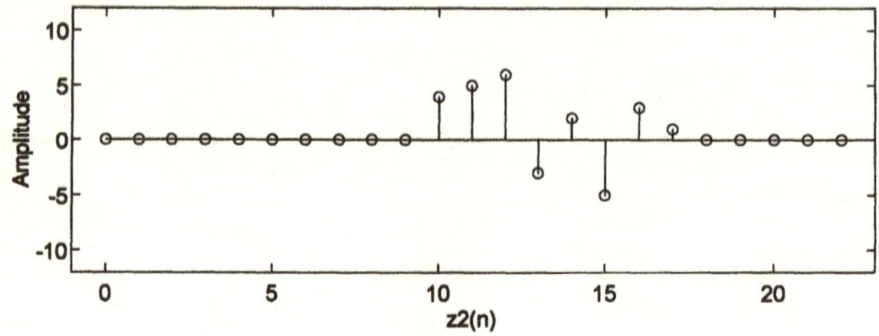

(c)

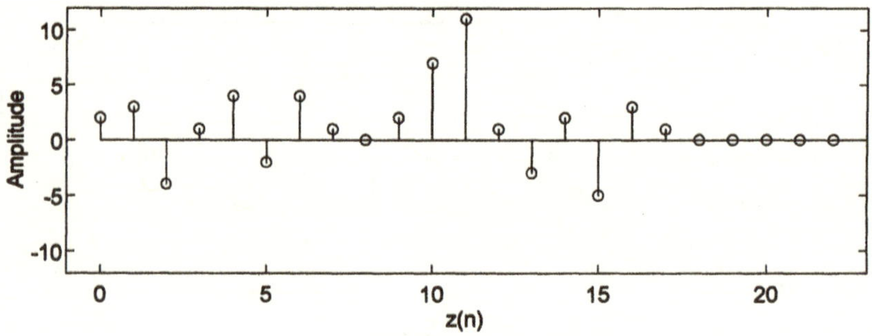

(d)

Figure 5.4 (continued).

The result is shown in Figure 5.5(b) and the peak occurs at $f = 0$. The first minima are at $\pm 1/T$.

However, if the outputs are from an FFT, the frequency components are discrete and will follow the contour of the sinc function as shown in Figure 5.6. In this figure the maximum frequency component of the power spectrum is not coincident with the maximum of the sinc function. The purpose of this discussion is to determine the true position of the maximum from these components. It should be noted that there are equal separations between these components and the distance is equal to $1/T$. Let us represent the amplitude of the frequency component by X_i: the highest one by X_0, the second highest one by X_1, and so on. From Figure 5.6, one can see that two of the highest amplitudes are in the mainlobe and the third one is in the first sidelobe.

Assume that k is the distance between the true peak and X_0. Since the distance between adjacent frequency components is $1/T$, this quantity can be used as the unit of distance. The positions of X_1 and X_2 are at $k - 1/T$, and $k + 1/T$, respectively. Their corresponding amplitudes can be written as

(a)

(b)

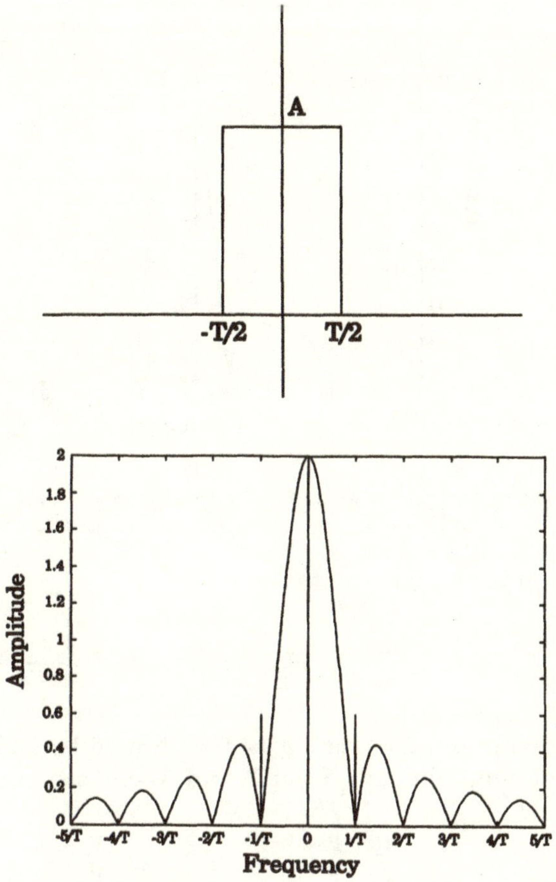

Figure 5.5 A rectangular function: (a) Time domain, (b) Frequency domain.

$$
\begin{aligned}
X_0 &= \frac{\sin(\pi T k)}{\pi k} \\
X_1 &= \frac{\sin\left[\pi T\left(k - \frac{1}{T}\right)\right]}{\pi\left(k - \frac{1}{T}\right)} = \frac{-\sin(\pi T k)}{\pi\left(k - \frac{1}{T}\right)} \\
X_2 &= \frac{\sin\left[\pi T\left(k + \frac{1}{T}\right)\right]}{\pi\left(k + \frac{1}{T}\right)} = \frac{\sin(\pi k)}{\pi\left(k + \frac{1}{T}\right)}
\end{aligned}
\qquad (5.13)
$$

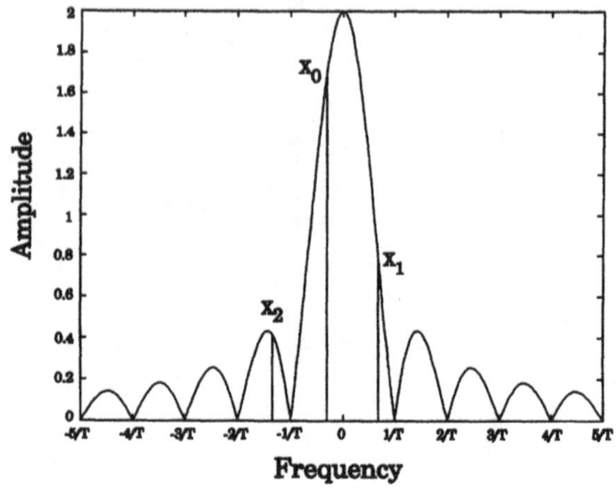

Figure 5.6 Frequency components from FFT of a rectangular window.

From these equations, one can find

$$\frac{X_1}{X_0} = \frac{k}{1-k} \quad \text{or} \quad k = \frac{X_1}{X_0 + X_1} \tag{5.14}$$

In the above equation, the relation $1/T = 1$ is used because the sampling interval is an assumed unity. Similarly, from X_1 and X_2 one can obtain

$$\frac{X_2}{X_1} = \frac{1-k}{1+k} \quad \text{or} \quad k = \frac{X_1 - X_2}{X_1 + X_2} \tag{5.15}$$

Either (5.14) or (5.15) can be used to find the center of the spectrum and each equation only uses two output components from the frequency domain. If the signal is noisy, both equations can be used to obtain k and the averaged value of the two can be used as the value of k.

If the frequency at X_0 and X_1, are k_0 and k_1 respectively, once k is found, the center frequency f_c can be considered as

$$f_c = k_0 \pm k \tag{5.16}$$

The positive sign is used if $k_1 > k_0$ and the negative if $k_1 < k_0$. Note that center frequency f_c is no longer an integer number.

The one important operation to obtain the above relations is the capability to eliminate the sine function by taking the ratio of the results in (5.13). Once the sine function is eliminated, the solution of k can be found easily.

5.5 PEAK POSITION ESTIMATION FOR HANNING WINDOW [5–11]

If a certain window is applied to the input data, the above approach can still be applicable. Let us use a Hanning window to demonstrate this idea. The Hanning window in the time domain can be written as

$$w(t) = \frac{1}{2} + \frac{1}{2}\cos\left(\frac{2\pi t}{T}\right) \tag{5.17}$$

where T is the total data length in time. The frequency response is shown in Figure 5.7.

The frequency response can be obtained from the Fourier transform of the constant and the cosine terms, which are discussed in Chapter 3. The highest four amplitudes of the frequency components are in the mainlobe. If the highest amplitude is k from the true maximum where $f = 0$, the amplitude can be written from (3.77) as

$$X_0 = \frac{\sin(\pi Tk)}{2\pi k} + \frac{1}{4}\left\{\frac{\sin\left[\pi T\left(k + \frac{1}{T}\right)\right]}{\pi\left(k + \frac{1}{T}\right)} + \frac{\sin\left[\pi T\left(k - \frac{1}{T}\right)\right]}{\pi\left(k - \frac{1}{T}\right)}\right\} \tag{5.18}$$

$$= \sin(\pi Tk)\left\{\frac{1}{2\pi k} - \frac{1}{4\pi\left(\frac{1}{k + T}\right)} - \frac{1}{4\pi\left(k - \frac{1}{T}\right)}\right\}$$

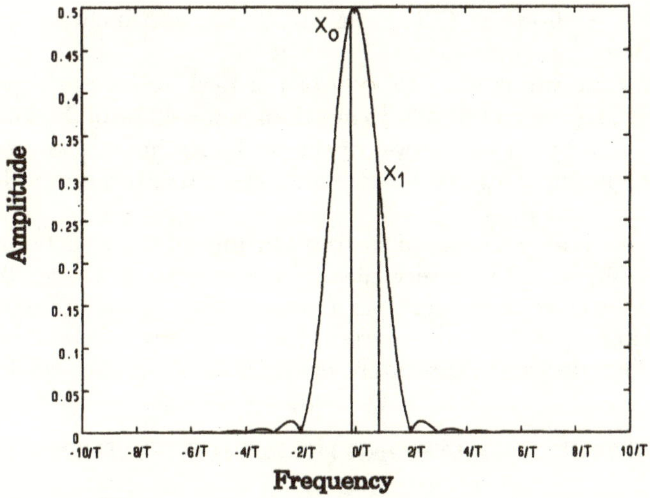

Figure 5.7 Frequency components from FFT of a Hanning window.

The second highest amplitude is $1/T$ away from k and can be written as

$$X_1 = \frac{\sin\left[\pi T\left(k - \frac{1}{T}\right)\right]}{2\pi\left(k - \frac{1}{T}\right)} + \frac{1}{4}\left\{\frac{\sin(\pi T k)]}{\pi k} + \frac{\sin\left[\pi T\left(k - \frac{2}{T}\right)\right]}{\pi\left(k - \frac{2}{T}\right)}\right\} \tag{5.19}$$

$$= \sin(\pi T k)\left\{\frac{-1}{2\pi\left(k - \frac{1}{T}\right)} + \frac{1}{4\pi k} + \frac{1}{4\pi\left(k - \frac{2}{T}\right)}\right\}$$

Similarly, if one finds the ratio of X_1 to X_0, the sine function can be canceled. If one assumes that $1/T = 1$, the result is

$$\frac{X_1}{X_0} = \frac{\frac{-1}{2(k-1)} + \frac{1}{4k} + \frac{1}{4(k-2)}}{\frac{1}{2k} - \frac{1}{4(k+1)} - \frac{1}{4\ (k-1)}} = \frac{k+1}{2-k}$$

or (5.20)

$$k = \frac{2X_1 - X_0}{X_0 + X_1}$$

Once the k value is found, the center frequency can be estimated by (5.16).

The above relations in (5.14) and (5.20) are also included in [9]. In that reference, some other windows are also included.

The spectrum estimation property of the peak selection is quite similar to that of zero padding. Both methods do not change the shape of the power spectrum, they only provide a better estimation of the peak location of the power spectrum. When there are multiple signals in the mainlobe, these two methods (in Sections 5.4 and 5.5) can cause error.

The two methods mentioned in this and the previous sections are sensitive to noise, especially when X_0 is very close to the true peak. Under this condition, X_1 and X_2 are very close to the minima and noise may reverse their amplitudes. If the amplitudes of X_1 and X_2 are reversed, the above equation will move the peak to the wrong direction and cause more error in the frequency reading.

5.6 PEAK POSITION ESTIMATION THROUGH ITERATION [11]

This approach is to generate X_0 and X_1 through iterations from the input data. The idea is to make $X_0 = X_1$ as shown in Figure 5.8. If one can find two frequencies with

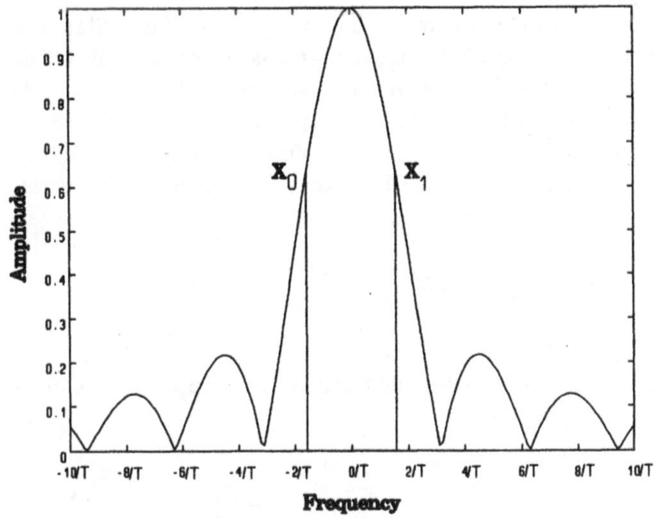

Figure 5.8 Desired relation of X_0 and X_1.

the same amplitudes close to the main peak, the center peak frequency will be between these two values. In order to keep this discussion simple, let us use an example with a rectangular window. When two frequencies are separated by $1/T$ and have the same amplitude, their amplitudes can be found as

$$\left|\frac{\sin(\pi T f)}{\pi f}\right|_{f=\pm\frac{1}{2T}} = \frac{2T}{\pi} \tag{5.21}$$

which is $2/\pi$ below the true peak. This value is very close to the peak and will not be affected much by noise. Thus, this approach is less noise-sensitive, but more computation is required.

The following procedure is used to find the two frequency components with approximately equal amplitudes. In order to keep this discussion clear, all the steps (including previously derived equations) will be included. Assume the input signal is $x(n)$, and X_0 and X_1 are the highest two components in the frequency domain. From these two components one can find k from (5.14) as

$$k^{(1)} = \frac{X_1}{X_0 + X_1} \tag{5.14}$$

The center frequency from (5.16) is

$$f_c^{(1)} = k_0 \pm k^{(1)} \tag{5.16}$$

In these equations superscripts are used to represent the order of iteration. From this frequency one component on each side is selected. The separation between them is chosen to be $1/T$ in order to use the relation in (5.14) again. If the separation is smaller than $1/T$, a linear approximation may be used to find the center frequency. However, there is a possibility that the two components are on the same side of the mainlobe and make the converging process slow. The two frequencies are

$$k_0^{(1)} = f_c^{(1)} - \frac{1}{2T} \qquad k_1^{(1)} = f_c^{(1)} + \frac{1}{2T} \tag{5.22}$$

From these two frequencies, their corresponding amplitudes can be found as

$$\begin{aligned} X_0^{(1)} &= \sum_{n=0}^{N-1} x(n)\, e^{\frac{-j2\pi n k_0^{(1)}}{N}} \\ X_1^{(1)} &= \sum_{n=0}^{N-1} x(n)\, e^{\frac{-j2\pi n k_1^{(1)}}{N}} \end{aligned} \tag{5.23}$$

It should be noted that these equations only calculate one frequency component at $k_0^{(1)}$ and $k_1^{(1)}$. Once $X_0^{(1)}$ and $X_1^{(1)}$ are found, they can be used to find $k^{(2)}$ from (5.14) and the same procedure repeats.

From simulated results, it appears that this process converges in about three cycles. In other words, once $X_0^{(3)}$ and $X_1^{(3)}$ are known, they can be used to find $k_0^{(4)}$ and $k_1^{(4)}$. The final frequency can be found as

$$f_c = \frac{k_0^{(4)} + k_1^{(4)}}{2} \tag{5.24}$$

Comparing this approach with the simple approach in (5.14), it appears that this method is less sensitive to noise.

This method can be applied to other windows. The only change required is (5.14). For simplicity, sometimes a linear relation can be used.

5.7 ACTUAL FREQUENCY DETERMINATION BY FAST FOURIER TRANSFORM [1, 2]

The purpose of this section is to find the actual frequency in hertz of the input signal after the FFT is performed. Parameters affecting this operation are the digitizing speed and the total number of data points. After the FFT, the frequency components should be in units of hertz. Let us start with the DFT. They are rewritten in the following forms:

$$X(k) = \sum_{n=0}^{N-1} x(n)\, e^{-j2\pi kn/N}$$
$$x(n) = \frac{1}{N}\sum_{k=0}^{N-1} X(k)\, e^{j2\pi kn/N} \tag{5.25}$$

Let the input signal be

$$x(t) = e^{j2\pi f_0 t} \tag{5.26}$$

After digitization, the input is

$$x(n) = e^{j2\pi f_0 n t_s} \tag{5.27}$$

where t_s is the sampling time. When this equation is substituted into (5.25), the result is

$$X(k) = \sum_{n=0}^{N-1} e^{j2\pi f_o n t_s}\, e^{-j2\pi kn/N}$$
$$= \sum_{n=0}^{N-1} e^{j2\pi n(f_o N t_s - k)/N} = \frac{1 - e^{j2\pi(Tf_o - k)}}{1 - e^{j2\pi(Tf_o - k)/N}} \tag{5.28}$$

where

$$T = Nt_s$$

and T is the total signal duration. It is interesting to note that the sampling interval t_s does not appear in the above equation explicitly, but only the total length of data T appears. The amplitude of the above equation can be written as

$$|X(k)| = \frac{\sin[\pi(Tf_o - k)]}{\sin\left[\frac{2\pi(Tf_o - k)}{N}\right]} \tag{5.29}$$

The peak of this equation occurs at $Tf_o = k$. However, k can only be an integer number. Thus, the frequency of the input signal is

$$f_o \approx \frac{k}{T} \tag{5.30}$$

Let us use a numerical example to illustrate the result. Assume the input is a sine wave with

$f_o = 200$ MHz $= 2 \times 10^8$ Hz,
$t_s = 10^{-9}$ sec,
$N = 64$, then $T = 64 \times 10^{-9}$ sec.

It is interesting to note that length of data from $x(0)$ to $x(63)$ only covers $63t_s$; however, the input to the FFT is periodic and the period is from $x(0)$ to $x(64)$. Thus, the length of data should be considered as $64t_s$. The signal can be written as

$$x(n) = \sin\left(\frac{2\pi f_o n t_s}{N}\right) \tag{5.31}$$

Applying the FFT, the result is shown in Figure 5.9. Since the input signal is real, the unambiguous band will be 500 MHz $(1/(2t_s))$ and only half of the spectrum components from 0 to 31 are plotted. Each frequency bin is $1/T = 15.625 \times 10^6$ Hz wide. The peak of the power spectrum is at $k = 13$, where the true frequency should be at $Tf_o = 12.8$ rather than 13. The corresponding frequency can be determined as $k/T = 13 \times 15.6 \times 10^6 = 203 \times 10^6$ Hz, which is close to the input frequency.

5.8 REAL INPUT COMPUTED BY A COMPLEX FFT OPERATOR [3]

Usually an FFT operator is designed to process complex data with a real and an imaginary part. However, in many implementations, the input is available only in

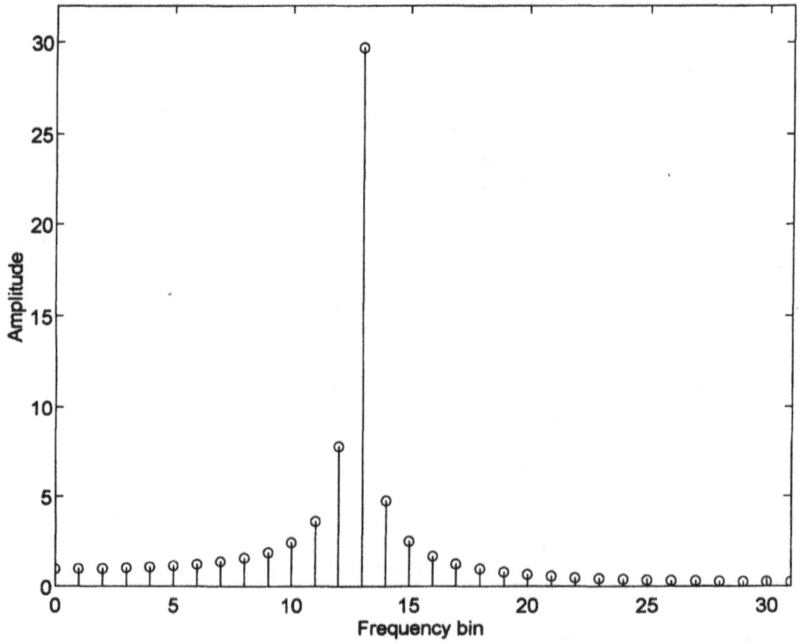

Figure 5.9 Frequency components from FFT of a sine wave.

real form because it is difficult to collect complex data with well-balanced in-phase (I) and quadrature phase (Q) outputs. The I and Q channels will be discussed in detail in Chapter 8. The real data can be used as the input to the real port of the FFT operator and zeros as input to the imaginary port; however, this arrangement does not use the resources efficiently.

This section will discuss the use of one FFT operator to process two sets of real data simultaneously. If $x(n)$ is real data, the FFT of the kth component is

$$X(k) = \sum_{n=0}^{N-1} x(n)\, e^{-j2\pi nk/N} \tag{5.32}$$

and the $(N-k)$th component can be written as

$$\begin{aligned} X(N-k) &= \sum_{n=0}^{N-1} x(n)\, e^{-j2\pi n(N-k)/N} \\ &= \sum_{n=0}^{N-1} x(n)\, e^{j2\pi nk/N} = \sum_{n=0}^{N-1} x^*(n)\,[e^{-j2\pi nk/N}]^* \end{aligned} \tag{5.33}$$

since $x(n) = x(n)^*$ where * is the complex conjugate. It is obvious that

$$\begin{aligned} X(k) &= X(N-k)^* \\ &\text{or} \\ Re[x(k)] &= Re[X(N-k)] \\ Im[X(k)] &= -Im[X(N-k)] \end{aligned} \tag{5.34}$$

where Re and Im represent the real and imaginary parts of the function respectively. The real part of the FFT is symmetric and the imaginary is antisymmetric with respect to $(N-1)/2$.

If $y(n)$ is another real function and has the same number of data points as $x(n)$, one can form a complex function $z(n)$ such that

$$\begin{aligned} z(n) &= x(n) + jy(n) \\ Z(k) &= X(k) + jY(k) \\ &= Re[X(k)] + jIm[X(k)] + j\{Re[Y(k)] + jIm[Y(k)]\} \\ &= Re[X(k)] - Im[Y(k)] + jIm[X(k)] + jRe[Y(k)] \end{aligned} \tag{5.35}$$

This equation can be written as

$$\begin{aligned} Re[Z(k)] &= Re[X(k)] - Im[Y(k)] \\ Im[Z(k)] &= Im[X(k)] + Re[Y(k)] \end{aligned} \tag{5.36}$$

Since $x(n)$ and $y(n)$ are real functions, the real parts of $X(k)$ and $Y(k)$ are symmetric with respect to $(N-1)/2$ and the imaginary parts are antisymmetric. The second half of the k components can be written in terms of the first half by using (5.36) and (5.34) as

$$\begin{aligned} Re[Z(N-k)] &= Re[X(N-k)] - Im[Y(N-k)] \\ &= Re[X(k)] + Im[Y(k)] \\ Im[Z(N-k)] &= Im[X(N-k)] + Re[Y(N-k)] \\ &= -Im[X(k)] + Re[Y(k)] \end{aligned} \tag{5.37}$$

From (5.36) and (5.37), it is easily seen that

$$\begin{aligned} Re[X(k)] &= \frac{Re[Z(k)] + Re[Z(N-k)]}{2} \\ Im[Y(k)] &= \frac{Re[Z(N-k)] - Re[Z(k)]}{2} \\ Re[Y(k)] &= \frac{Im[Z(k)] + Im[Z(N-k)]}{2} \\ Im[X(k)] &= \frac{Im[Z(k)] - Im[Z(N-k)]}{2} \end{aligned} \tag{5.38}$$

This is the desired result.

One can divide the real data $x(n)$ into two sets, $x_1(n)$ and $x_2(n)$, and apply $x_1(n)$ to the real input part of the FFT and $x_2(n)$ to the imaginary part. The output of the FFT is represented by $Z(k)$. The real and imaginary parts of $X_1(k)$ and $X_2(k)$ can be found from the above equation.

5.9 AUTOCORRELATION [1–4]

In this section, the definitions of autocovariance and autocorrelation will be discussed because they can be used in spectrum estimation. The difference between autocorrelation and convolution will also be discussed.

Autocorrelation is often used in spectrum estimation. Autocorrelation is quite similar to convolution. If there are N points of input data represented by $x(n)$ where n is from 0 to $N-1$, the autocorrelation is defined as

$$R(m) = \frac{1}{N}\sum_{n=0}^{N-m-1} x(n)x(n+m) \tag{5.39}$$

where m is referred to as the lag of the autocorrelation. Strictly speaking, this quantity should be referred to as the "sample" autocorrelation, which approximates

$E[x(n)x(n+m)]$ where $E[\ ]$ represents expectation value. The value of m can be either positive or negative. When the argument m is negative, the autocorrelation can be related to the one with positive m as

$$R(-m) = R(m)^* \tag{5.40}$$

if $R(k)$ is complex. If $R(m)$ is real, then $R(-m) = R(m)$.

The autocorrelation can be explained with a simple example. If the lag is m, the input data from 0 to $N-1$ is divided into two groups of same length: one from 0 to $N-m-1$, the other one from m to $N-1$. The elements in these two groups are multiplied term by term as shown in Figure 5.10. The sum of all the product terms is equal to N times the autocorrelation $R(m)$.

The autocorrelation can be considered as the measurement of the similarity between the two groups of data. If the two groups are alike, the autocorrelation value is high; otherwise, it is low. When $m = 0$, the two groups of data are identical; therefore, $R(0)$ produces the highest value among the autocorrelations with different lags. When the m value is large, only a few terms can be obtained in the summation, but the denominator N in (5.39) is a fixed constant. Therefore, the magnitude of $R(m)$ is usually small because the summation is divided by N; however, theoretically this value could be large. The autocorrelation defined in (5.39) is referred to as the biased form.

One can define the unbiased autocorrelation as

$$R_u(m) = \frac{1}{N-m}\sum_{n=0}^{N-m-1} x(n)x(n+m) \tag{5.41}$$

In this definition, when m is large, the denominator is small and it is equal to the number of summation. This form is seldom used in spectrum estimation because it may lead to a negative power spectrum [2]. If $R(m)$ is small, it has less effect on the power spectrum calculated. It is desired to keep the value of $R(m)$ small because it contains only a few data points, which should have less effect on the power spectrum.

Another quantity which is very close to autocorrelation is called autocovariance. Autocovariance can be written as

		x(0)	x(1)	x(2)	. . . .	x(N–m–1)
		x(m)	x(m+1)	x(m+2)	. . . .	x(N–1)

NR(m) = x(0)x(m) + x(1)x(m+1) + x(2)x(m+2) + . . . + x(N–m–1)x(N–1)

Figure 5.10 Representation of $R(m)$ calculation.

$$\Gamma(m) = \frac{1}{N}\sum_{n=0}^{N-m-1}[x(n) - \mu][x(n+m) - \mu] \tag{5.42}$$

where μ is the "sample" mean of the input data $x(n)$ and can be written as

$$\mu = \frac{1}{N}\sum_{n=0}^{N-1}x(n) \tag{5.43}$$

Strictly speaking, the result in (5.42) should be called "sample" autocovariance. The only difference between the autocovariance and the autocorrelation is that in the autocovariance the mean is subtracted from the input data. If the mean is zero, then these two quantities are identical.

The convolution operation discussed in Section 3.7 is mathematically somewhat similar to the autocorrelation, but they have different meanings. If an input signal $x(n)$ passes through a linear system, the output signal can be obtained from convolution. Therefore, physical interpretation of convolution and autocorrelation is quite different.

The mathematical representation of convolution in digital form can be written as

$$y(m) = \sum_{n=0}^{N-m-1}h(n)x(m-n) = \sum_{n=0}^{N-m-1}h(m-n)x(n) \tag{5.44}$$

where $x(n)$ is the input data, $h(n)$ is the impulse response of the linear function, and $y(m)$ is the output. The major difference between this equation and (5.39) is that there is a minus sign in either $x(n)$ or in $h(n)$, but there is no minus sign in (5.39). Physically, the minus sign means that there is a direction change in either $x(n)$ or $h(n)$ as discussed in Section 3.7. In autocorrelation the result represents the similarity of two data sets, while in convolution the result represents the output of a linear system.

5.10 AUTOCORRELATION (BLACKMAN-TUKEY) SPECTRUM ESTIMATION [1–4, 12]

In this section, the autocorrelation is used to find the power spectrum of the input signals. The power spectrum can be estimated from two approaches. First, it can be estimated through FFT of the input data in time domain and then squaring the result. This approach is called the periodogram and is discussed in Chapter 4. This result can be obtained directly from the FFT operation. The power spectrum is obtained from the autocorrelation of the input data. It can be shown that [12]

$$\lim_{T\to\infty}\frac{1}{T}E\left[\left|\int_{-T/2}^{T/2} x(t)\; e^{-j2\pi ft}dt\right|^2\right] = \lim_{T\to\infty}\frac{1}{T}\int_{-T/2}^{T/2}R(\tau)\; e^{-j2\pi f\tau}d\tau \tag{5.45}$$

where $E[\;]$ represents the expectation value.

The power spectrum $P(k)$ from limited data can be obtained from the above equation as

$$P(k) = \sum_{m=-M}^{M} R(m)\ e^{-j2\pi kmt_s} \tag{5.46}$$

Where the m value is from $-M$ to M, k is the frequency component, and t_s is the unit of sampling time. This equation is usually referred to as the Blackman-Tukey method. The total number in the summation is $2M + 1$ because it includes $m = 0$. If the biased autocorrelation is used in the above equation, the result is equal to the periodogram as discussed in [2].

In general, the biased autocorrelation $R(m)$ is used in the above equation in order to produce a positive power spectrum. The biased $R(m)$ can be considered as a windowed function. When m is large, $R(m)$ is usually low. Sometimes, the summation of m is limited to approximately $m = -N/10$ to $N/10$, and a maximum value of $-N/5$ to $N/5$ is often recommended. However, a window function can also be added to the biased autocorrelation to further reduce the sidelobes. With an additional window, the power spectrum can be written as

$$P(k) = \sum_{m=-M}^{M} W(m)R(m)\ e^{-j2\pi km} \tag{5.47}$$

where $W(m)$is the window function. This window must be symmetrical in order to produce an even function of $P(k)$.

The expected value of $R(m)$ can be found as follows. Since $R(m) = R(-m)$, $R(m)$ can be written as

$$R(m) = \frac{1}{N}\sum_{m=0}^{N-1} x(n)x(n+m) = \begin{cases} \dfrac{1}{N}\displaystyle\sum_{n=0}^{N-|m|-1} x(n)x(n+m) & \text{for } |m| \le N-1, \\ 0 & \text{otherwise} \end{cases} \tag{5.48}$$

The expectation value of $R(m)$ can be written as

$$E[R(m)] = \frac{1}{N}\sum_{n=0}^{N-|m|-1} E[x(n)x(n+m)] = \frac{1}{N}\sum_{n=0}^{N-|m|-1} R(m) = \frac{R(m)}{N}\sum_{n=0}^{N-|m|-1} 1 \tag{5.49}$$

since $R(m)$ is independent of n. The summation equals to $N - |m|$; thus, the above equation can be written as

$$E[R(m)] = \frac{N-|m|}{N}R(m) \underset{\text{as } N\to\infty}{\to} R(m) \tag{5.50}$$

The term $(N - |m|)/N$ represents a triangular window. Therefore, one can argue that the Blackman-Tukey method is inherently limited by an effective window and the value of $R(m)$ with $m > N$ is arbitrarily assumed zero.

From the DFT and periodogram discussed in the previous chapter, the frequency resolution equals to f_s/N where f_s is the sampling frequency and N is the total number of data points. If two frequencies f_{i1} and f_{i2} are separated by less than the frequency resolution cell (i.e., $|f_{i1} - f_{i2}| \leq f_s/N$), the periodogram and the Blackman-Tukey methods cannot distinguish them. Under this condition, zero padding will not help either. The only way to distinguish them through these two methods is to increase the actual data duration ($T = Nt_s$).

5.11 APPLICATION OF FFT TO SPECTRUM ESTIMATION FROM AUTOCORRELATION FUNCTION [3]

In this section, it is intended to use the FFT to calculate the result from (5.46). Equation (5.46) is similar to the DFT, but the k value can be arbitrarily assigned. For example, any given k value can be calculated from the above equation. If k is restricted to exactly match the DFT, the FFT algorithm can be used to save calculation time. In order to adopt the FFT, (5.46) must be changed into the proper form. A DFT can be written as

$$X(k) = \sum_{n=0}^{N-1} x(n)\, e^{\frac{-j2\pi kn}{N}} \tag{5.51}$$

where $X(k)$ is the frequency domain response and $x(n)$ is the time domain sampled data point. In this equation, both n and k are discrete and N is usually a power of 2 number.

The following procedures are needed to change (5.46) into the proper form.

1. The notation of t_s must be changed. By comparing (5.46) and (5.51), one can assign that

$$t_s = \frac{1}{N} \tag{5.52}$$

   where N is a number of the power of 2. Comparing with (5.28), this implies that T is assumed to be unity.
2. Equation (5.46) has to sum $2M + 1$ terms, which cannot be a power of 2 number. In order to use the FFT, the total terms under the summation sign must be changed to a power of 2 number. To accomplish this, (5.46) must be padded with zeros.
3. Equation (5.46) sums values from $-M$ to M, but in (5.51), the summation starts from 0. Therefore, the summation in (5.46) must be rearranged to start from 0.

In order to accomplish step 1, choose the N value as

$$2M + 1 < N < 2(2M + 1) \tag{5.53}$$

but N must be a power of 2 number.

Separate the summation of (5.46) as

$$\begin{aligned} P(k) &= \sum_{m=-M}^{M} R(m)\, e^{\frac{-j2\pi mk}{N}} \\ &= \sum_{m=0}^{M} R(m)\, e^{\frac{-j2\pi mk}{N}} + \sum_{m=-M}^{-1} R(m)\, e^{\frac{-j2\pi mk}{N}} \end{aligned} \tag{5.54}$$

On the second summation, make the change of variable as

$$m = m' + N \tag{5.55}$$

Then (5.54) can be written as

$$\begin{aligned} P(k) &= \sum_{m=0}^{M} R(m)\, e^{\frac{-j2\pi km}{N}} + \sum_{m'=N-M}^{N-1} R(m'+N)\, e^{\frac{-j2\pi k(m'+N)}{N}} \\ &= \sum_{m=0}^{M} R(m)\, e^{\frac{-j2\pi km}{N}} + \sum_{m'=N-M}^{N-1} R(m'+N)\, e^{\frac{-j2\pi km'}{N}} \end{aligned} \tag{5.56}$$

In the above equation, the relation of

$$e^{-j2\pi k} = 1 \tag{5.57}$$

where k = integer is used.

Since both m and m' are dummy variables, they can be replaced by n. The result can be written as

$$P(k) = \sum_{n=0}^{M} R(n)\, e^{\frac{-j2\pi kn}{N}} + \sum_{n=M+1}^{N-M-1} 0 + \sum_{n=N-M}^{N-1} R(n+N)\, e^{\frac{-j2\pi kn}{N}} \tag{5.58}$$

In this equation, two important facts should be noted. First, zeros are added from $n = M + 1$ to $N - M - 1$. The total terms on the right equal N, a number of the power of 2. Second, the $R(n)$ must be properly arranged. The $R(n)$ for $n = 0$ to M stay the same. Under the third summation sign, the $R(n)$ should be changed from the range of $n = -M$ to -1 (5.54) to the range from $n = 2N - M$ to $2N - 1$; thus, the above equation can be written as

$$
\begin{aligned}
P(k) &= \sum_{n=0}^{M} R(n)\, e^{\frac{-j2\pi nk}{N}} + \sum_{n=M+1}^{N-M-1} 0 + \sum_{n=N-M}^{N-1} R(n+N)\, e^{\frac{-j2\pi kn}{N}} \\
&= \sum_{n=0}^{N-1} R(n)\, e^{\frac{-j2\pi kn}{N}}
\end{aligned}
\tag{5.59}
$$

which is exactly the same form as (5.51). Therefore, FFT can be used to perform the calculation.

Now, let us use an example to demonstrate this operation. Let $R(n)$ be from -4 to 4 ($M = 4$), as shown in Figure 5.11(a). There are total of 9 terms. From (5.53), choose N between 9 and 18 (and N should be the number of power of 2); thus, $N = 16$. The rearranged $R(n)$ is shown in Figure 5.11(b). After this rearrangement, FFT can be used to calculate $P(k)$.

In the above discussion, (5.46) is directly transformed into the form of (5.51); therefore, the correct phase relation is kept. The result obtained for $P(k)$ is usually a complex value. If the power spectrum is desired, the absolute value will be used. The $R(n)$ can be rearranged slightly differently and the same results will be obtained. This approach is to shift the entire $R(n)$ from $n = -M$ to M to $n = 0$ to $2M$ and pad with zeros at the end of the data string as shown in Figure 5.11(c). This arrangement will not generate the same result as (5.51) and there is a phase difference. However, the absolute value of $P(k)$ is the same, which is important for power spectrum estimation. This new way of shifting $R(n)$ might be somewhat easier to accomplish.

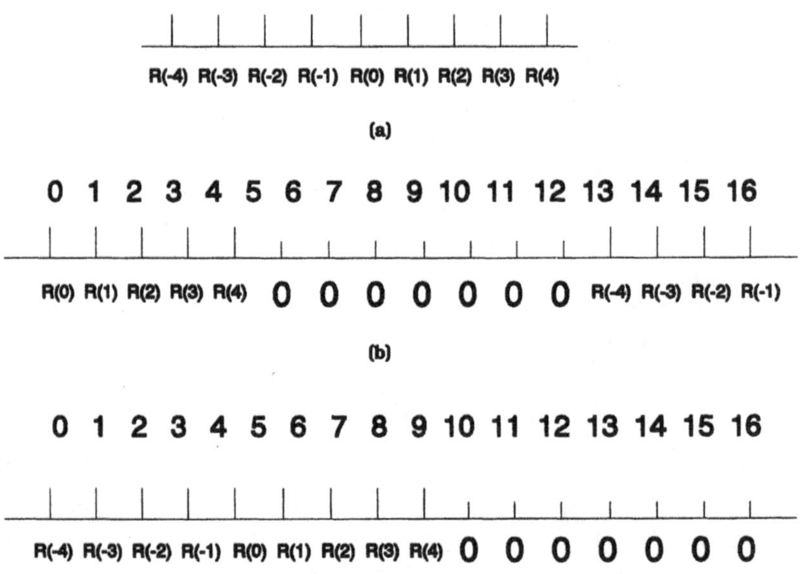

Figure 5.11 Rearranged $R(n)$.

5.12 BASIC IDEA OF SUB-NYQUIST SAMPLING SCHEME [13–18]

In the next three sections, the concept of using sub-Nyquist sampling to design an EW receiver will be discussed. This section discusses the basic idea. The next section will discuss the difference in phase relation between an analog and a digital system because it is important in the sub-Nyquist sampling method. In Section 5.14, some potential problems and solutions, which are unique in the sub-Nyquist approach, will be presented.

The sub-Nyquist method may have two impacts that might be of interest to EW receiver designers. First, this method can increase the instantaneous bandwidth of the receiver. Second, for a given bandwidth, the FFT processing speed need not match the instantaneous bandwidth. Actually, these two impacts are closely related. The basic sub-Nyquist sampling scheme is very similar to the instantaneous frequency measurement (IFM) receiver briefly discussed in Chapter 2.

The main part of an IFM receiver is a correlator. The input to the correlator consists of a delayed and an undelayed copy of the same signal. The outputs from the correlator are

$$E = \sin(2\pi f\tau) \qquad \text{and} \qquad F = \cos(2\pi f\tau) \tag{5.60}$$

where f is the frequency of the input signal and τ is the delay time of the delayed path. In the above equation, the amplitude of signal is neglected for simplicity. Since the delay time is known, the frequency can be obtained from

$$\theta = \tan^{-1}\left(\frac{E}{F}\right) = 2\pi f\tau \tag{5.61}$$

The only restriction in the above equation is that $\theta < 2\pi$. If $\theta > 2\pi$, there is an ambiguity problem. The maximum unambiguous bandwidth obtained from this equation is

$$\Delta B = \frac{2\pi}{2\pi\tau} = \frac{1}{\tau} \tag{5.62}$$

This relation only limits the bandwidth, but not the center frequency; therefore, the center frequency can be any value. For example, if $\tau = 0.5$ ns, the unambiguous bandwidth is 2 GHz. The frequency range can be either from 0 to 2 GHz or from 2 to 4 GHz, or any other values as long as the bandwidth is 2 GHz.

The sub-Nyquist sampling scheme can be considered as an IFM receiver implemented through digital techniques. An important difference is that a conventional IFM receiver can process only one signal at a time, but a digital IFM receiver can process simultaneous signals because of the FFT operation.

In the sub-Nyquist approach discussed in this chapter, the input signal is divided into two paths—an undelayed one and a delayed one—and A/D converters are used to digitize the signal as shown in Figure 5.12. In actual design, the delay can be introduced in the clock pulse rather than in the radio frequency (RF) circuit.

The digitized outputs can be processed through an FFT operation. The FFT will generate real and imaginary parts in the frequency domain. Let $X_{ru}(k)$ and $X_{iu}(k)$ represent the real and imaginary components of the undelayed case and $X_{rd}(k)$ and $X_{id}(k)$ the delayed case. The amplitude of the FFT output can be calculated from the undelayed output as

$$X_u(k) = [X_{ru}(k)^2 + X_{iu}(k)^2]^{1/2} \tag{5.63}$$

The delayed path has the same amplitude components. Let $X_u(k_m)$ represent the maximum amplitude of the frequency component from the undelayed path; thus, $X_u(k_m)$ can represent the input frequency. It should be noted that, as in the FFT operation, the input frequency need not be exactly on any frequency bin. The phase difference between the delay and undelayed path can be written as

$$\theta = \theta_d - \theta_u = 2\pi f \tau \tag{5.64}$$

where

$$\theta_d = \tan^{-1}\left[\frac{X_{id}(k_m)}{X_{rd}(k_m)}\right]$$
$$\theta_u = \tan^{-1}\left[\frac{X_{iu}(k_m)}{X_{ru}(k_m)}\right] \tag{5.65}$$

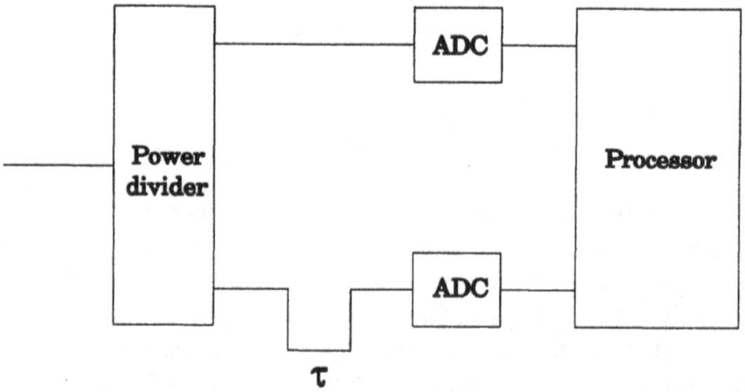

Figure 5.12 Basic sub-Nyquist sampling.

From this phase difference θ, the frequency of the input signal can be obtained because τ is known.

As long as the input frequencies (or multiple FFT peaks) are sufficiently separated, the input frequencies can be identified by observing those frequency bins whose magnitude exceeds a threshold. This is why this approach can process simultaneous signals.

In general, if the signal is sampled below the Nyquist rate, there is ambiguity in the FFT output. This ambiguity can be resolved by the phase difference in (5.64). Although this approach is named sub-Nyquist because the uniform sampling rate is lower than the Nyquist rate, it does not violate the Nyquist sampling theorem. The closest sampling time is the delay time τ. When the input bandwidth is less than $1/\tau$, the Nyquist sampling criterion is not violated.

Let us use an example to illustrate this approach. If an ADC can only operate at 250 MHz, the maximum unambiguous bandwidth is 125 MHz. If the desired input bandwidth is 1,000 MHz and this ADC is used to collect data, the entire input bandwidth will fold into a 125-MHz output band as shown in Figure 5.13. In this figure $f_s = 250$ MHz. This means there are eight ambiguity zones. The FFT can only determine the input frequency within 125 MHz without ambiguity. If the delay line τ is chosen to be less than 1 ns (say 0.8 ns) from (5.62), the corresponding unambiguous frequency band is over 1,250 MHz. Thus, any frequency within the bandwidth of 1,000 MHz can be determined uniquely by the phase difference in (5.64). For example, if the measured frequency is 40 MHz from the FFT operation, one does not know the input frequency because it can be in any one of the eight zones. The input could be 40, 210(250 – 40), 290(250 + 40), 460(500 – 40), 540(500 + 40), 710(750 – 40), 790(750 + 40), and 960(1,000 – 40)MHz as shown in Figure 5.13. The frequency obtained from the phase difference at this peak value (close to 40 MHz) can determine which zone this signal is in. From the example, if the frequency measured by the phase difference is close to 460 MHz, the true input frequency will be 460 MHz. Thus, the fine frequency resolution can be determined by the FFT and the phase difference can be used to resolve the ambiguity zone.

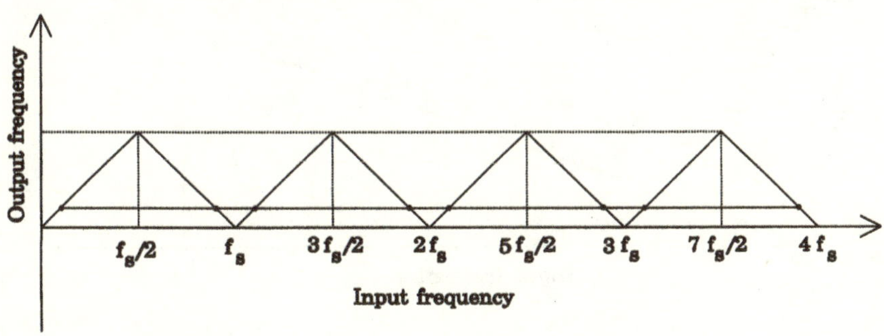

Figure 5.13 Input band versus output band.

From the above discussion, it may appear that (5.64) can be used to eliminate ambiguity frequency problems. If the phase is obtained from a continuous system, such as in an analog IFM receiver, this is true. However, the phase is not continuous in the sampled case and this problem will be discussed in the next two sections. As a result, when the input frequency is close to the multiple of $f_s/2$, the phase measurement could be erroneous. Therefore, there are regions where the frequency of the input signal cannot be obtained. In order to eliminate these regions, additional hardware is required.

5.13 PHASE RELATION IN A SUB-NYQUIST SAMPLING SYSTEM [13, 14]

In this section, the phase difference in an analog and a digital system will be compared. The phase difference in a digital system will cause some receiver design changes from a conventional analog system.

In a conventional IFM receiver, the phase relation is obtained from the I and Q channel outputs. The phase versus frequency is continuous. Thus, the phase difference between the delayed and undelayed phase is continuous. The phase versus frequency of an IFM receiver is shown in Figure 5.14 where f_1 and f_2 are the low and high frequencies, respectively.

The difference phase will increase monotonically, except that an abrupt 2π phase change (from 2π drops to 0) will occur at a certain frequency. For simplicity, let us limit the phase shift within 2π as shown in Figure 5.14.

This idea is further illustrated mathematically by using a sine function. In the analog case, a sinusoidal wave can be written as

$$s(t) = \cos[2\pi f_o(t - \tau) - \phi] \tag{5.66}$$

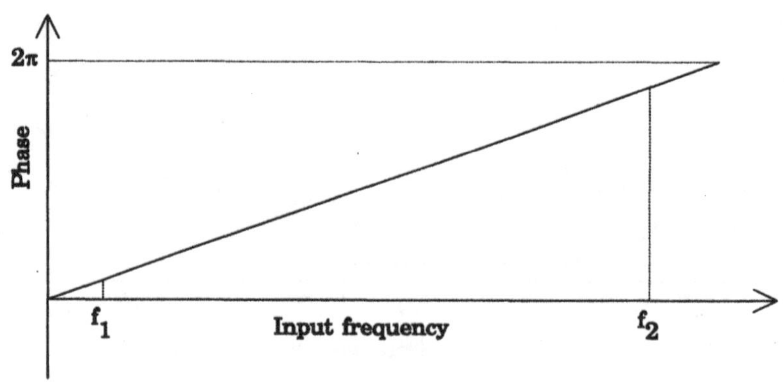

Figure 5.14 Phase versus frequency from an IFM correlator.

where f_0 is the input frequency, τ is a time delay, and ϕ is the initial phase angle. As discussed in Chapter 3, the Fourier transform of this signal can be represented by a pair of delta functions at plus and minus f_o with positive and negative phase shifts as

$$S(f) = \frac{1}{2}[\delta(f - f_o)\, e^{j\theta} + \delta(f + f_o)\, e^{-j\theta}] \tag{5.67}$$

where

$$\theta = \phi + 2\pi f_o \tau \tag{5.68}$$

Note that in the above equation, the phase for the two delta functions has opposite signs.

If the signal is truncated in time using a window function $w(t)$, the delta functions are convolved with the function $W(f)$ and the transform becomes

$$S(f) = \frac{1}{2}[W(f - f_o)\, e^{j\theta} + W(f + f_o)\, e^{-j\theta}] \tag{5.69}$$

The window function $W(f)$ may modify the phase of the input signal; however, the phase of the positive and negative spectra still has the same amplitude (but opposite signs).

In a digital system, the sine signal is sampled at frequency f_s and the Fourier integral is replaced by DFT. The output $S_p(k)$ in the frequency domain is periodically replicated as

$$S_p(k) = S(k) \star \sum_k \delta(k - nf_s) \tag{5.70}$$

where k is an integer. An infinite number of zones appear, each of width f_s and containing a replica (or alias) of either the positive or negative power spectrum. If one considers the positive and negative frequencies separately, the zone width becomes $f_s/2$.

If the signal in a zone from nf_s to $(n + 1/2) f_s$ is the positive frequency, the signal in an adjacent zone from $(n + 1/2) f_s$ to $(N + 1) f_s$ is the negative frequency. Thus, their phases are opposite in sign. As the input frequency increases to pass a zone boundary, the phase changes sign. The phase versus frequency of the sampled data is shown in Figure 5.15. At every multiple of $f_s/2$, there is a phase reversal.

If the phase in Figure 5.15 covers the entire range from -2π to 2π, the phase measurement can be ambiguous. The explanation is as follows. Since the real and the imaginary parts are calculated from FFT, the θ angle calculated from (5.64) is within $-\pi$ to π. The phase difference calculated from this relation has two possible values.

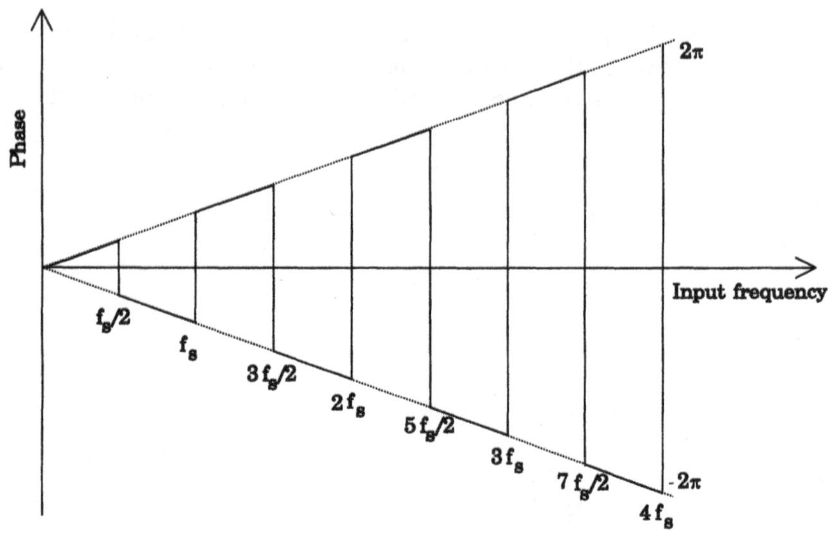

Figure 5.15 Phase versus frequency of a sub-Nyquist sampling scheme.

Let us use two examples to demonstrate the calculation of phase shifts. First, let $\theta_d = \pi/6$ and $\theta_u = 3\pi/4$. The phase difference can be either $\theta_1 = \theta_d - \theta_u = -7\pi/12$ or $\theta_1' = \theta_d - \theta_u = (\pi/6 + 2\pi) - 3\pi/4 = 17\pi/12$ because both the calculated angles are within $\pm 2\pi$. Second, if $\theta_d = \pi/6$ and $\theta_u = -3\pi/4$, the phase difference can be either $\theta_1 = \theta_d - \theta_u = 11\pi/12$ or $\theta_1' = \theta_d - \theta_u = \pi/6 - (-3\pi/4 + 2\pi) = -13\pi/12$. In each example, there are two different phase angles and one cannot determine which one is the true phase difference.

From these examples, it is obvious that the two possible phase shifts have the following relations:

$$|\theta_1| + |\theta_1'| = 2\pi \tag{5.71}$$

Both θ_1 and θ_1' have an absolute value of less than 2π, but they can be either positive or negative.

If one desires to use the entire phase shift from -2π to 2π for a digital receiver design, there are ambiguous frequency ranges. There are two ways to eliminate these ambiguous range pairs. One is to limit the input bandwidth from $\theta = -\pi$ to π, and this approach is obvious. For instance, in the previous example, to cover 1,000-MHz bandwidth by using a 250-MHz ADC, one can choose the delay time to be 0.4 ns rather than 0.8 ns. With this short delay, the overall phase difference can be restricted within $\pm\pi$ and there is no ambiguity.

The other approach is to restrict the choice of the sampling frequency. One can choose the delay time with the following relation:

$$\left(n + \frac{1}{2}\right) f_s \tau = 1 \tag{5.72}$$

where n is an integer. This approach divides the input bandwidth $(1/\tau)$ into an odd number of sampling bandwidth $(f_s/2)$ intervals. Under this condition the phase angles θ_1 and θ_1' have the same sign with respect to the vertical line passing π, as shown in Figure 5.16. If both phases have the same sign, the condition in (5.71) can no longer be fulfilled. For example, in the range 0 to $f_s/2$, a small positive-valued θ is the correct angle. The corresponding θ' ($\theta - 2\pi$, a negative value) is in the range $3f_s$ to $7f_s/2$ because of the symmetry. However, θ' must have a positive value in this frequency range, as shown in Figure 5.16. Thus, θ' is not an acceptable answer and θ is the only answer. Therefore, there is no ambiguous range if the relation in (5.72) is used.

5.14 PROBLEMS AND POTENTIAL SOLUTIONS OF SUB-NYQUIST SAMPLING SCHEME

If the signal frequency is close to the multiple of $f_s/2$, the subsampling approach fails. The reason and some possible solutions will be discussed in this section.

As discussed in the previous section, if the input is real, the spectrum will appear in pairs: one at f_i, the other at $f_s - f_i$. Because of the limited data length in time domain, the spectra has sidelobes. If the input signal frequency is close to

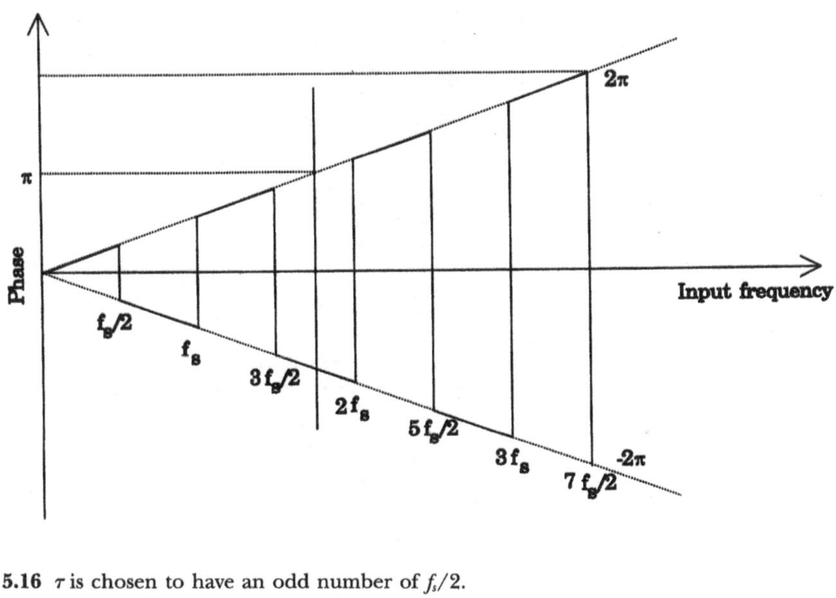

Figure 5.16 τ is chosen to have an odd number of $f_s/2$.

$nf_s/2$, both spectra are close to $nf_s/2$ as shown in Figure 5.17. The sidelobe of one spectrum will interfere with the other one. This interference will distort the phase measurement. The wrong phase information may cause frequency measured to be assigned to the wrong subband, and a catastrophic error in the frequency reading would result.

There are three possible ways to remedy this problem, as follows.

1. One straightforward approach is to reduce the amplitude of the sidelobes. As mentioned in the previous chapter, a proper windowing function will increase the width of the mainlobe but reduce the sidelobes. When the sidelobes are low, the interference from neighboring zones will be reduced. Although windowing can narrow the region of interference, it cannot eliminate them. A receiver using this approach will have "holes" near the multiple of $nf_s/2$.
2. In order to reduce these "holes," a second channel can be built. The second channel has a sampling rate f'_s, which is relatively prime to the first sampling rate f_s. The arrangement is shown in Figure 5.18. When the input signal is near the multiple of $f_s/2$, it should be far away from the multiple of $f'_s/2$. Thus, when the first channel generates wrong frequency information, the second channel will produce the correct one. When the signal is close to a multiple of $f'_s/2$, the first channel should be used to read the frequency. In other words, between the two channels, one of the frequency readings must be correct.

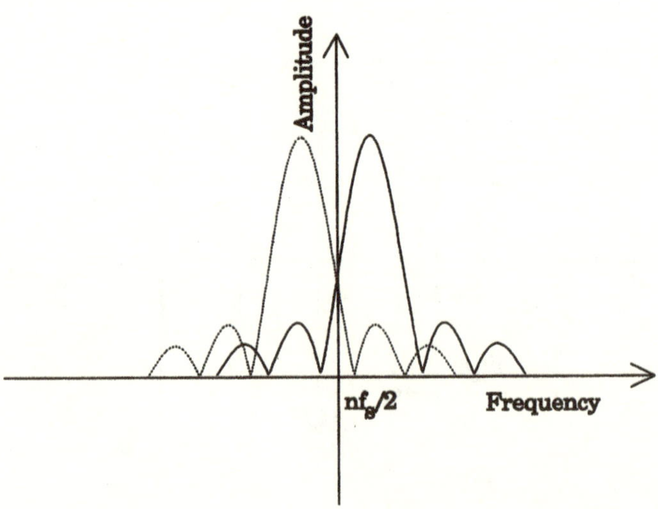

Figure 5.17 Spectra close $nf_s/2$.

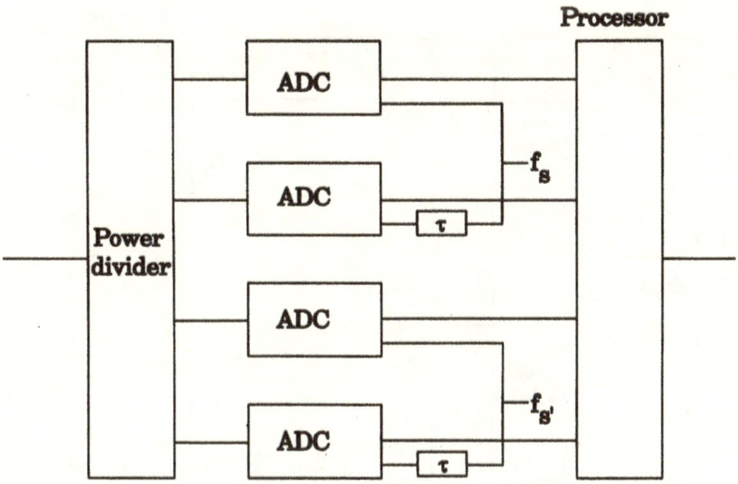

Figure 5.18 Sub-Nyquist scheme with two sampling rates.

3. Another way to eliminate the "holes" in the sub-Nyquist sampling scheme is to use the *I* and *Q* channels. The *I* and *Q* channels are used in both the delayed and undelayed paths as shown in Figure 5.19. All four ADCs operate at the same sampling speed f_s. In both Figures 5.18 and 5.19, the delay is introduced in the clock rather than in the RF chain.

   This arrangement generates a complex signal. As mentioned in Chapter 3, when the input signal is complex, the frequency component exists only on one side of the frequency axis—either on the positive or on the negative side. Let us assume that the spectrum is positive. Under this condition, when the spectrum is at the multiple number of $f_s/2$, there is no negative component to interfere with it. The phase relation measured through this approach is always correct and there is no frequency error at multiples of $f_s/2$.

   Although it appears that the *I* and *Q* channel approach has obvious advantages, it has a practical limitation (i.e., the channel imbalance), especially for a wide frequency band system. If the two channels are perfectly balanced in both amplitude and phase, this is the preferred approach. If the two channels cannot be perfectly balanced, the negative frequency cannot be totally canceled, and the amplitude of the negative frequency depends on how well the two channels are balanced. The negative frequency limits the dynamic range of the receiver, and this will be discussed in detail in Chapter 8.

   If two signals are separated by f_s, they will fold into the same peak and the sub-Nyquist sampling scheme cannot solve them easily.

   The sub-Nyquist sampling approach can be considered as a special way to fold a wideband input into a narrow output band. This concept is used

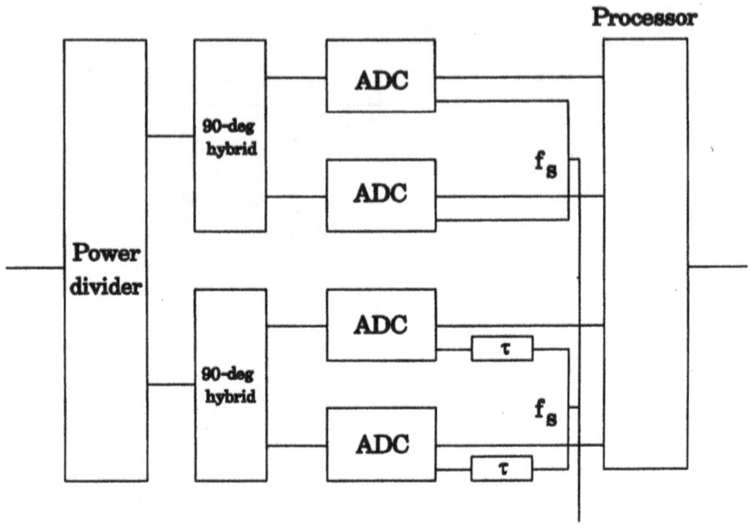

Figure 5.19 Sub-Nyquist sampling with I and Q channels.

often in analog EW receivers. One of the prices one pays is that the noise in the input bandwidth will fold into the output band. For example, folding eight input bands into one will increase the noise by 9 dB(10 log 8). This increase of noise will reduce the sensitivity of the receiver.

5.15 DISCRETE FOURIER TRANSFORM THROUGH DECIMATION [19, 20]

This section will discuss the decimation scheme of DFT. The discussion is based on [20] and only the final results will be presented here. It is anticipated that with modification, this approach might be applicable to real-time signal processing. This scheme uses many parallel FFT operators to perform many individual FFTs and combines them into a single FFT output. The basic approach will be discussed in this section.

If there are N total points, the data can be divided into r subgroups and each group contains s data points. FFT will be performed on each group and the results will be combined to obtain the desired results. The original data points can be represented by $x(n)$ where n is from $n = 0$ to $N - 1$. After regrouping there are r groups, where each one can be represented as $x_i(n)$ where i is from 0 to $r - 1$. The data point n can be written as

$$n = lr + i \qquad i = 0, 1, \ldots r - 1; \quad l = 0, 1, \ldots s - 1 \tag{5.73}$$

This processing is referred to as decimation, or the input data is decimated by r.

Let us use an example to illustrate this operation. Assume $N = 128$ and the data are divided into four groups ($r = 4$), and each group contains 32 data points ($s = 32$). Under this condition, $i = 0, 1, 2, 3$ and $l = 0, 1, 2, \ldots 31$. If $i = 0$, the data points consist of $n = 0, 4, 8, 12, \ldots 124$, which corresponds to $l = 0$ to 31. These results mean that $x_0(0) = x(0)$, $x_0(1) = x(4)$, $x_0(2) = x(8)$, $\ldots$ $x_0(31) = x(124)$. A similar argument can be applied to the following: for $i = 1$, the data points consist of $x(n)$ with $n = 1, 5, 9, 13, \ldots 125$; for $i = 2$, $n = 2, 6, 10, 14, \ldots 126$; and for $i = 3$, $n = 3, 7, 11, 15, \ldots 127$. Figure 5.20 shows only the first few points of the data.

From the four subgroups, four DFTs will be performed. The result is

$$X_i(k) = \sum_{n=0}^{s-1} x_i(n) \; e^{-j2\pi kn/s} \tag{5.74}$$

where n is given by (5.73). In the above equation, each X_i contains s (for this example $s = 32$) frequency components or k ranges from 0 to $s - 1$ (0 to 31). The final DFT of the N points can be obtained by combining the individual results from (5.74) as

$$X(k) = \frac{1}{r}\sum_{i=0}^{r-1} X_i(k \bmod s) \; e^{-j2\pi ki/N} \tag{5.75}$$

where (k mod s) means the remainder of k divided by s. If $k = 68$, since $s = 32$, $k/s = 68/32 = 2 + 4/32$ and the remainder is 4.

Since $X(k)$ is the DFT of all the points $x(n)$, there should be N frequency components or the k value of $X(k)$ is from 0 to 127 (for this example). However, the value of k in $X_i(k)$ is from 0 to 31; thus, the (k mod s) operation is used to change the k value. Let us use the above example to demonstrate this point. If $k = 2$, then (k mod s) = (2 mod 32) = 2; thus,

$$X(2) = \frac{1}{4}\sum_{i=0}^{3} X_i(2) \; e^{-j2\pi 2i/N} \tag{5.76}$$

It takes four complex multiplications to obtain the above result. For a component at $k = 34$, (k mod s) = (34 mod 32) = 2; thus,

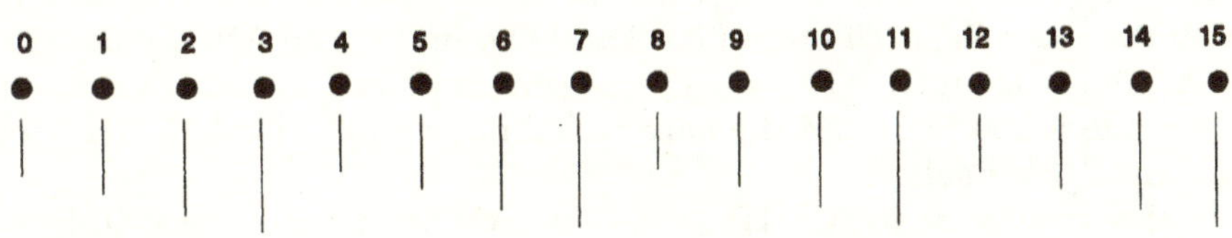

Figure 5.20 Data divided into four subgroups.

$$X(34) = \frac{1}{4}\sum_{i=0}^{3} X_i(2)\, e^{-j2\pi 34i/N} \tag{5.77}$$

It should be noted that in the calculation at $k = 34$, the same $X_i(2)$ is used because there is no $X_i(34)$ available from (5.75). The only difference between (5.76) and (5.77) is that as a kernel function, the former uses $k = 2$ whereas the latter one uses $k = 34$.

If one wants to calculate all the 128 points from this method, it appears that the last calculation as shown in (5.75) will take $N \times r$ complex multiplications. In this example, it will take 512 (4×128) operations, which is more complicated than the straightforward FFT approach that takes approximately $(128/2)\log_2(128)$ or 448 operations. In addition to the last step operation, there are four separate FFT operations to find the four $X_i(k)$ operations. Therefore, if one uses this approach to find all the components of the DFT, it may not be very attractive. Although the input can be processed in parallel, to combine all four individual operations is also rather complex. However, the operation in (5.75) should be further investigated to see whether it can be simplified.

5.16 APPLICATIONS OF DECIMATION METHOD TO EW RECEIVERS

The basic idea in this section is to calculate all the FFT in the subgroups and find the peaks. From these peak values, one needs only to calculate the frequencies of interest. This operation uses less calculation than to find all the frequency components. Although the above decimated method to obtain FFT may not be very attractive to generate the FFT data, this modification might be applicable to EW receivers. Since an EW receiver is required to find only a few signals in a short time interval, as discussed in Chapter 2, all the frequency components from the FFT may not be needed.

Let us assume that the input signal is downconverted to complex outputs and four pairs of ADCs are used to digitize the output. If each ADC is operating at 250 MHz and staggered by 1 ns, this arrangement is equivalent to sampling at 1 GHz. If 128 data points are used for the calculation, the total data collection time is 128 ns. The probability of intercepting a large number of simultaneous signals is very low. Therefore, the receiver can be designed to process only a few signals.

Let us use an example to illustrate this idea. Figure 5.21 shows the arrangement. In this example, each output has I and Q outputs. The ADC operates at 250 MHz and the delay time τ is 1 ns. If one considers only one pair of ADCs, the sampling rate is 250 MHz and the corresponding bandwidth is 250 MHz because of the I and Q channels.

In this arrangement, $N = 128$, $r = 4$, and $s = 32$. If there are two sinusoidal signals contained in the data, the power spectrum obtained from any one of the four FFT outputs should have two distinct peaks. First, all the four FFTs are used

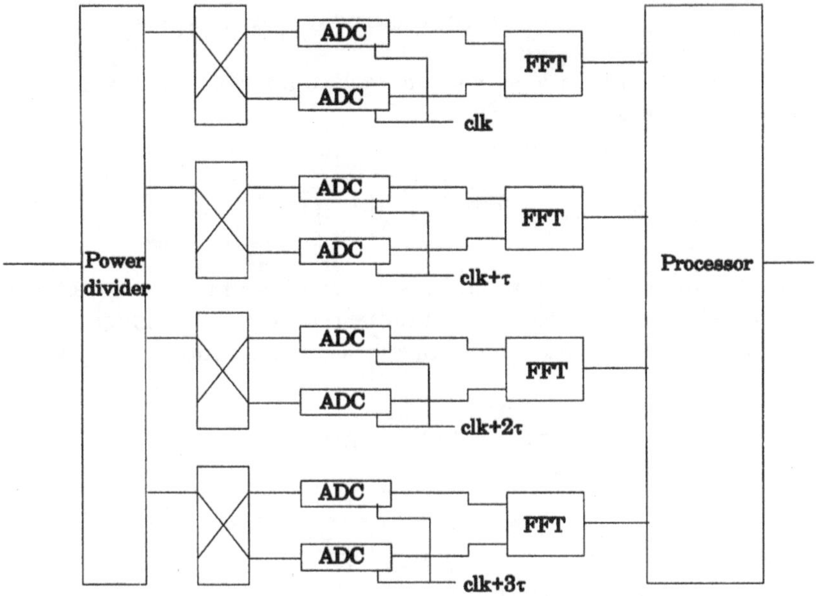

Figure 5.21 Input signals divided into four parallel channels.

to calculate the four sets of outputs $X_i(k)$ for $i = 0$ to 3. Let us represent these two peaks by the amplitudes of $X_0(k_1)$ and $X_0(k_2)$. It should be noted that the peaks in all the four outputs of $X_i(k)$ for $i = 0$ to 3 are at the same k values and have the same amplitudes. In other words, once k_1 and k_2 are found from the amplitude of $|X_0(k)|$, all the rest of the three $X_i(k)$ (for $i = 1, 2, 3$) with these k values will be used in (5.75) to find the correct frequency.

For each peak (or each k value) in $X_i(k)$, there are four possible values in the overall FFT output and they are $X(k)$, $X(k + s)$, $X(k + 2s)$, and $X(k + 3s)$ where $s =$ 32 in this example. One needs to find all these four components to determine the true peak. To find one of the $X(k + is)$ ($i = 0, 1, 2,$ and 3) values from (5.75) requires four complex multiplications, and a total of 16 operations are required. Two signals require 32 operations. The total operations including the four sets of FFT are $4(32/2)\log_2(32) + 32 = 352$, which is less than the 448 operations required for straightforward 128 point FFT.

When the two signals are separated by an integer multiple of 250 MHz, the two input frequencies will fold into one peak. Under this condition, the peaks from each $X_i(k)$ are different in amplitude because of the different phases in the two signals. This is an important sign that two signals are folding into one peak. The highest peak can be used to choose the k value. Once the k value is selected, $X(k)$, $X(k + s)$, $X(k + 2s)$, and $X(k + 3s)$ will be calculated. Two of these $X(k + is)$ values should have peaks and they represent the true frequencies. This requires only 16 complex multiplications.

In general, if there are p signals, the total number of complex multiplications (N_c) can be approximated as

$$N_c = r\left(\frac{s}{2}\right)\log_2 s + pr^2 \tag{5.78}$$

where r is total number of groups and s is the number of points in each subgroup. When this number is equal to the operations required for a straightforward FFT, the advantage of this method will disappear. This condition is

$$\frac{rs}{2}\log_2(rs) = \frac{rs}{2}\log_2 s + pr^2$$

or

$$p = \frac{s}{2r}\log_2 r \tag{5.79}$$

For the above example, p can be as large as 32.

Of course, the calculation can be further simplified by finding $X_0(k)$ and its peak values at k_1 and k_2 and so forth. Then find the other three $X_i(k)$ at these k values. This way, one needs to use the FFT once instead of four times. However, there is a chance when the two signals are separated by 250 MHz that the amplitude in $X_0(k)$ may be very low and one will miss the signal. If all four $X_i(k)$values are calculated, the four sets of outputs have different amplitudes and the signal can always be recognized.

5.17 SIMPLIFIED DECIMATION METHOD [21]

In this section, the decimation method is further simplified by reducing the hardware. This basic approach is somewhat similar to the sub-Nyquist sampling scheme discussed in Section 5.12.

Let us continue to use the above example to demonstrate the approach. The basic idea is to use two sets of data delayed by t to find the peaks and the corresponding phase differences through $X_0(k)$ and $X_1(k)$. From this phase difference, the value of $X_2(k)$ and $X_3(k)$ can be estimated. When all the $X_i(k)$ values are obtained, the true frequency can be found through (5.75).

In Figure 5.22, only the first two pairs of samplers are kept. The sampling rate is maintained at 250 MHz and the delay time τ is still 1 ns. Under this arrangement, only half of the data shown in Figure 5.21 will be obtained. A total of 64 points of complex data will be collected. In Figure 5.20, data points 0, 1, 4, 5, 8,

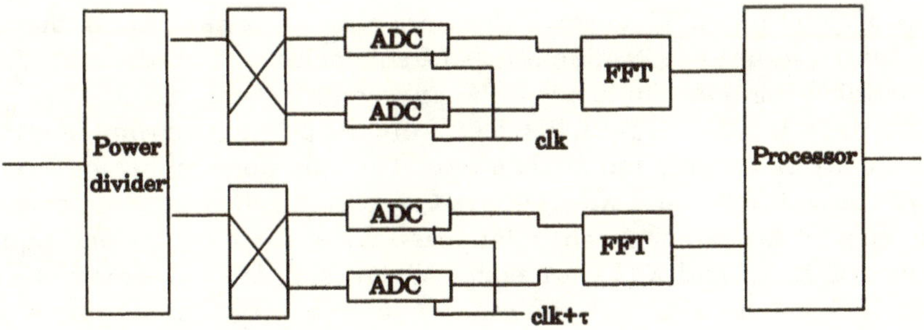

Figure 5.22 Simplified decimation method.

9, . . . are available, but the other half consisting of 2, 3, 6, 7, 10, 11, . . . will not be obtained. From the first group of 32 points, one can calculate $X_0(k)$ and from the amplitude of $|X_0(k)|$ one can find two peaks k_1 and k_2. From the second group of 32 points, one can find $X_1(k_1)$ and $X_1(k_2)$. From $X_0(k_m)$ and $X_1(k_m)$ where $m = 1$, 2 for this case, one can find the phase difference, which can be written as

$$\theta_m = \tan^{-1}\left\{\frac{\mathrm{Im}[X_1(k_m)]}{\mathrm{Re}[X_1(k_m)]}\right\} - \tan^{-1}\left\{\frac{\mathrm{Im}[X_0(k_m)]}{\mathrm{Re}[X_0(k_m)]}\right\} \tag{5.80}$$

From θ_m one can find the frequency of the input signal as in the phase delay method. Now, let us use the θ_m values to generate $X_2(k_m)$ and $X_3(k_m)$ because the phase difference from one set of data to another set can be expressed as

$$X_{i+1}(k_m) = X_i(k_m)\, e^{-j\theta_m} \tag{5.81}$$

From this equation one can find that

$$\begin{aligned} X_2(k_m) &= X_0(k_m)\, e^{-j2\theta_m} \\ X_3(k_m) &= X_0(k_m)\, e^{-j3\theta_m} \end{aligned} \tag{5.82}$$

Once all the $X_i(k_m)$ values are found, the quantities of $X(k)$, $X(k + s)$, $X(k + 2s)$, and $X(k + 3s)$ can be calculated from (5.75).

The major advantages of this approach are that less hardware is used and the processing can be relatively simple because some of the $X_i(k_m)$ values are calculated by a simple phase shift.

One of the possible disadvantages is that this approach may lose a signal. When two signals are separated by a multiple of f_s, the two peaks in the power spectrum from the two input signals will be folded into a single peak. If all the information is available, such as described in Section 5.15, the input signals can be uniquely determined. However, if there are only two pairs of ADCs, there is not enough information to determine the two frequencies.

The following question needs to be answered: If there are only two pairs of ADCs and there is only one peak detected from the power spectrum, how can one tell if there are two signals rather than one? This question can be answered from the amplitudes of $X_0(k_m)$ and $X_1(k_m)$ where k_m represents the frequency component that the peak of the power spectrum appears. When there is only one signal, the amplitudes of $X_0(k_m)$ and $X_1(k_m)$ are equal. When there are two signals, the amplitudes of $X_0(k_m)$ and $X_1(k_m)$ are different. The reason of this argument can be expressed as follows. For one signal

$$X_0(k_m) = c_1 \quad \text{and} \quad X_1(k_m) = c_1 e^{-j\theta_1} = c_1 e^{-j2\pi f_1 \tau} \tag{5.83}$$

where c_1 is a constant, θ_1 is the phase shift caused by the delay, f_1 is the input frequency, and τ is the delay time. It is obvious that the amplitudes of $X_0(k_m)$ and $X_1(k_m)$ are equal.

For two signals, the results can be written as

$$\begin{aligned} X_0(k_m) &= c_1 + c_2 \\ X_1(k_m) &= c_1 e^{-j\theta_1} + c_2 e^{-j\theta_2} \\ &= c_1 e^{-j2\pi f_1 \tau} + c_2 e^{-j2\pi f_2 \tau} \end{aligned} \tag{5.84}$$

where c_2 is a constant, and θ_2 and f_2 are the phase shift and frequency of the second signal, respectively. In general, the amplitude of $X_0(k_m)$ and $X_1(k_m)$ are not equal. From the relative amplitude of $X_i(k_m)$, one can find there is more than one signal. However, there is not enough information to determine their frequencies. Additional information is required to determine the frequency of more than one signal.

REFERENCES

[1] Marple, S. L., Jr. *Digital Spectral Analysis With Applications*, Englewood Cliffs, NJ: Prentice Hall, 1987.

[2] Kay, S. M. *Modern Spectral Estimation, Theory and Application*, Englewood Cliffs, NJ: Prentice Hall, 1986.

[3] Rabiner, L. R., and Gold, B. *Theory and Application of Digital Signal Processing*, Englewood Cliffs, NJ: Prentice Hall, 1975.

[4] Oppenheim, A. V., and Schafer, R. W. *Digital Signal Processing*, Englewood Cliffs, NJ: Prentice Hall, 1975.

[5] Kleinrock, L. "Detection of the Peak of an Arbitrary Spectrum," *IEEE Trans. Information Theory*, Vol. IT-10, July 1964, pp. 215–221.

[6] Palmer, L. C. "Coarse Frequency Estimation Using the Discrete Fourier Transform," *IEEE Trans. Information Theory*, Vol. IT-20, Jan. 1974, pp. 104–109.

[7] Rife, D. C., and Boorstyn, R. R. "Single-Tone Parameter Estimation From Discrete-Time Observations," *IEEE Trans. Information Theory*, Vol. IT-20, Sept. 1974, pp. 591–598.

[8] Rife, D. C., and Vincent, G. A. "Use of the Discrete Fourier Transform in the Measurement of Frequencies and Levels of Tones," *The Bell System Technical Journal*, Feb. 1970, pp. 197–228.

[9] Rife, D. C. and Boorstyn, R. R. "Multiple Tone Parameter Estimation From Discrete Time Observations," *The Bell System Technical Journal*, Nov. 1976, pp. 1389–1410.

[10] Ng, S. S. "A Technique for Spectral Component Location Within a FFT Resolution Cell," *IEEE International Conference on Acoustics, Speech, and Signal Processing*, Vol. 3, March 19–21, 1984, pp. 38.8.1–38.8.3.

[11] Pasala, K. University of Dayton, Private communication.

[12] Papoulis, A. *The Fourier Integral and its Applications*, New York, NY: McGraw-Hill Book Co., 1962.

[13] Rader, C. M. "Recovery of Undersampled Periodic Waveforms," *IEEE Trans. Acoustics, Speech, and Signal Proc.*, Vol. ASSP-25, June 1977, pp. 242–249.

[14] Sanderson, R. B., Tsui, J. B. Y., and Freese, N. "Reduction of Aliasing Ambiguities Through Phase Relations," *IEEE Trans. Aerospace and Electronic Systems*, Vol. 28, Oct. 1992, pp. 950–956.

[15] Shapiro, H. S., and Silverman, R. A. "Alias Sampling of Random Noise," *J. Soc. Indust. Appl. Math.*, Vol. 8, June 1960, pp. 225–248.

[16] Beutler, F. J. "Error-Free Recovery of Signals From Irregularly Spaced Samples," *SIAM Review*, Vol. 8, July 1966, pp. 328–335.

[17] Beutler, F. J. "Alias-Free Randomly Timed Sampling of Stochastic Processes," *IEEE Trans. Information Theory*, Vol. IT-16, March 1970, pp. 147–152.

[18] Jenq, Y. C. "Digital spectra of nonuniformly sampled signals: fundamentals and high-speed waveform digitizers," *IEEE Trans. Instrumentation and Measurement*, Vol. 37, June 1988, pp. 245–251.

[19] Vaidyanathan, P. P. *Multirate systems and filter banks*, Englewood Cliffs, NJ: Prentice Hall, 1992.

[20] Cooley, J. W., Lewis, P. A. W., and Welch, P. D. "The Finite Fourier Transform," *IEEE Trans. Audio Electroacoustics*, Vol. AU-17, June 1969, pp. 77–85.

[21] Choate, D. B., and Tsui, J. B. Y. "Note on Prony's Method," *IEEE Proc.-F*, Vol. 140, April 1993, pp. 103–106.

CHAPTER 6

Analog-to-Digital Converters

6.1 INTRODUCTION

In this chapter, the performance of analog-to-digital converters (ADCs) will be discussed. The discussion will concentrate on the impact of ADCs on the performance of receivers; therefore, the discussion will be emphasized from a system point of view. The important parameters of the ADC related to receiver performance are number of bits, number of effective bits, maximum sampling frequency, and input bandwidth.

Unfortunately, the performance of an ADC will not be perfect. There are, however, several different ways to represent these deficiencies. The performance of an ideal ADC will be presented first, then the imperfection will be added on. The most significant effect of an ADC is on the dynamic range of the receiver, which is closely related to the sensitivity of the receiver because the sensitivity is the lower limit of the dynamic range of the receiver. There are several ways to consider the dynamic range, and each approach will lead to a slightly different result. All these approaches will be discussed. Since the ADC is basically a nonlinear device, the analysis that can be performed is rather limited and sometimes very difficult. Some of the results are obtained from numerical simulations.

Before starting the discussion on ADC performance, two types of ADCs will be discussed. One is the folding system, which can operate at very high speed. The second type is the sigma-delta ADC. This type of ADC can trade operating speed for number of bits. There are many different ways to build ADCs [1–11]. Some of the approaches can even be optical.

6.2 ADC THROUGH FOLDING TECHNIQUE [12–16]

The most common high-speed ADC is a flash converter. In this type of ADC, a large number of comparators are needed. If the ADC has b bits, the number of comparators N is

$$N = 2^b \tag{6.1}$$

If the ADC has 8 bits, using this equation, 256 comparators are required. It is not only complicated to build this large number of comparators, but it is also difficult to drive it because of the low input impedance due to many comparators in parallel. Another difficulty is to build the levels of comparators very close to each other. In order to save hardware, the folding technique is used.

Before the discussion of the folding technique, let us discuss a two-stage approach. A block diagram is shown in Figure 6.1. There are two 4-bit ADCs (a coarse one and a fine one), a digital-to-analog converter (DAC), and a comparator. The input signal is digitized by the coarse ADC to generate the most significant bits (MSBs). The MSBs are converted back into an analog signal through the DAC and subtracted from the input signal through an analog comparator. The output from the comparator is digitized again to generate the least significant bits (LSBs). In this arrangement, two 4-bit ADCs only require 32 comparators (2 × 16) to generate 8 bits of digitized data.

The folding technique is somewhat similar to the two-stage approach. Instead of using the DAC and the analog comparator, a folding circuit is used. A block is shown in Figure 6.2. In this figure, there are also two 4-bit ADCs (again, a coarse one and a fine one). The coarse ADC provides the MSB. The input signal passes through the folding circuit. The folding circuit can be considered as performing a somewhat similar function of the DAC and the analog comparator as in the two stage ADC.

The input versus output of the folding circuit is shown in Figure 6.2(b). The folding circuit is analog. Both the input and the output are shown on the y-axis. The x-axis represents time. The input versus time is a linear function. The output versus time is represented by many triangle-shaped functions. The output can be found from the input level through the time axis, as shown in Figure 6.2(b). In this example, the input is V_i and the output is V_o. The output from the folding circuit is digitized to obtain the LSB.

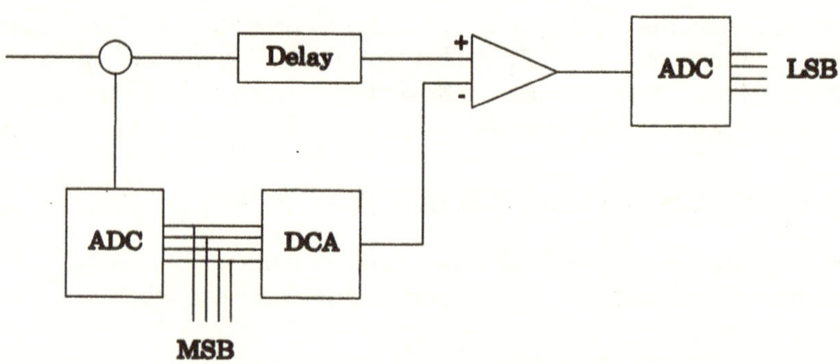

Figure 6.1 A two-stage ADC.

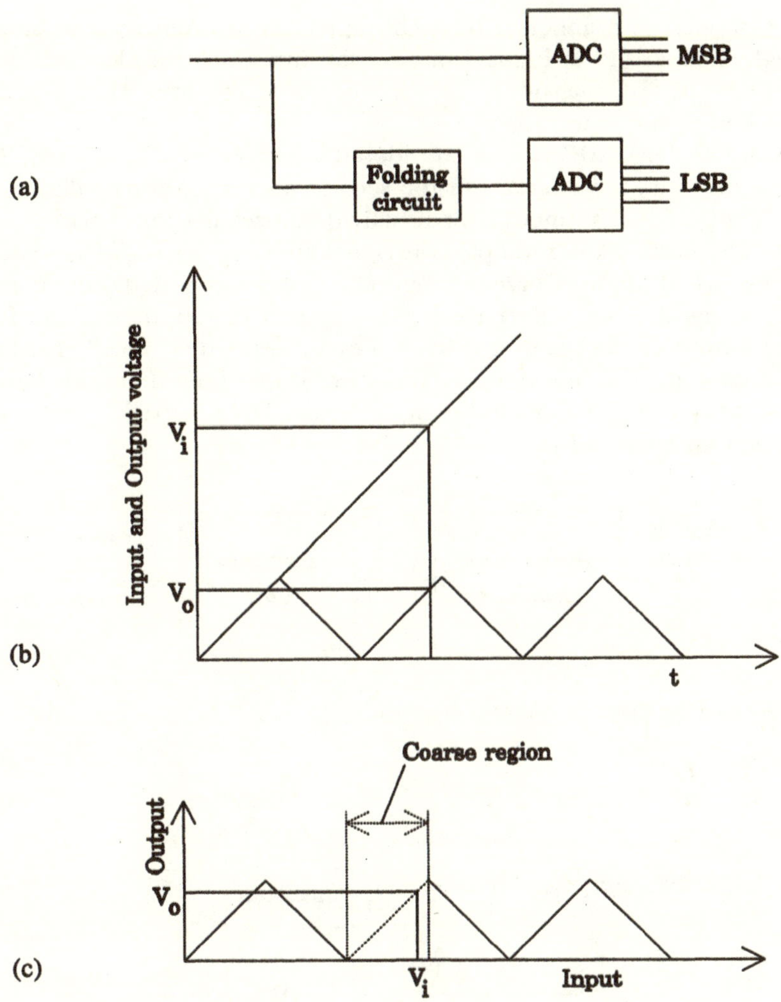

Figure 6.2 Folding technique ADC: (a) Functional diagram, (b) input, output versus time, (c) input versus output.

The MSB will determine the coarse amplitude of the input signal. The coarse amplitude is determined by the triangular region where the input signal is (Figure 6.2(c)). The LSB will provide the fine amplitude information of the input signal.

6.3 ADC THROUGH SIGMA-DELTA MODULATION [17–26]

The basic idea of the sigma-delta ADC is to trade digitizing speed for number of bits. A high-speed ADC with a few number of bits (usually one bit) is used to make an ADC with a large number of bits. This approach is often used to build ADCs with a high number (10 or higher) of bits, but with a relatively low sampling rate. Thus, this kind of ADC may not be used in microwave receivers. However, the basic operation idea is very interesting.

The sigma-delta ADC can be considered as consisting of two parts. The first part is a sigma-delta modulator and the second part is a low-pass digital filter as shown in Figure 6.3. The modulator usually quantizes the input signal into ones and zeros. The lowpass filter will pass the signal and stop the high-frequency noise.

Figure 6.4(a) shows a basic configuration of a sigma delta modulator. The integrator integrates the difference signal between the input and the feedback loop. The output of the integrator feeds a quantizer, which usually has only one bit (a comparator). The output of the quantizer is the desired output. This output is also fed back through a DAC to the input signal. The feedback signal is stronger than the highest expected input.

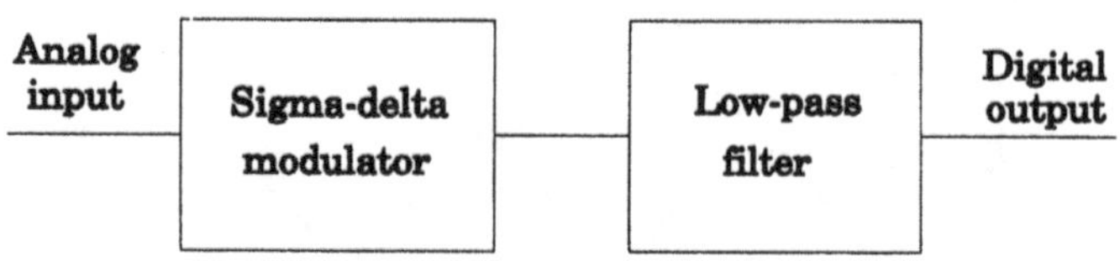

Figure 6.3 A functional diagram of a sigma-delta ADC.

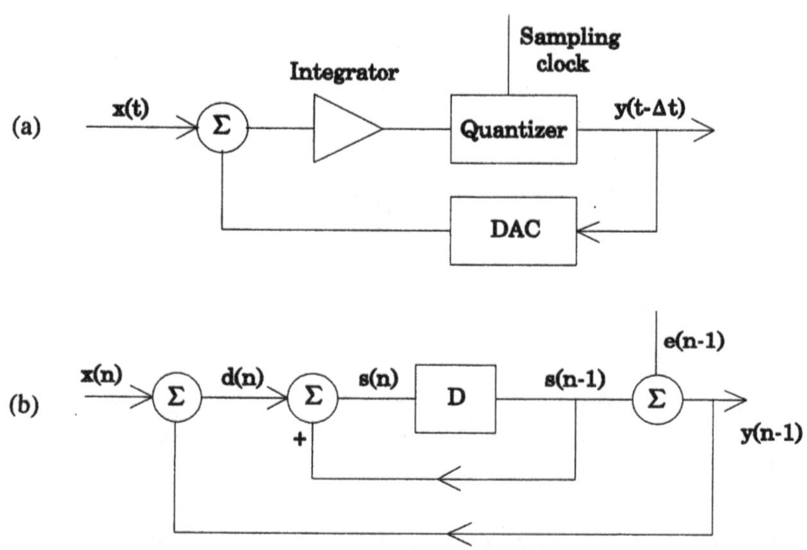

Figure 6.4 A sigma-delta modulator: (a) simple circuit, (b) signal flow diagram.

Let us use an input signal with constant amplitude to illustrate the output of the quantizer. Let us arbitrarily assume that the threshold of the quantizer is at 0.5V, the input is less than 1V, and the feedback voltage is 1V. If the voltage at the input of the quantizer is greater than or equal to 0.5V, the output is 1; otherwise, it is equal to 0. If the input signal to the sigma-delta modulator is a constant voltage at 1/7V and initial output of the integrator is assumed 0, the integrator output at each clock cycle will be 0, 1/7, 2/7, 3/7, 4/7, −2/7, −1/7, 0, 1/7, 2/7, 3/7, 4/7, −2/7, −1/7, 0, 1/7, 2/7, 3/7, 4/7, −2/7, and −1/7, and this pattern continues. These outputs can be obtained as follows. When the output of the integrator is less than 0.5V, there is no feedback signal and the integrator will add the input to its previous value. When the integrator output is greater than or equal to 0.5V, a 1V signal will be fed back to the input. The integrator output is the sum of the input signal and its previous value minus 1.

With these integrator outputs, the corresponding quantizer output will be 000010000001000000100, and this pattern continues. It is interesting to see that the quantized output has a 1 in every 7 sequential outputs. Thus the average is close to 1/7, the input signal. If the threshold is not 0.5V or the initial condition of the integrator is not zero, the quantized output has similar pattern (one 1 in 7 sequential outputs), but the initial value is different. This is different from a conventional sigma-delta modulator in which the output is the difference between two stages.

This output passes through a low-pass filter and has an output every M clock cycles. The averaged output is close to the input level. For example, if $M = 8$, the output could contain one 1 or two 1's. If there is one 1, the output will be interpreted as 1/8. If there are two 1's, the output will be interpreted as 2/8 = 1/4. If M is a larger value, the interpreted value will be closer to 1/7. For example, if $M = 32$, there will be either four or five 1's. The interpreted value can be either 4/32 = 1/8 or 5/32, which is closer to 1/7. If M is very large, the result can be rather accurate (high number of bits). A 10-bit ADC requires $M = 1{,}024$. In general, this value can represent the average amplitude of the input in these M cycles.

Every M clock cycles generate an output equivalent to a sampling time of M clock periods. The input frequency must be low enough not to violate the Nyquist sampling rate ($1/M$ cycles). These binary data can be converted into a power of 2 number every M cycles. This explains how a sigma-delta ADC works.

Now let us show how a sigma-delta modulator reshapes the noise spectrum and translates it to high frequency. From Figure 6.4(b), one can write the following equations:

$$\begin{aligned} d(n) &= x(n) - y(n-1) \\ s(n) &= d(n) + s(n-1) \\ y(n-1) &= e(n-1) + s(n-1) \end{aligned} \tag{6.2}$$

where $x(n)$, $y(n)$, and $e(n)$ are the input, output, and quantization noise, respectively, and $s(n)$ and $d(n)$ are shown in Figure 6.4(b). From these equations one can express the output in terms of the input $x(n)$ and quantization noise $e(n)$. From the first two equations, one can obtain

$$s(n) - s(n-1) = x(n) - y(n-1) \tag{6.3}$$

Replace $s(n-1)$, $y(n-1)$, and $e(n-1)$ by $s(n)$, $y(n)$, and $e(n)$ in the third equation of (6.2). Using (6.3), the result is

$$y(n) = x(n) + e(n) - e(n-1) \tag{6.4}$$

Taking the z transform, the result is

$$Y(z) = X(z) + (1 - z^{-1})E(z) \equiv X(z) + N(z) \tag{6.5}$$

This equation shows that the output $Y(z)$ is equal to the input $X(z)$ and noise $N(z)$. This noise is related to the quantization noise $E(z)$. By replacing z by $e^{j2\pi ft_s}$, the frequency response of the noise $N(f)$ can be written as

$$\begin{aligned} N(f) &= |(1 - e^{-j2\pi ft_s})E(f)|^2 \\ &= 2[1 - \cos(2\pi ft_s)]|E(f)|^2 \\ &\equiv 2\left[1 - \cos\left(\frac{2\pi f}{f_s}\right)\right]A \end{aligned} \tag{6.6}$$

where $E(f)$ is the quantization noise (which is replaced by a constant A), t_s is the sampling time, and f_s is the sampling frequency. It will be shown later in Section 6.7 that quantization noise $E(f)$ is not a function of frequency. One can see that $N(f)$ will increase with increasing f, as shown in Figure 6.5. The high frequency will fold back into the digital output. Thus, a digital low-pass filter is usually used to attenuate the out-of-band noise power at the high frequency in building a sigma-delta ADC.

6.4 BASIC SAMPLE AND HOLD CIRCUIT [1–3]

In order to quantize an input signal at a certain instant of time, the signal should be held constant at that instant. If the input signal changes rapidly and the digitizing process is slow, the accuracy of the output data will be poor. One approach is to place a sample-and-hold circuit in front of a quantizer. A sample-and-hold circuit can create a very narrow aperture window, which will take the input at the desired instant and keep the voltage constant for a relatively long period of time, such that the digitizing circuit can operate properly.

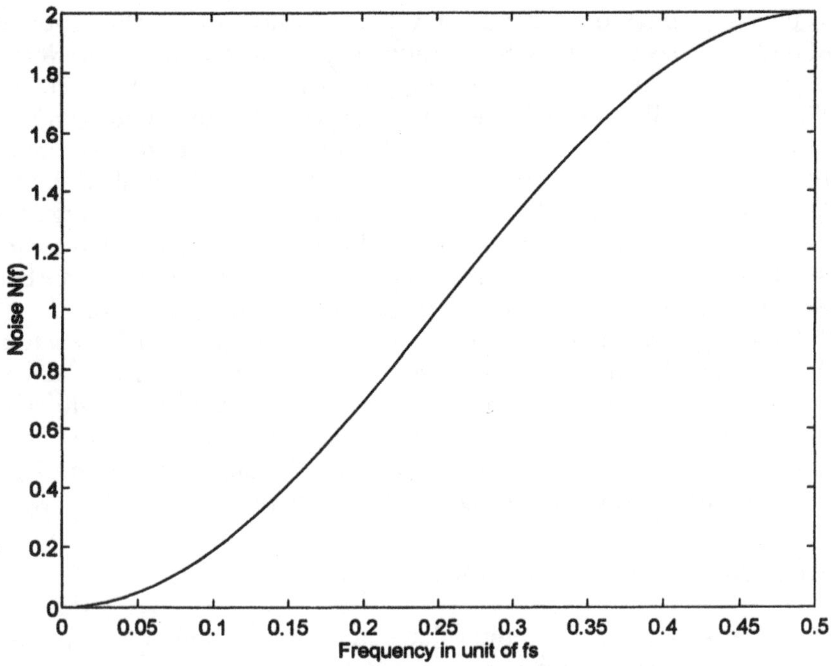

Figure 6.5 Noise spectrum density of sigma-delta modulator.

A simple sample-and-hold circuit is shown in Figure 6.6. It consists of a sample/hold switch and a holding capacitor. The two amplifiers are used for impedance matching.

When the switch is closed, the sample and hold is in the sample mode and the voltage on the capacitor follows the input voltage. When the switch is open, the voltage on the capacitor stays at a constant value and is called the hold mode.

In the sample mode, the operation can be divided into two time periods: acquisition and tracking. When the switch is closed, the voltage on the capacitor

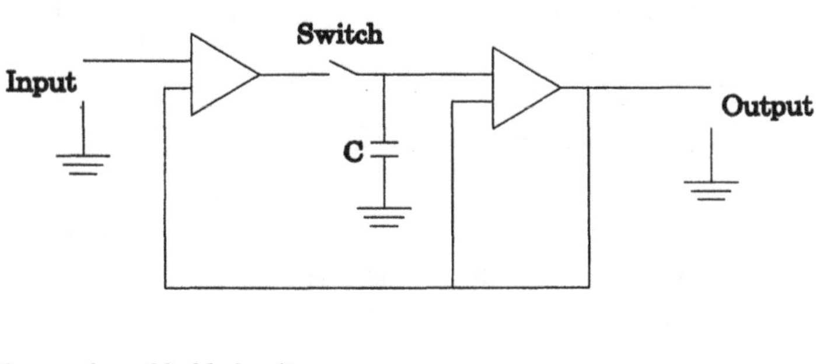

Figure 6.6 Basic sample-and-hold circuit.

starts to change rapidly from the previously held value toward the input voltage until it finally reaches it. This period is called the acquisition time. The voltage on the capacitor follows the input voltage and this period is called the tracking time.

When the switch is opened, the voltage on the capacitor should stay constant and this is the hold mode. However, after the switch is open, the voltage on the capacitor usually oscillates slightly due to transient effect and is called the settling time. After the settling time, the voltage will drop slightly due to charge leakage through the finite input impedance of amplifier A_2. Figure 6.7 shows the time response of a sample-and-hold circuit. The aperture time is the elapsed time between initiation and completion of the mode transition from sample to hold.

The aperture uncertainty is the variance of the aperture time. This time can be very short, usually in the picoseconds range for high-frequency ADCs. The aperture jitter is the variation in the effective sample instant due to the aperture uncertainty. The sampling time offset is the time interval between the sample-to-hold mode transition command and the actual initiation of the transition. The sampling time uncertainty is the variance of the sampling time offset.

6.5 BASIC ADC PERFORMANCE AND INPUT BANDWIDTH

An ADC converts a continuous input voltage into discrete output levels, which can be represented by binary coded words. The smallest discrete step size is called the

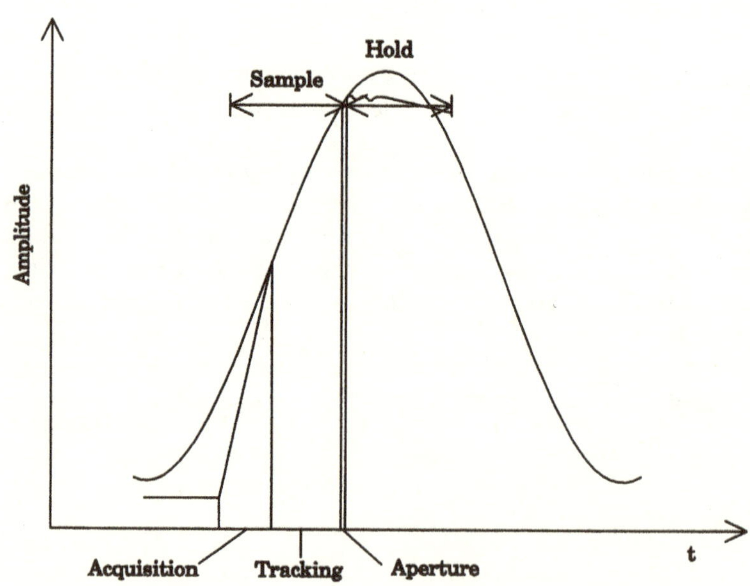

Figure 6.7 Time domain response of a sample-and-hold circuit.

quantization level. The conversion usually occurs at uniformly spaced time intervals. This is often referred to as the sampling time. The transfer function representing the input versus output of an ADC is shown in Figure 6.8. Figures 6.8(a, b) show the 3-bit midtread and midriser approaches, respectively. The x-axis is the analog input and the y-axis represents the digital output. In the midtread configuration, there is a zero level, but the number of positive levels do not equal the negative levels since the total number of levels is usually a power of 2 number. In this figure, there is one more negative level than the number of positive levels. It is obvious that the midtread approach has an unsymmetrical output.

In the midriser approach, there is no zero level, but they have an equal number of positive and negative levels. Thus, the output is symmetrical. In testing high-frequency ADCs, a sine wave is often used. Since the sine wave is symmetrical, the midriser model is often applied.

Figure 6.9 shows the transfer characteristic of an ideal ADC. If the input increases linearly with respect to time, the output and the quantization error are as shown in Figure 6.9. It is obvious that the quantization process is a nonlinear one, making it difficult to analyze mathematically. In a practical ADC, it is also difficult to make the quantization level uniform; thus, the quantization error is worse than in the ideal case.

It is often assumed that the input frequency to an ADC is half the sampling frequency in order to fulfill the Nyquist sampling criterion, but this is not always true. In order to avoid aliasing ambiguity, it is required that the input bandwidth (not necessarily the input frequency) of the ADC be less than half of the sampling frequency. It is often desirable to have an input frequency higher than the maximum sampling frequency because the input bandwidth does not necessarily start from a direct current (dc) level. For example, if the maximum sampling frequency of the ADC is 200 MHz, the unambiguous bandwidth is 100 MHz. The input frequency spectrum does not necessarily extend from dc to 100 MHz. It can be from 120 to 220 MHz, and this selection keeps the input bandwidth below an octave. An input bandwidth of less than one octave will eliminate the second harmonic, which can be generated by an analog front end or by the ADC nonlinear transfer characteristic.

Sometimes ADCs are used in parallel to increase sampling speed, as discussed in the previous chapter. If the input frequency of the ADC is high, these ADCs can be used to increase the bandwidth of the system. If the input frequency of the ADC is limited to half the sampling frequency, a parallel approach cannot be adopted because the ADC cannot receive high-frequency input signals.

6.6 APPARENT MAXIMUM AND MINIMUM SIGNALS TO AN ADC

The maximum signal to an ADC is often defined as a sine wave with an amplitude that matches the highest level of the ADC. If a signal is stronger than this maximum level, the output waveform will be clipped. If an input signal is lower than this

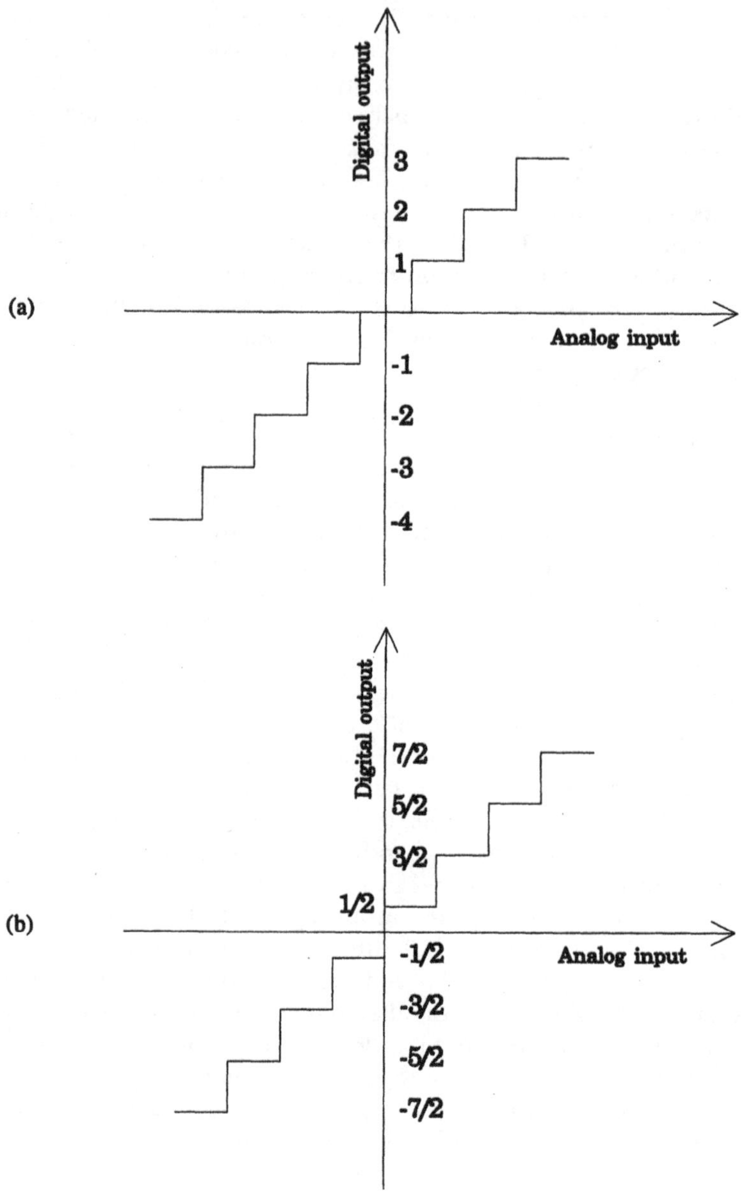

Figure 6.8 Transfer function of an ADC: (a) midtread, (b) midriser.

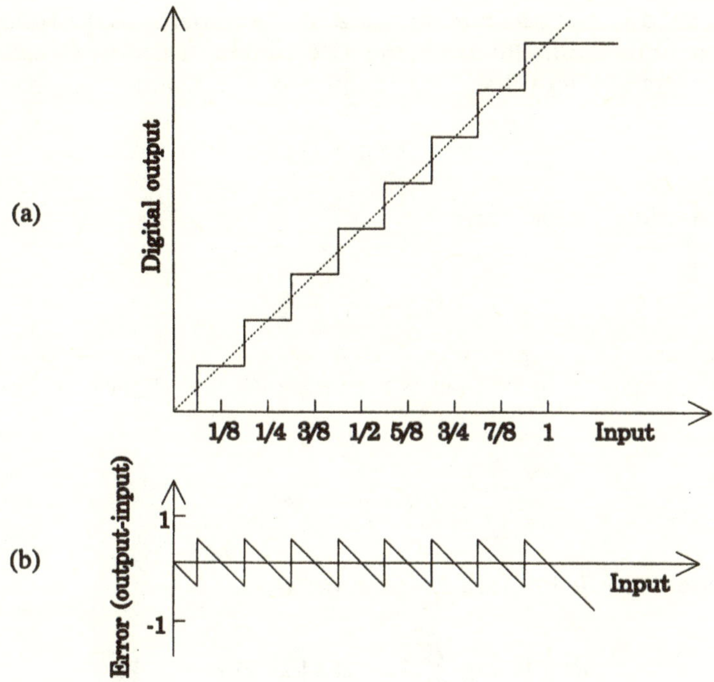

Figure 6.9 Performance of an ideal ADC: (a) input versus output, (b) quantization error.

signal, all the number of bits may not be exercised. The maximum level is often used to determine the upper limit of the dynamic range. If there is no noise and the input voltage matches the maximum range of the ADC, the voltage V_{max} is

$$2V_{max} = 2^b Q \qquad \text{or} \qquad V_{max} = 2^{(b-1)}Q \tag{6.7}$$

where b is the number of bits and Q is the voltage per quantization level. This equation shows that the sine wave reaches the top of the highest as well as the bottom of the lowest quantization levels.

A sine wave with amplitude matching this maximum voltage has a power of

$$P_{max} = \frac{V_{max}^2}{2} = \frac{2^{2(b-1)}Q^2}{2} = \frac{2^{2b}Q^2}{8} \tag{6.8}$$

In the above equation, the input impedance is assumed to be unity. In later discussions, the power ratio is often of interest and hence the impedance will be canceled when the ratio of powers is taken. However, in some calculations, it is necessary to include the impedance in order to have the actual power.

If there is no noise, the minimum signal is considered as the voltage that can cause change in the LSB. Otherwise, the ADC cannot detect the signal. Under this condition, the minimum voltage V_{min} is equal to one quantization level, or

$$2V_{min} = Q \tag{6.9}$$

The corresponding power level is

$$P_{min} = \frac{V_{min}^2}{2} = \frac{Q^2}{8} \tag{6.10}$$

The dynamic range can be defined as the ratio of P_{max} to P_{min}, which can be written as

$$DR = \frac{P_{max}}{P_{min}} = 2^{2b} \tag{6.11}$$

which is often expressed in logarithmic form as

$$DR = 10 \log\left(\frac{P_{max}}{P_{min}}\right) = 20b \log(2) \approx 6b \quad \text{dB} \tag{6.12}$$

This is why the dynamic range of an ADC is often referred to as 6-dB per bit.

However, the dynamic range of a receiver also depends on the amplifier performance in front of the receiver and the ADC. This problem will be discussed in the next chapter.

6.7 QUANTIZATION NOISE OF AN IDEAL ADC [1–3, 27–32]

An ADC converts the input signal from analog to digital form, and this process is nonlinear. For example, a 1-bit ADC is equivalent to a hard limiter, which is a nonlinear device. As shown in Figure 6.10, a sine wave is converted point by point into two different output levels. There is a difference (or error) between the true value of the sine wave and the quantized one. Because the error can be any value within the quantization level, it is reasonable to assume that the probability the error is uniformly distributed over the quantization level Q. Thus, the probability density function of the amplitude is $1/Q$. The quantization noise power can be calculated from the error as

$$N_b = \frac{1}{Q}\int_{-Q/2}^{Q/2} x^2 dx = \frac{Q^2}{12} \tag{6.13}$$

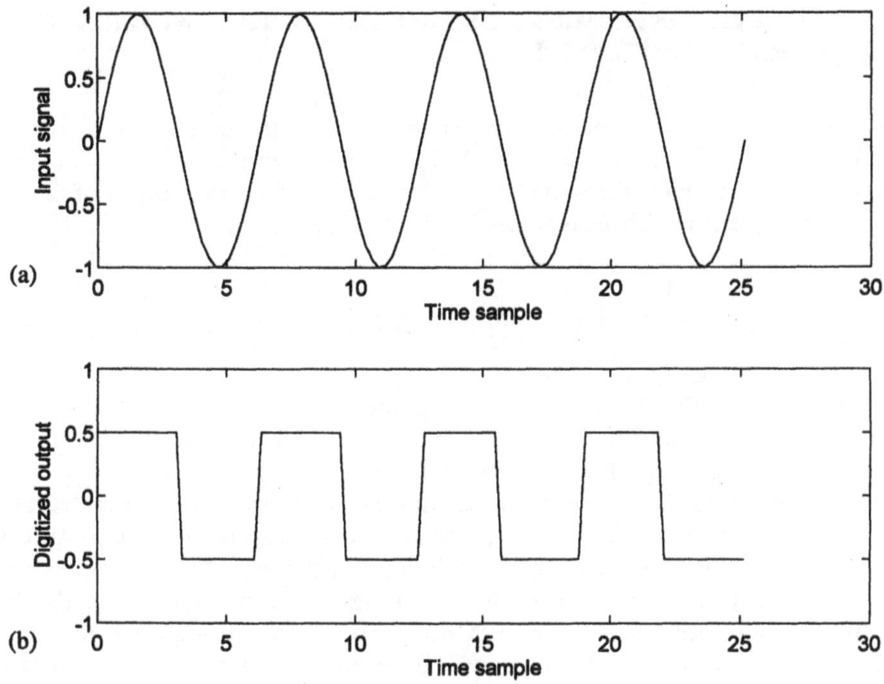

Figure 6.10 Sine wave and one bit quantizer: (a) input signal, (b) digitized output.

This quantity is sometimes used as the sensitivity level of the receiver. Under this condition, the maximum signal-to-noise ratio (S/N), by combining (6.8), can be expressed as

$$\left(\frac{S}{N}\right)_{\max} = \frac{P_{\max}}{N_b} = \frac{3}{2}2^{2b} \tag{6.14}$$

This quantity can be expressed in logarithmic form as

$$DR = 10\log\left(\frac{P_{\max}}{N_b}\right) = 10\log(1.5) + 20b\log(2) = 1.76 + 6.02b \quad \text{dB} \tag{6.15}$$

This difference between (6.12) and (6.14) is the factor 1.76 because the two lower limits are different.

6.8 NOISE LEVEL DETERMINED BY PROCESSING BANDWIDTH AND DITHERING EFFECT [33, 34]

The lower limit (noise level) of the dynamic range depends on the processing bandwidth. The processing bandwidth mentioned here is usually data length dependent. If one determines to perform N points DFT on the output of the ADC, the processing bandwidth B_V is related to the DFT length as

$$B_V = \frac{f_s}{N} = \frac{1}{Nt_s} \tag{6.16}$$

where f_s and t_s are the ADC sampling frequency and sampling time, respectively. Thus, a longer DFT operation creates a narrower processing bandwidth. The narrower the bandwidth, the lower the noise level.

Since the signal is coherent and the noise is incoherent, the signal level from the DFT output is proportional to N and the noise level is proportional to $N^{1/2}$. Therefore, when the length of the DFT increases, the signal amplitude increases faster than the noise level. As a result, a weak signal can be detected by the ADC. However, as mentioned before, the signal must be strong enough to cross the first level of the ADC in order to be detected. However, one can add noise to the signal to be detected by the ADC and at the same time increase the FFT length to improve the S/N. This approach can be used to detect a very weak signal. This processing is often referred to as noise dithering. The purpose of a dithering is to make a weak signal cross quantization level in an ADC.

Figure 6.11 shows an example of noise dithering. In this example, the input signal alone is too weak to cross the LSB and the output is at a constant level of 0.5. If noise is added, the noise can be sensed by the ADC. In this case the S/N is set at −10 dB. The signal can be identified at the 100th frequency bin. The dc component represents the bias of the ADC output.

If there are two input signals of different amplitude and frequency, an ideal EW receiver will receive both the strong and the weak signals. The difference between the strong and weak signals is referred to as the instantaneous dynamic range of the receiver.

If there is only one weak signal, which is not strong enough to cross the first quantization level, the ADC cannot sense the signal. If there is a strong signal, the weak signal may not need to cross the first quantization level to be detected. The strong signal will cross different quantization levels and, since the weak signal is superimposed on the strong one, both signals may be detected. In this case, the strong signal can be considered as the dithering signal.

In addition to the noise floor, the spurious responses (sometimes referred to as spurs) often limit the lower limit of the dynamic range. The spurious responses in an ADC will be discussed in the next section.

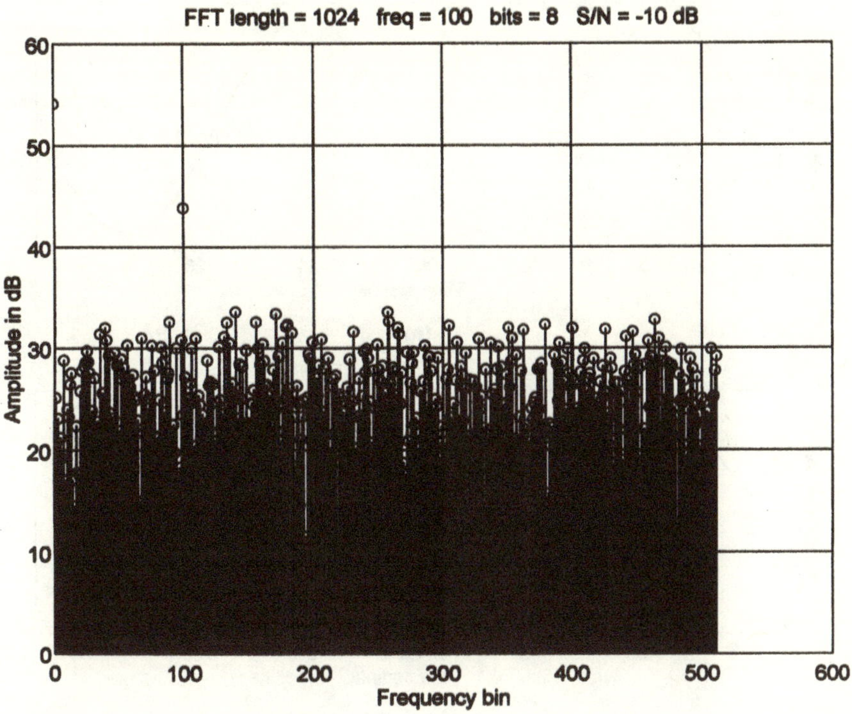

Figure 6.11 Signal detected with noise dithering.

6.9 SPURIOUS RESPONSES

If an input signal with arbitrary frequency is applied to the input of the ADC, the error between the true signal and the digitized value cannot be predicted. Thus, it is reasonable to assume the error is uniformly distributed. However, if the input signal frequency is commensurate with respect to the sampling frequency f_s, the error function is highly correlated. Under this condition, a uniform distribution is no longer a good assumption.

For example, if the input frequency f_i is related to the sampling frequency f_s by

$$f_s = nf_i \tag{6.17}$$

where n is an integer, the error will exhibit a repetitive pattern from one cycle to the next, as shown in Figure 6.12. From this consideration, the assumption of uniform distribution of error will no longer hold. Figure 6.12(a) shows a sine wave

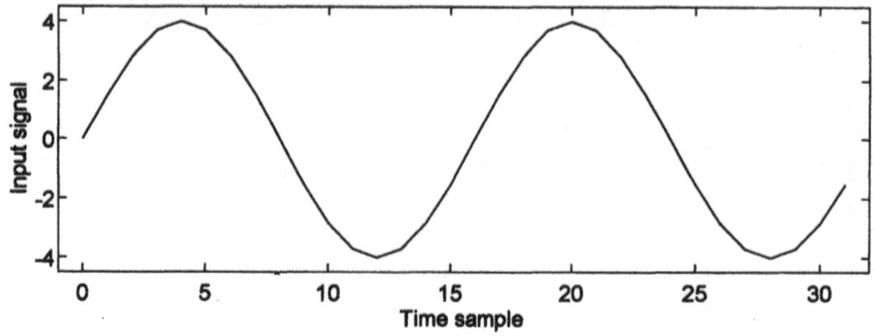

(a)

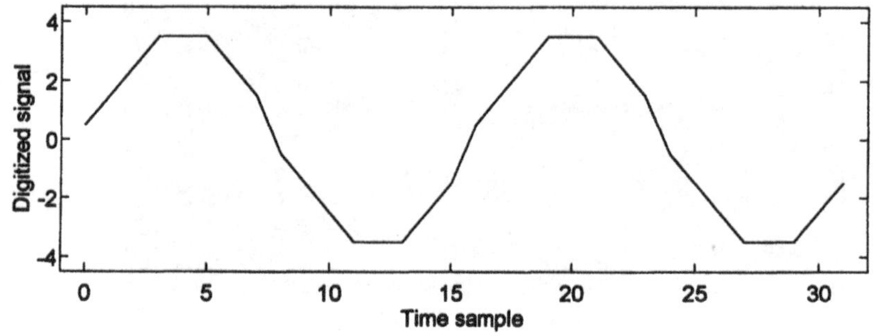

(b)

Figure 6.12 Coherent digitizing error: (a) input signal, (b) digitized output, (c) error signal, (d) FFT of input signal, (e) FFT of digitized output, (f) FFT of digitized output with 64 points.

with two cycles and is sampled at 32 points. Figure 6.12(b) shows the quantized version with 3 bits. This case can be considered as $f_s = 32$ and $f_i = 2$. The error in time domain is shown in Figure 6.12(c). One should note that the errors from points 0 to 15 are the same as those at 16 to 32. If one increases the number of points, the errors will just repeat themselves. Figure 6.12(d) shows the FFT results of the sine wave. Since the input frequency matches one of the frequency bins ($n = 2$) of the FFT output, there are no sidelobes. Figure 6.12(e) shows the FFT result of the quantized version, which contains spurs.

Since the error data repeats every input cycle, increasing the FFT length will not change the levels of the spurs. Figure 6.12(f) shows the results of 64 points of FFT outputs with a signal of four cycles. The spur levels are identical to the results in Figure 6.12(e). Thus, the spur levels do not decrease with increasing of the FFT length. It should be noted that in producing Figure 6.12, the input signal starts from $t = 1 \times 10^{-8}$ to $t = 4\pi + 1 \times 10^{-8}$. If the input signal passes exactly at $t = 0$, the computer error will disturb the quantized data.

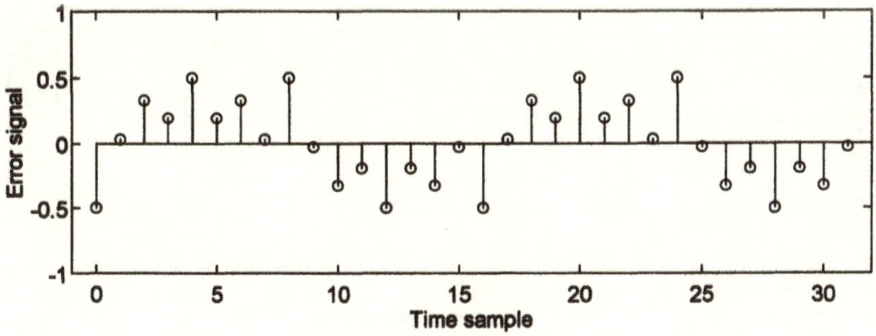

(c)

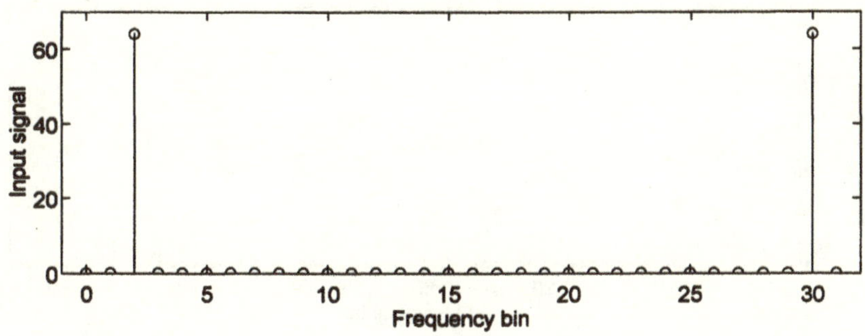

(d)

Figure 6.12 (continued).

These spurs are caused by the quantization error. The DFT of the quantized data can be written as

$$X_d(k) = \sum_{n=0}^{N-1} x_d(n)\, e^{\frac{-j2\pi nk}{N}} = \sum_{n=0}^{N-1} [x(n) - x_e(n)]\, e^{\frac{-j2\pi nk}{N}} \tag{6.18}$$

where $x(n)$ is the input in time domain, $x_d(n)$ is the corresponding digitized data, and $x_e(n)$ represents the error function. The error in the frequency domain is caused by the DFT of $x_e(n)$.

It has been demonstrated that errors in the digitized signal can be periodic in nature. In (6.17), even if n is not an integer, the output can still be periodic. For example, if $f_s = 32$ Hz and $f_i = 0.5$ or 1.5 Hz, the output will repeat itself every 64 points instead of every 32 points. This section illustrates that some of the spurs cannot be reduced by increasing the FFT length.

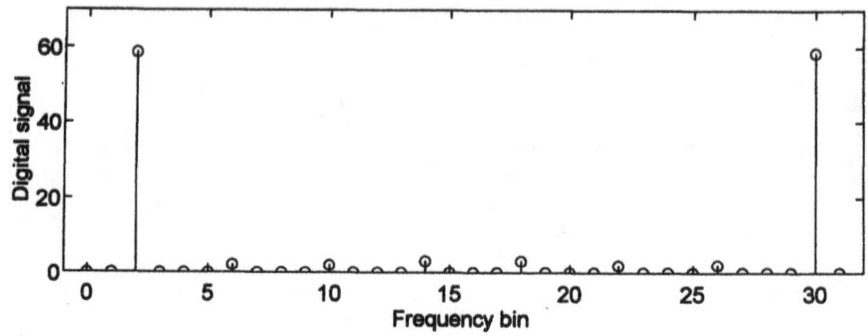

(e)

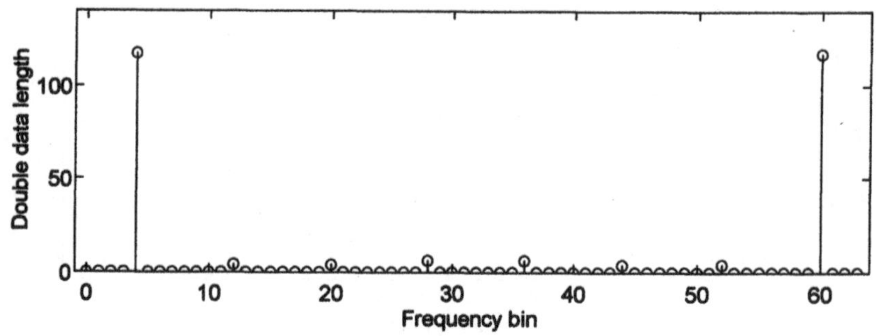

(f)

Figure 6.12 (continued).

6.10 ANALYSIS ON SPUR AMPLITUDES [35–37]

In the previous section, it was demonstrated that digitized signals generate spurious responses. Some of spurs cannot be reduced, even if longer FFT is used. It is desirable to find the maximum of the spur level so that the dynamic range of the receiver can be determined. Spurs can be generated through other processes (i.e., through the nonideal characteristics of ADCs). Sometimes, sidelobes can be considered as spurs because they too will limit the instantaneous dynamic range of a receiver.

This section is a study of the amplitude of the spurs based on [36]. Since digitization is a nonlinear process, a general analysis is not achievable. The following discussion can be considered as a special case, but it does provide some interesting results. If x is the input signal to an ADC, x_d is the output signal and g is the transfer characteristic function of the ADC. Then,

$$x_d(t) = g(x(t)) \tag{6.19}$$

If the input is a cosine wave, then

$$x(t) = \cos(2\pi ft + \theta) \equiv \cos\phi(t)$$
$$\text{and} \tag{6.20}$$
$$x_d(t) = g[\cos\phi(t)]$$

Since the digitized signals are inherently periodic, they can be represented by the Fourier series as

$$x_d(t) = A_0 + \sum_{n=1}^{\infty} A_n \cos(n\phi) \tag{6.21}$$

where A_1 is the amplitude of the fundamental frequency and the higher orders of A_n represent the spurious responses. The values of A can be found as

$$A_0 = \frac{1}{\pi}\int_0^{\pi} g(\phi)\, d\phi$$
$$A_n = \frac{2}{\pi}\int_0^{\pi} g(\phi)\cos(n\phi)\, d\phi \tag{6.22}$$

Let us use an example to demonstrate this approach. Figure 6.13 shows that a cosine wave is digitized by a 3-bit ADC into eight levels and the amplitude of the signal matches the full level of the ADC. The function of $g(\phi)$ can be written as

$$g(\phi) = \begin{bmatrix} 7/8 & \cos^{-1}(1) \ge \phi \ge \cos^{-1}(3/4) \\ 5/8 & \cos^{-1}(3/4) \ge \phi \ge \cos^{-1}(1/2) \\ 3/8 & \cos^{-1}(1/2) \ge \phi \ge \cos^{-1}(1/4) \\ 1/8 & \cos^{-1}(1/4) \ge \phi \ge \cos^{-1}(0) \\ -1/8 & \cos^{-1}(0) \ge \phi \ge \cos^{-1}(-1/4) \\ -3/8 & \cos^{-1}(-1/4) \ge \phi \ge \cos^{-1}(-1/2) \\ -5/8 & \cos^{-1}(-1/2) \ge \phi \ge \cos^{-1}(-3/4) \\ -7/8 & \cos^{-1}(-3/4) \ge \phi \ge \cos^{-1}(-1) \end{bmatrix} \tag{6.23}$$

In this equation, the ADC is a midrise type and the output is normalized to unity. To find the constant A_n, one can substitute $g(f)$ into (6.22) and the result is

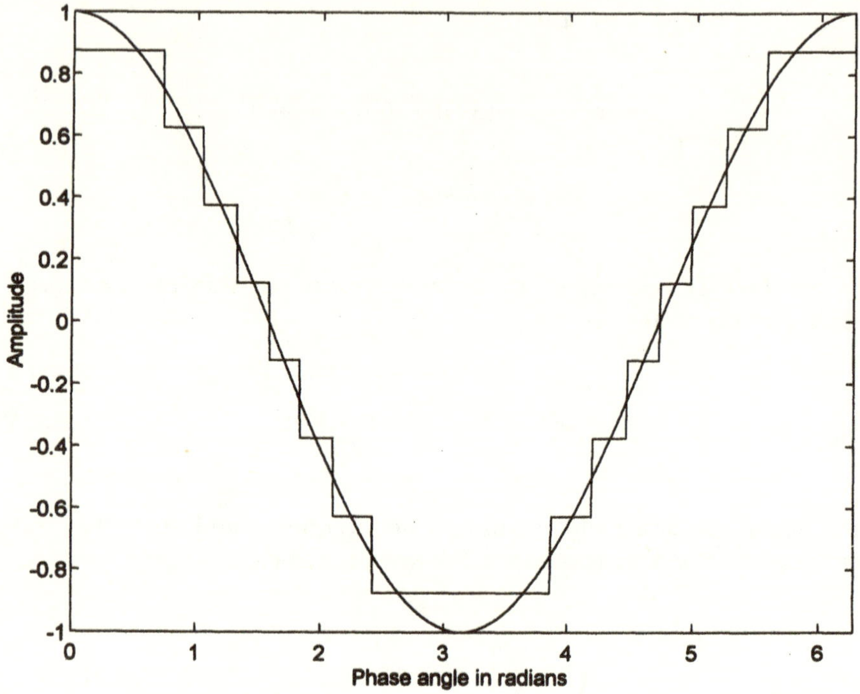

Figure 6.13 A cosine wave digitized by a 3-bit ADC.

$$
\begin{aligned}
A_n &= \frac{2}{\pi}\int_0^{\pi} g(\phi)\ \cos(n\phi)\, d\phi \\
&= \frac{2}{\pi}\left[\frac{7}{8n}\sin(n\phi)\right]_{\cos^{-1}(1)}^{\cos^{-}\left(\frac{3}{4}\right)} + \frac{2}{\pi}\left[\frac{5}{8n}\sin(n\phi)\right]_{\cos^{-1}\left(\frac{3}{4}\right)}^{\cos^{-1}\left(\frac{1}{2}\right)} + \ldots \\
&\quad + \frac{2}{\pi}\left[\frac{-7}{8}\sin(n\phi)\right]_{\cos^{-1}(-1)}^{\cos^{-1}\left(\frac{-3}{4}\right)} \qquad (6.24) \\
&= \frac{1}{2n\pi}\left\{1 + 2\sin\left[n\cos^{-1}\left(\frac{1}{4}\right)\right] + 2\sin\left[n\cos^{-1}\left(\frac{1}{2}\right)\right]\right. \\
&\quad \left. + 2\sin\left[n\cos^{-1}\left(\frac{3}{4}\right)\right]\right\}
\end{aligned}
$$

In general, this approach can be extended to b bits. If an ADC has b bits, the amplitude of the n-th harmonics can be found in a similar way as

$$A_n = \frac{(-4)\, j^{n+1}}{2^b n\pi}\left[1 + 2\sum_{k=1}^{\frac{2^b}{2}-1} \sin\left[\left[n \cos^{-1}\left(\frac{2k}{2^b}\right)\right]\right]\right] \tag{6.25}$$

In this equation, A_1 is the amplitude of the fundamental component and A_3 is the third harmonic, which is usually the strongest harmonic because it is the lowest ordered odd harmonic. Thus, the ratio of the fundamental to the highest harmonic in decibels can be written as

$$\left(\frac{A_3}{A_1}\right)_{\mathrm{dB}} = 20 \log\left(\frac{A_3}{A_1}\right) \tag{6.26}$$

The results of (6.25) and (6.26) are listed in Table 6.1. From these values, it is found that the highest spurious response is approximately 9b-dB below the fundamental component. If b = 8, the strongest spur is about 72-dB below the fundamental component. From this discussion, it appears that the maximum dynamic range is 9-dB per bit rather than 6 dB as predicted in (6.12).

It should be noted that the above discussion is based on the assumption that the time domain data is continuous. Under this situation, if the input signal is continuous, all the levels of the ADC are exercised and at each level there are many data points. This situation does not occur in the sampled data, especially for high frequency signals.

Some simulated results are used to demonstrate the spur analysis. Figure 6.14 shows the results of the time and frequency domain of ADC with 3 to 5 bits. In

Table 6.1
Highest Spur Level Versus Number of Bits

Number of Bits	*Largest Spur (dBc)*
1	–9.5
2	–18.3
3	–27.0
4	–35.9
5	–44.8
6	–53.8
7	–62.8
8	–71.8

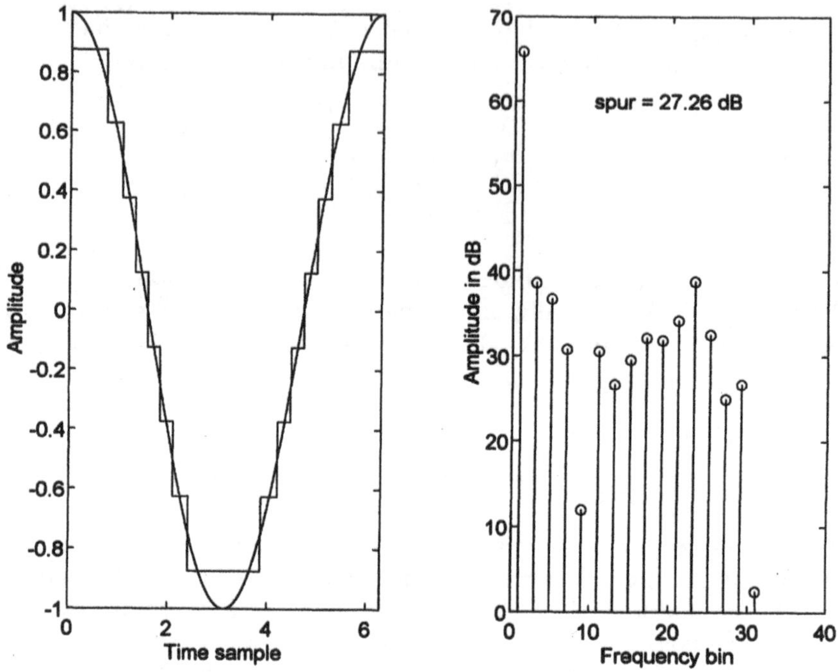

(a)

Figure 6.14 Time and frequency domain of ADC outputs: (a) 3 bits, (b) 4 bits, (c) 5 bits.

these simulations, the input signal is a cosine wave with one complete cycle and 1,024 points of data are obtained in the time domain. Since the data is real, only 512 frequency components are independent. However, in order to show the first few individual frequency components clearly, only the first 32 components are shown. In these figures, the fundamental frequency is at location 1, which is the highest. The third harmonic is 3, which is next to the fundamental since the second harmonic is zero. It appears that the third harmonic is close to the highest one, and it is approximately 9-dB per bit below the input signal. However, these special cases may not represent the worst situation in practice.

6.11 FURTHER DISCUSSION ON SPUR AMPLITUDES

In a digital receiver, the length of the FFT might be limited by the minimum pulse width. If the input frequency is relatively high compared to the sampling frequency, all the ADC output levels may not be exercised. Under this condition, the third frequency component might not be the highest one and the amplitude of the

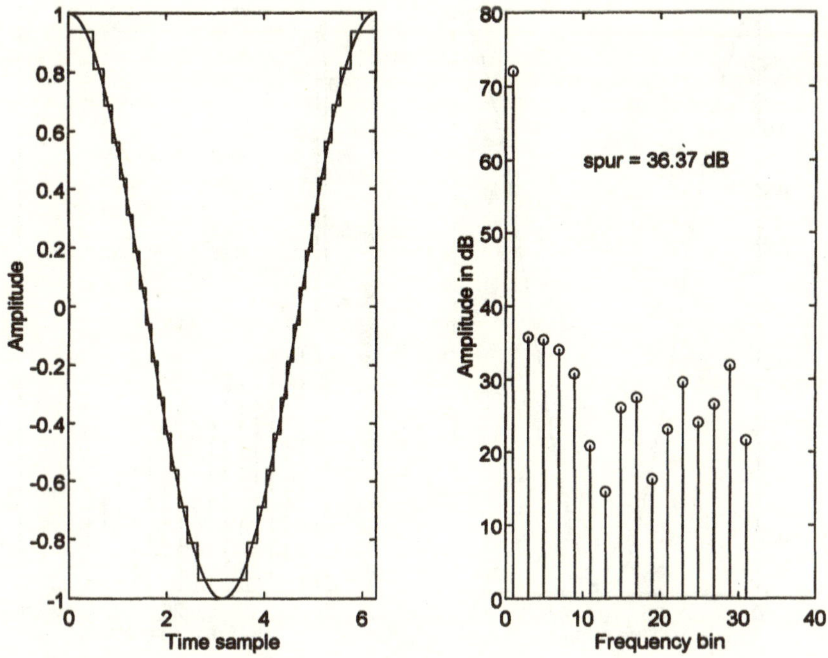

(b)

Figure 6.14 (continued).

highest spur will be difficult to predict. In other words, the analysis discussed above may no longer be applicable. Figures 6.15 and 6.16 show the time and frequency responses of a 3-bit ADC with 64 points of FFT. In Figure 6.15(a), all the digitization levels are exercised, the frequency output has (Figure 6.15(b)) approximately 25 dB of dynamic range, but the highest spur is not the third harmonic of the fundamental. In Figure 6.16(a), many quantization levels are missing and the highest output in the frequency domain (Figure 6.16(b)) is about 30 dB, which is higher than the expected value of 27 dB.

In all the above simulated results, the input frequency equals one of the frequency bins of the FFT output. Under this condition, no sidelobes are generated. If the input frequency does not match one of the FFT output frequency bins, the output has high sidelobes and it is difficult to distinguish the sidelobes from the spurs. Figure 6.17 shows these results for a 1.5-cycle sine wave. The frequency plot is shown in Figure 6.17(b), and it is difficult to separate the spurs from the sidelobes.

Finally, the spur levels are evaluated through simulated data. The input conditions are as follows. A Blackman window, which has a maximum sidelobe of –58 dB, is used to suppress the sidelobes. The length of the FFT is arbitrarily chosen

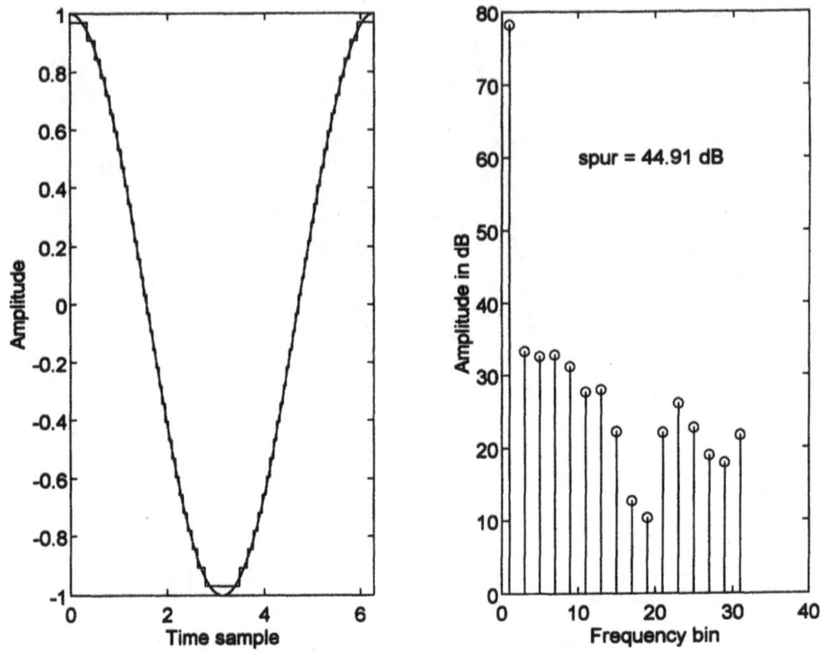

(c)

Figure 6.14 (continued).

to be 128 points and the ADCs have from 3 to 12 bits. For each ADC, 1,000 sine waves with random frequencies are used as input. There are 64 frequency bins. The input frequency is limited from the second frequency bin to the 62nd frequency bin because if the input frequency is too close to the ends, it is difficult to find the proper dynamic range. The input signal passing through the Blackman filter is digitized and 128 data points are collected. First, take the absolute values of the FFT outputs, then take the logarithm of them. A peak is defined as a frequency component that is higher than both its adjacent neighbors and at least 1.5-dB higher than one of them to avoid a false peak. The dynamic range is defined as the distance between the highest and the second highest peaks. There are 1,000 dynamic ranges corresponding to the 1,000 input frequencies and the lowest value is considered as the desired dynamic range. These results are shown in Figure 6.18 as the "*" marked curve. In this figure, it appears that the dynamic range increases by approximately 6-dB per bit from 3 to 7 bits. It starts to saturate at about 8 bits and approaches 58 dB, which is the dynamic range of the Blackman window.

A similar simulation is carried out for a 512-point FFT. The result is shown in Figure 6.18 and is represented by the "o" curve. The result is quite similar to

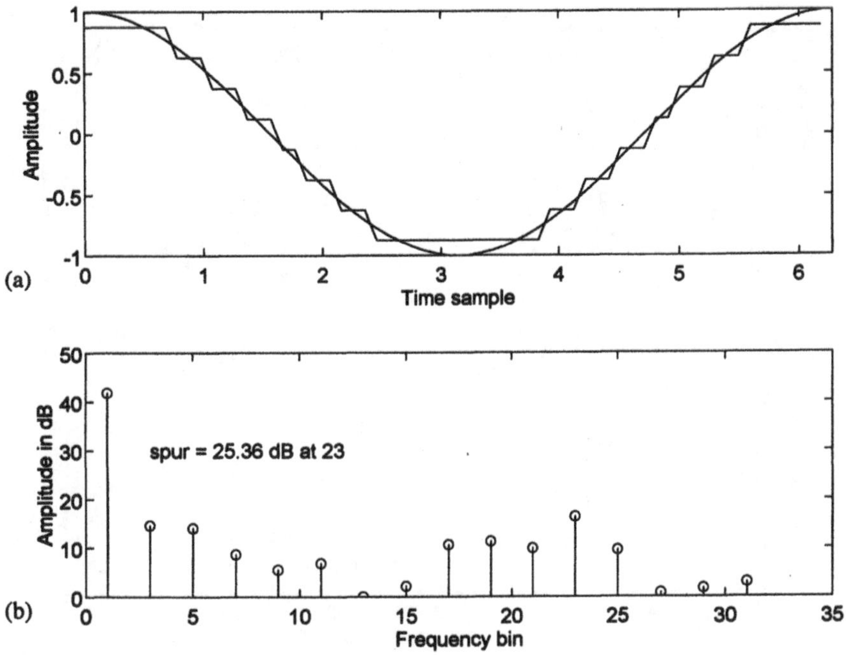

Figure 6.15 Outputs of a 3-bit ADC with no levels missing: (a) time domain, (b) frequency.

the 128-point FFT. However, from 3 to 8 bits, the dynamic ranges are slightly higher. This is probably due to the finer frequency resolution to obtain the main peak. The maximum dynamic range is approaching 58 dB. This is one way to obtain an approximate result for the maximum dynamic range expected from a certain ADC.

The above discussion still represents the best results because the amplitude of the input signal matches the maximum level of the ADC. If the input signal does not match the full scale of the ADC, the spurs will be higher than in the ideal case. Assume that the input signal is less than the optimum input. Therefore, not all the possible ADC levels will have outputs. Under this condition, the number of bits is less than the maximum available bits and the spurious responses will be higher than the ideal case. If the signal is stronger than the maximum input level, the digitized output will show a saturation effect. As a result, the maximum spur level will increase also.

6.12 NOISE EFFECTS IN ADC [33, 34]

The ADCs discussed in the previous sections are assumed to be ideal. However, the performance of almost all the ADCs is not perfect. For example, the quantization

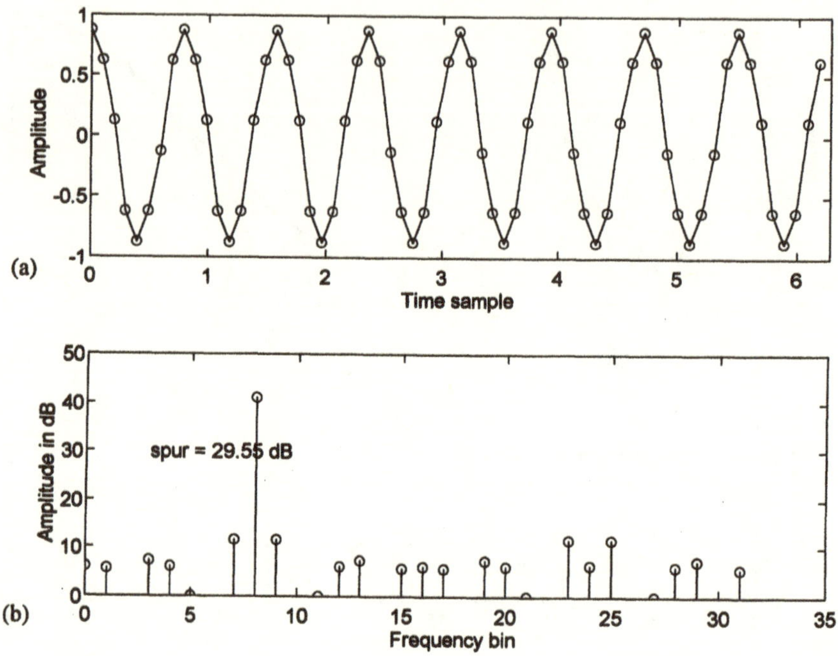

Figure 6.16 Outputs of a 3-bit ADC with many levels missing: (a) time domain, (b) frequency.

steps may not be uniform. Some steps are wider and some are narrower. In the extreme case, a certain quantization level is so narrow that it may never generate an output, and this is referred to as the missing bits. The sampling window is not always stable and the window will jitter, and this will have an adverse effect. There is also noise in the ADC circuit. For some ADCs, even without an input signal, the least significant bit may toggle in a random manner. The effect of noise will be discussed in this section.

Common sense tells us that noise will affect the sensitivity of a receiver. In many narrowband receivers (e.g., those used for communications), the noise level is kept as low as possible. In an EW receiver, the noise level is not the only concern; the dynamic range is also of concern. High sensitivity (low noise) usually means low dynamic range. This is true in a digital receiver also.

In an ADC, the noise sometimes has a positive effect. For instance, noise may reduce the spurs generated from the coherent digitization error. Some of the spurs are generated from coherent error, but the noise is incoherent. When noise is added into the input signal, the digitization coherence will be reduced; thus, the spurs are usually reduced or may even disappear. Figure 6.19 shows this effect through simulated data. Figure 6.19(a) shows the FFT output of the digitized sine wave without noise. The data contains two sine waves and the ADC has 8 bits. In

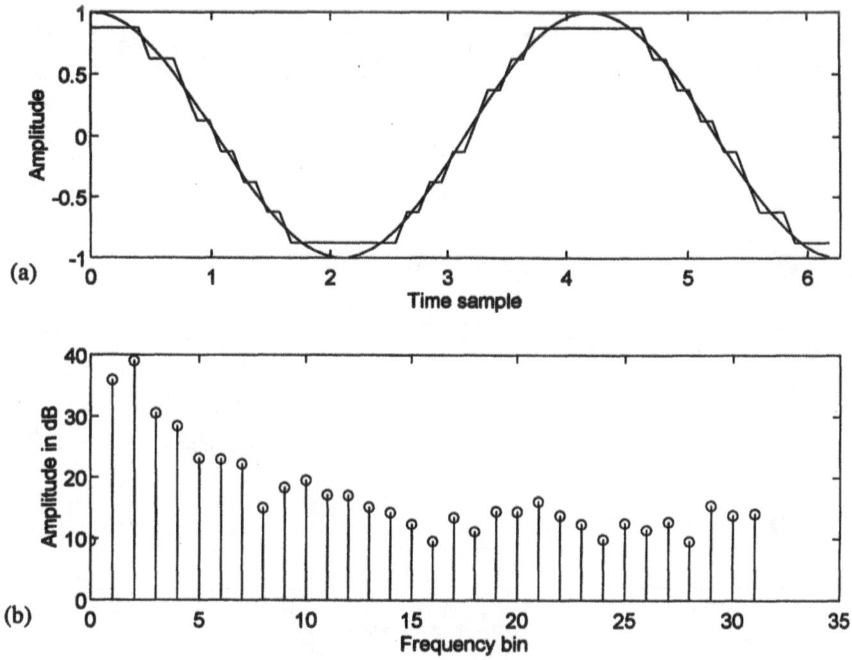

Figure 6.17 Output of 3-bit ADC with input frequency between bins: (a) time domain, (b) frequency.

order to keep this discussion simple, the frequencies of the two signals are selected to match two frequency bins. The first signal is at the 100th frequency bin and the second is at the 300th frequency bin with an amplitude 59 dB below the first one. From this figure, it is difficult to determine the position of the second signal because of the high spur level. If a receiver is designed to process simultaneous signals, these spurs will limit the instantaneous dynamic range because the weak signal must be higher than the spurs to be detected.

Figure 6.19(b) uses the same input signals, but with added noise. In this case, the *S/N* is 50 dB with respect to the strong signal. The amplitude of the second signal is clearly shown in this figure.

In general, the noise will reduce most of the spurs. Since the noise is random, there is a chance under certain conditions that a spur may increase in amplitude. If a spur is generated entirely from digitizing effect and there is no signal at the spur frequency, noise can disperse them easier than a true signal because the noise has a zero mean and its effect on the signal will be averaged out for a longer period of time. It can be easily shown that when the noise power increases, it will mask all the spurs, but it also reduces the sensitivity of the receiver. Thus, a little noise will reduce the spur levels. It might improve the dynamic range slightly (at least there is no adverse effect), but more noise will reduce the sensitivity. A general

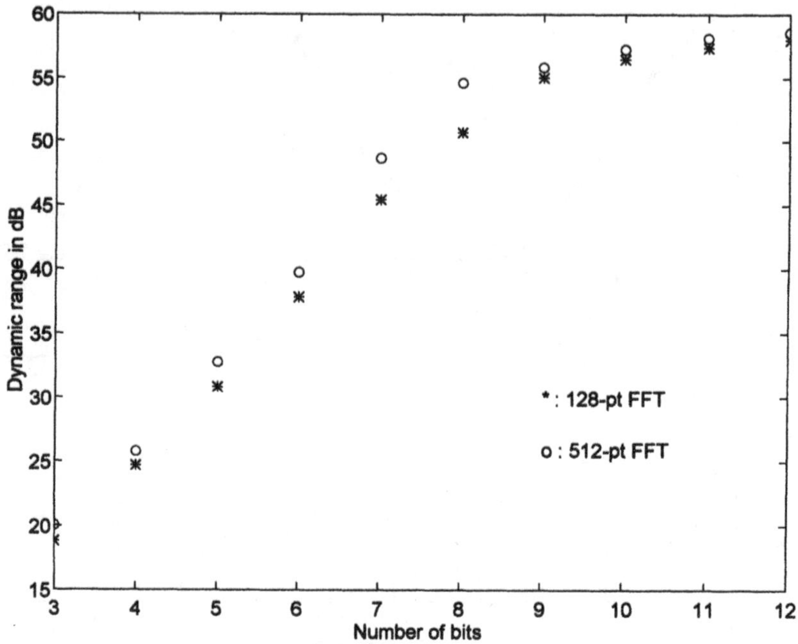

Figure 6.18 Dynamic range versus bits of ADC.

analysis of noise effect might be difficult. Noise floor will be further discussed in the next chapter.

6.13 SAMPLING WINDOW JITTERING EFFECT [37, 38]

As mentioned in Section 6.4, in a sample-and-hold circuit the sampling window (or aperture) has an unavoidable uncertainty period, which is called the sampling window jittering. This effect can be illustrated as follows.

Let us consider the following case. If the input is a constant voltage, the jittering effect will not affect the output because the input does not change with time. If the input voltage changes rapidly, a small jittering of the sampling window will have a prominent effect on the output. If the input is sinusoidal, the ADC output also depends on the amplitude of the signal because the larger the amplitude, the larger the change with respect to time. This effect is shown in Figure 6.20. Figure 6.20(a) shows a signal with low amplitude and low frequency and Figure 6.20(b) shows a signal with high amplitude and high frequency. If the ADC has the same amount of jitter, the signal with high amplitude and high frequency has a higher output change.

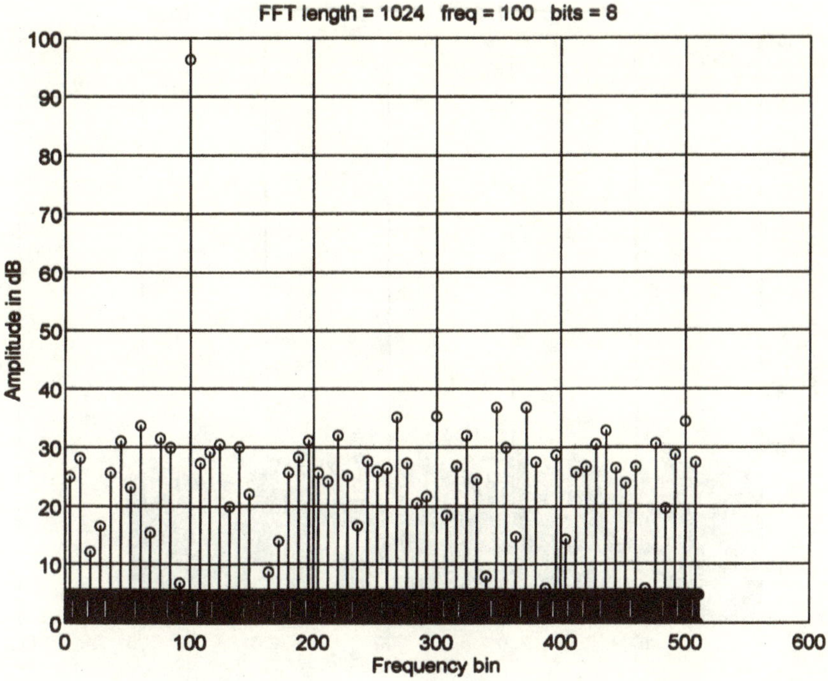

(a)

Figure 6.19 Noise effect on spurs and signals: (a) no noise, (b) noisy case.

The jitter effect can be measured in terms of *S/N*. This effect can be studied as follows. If the input signal is

$$v(t) = A\sin(2\pi ft) \tag{6.27}$$

where *A* and *f* are the amplitude and the frequency of the signal, respectively, then the derivative is

$$\frac{dv}{dt} = 2\pi fA\cos(2\pi ft) \tag{6.28}$$

The root mean square (rms) value of the derivative is

$$\left.\frac{dv}{dt}\right|_{\text{rms}} = \frac{2\pi fA}{\sqrt{2}} \tag{6.29}$$

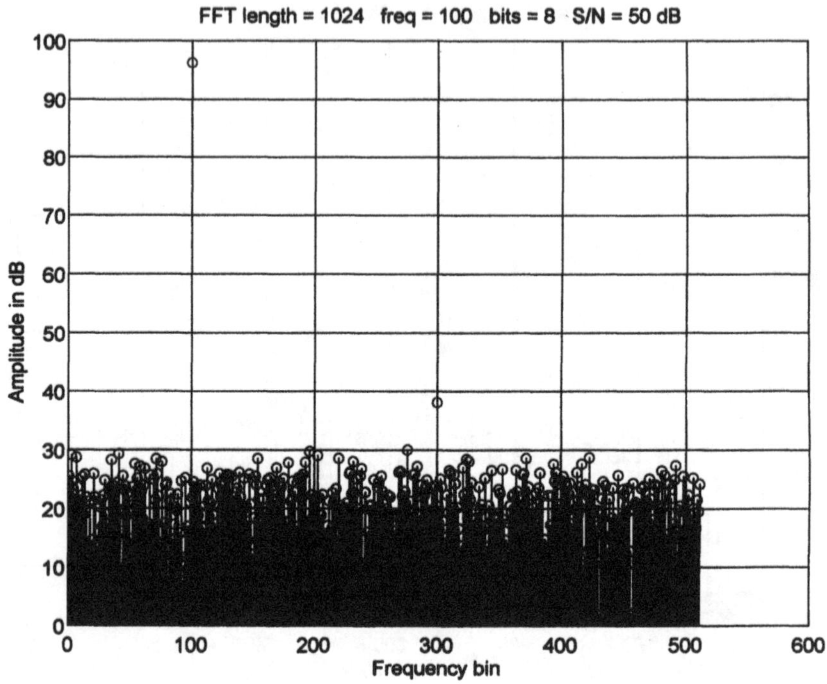

(b)

Figure 6.19 (continued).

This relation can be used to relate the rms error voltage and rms aperture. This relation can be written as

$$\frac{\Delta V_{rms}}{t_a} = \frac{2\pi f A}{\sqrt{2}} \tag{6.30}$$

where t_a represents the rms value jitter time. This equation can be written in terms of signal-to-noise as

$$\left(\frac{S}{N}\right)_{dB} = 20 \log\left[\frac{A/\sqrt{2}}{\Delta V_{rms}}\right] = 20 \log\left[\frac{1}{2\pi f t_a}\right] \tag{6.31}$$

where $A/\sqrt{2}$ is the input amplitude and ΔV_{rms} is considered as noise. It is assumed that jitter has a normal distribution with zero mean and a variance of σ^2 or $(\sigma = t_a)$. Under this condition, (6.30) can be written as

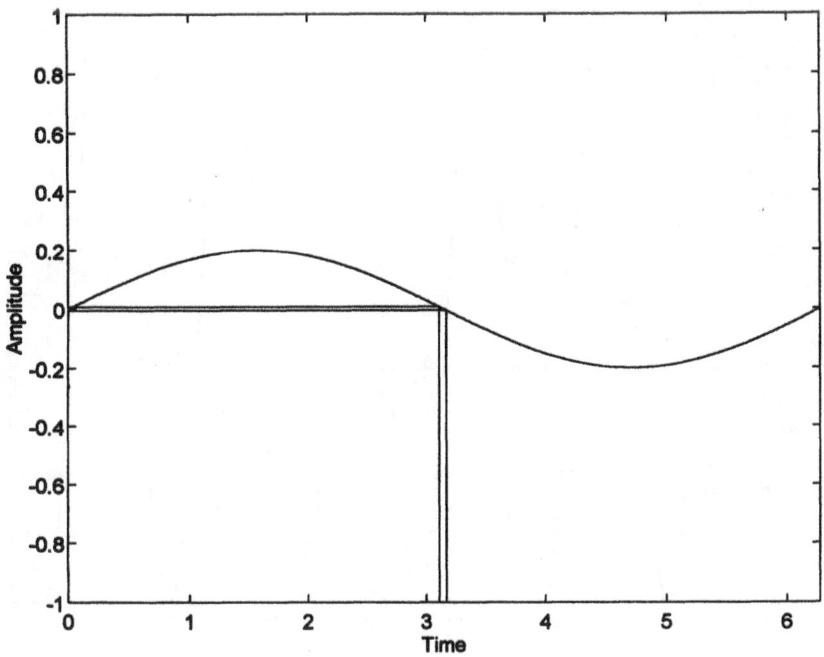

(a)

Figure 6.20 Sample window jitter effect on signal output: (a) signal with low amplitude and low frequency, (b) signal with high amplitude and high frequency.

$$\Delta V_{\text{rms}} = \frac{2A\pi f\sigma}{\sqrt{2}} \tag{6.32}$$

If one considers only the jittering and the quantization noise, the signal-to-signal ratio can be found from the sum of the two noise powers. The total noise N_{qj} is

$$N_{qj} = \frac{Q^2}{12} + \frac{(A2\pi f\sigma)^2}{2} \tag{6.33}$$

The S/N can be written as

$$\frac{S}{N} = \frac{A^2}{2N_{qj}} = \frac{A^2}{2\left[\frac{Q^2}{12} + \frac{(2\pi fA\sigma)^2}{2}\right]} \tag{6.34}$$

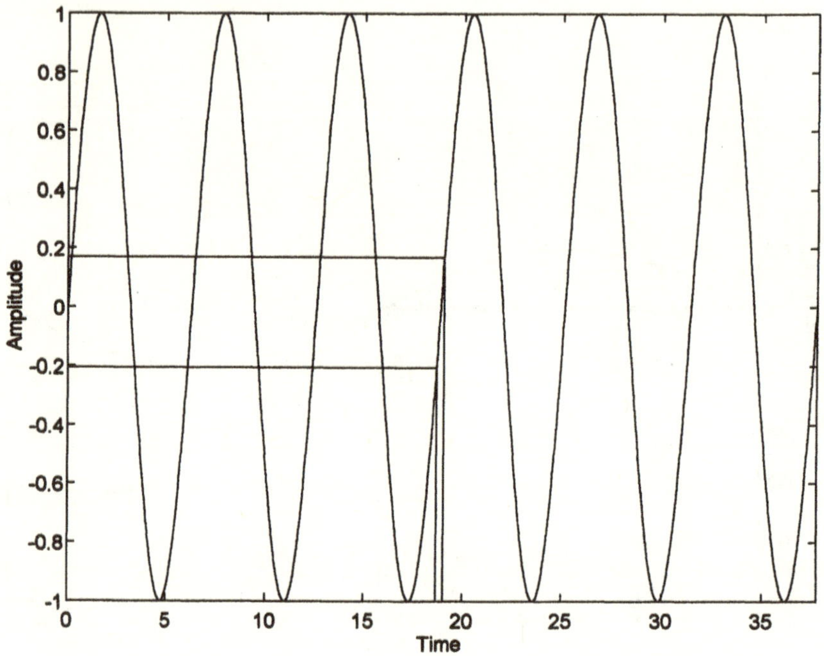

(b)

Figure 6.20 (continued).

From (6.7), this equation can be written in decibels as

$$\frac{S}{N} = 10 \log\left[\frac{3\ 2^{2b}}{2 + 3\ (2^b 2\pi f\sigma)^2}\right] \text{ dB} \tag{6.35}$$

This result can be plotted in Figure 6.21. In this figure, $b = 8$ and $t_a = \sigma = 2.5$ ps. This S/N drops from about 50 dB at low frequency to about 36 dB at 1 GHz. If the S/N response of an ADC can be measured, this type of curve can be used to determine the sample window jittering time.

6.14 ADC TEST THROUGH HISTOGRAM [28, 31, 32, 37–51]

There are several ways to evaluate an ADC. Some of the most common approaches are: histogram, sine wave curve fitting, and FFT. A pure sine wave with amplitude matching the maximum input of an ADC is used for all three tests. A pure sine wave is often generated by passing a sine wave (in general, nonideal) from a signal

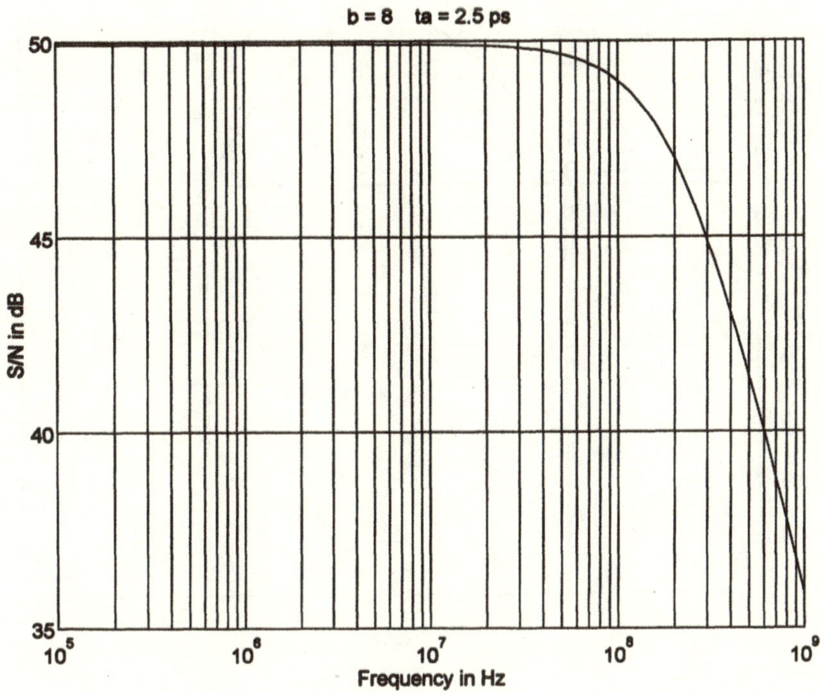

Figure 6.21 Signal-to-noise ratio versus frequency of an ADC.

generator through a very narrowband filter. The filter will reduce the harmonics in the nonideal sine wave. For a sine wave of amplitude A, the probability density of occurrence at a certain voltage V can be found with the help of Figure 6.22. This figure shows a half-cycle of a sine wave. Since there are two values of dx for every value of dv,

$$p(V)\ dV = 2p(x)\,dx \quad \text{or} \quad p(V) = \frac{2p(x)}{\dfrac{dV}{dx}} \tag{6.36}$$

where

$$V = A \sin x \quad \text{and} \quad \frac{dV}{dx} = A \cos x = A\sqrt{1 - \sin^2 x} = \sqrt{A^2 - V^2} \tag{6.37}$$

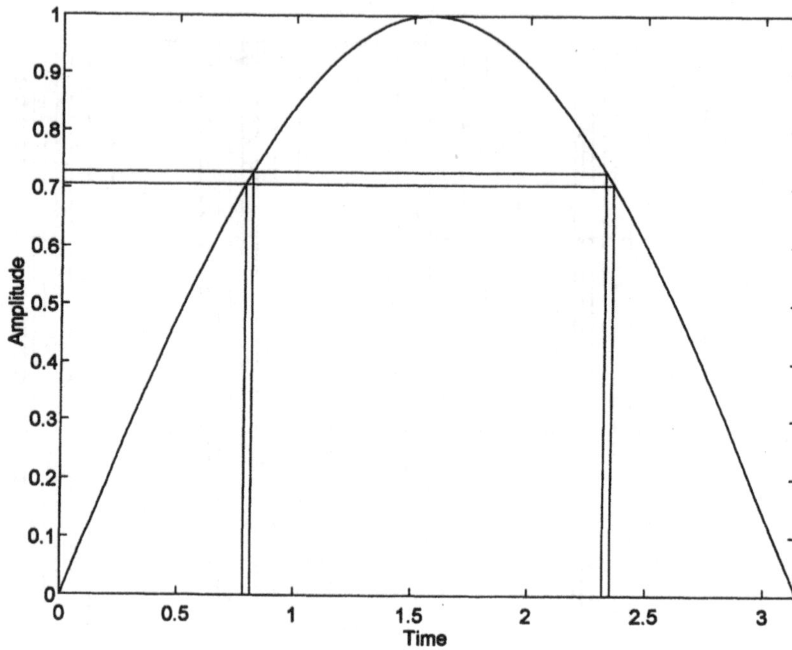

Figure 6.22 A half-cycle sine wave.

and $p(x)$ is uniformly distributed over 2π or $p(x) = 1/2\pi$. The probability density $p(v)$ of the data can be written as

$$p(V) = \frac{1}{\pi\sqrt{A^2 - V^2}} \tag{6.38}$$

This equation can be plotted as shown in Figure 6.23, where missing bits will show zero counts. Figure 6.24(a) shows a histogram of an ADC with small differential nonlinearity and no missing bits. Figure 5.24(b) shows the result of an ADC with large differential nonlinearity and many missing bits.

6.15 ADC TEST THROUGH SINE CURVE FITTING [28–32, 37–51]

A popular way to evaluate the performance of an ADC is through sine curve fitting. If the input signal is a pure sine wave and the ADC is ideal, a certain pattern will be generated. In general, a practical ADC will generate a pattern that will be different from the ideal case. In addition, since some noise would always be present, the code level transition is probabilistic. In order to reduce the uncertainty, a large

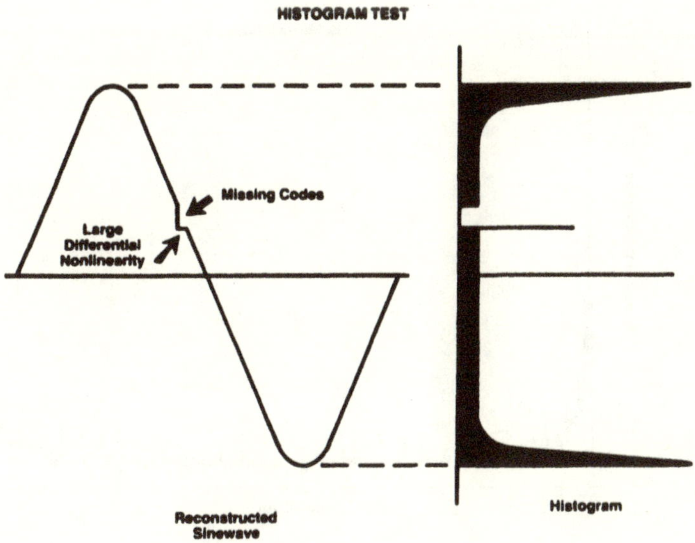

Figure 6.23 Histogram of a sine wave. (Courtesy of Hewlett Packard Co. [28].)

number of data points are needed. Assuming the noise has a zero mean and normal distribution, the estimation accuracy can be computed. Table 6.2 gives the record length versus a 99.9% (3σ) precision confidence level, expressed as a percentage of the rms noise value (σ)[30].

A record of data is taken with a sine wave input with specific parameters. Fit a sine wave function to the record by adjusting the phase, amplitude, dc value, and frequency to minimize the sum of the squared difference between the sine function and the data points. There are two algorithms to fit the curve. One is for known frequency, which means the sampling clock and the input frequency are known and stable. The other algorithm is for unknown frequency, which means the frequency is not known exactly. Equations are available in [30]. The final result is usually converted into effective bits. The effective bits (b_{eff}) are defined as [41]

$$b_{eff} = b - \log_2\left\{\frac{\text{rms error (actual)}}{\text{rms error (ideal)}}\right\} \tag{6.39}$$

where b is the number of bits of the perfect ADC.

6.16 ADC TEST THROUGH FFT OPERATION [52–56]

Another popular way to evaluate the ADC is through an FFT operation. Because the present approach to build a wideband digital receiver is through an FFT

(a)

(b)

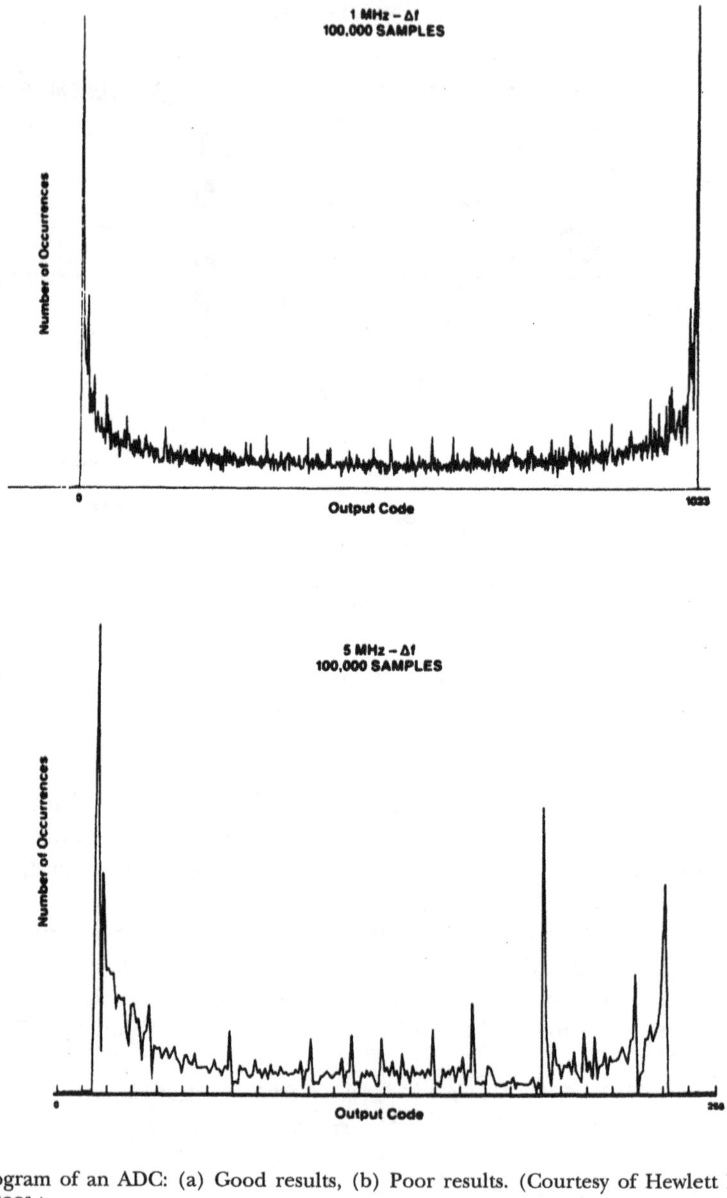

Figure 6.24 Histogram of an ADC: (a) Good results, (b) Poor results. (Courtesy of Hewlett Packard Co. [28].)

Table 6.2
Record Length Versus Confidence Level

Record Length (points of data)	*Precision of Estimates of Code Transition Level (times σ)*
64	45%
256	23%
1,024	12%
4,096	6%

operation, as discussed in Chapter 11, this test should provide results closely related to receiver performance. One must keep in mind, however, that the ADC performance does not represent the performance of the receiver. The receiver performance depends on the design of the entire receiver. For example, one of the primary requirements of an electronic warfare (EW) receiver is the minimum pulse width. The minimum pulse width usually determines the length of the FFT, which in turn determines the frequency data resolution and the sensitivity of the receiver. If the length of the FFT is longer than the minimum pulse width, it will affect the sensitivity of the receiver as discussed in Section 2.19. An FFT length longer than the pulse width may also complicate the parameter encoder design because one short pulse can appear in many output frequency bins. The pulse amplitude, pulse width, and the carrier frequency of the short pulse might be severely distorted by a narrowband filter and the encoder may produce erroneous information. Therefore, the test of the ADC should be independent of the receiver design. In general, a long FFT is usually used to evaluate ADC performance.

In order to test the ADC independently of the receiver design, the following provisions should be observed. The input must be a pure sinusoidal wave with frequency matched to one of the output FFT frequency bins. An input frequency matching to a frequency bin does not have any sidelobes as discussed in Section 4.5.2, therefore, a rectangular window can be used. The performance of the ADC should be independent of the FFT length. Usually, the longer the FFT the narrower the output frequency bins, and therefore, the lower the noise level. There are, however, spurs in the frequency domain. The spur levels may be independent of the FFT length. In other words, long FFT may not decrease the spur levels. If an ADC has n bits, there are 2^n output levels. Usually all the levels should be exercised. Using a sinusoidal wave to exercise all the levels, the minimum length N_m of the FFT should be [55]

$$N_m = \pi 2^n \tag{6.40}$$

For an 8-bit ADC, N_m is about 804. Of course, the length of the FFT should be a base-2 number. A relatively long FFT is usually used to measure the ADC performance because it can better reveal the fine structure of the spur responses. Thus, this discussion is different from the discussion in Section 6.11 where the results are dependent on FFT length and window shape. Due to the minimum pulse width requirement, the FFT used in a receiver design is usually limited to a maximum of about 512 points. The real-time processing requirement in a receiver also limits the length of FFT used in a receiver design.

If the sampling frequency is f_s and the length of the FFT is N, the frequency data resolution is f_s/N. The input frequency must be

$$f_s = \frac{Mf_s}{N} \tag{6.41}$$

where N is the length of the FFT and also a base-2 number, and M is an integer. If M is an odd number, the input is referred to as the optimum input frequency. Under this condition there are more levels to be exercised. Figure 6.25 illustrates this phenomenon. To generate these figures $N = 64$, $f_s = 128$ are used. In Figure 6.25(a) ($M = 3$) there are three cycles and each cycle starts with a different phase relative to the sampling clock. In Figure 6.25(b) ($M = 4$) there are four cycles and all cycles start with the same phase. In Figure 6.25(c) and (d), the sorted outputs are plotted and there are more output values for $M = 3$ than $M = 4$. Figure 6.25(e) and (f) show the phase angle (or the x values in Figure 6.25(a) and (b)) and these values are adjusted to be less than 2π. Figure 6.25(e) shows that the phase values are uniformly distributed between 0 and 2π, but Figure 6.25(f) does not have this property.

In order to generate a pure sinusoidal wave, the input signal to the ADC is properly filtered to reduce the unwanted harmonics. The input and the sampling frequency should be phase locked such that the input can be put in one of the frequency bins accurately. Using this arrangement, the relative frequency inaccuracies between the input signal and the sampling frequency are minimized.

The single frequency test results of an 8-bit ADC are shown in Figure 6.26. The FFT length is 16,384 (2^{14}). Because the input signal is real, only half of the outputs (8192) need to be displayed. In this figure the input is a single frequency in one frequency bin. The amplitude of the input matches the maximum input level of the ADC. In other words, the input signal exercises all of the ADC levels without saturating it.

A harmonic response is an output frequency f_o, which can be written as

$$f_o = mf_i \tag{6.42}$$

where m is a positive integer and f_i is the input frequency. Higher f_o frequencies can alias into the baseband through sampling. Several of the spurious responses

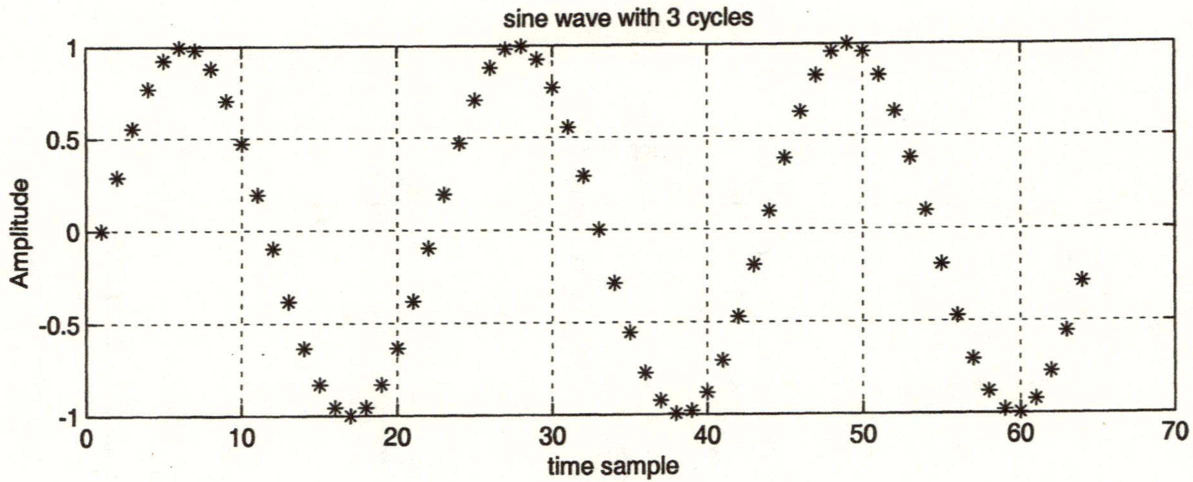

(a)

Figure 6.25 Illustration of digitized points: (a) output in time domain $M = 3$, (b) output in time domain $M = 4$, (c) sorted output $M = 3$, (d) sorted output $M = 4$, (d) sorted phase values $M = 3$, (f) sorted phase values $M = 4$.

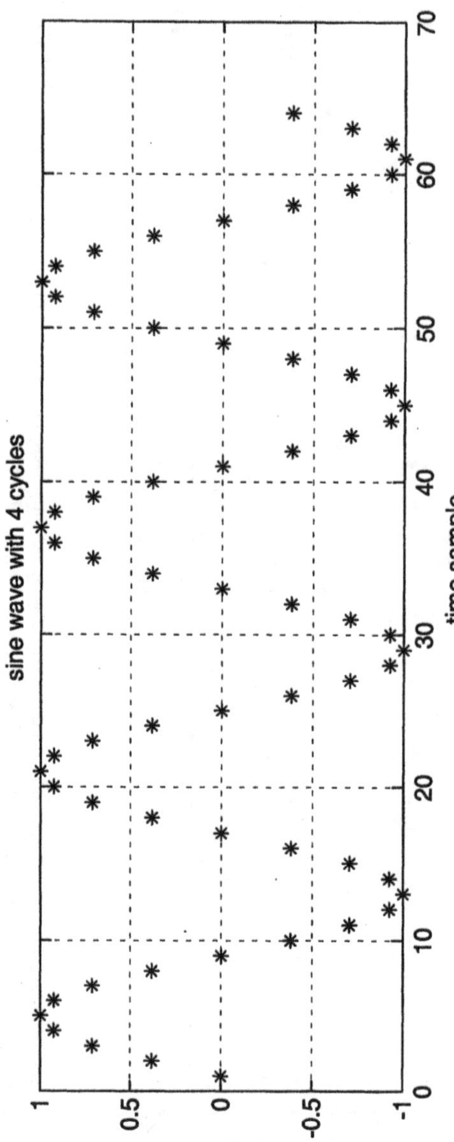

(b)

Figure 6.25 (continued).

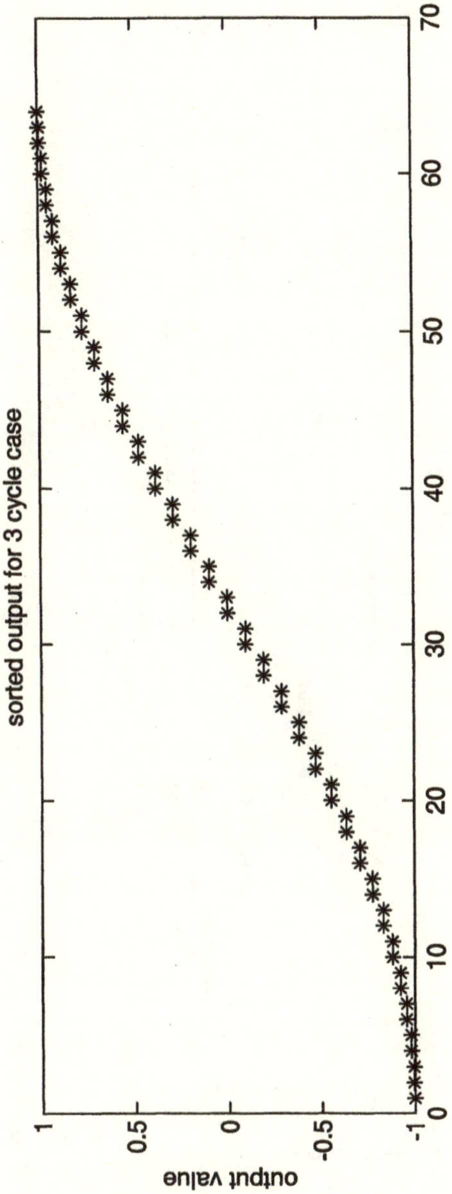

(c)

Figure 6.25 (continued).

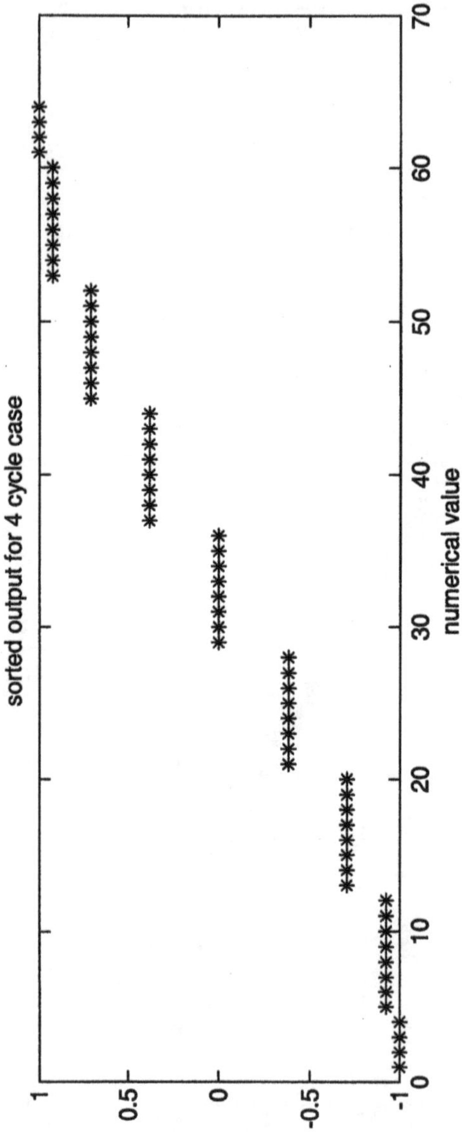

(d)

Figure 6.25 (continued).

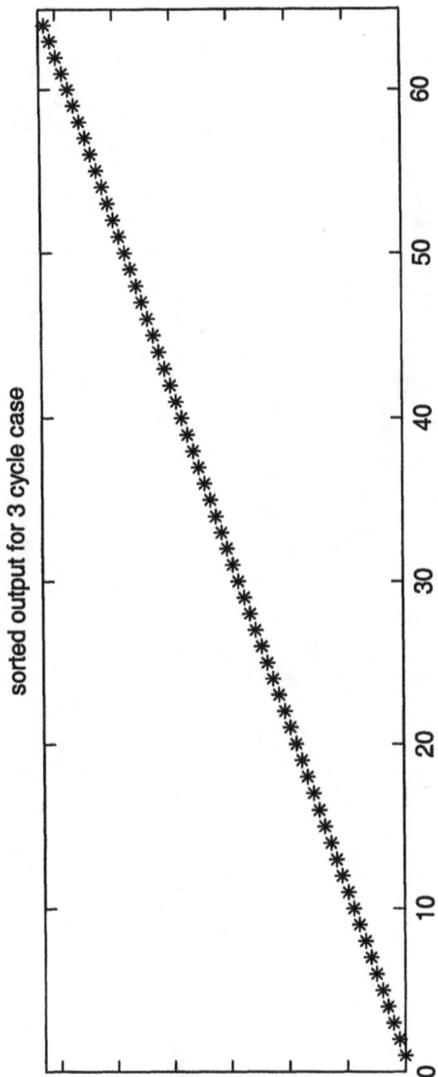

(e)

Figure 6.25 (continued).

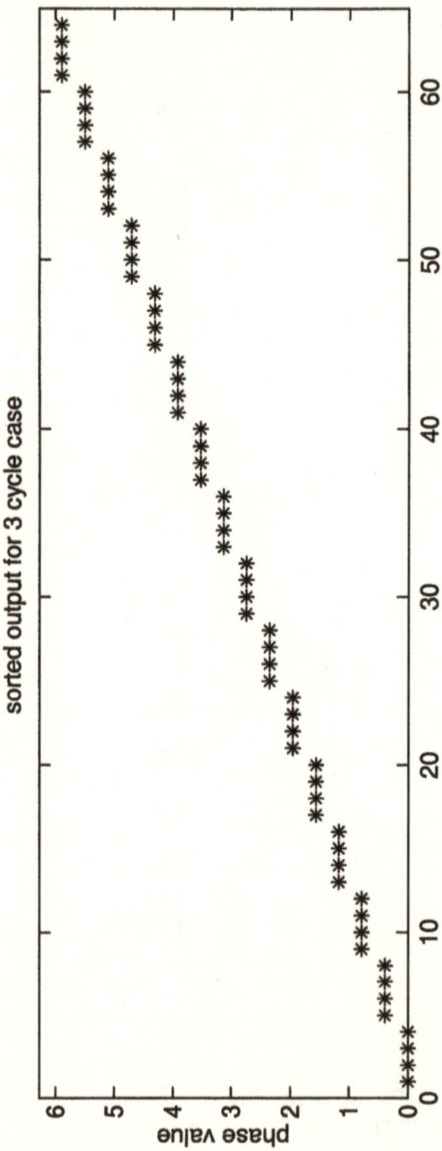

(f)

Figure 6.25 (continued).

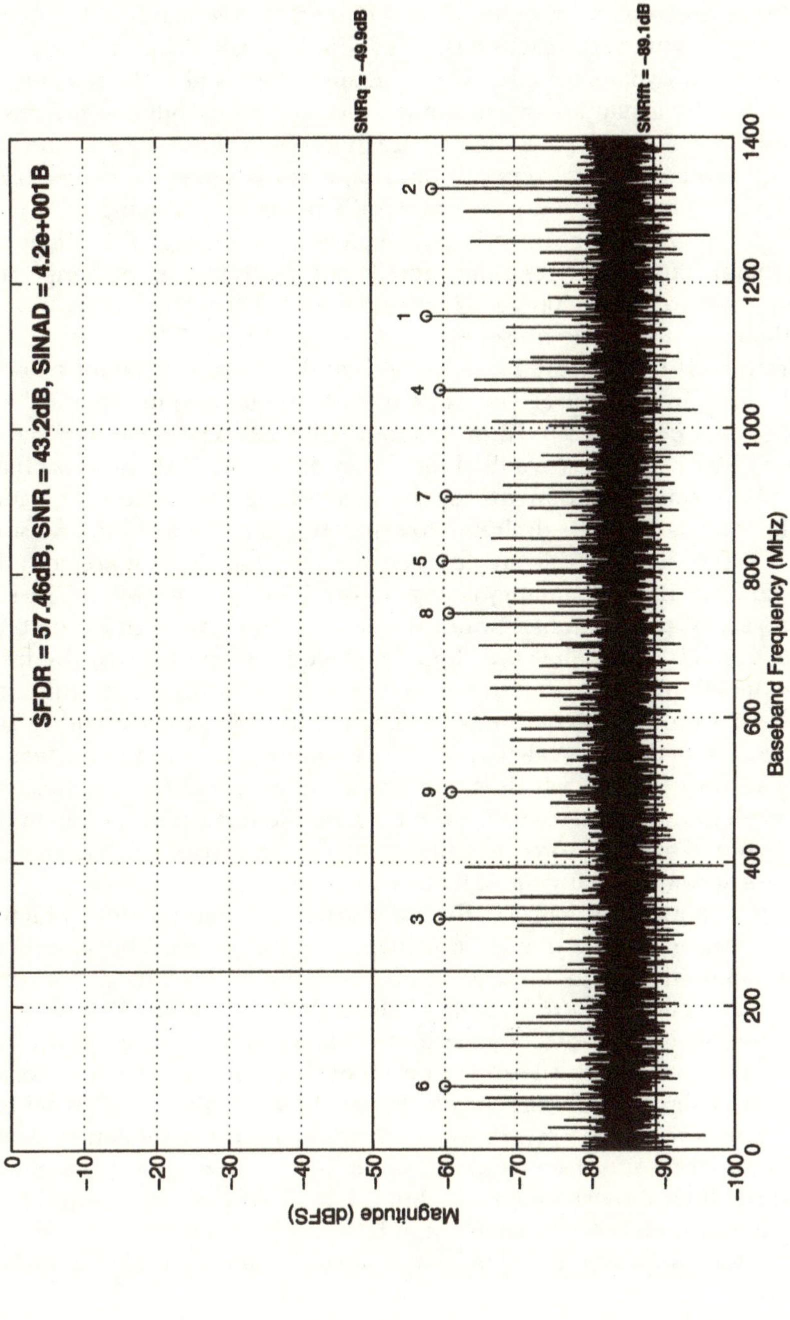

Figure 6.26 FFT output of an 8-bit ADC.

can be identified as harmonics. Spur numbers 9 and 8 are the second and third harmonics of the input signal. There are also many spurious responses that cannot be identified as probable harmonics. From Figure 2.26 one can see that from the peak of the input signal to the highest spur is about 57 dB, which can be considered as the single signal dynamic range. This dynamic range is also the instantaneous dynamic range. The instantaneous dynamic range is the capability to process both a strong and a weak signal presented simultaneously at the receiver. The single-tone spur-free dynamic range versus input frequency is shown in Figure 6.27. In Figure 6.27(a) the input signal power is 0.5 dB below the clipping of the ADC, while in Figure 6.27(b), the input is 1 dB below the clipping. The single-signal spur-free dynamic range changes quite a few decibels with the input change of only 0.5 dB. The input frequency range is from 1,500 to 3,000 MHz.

The theoretical signal-to-noise ratio of an 8-bit ADC is 49.9 dB, which can be obtained from (6.15). Since the input signal is at 0 dB, the equivalent noise level is –49.9 dB shown in the figure. The noise in each frequency bin is about 39.1 dB [10log(16,384/2)] below –49.9 dB or at about –89.1 dB. The measured signal-to-noise ratio is 43.2 dB, which is less than the theoretical value. This value is obtained by adding all the noise components together excluding the highest nine components. In this calculation, it is desirable to separate the spurs from the noise. The highest nine responses are arbitrarily chosen as spurs, and the rest are considered as noise spikes. The input frequency changes from 1,500 to 3,000 MHz. The signal-to-noise ratio across the input bandwidth is shown in Figure 6.28. In Figure 6.28(a) the input power is 0.5 dB below the clipping, while in Figure 6.28(b) the input is 1 dB below the clipping. The difference between these two figures is quite small.

Figure 6.29 shows similar results as Figure 6.28. The only difference is that the spurs are included in the calculation. The quantity is referred to as the signal-to-noise-and-distortion ratio, which is often abbreviated as SINAD. The results are slightly lower than the signal-to-noise ratio because the distortions are included in the denominator. This is a more meaningful quantity because one cannot neglect the spur distortion when using an ADC in a receiver.

Another important property is the third-order intermodulation, which provides information on two simultaneous signals of equal power. This parameter is discussed in detail in Section 7.5, and its effect on the performance of a receiver in Section 16.16. The measured result of the ADC is shown in Figure 6.30. In this figure, the third-order intermodulation products are labeled as 4 and 5. In a conventional receiver the third-order intermodulation products limit the lower level of the input signal. The range from the amplitude of the input signal to the intermodulation level is referred to as the third-order intermodulation dynamic range or the two-tone spur-free dynamic range. In this case the intermodulation products do not limit the dynamic range but the highest spurious response, which limits the two-tone spur-free dynamic range to about 51.3 dB.

The two-tone spur-free dynamic range versus frequency is shown in Figure 6.31.

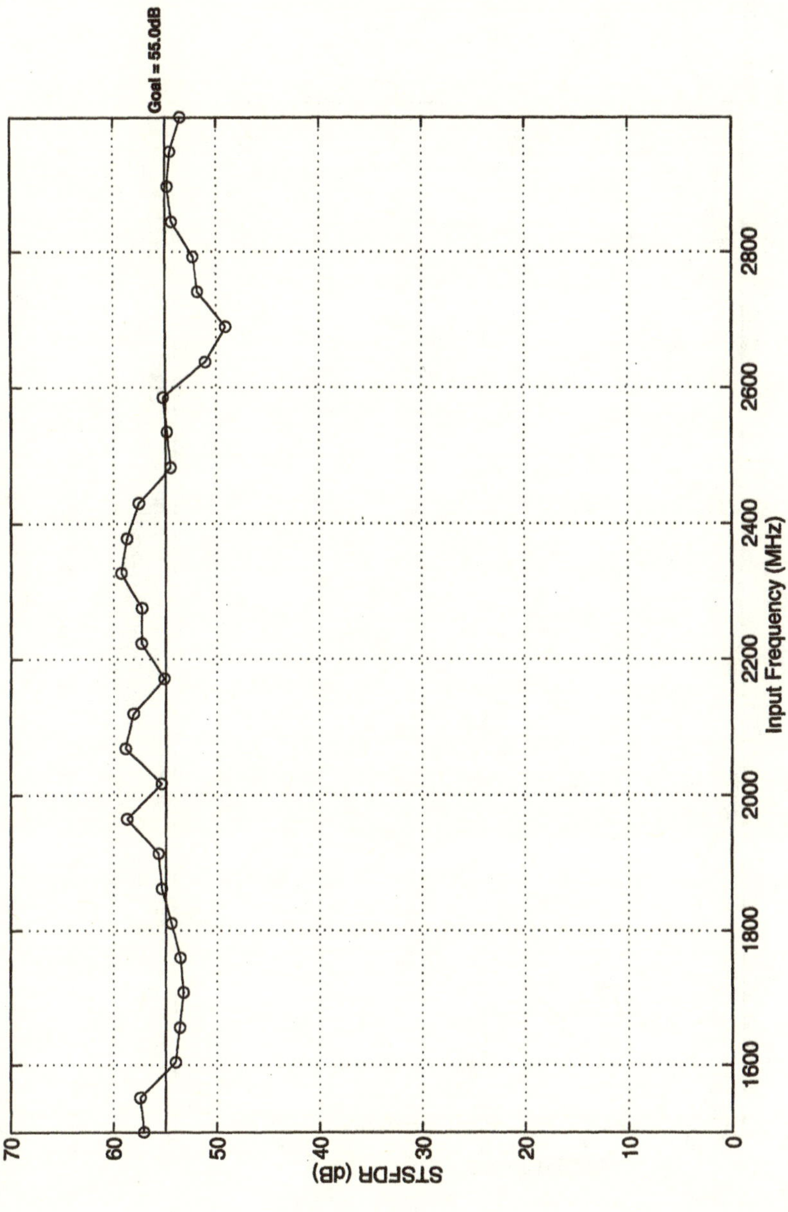

(a)

Figure 6.27 Single-signal spur-free dynamic range versus input frequency: (a) input 0.5 dB below clipping, (b) input 1 dB below clipping.

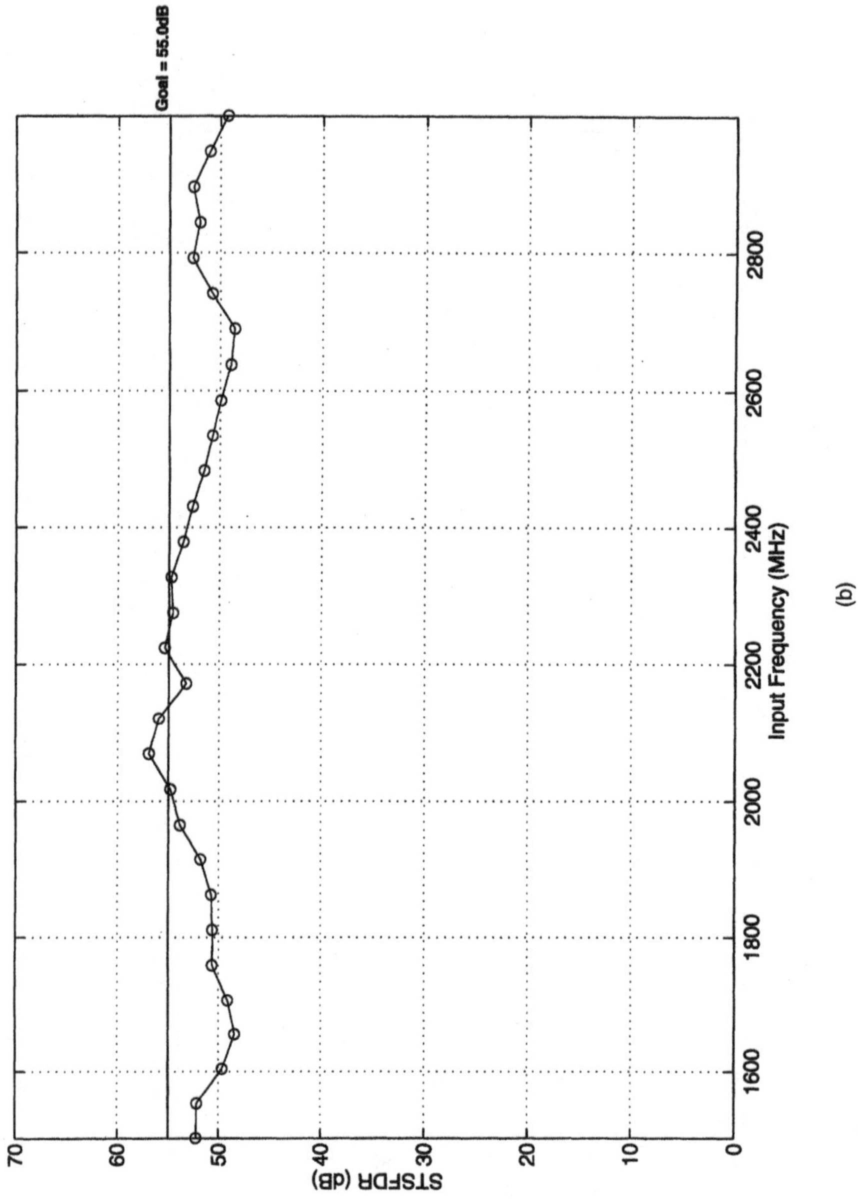

Figure 6.27 (continued).

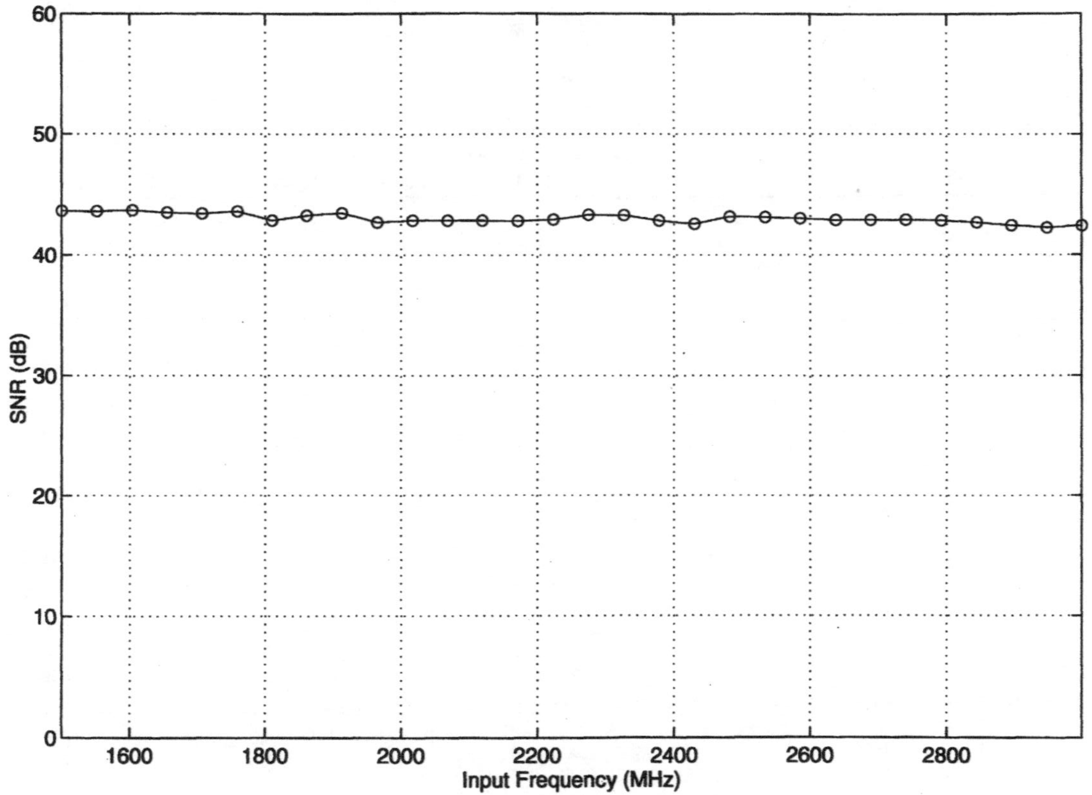

(a)

Figure 6.28 Signal-to-noise ratio versus the input frequency range: (a) input 0.5 dB below clipping, (b) input 1 dB below clipping.

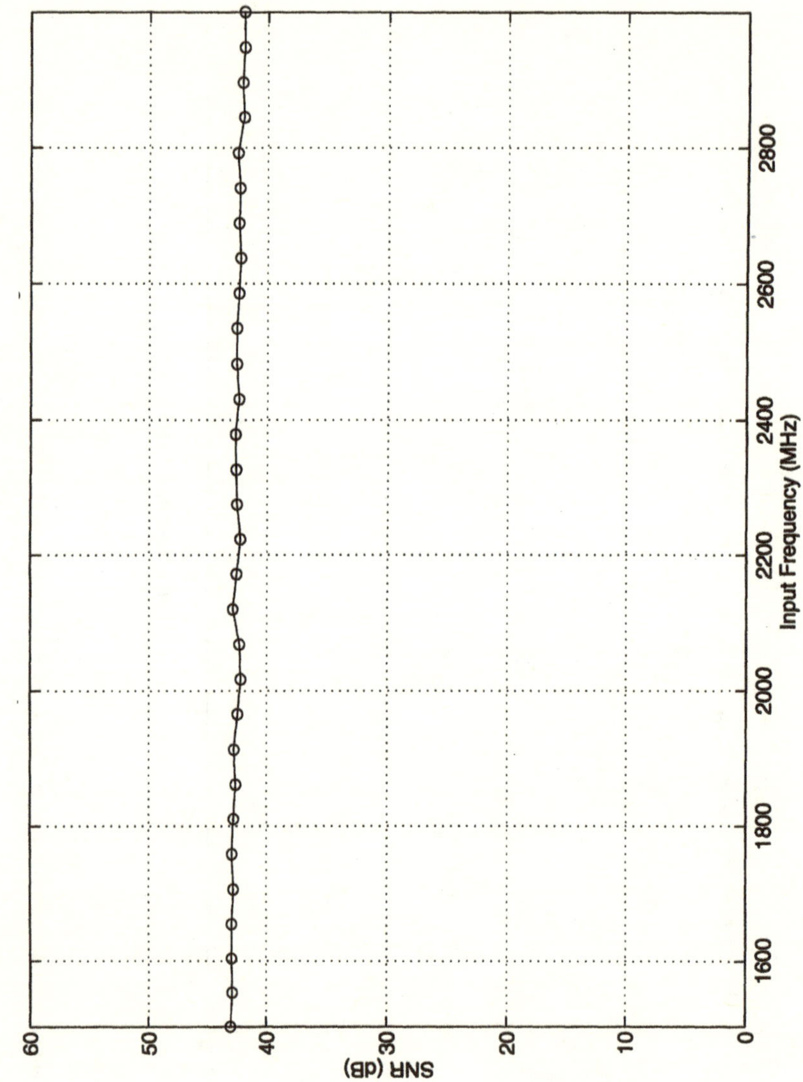

(b)

Figure 6.28 (continued).

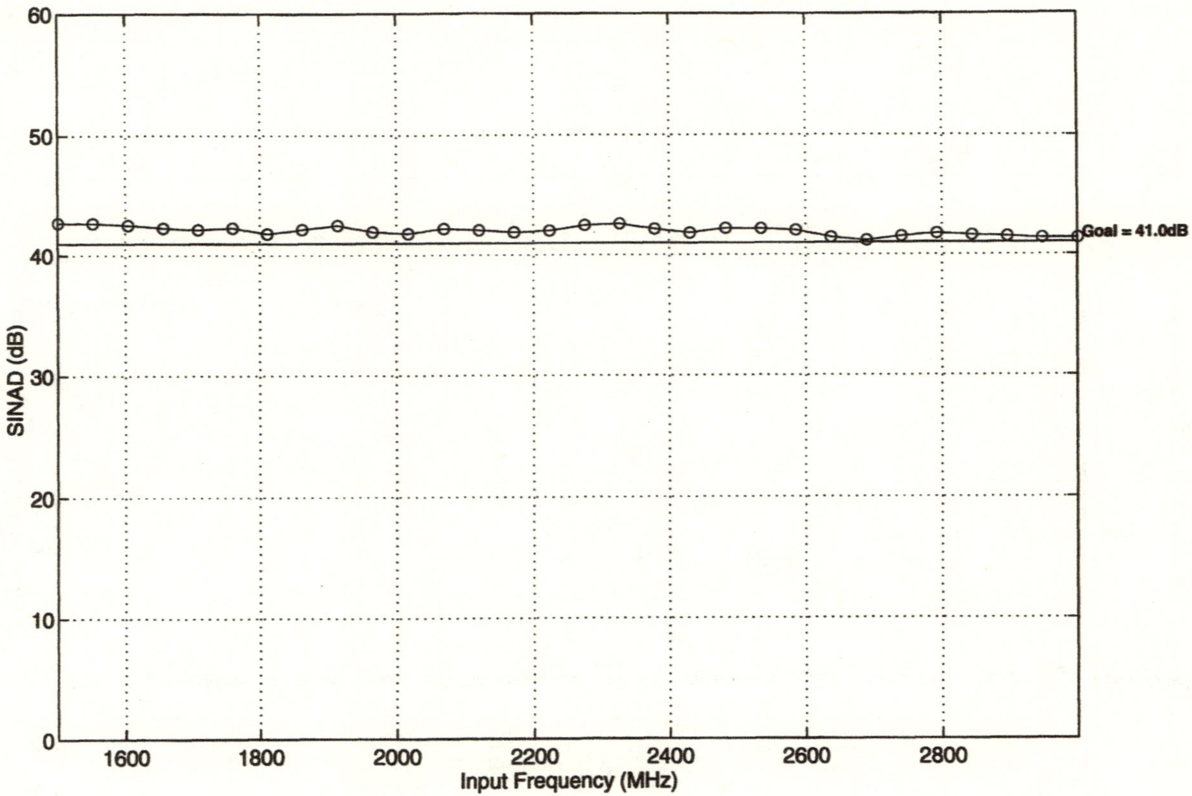

(a)

Figure 6.29 Signal-to-noise-and-distortion ratio versus the input frequency range: (a) input 0.05 dB below clipping, (b) input 1 dB below clipping.

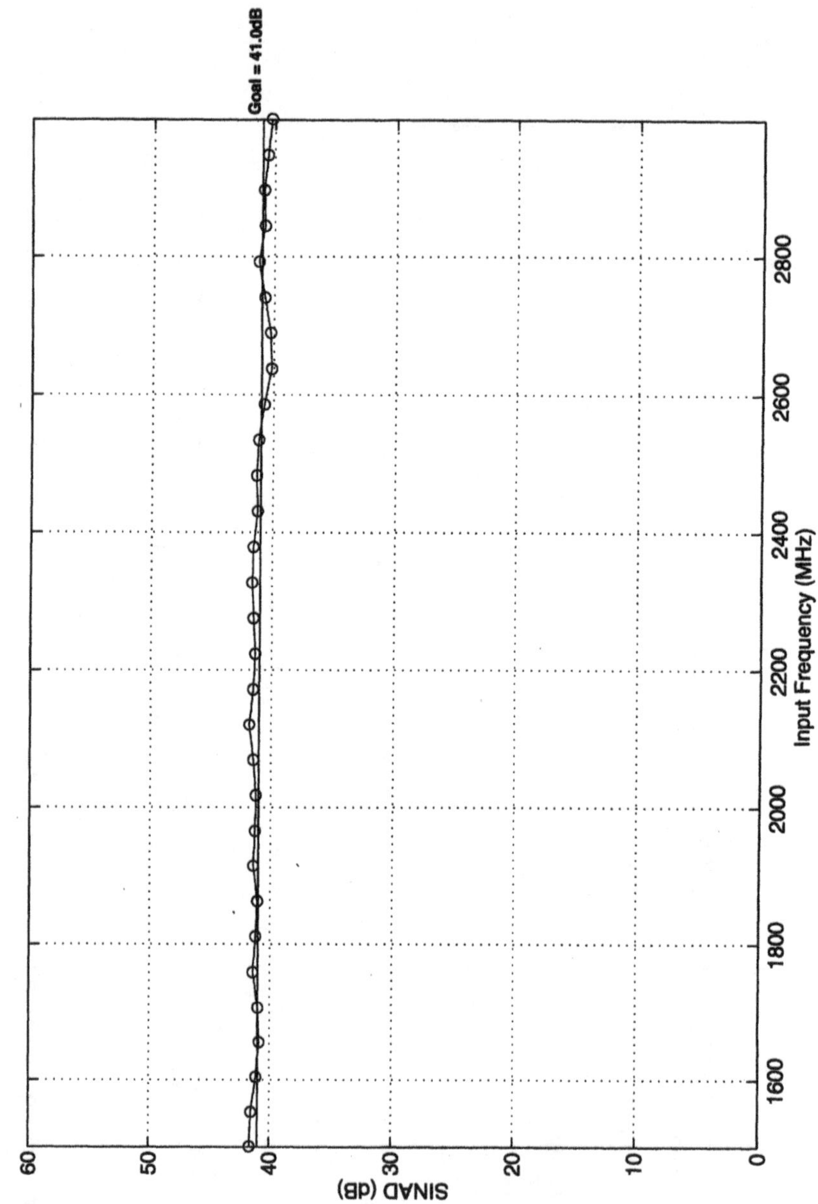

(b)

Figure 6.29 (continued).

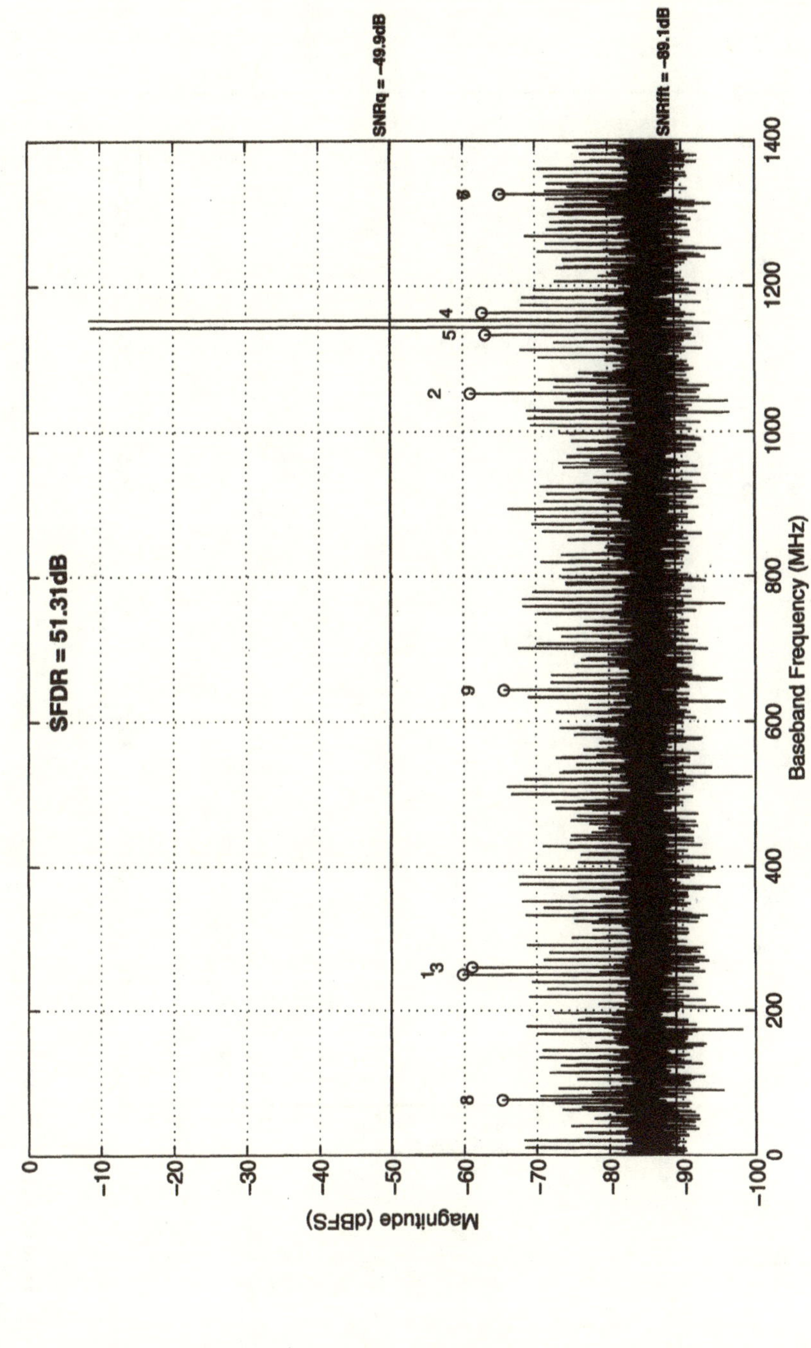

Figure 6.30 Third-order intermodulation of the 8-bit ADC.

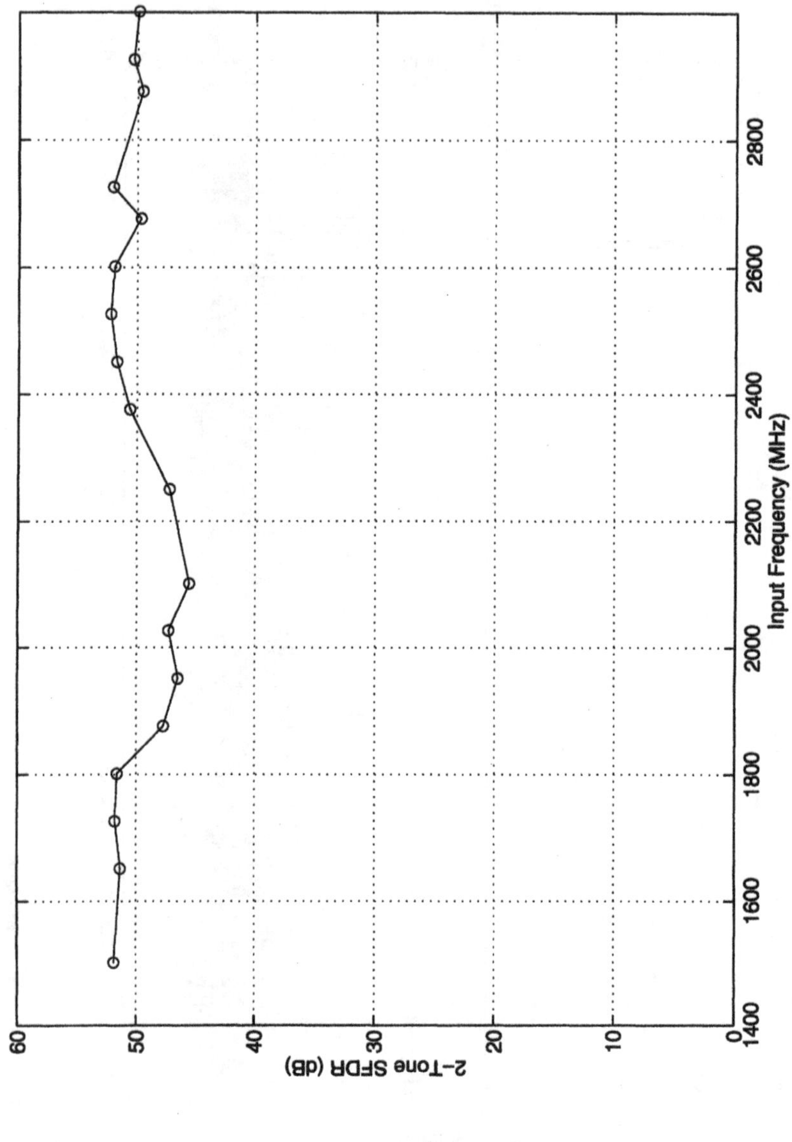

Figure 6.31 Two-tone spur-free dynamic range versus input frequency.

The effective number of bits can be obtained from the signal-to-noise ratio or from the signal-to-noise-and-distortion ratio. From (6.14) and (6.15), the effective number of bits can be written as

$$b_{eff} = \frac{(S/N)_{dB} - 1.76}{6.02} \tag{6.43}$$

or

$$b_{eff} = \frac{(\text{SINAD})_{dB} - 1.76}{6.02} \tag{6.44}$$

In this equation the signal-to-noise ratio is replaced by the signal-to-noise-and-distortion ratio. In both equations the signal-to-noise and the signal-to-noise-and-distortion ratio are expressed in decibels. The result from (6.44) is a more conservative estimation. The effective number of bits versus input frequency is shown in Figure 6.32. For the 8-bit ADC the effective number of bits is above 6.5 bits.

6.17 REQUIREMENTS ON ADC

The performance of a receiver depends on the receiver design and the performance of the ADC. The receiver performance cannot surpass the performance of the ADC used in the receiver. For example, in Figure 6.26 the single-signal spur-free dynamic range is about 57 dB. One cannot expect to build a receiver using this ADC to obtain a single signal dynamic range of more than 57 dB, which is obtained from the 16,384-point FFT. If the length of the FFT in a receiver is 256 points instead of 16,384, the noise will increase about 18 dB [10log(16,384/256)]. Under this condition the noise spikes may still limit the lower level of the dynamic range. One can conclude, however, that the dynamic range of a receiver should be less than the dynamic range obtained from the ADC tests.

The RF input bandwidth of a receiver should be less than an octave to avoid in-band second harmonic. Octave bandwidth means the highest frequency is double the lowest frequency, such as from 1 to 2 GHz. For a wide bandwidth receiver it is usually impossible to achieve this design goal in the baseband from 0 to $f_s/2$ as shown in Figure 6.33. In an intermediate frequency band such as from $f_s/2$ to f_s, this goal can be achieved. It is common practice to build the RF channel of a wideband receiver in the second frequency zone. Under this condition, the ADC sampling the RF must be able to accommodate the input frequency. Therefore, if the maximum sampling frequency of the ADC is f_s, the input frequency should be able to reach f_s also. The ADC illustrated in the last section can operate at a sampling frequency of 3 GHz, and the input frequency can also reach 3 GHz. Thus it is suitable for wideband receiver operation.

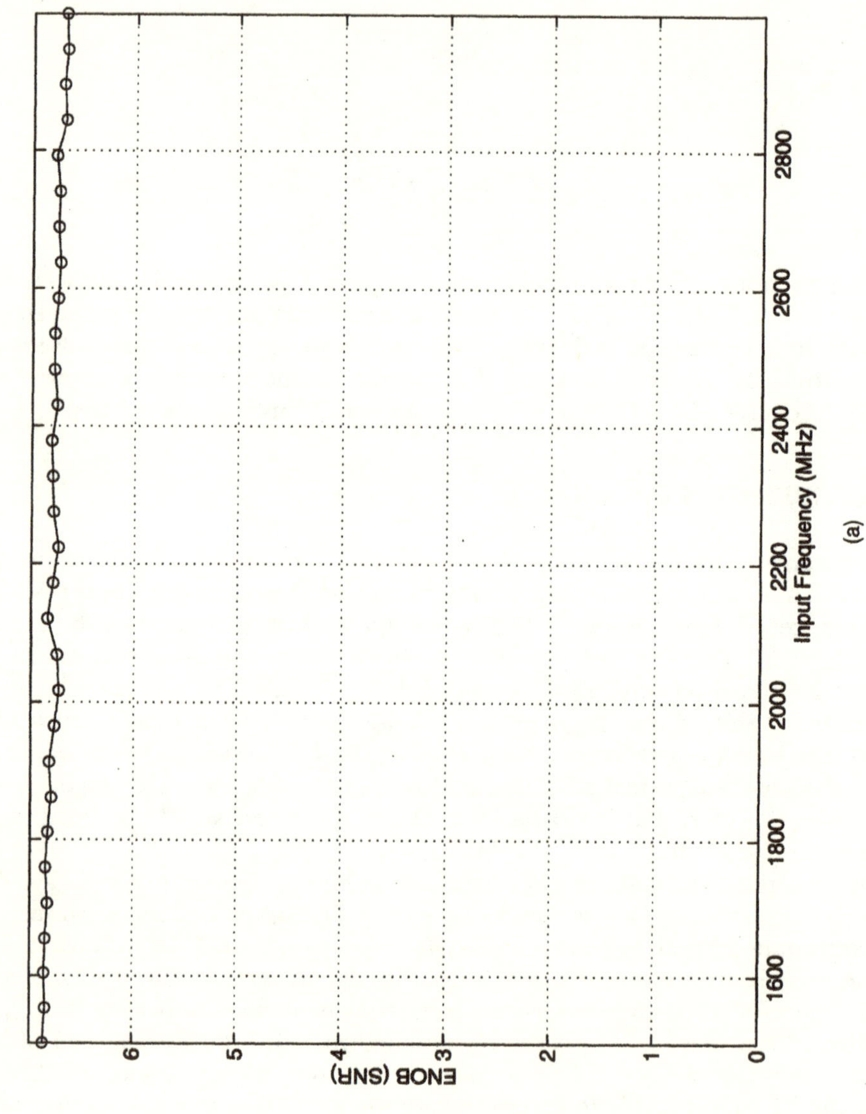

Figure 6.32 Number of effective bits versus input frequency: (a) against signal-to-noise ratio, (b) against SINAD.

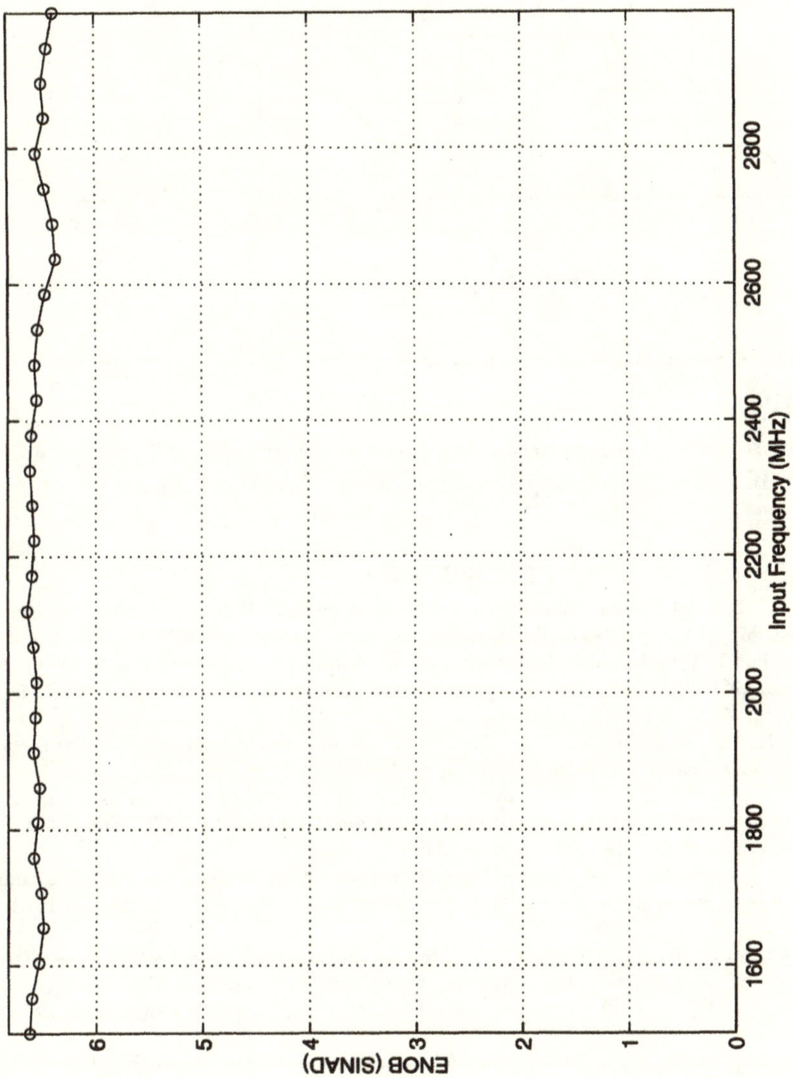

(b)

Figure 6.32 (continued).

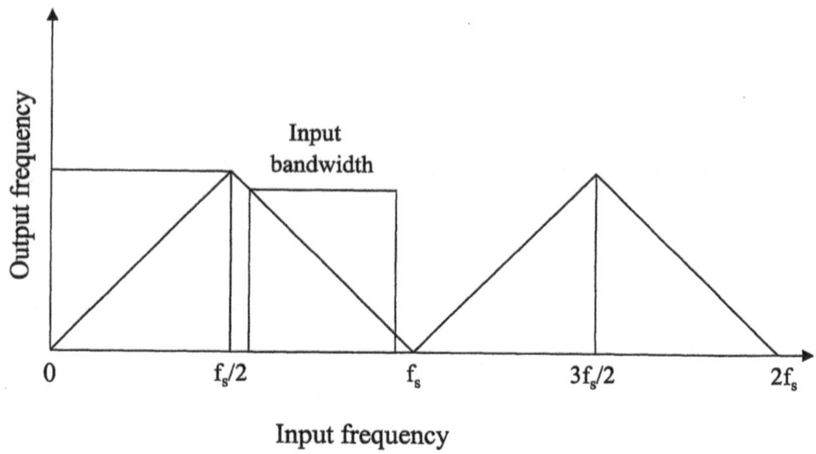

Figure 6.33 Input versus output frequency of band aliasing.

In order to achieve the performance goals listed in Table 2.2, it appears that an ADC with 10 bits and sampling frequency above 2.5 GHz is needed.

REFERENCES

[1] Jaeger, R. C. "Tutorial: Analog Data Acquisition Technology Part I Digital-to-Analog Conversion," *IEEE MICRO*, May 1982, pp. 20–37, "Part II Analog-to-Digital Conversion," IEEE MICRO, Aug. 1982, pp. 46–56, "Part III Sample-and-Hold, Instrumentation Amplifiers, and Analog Multiplexers," *IEEE MICRO*, Nov. 1982, pp. 20–35, "Part IV System Design, Analysis and Performance," *IEEE MICRO*, Feb. 1983, pp. 52–61.

[2] Stafford, K. R., Cray, P. R., and Blanchard, R. A. "A Complete Monolithic Sample/Hold Amplifier," *IEEE J. Solid-State Circuits*, Vol. SC-9, Dec. 1974, pp. 381–387.

[3] Tewksbury, S. K., Meyer, F. C., Rollenhagen, D. C., Schoenwetter, H. K., and Souders, T. M. "Terminology Related to the Performance S/H, A/D, and D/A Circuits," *IEEE Trans. on Circuits and Systems*, Vol. CAS-25, July 1978, pp. 419–426.

[4] Gordon, B. M. "Linear Electronic Analog/Digital Conversion Architectures, Their Origins, Parameters, Limitations, and Applications," *IEEE Trans. Circuits and Systems*, Vol. CAS-25, July 1978, pp. 6–33.

[5] Erdi, G., and Henneuse, P. R. "A Precision FET-less Sample-and-Hold With High Charge-to-Droop Current Ration," *IEEE J. Solid-State Circuits*, Vol. SC-13, Dec. 1978, pp. 864–873.

[6] Bucklen, W. K. "Digitize With Monolithic Flash a-d Converters," *Electronic Design*, Sept. 1980, pp. 93–98.

[7] Becker, R. A., Woodward, C. E., Leonberger, F. J., and Williamson, R. C. "Wide-Band Electrooptic Guided-Wave Analog-to-Digital Converters," *IEEE Proc.*, Vol. 72, July 1984, pp. 802–819.

[8] Porter, A. "Sampling Sees Skinny Signals," *Electronics Week*, Jan. 7, 1985, pp. 41–44.

[9] Mavor, J., and Grant, P. M. "Operating Principles and Recent Developments in Analog and Digital Signal Processing Hardware," *IEEE Proc.*, Vol. 134, July 1987, pp. 305–334.

[10] Rush, K., and Byrne, P. "A 2 GHz 8b Data Acquisition System," *IEEE International Solid-State Circuits Conference*, Vol. 34, Feb. 14, pp. 176–177, 1991.

[11] Schiller, C., and Byrne, P. "A 4-GHz 8-b ADC System," *IEEE Journal on Solid-State Circuits*, Vol. 26, Dec. 1991, pp. 1781–1789.

[12] Fiedler, U., and Seitzer, D. "A High Speed 8 bit A/D Converter Based on a Gray-Code Multiple Folding Circuit," *IEEE Journal on Solid-State Circuits*, Vol. SC-14, June 1979, pp. 547–551.

[13] Van De Grift, R. E., and Van De Plassche, R. J. "A High Speed 7 bit A/D Converter," *IEEE Journal on Solid-State Circuits*, Vol. SC-14, Dec. 1979, pp. 938–943.

[14] Van De Grift, R. E., and Van De Plassche, R. J. "A Monolithic 8-Bit Video A/D Converter," *IEEE Journal on Solid-State Circuits*, Vol. SC-19, June 1984, pp. 374–378.

[15] Shimizu, T., Hotta, M., and Maio, K. "A 10 Bit 20 MHz Two Step Parallel ADC With Internal S/H," *IEEE International Solid-State Circuit Conference*, Feb. 1988, pp. 224–225.

[16] Van De Plassche, R. J., and Baltus, P. "An 8-Bit 100 MHz Full Nyquist Analog-to-Digital Converter," *IEEE Journal on Solid-State Circuits*, Vol. 23, Dec. 1988, pp. 1334–1344.

[17] Goodman, D. J. "The Application of Delta Modulation to Analog-to-Digital PCM Encoding," *Bell Syst. Tech. J.*, Vol. 48, Feb. 1969, pp. 321–343.

[18] Agrawal, B. P., and Shenoi, K. "Design Methodology for ΣΔM," *IEEE Trans. Communications*, Vol. COM-31, March 1983, pp. 360–370.

[19] Candy, J. C. "A Use of Double Integration in Sigma Delta Modulation," *IEEE Trans. Communication*, Vol. COM-33, March 1985, pp. 249–258.

[20] Gary, R. M. "Oversampled Sigma-Delta Modulation," *IEEE Trans. Communications*, Vol. COM-35, May 1987, pp. 481–489.

[21] Boser, B. E., and Wooley, B. A. "The Design of Sigma-Delta Modulation Analog-to-Digital Converters," *IEEE J. Solid-State Circuits*, Vol. SC-23, Dec. 1988, pp. 1298–1308.

[22] Temes, G. C., and Candy, J. C. "A Tutorial Discussion of the Oversampling Method for A/D and D/A Conversion," *IEEE International Symposium on Circuits and Systems 90*, May 1–3, 1990, pp. 910–913.

[23] He, N., Kuhlmann, F., and Buzo, A. "Double-Loop Sigma Delta Modulation With DC Input," *IEEE Trans. Communications*, Vol. COM-38, April 1990, pp. 487–495.

[24] Leung, B. "The Oversampling Technique for Analog to Digital Conversion: a Tutorial Overview," *Analog Integrated Circuits and Signal Processing 1*, Boston, MA: Kluwer Academic Publishers, 1991, pp. 65–74.

[25] Candy, J. C., and Temes, G. C., Editor. *Oversampling Delta-Sigma Converters: Theory, Design and Simulation*, Piscataway, NJ: IEEE Press, 1992.

[26] Poole, M. A., Mitre Corp., Bedford, MA, Private communication.

[27] Naylor, J. R. "Testing Digital/Analog and Analog/Digital Converters," IEEE Trans. *Circuits and Systems*, Vol. CAS-25, July 1978, pp. 526–538.

[28] "Dynamic Performance Testing of A to D Converters," Hewlett Packard, Product note 5180A-2.

[29] Carrier, P. "A Microprocessor Based Method for Testing Transition Noise in Analog to Digital Converters," *Proc. 1983 IEEE International Test Conference*, Oct. 1983.

[30] Doernberg, J., Lee, H. S., and Hodges, D. A. "Full-Speed Testing of A/D Converters," *IEEE Journal of Solid State Circuits*, Vol. SC-19, Dec. 1984, pp. 820–827.

[31] "IEEE Trial-Use Standard for Digitizing Waveform Recorders," IEEE Std 1057, for trial use, July 1989.

[32] "A Guide to Waveform Recorder Testing," Prepared by The Waveform Measurement and Analysis Committee of the IEEE Instrumentation and Measurement Society, April 1990. This guide contains four papers: 1) Linnenbrink, T. E. "Introduction to Waveform Recorder Testing," 2) Green, P. J. "Effective Waveform Recorder Evaluation Procedures," 3) Souders, T. M., and Flach, D. R. "Step and Frequency Response Testing of Waveform Recorders," 4) Grosby, P. S. "Waveform Recorder Sine Wave Testing-Selecting a Generator."

[33] Sklar, B. *Digital Communications: Fundamentals and Applications*, Englewood Cliffs, NJ: Prentice Hall, 1988.

[34] Wong, P. W. "Quantization Noise, Fixed-Point Multiplicative Roundoff Noise, and Dithering," *IEEE Trans. Acoustics, Speech, and Signal Proc.*, Vol. 38, Feb. 1990, pp. 286–300.

[35] Davenport, W. B., and Root, W. L. *An Introduction to the Theory of Random Signals and Noise*, New York, NY: McGraw-Hill Book Co., 1958, Reprinted in 1987.
[36] West, P. D. Georgia Tech Research Institute, Private communication.
[37] Larson, E. L. "High-Speed Analog-to-Digital Conversion With GaAs Technology: Prospects, Trends and Obstacles," *IEEE International Symposium on Circuits and Systems*, 1988, pp. 2871–2878.
[38] Walter, K. "Test Video A/D Converters Under Dynamic Condition," *EDN*, Aug. 1982 pp. 103–112.
[39] Hamming, R. W. *Digital Filters*, 2nd Edition, Englewood Cliffs, NJ: Prentice Hall, 1983.
[40] Harris, F. J. "On the Use of Windows for Harmonic Analysis With the Discrete Fourier Transform," *Proc. of the IEEE*, Vol. 66, Jan. 1978, pp. 51–83.
[41] Kuffel, J., McComb, T. R., and Malewski, R. "Comparative Evaluation of Computer Methods for Calculating the Best Fit Sinusoid to the High Purity Sine Wave," *IEEE Trans. Instrumentation and Measurement*, Vol. IM-36, June 1987, pp. 418–422.
[42] Linnenbrink, T. "Effective Bits: Is That All There Is," *IEEE Trans. Instrumentation and Measurement*, Vol. IM-33, Sept. 1984, pp. 184–187.
[43] Patstone, W., and Dunbar, C. "Choosing a Sample-and-Hold Amplifier Is Not as Simple as it Used To Be," *Electronics*, Aug. 2, 1973, pp. 101–104.
[44] Peetz, B. E., Muto, A. S., and Neil, J. M. "Measuring Waveform Recorder Performance," *Hewlett Packard Journal*, Dec. 1982, pp. 21–29.
[45] Peetz, B. E. "Dynamic Testing of Waveform Recorders," *IEEE Trans. on Instrumentation and Measurement*, Vol. IM-32, March 1983, pp. 12–16.
[46] Rosenbaum, M. J. "Correct Timing of Sample-and-Hold System Improves Leveling Performance," *Microwave Journal*, May 1987, p. 325.
[47] Morgan, D. R. "Finite Limiting Effects for a Band-Limited Gaussian Random Process With Applications to A/D Conversion, *IEEE Trans. on Acoustics, Speech and Signal Processing*, Vol. 36, July 1988, pp. 1011–1016.
[48] Jenq, Y. C. "Measuring Harmonic Distortion and Noise Floor of an A/D Converter Using Spectral Averaging," *IEEE Trans. on Instrumentation and Measurement*, Vol. 37, Dec. 1988, pp. 525–528.
[49] Jenq, Y. C., and Crosby, P. B. "Sinewave Parameter Estimation Algorithm With Application to Waveform Digitizer Effective Bits Measurement," *IEEE Trans. on Instrumentation and Measurement*, Vol. 37, Dec. 1988, pp. 529–532.
[50] White, D. R. "The Noise Bandwidth of Sampled Data Systems," *IEEE Trans. on Instrumentation and Measurement*, Vol. 37, Dec. 1989, pp. 1036–1043.
[51] Shinagawa, M. S., Akazawa, Y., and Wakimoto, T. "Jitter Analysis of High Speed Sampling Systems," *IEEE Journal of Solid-State Circuits*, Vol. 25, Feb. 1990, pp. 220–224.
[52] Sharpin, D. Wright Laboratory, Wright Patterson Air Force Base, OH, Private communication.
[53] Moulin, D., Mitre Corp., Bedford MA, Private communication.
[54] Colleran, W. T., and Abidi, A. A. "A 10-bit, 75-MHz Two-Stage Pipelined Bipolar A/D Converter," *IEEE Journal of Solid-State Circuits*, Vol. 28, Dec. 1993, pp. 1187–1199.
[55] IEEE Std 1241, edited December 6, 1999.
[56] Kien, D., Engineer, Veridian, Dayton, OH, private communication.

APPENDIX 6.A

```
% df6_18a.m simulates an ideal ADC and find spur levels with automatic change
% of input frequencies.
% JT 26 March 92
% ******** USER INPUT ********
clear
```

```
input(' # of bits = ');
bits = ans;
kk = input('Do you want window? y/n: ', 's');
if kk == 'y',
      ws = 'Blackman window';
   else
      ws = 'square window';
end
snr = 20*log10((2^(bits-1))/sqrt(2));
f_samp = 250;
ts = 1/f_samp;
% ****************
n = 516;
points = n;
time = [0:n-1];
amp_no = 1;
% ******** CALCULATION OF CONSTANTS ********
amp1 = sqrt(2*amp_no)*(10^(snr/20));
% ******** START LOOP CHANGE INPUT FREQUENCY ********
To = n*ts;
for k = 1:1000;
kn = randn(1,1);
fi1 = (kn*240+8)/To;
x = amp1*cos(2*pi.*fi1*ts*time);
% ******** QUANTIZATION ********
x_q1 = quantiz(x, bits);
% ******** WINDOW ********
win = blackman(n);
if kk == 'y',
      x_q = x_q1 .* win';
   else
      x_q = x_q1;
end
% ******** FFT ********
x_qf = fft(x_q);
y = abs(x_qf);
y_log = 20*log10(y);
% ******** FIND MAX SPURS ********
[p1 i1 p2 i2 dr(1,k)] = peak(y_log(1:n/2));
end
% ******** END OF LOOP ********
m_min = min(dr)
```

APPENDIX 6.B

```
% QUANTIZ simulates an ideal mid-rise qunatizer
% JT 6 April 92
% x : input data to exercise all bits xmax=2^(bits-1)
% bits : number of bits
% x_q: output
% ****************
function x_q = quantiz(x,bits)
% ****************
q_levels = 2^bits;
q_max = 2^(bits-1);
q_min = -2^(bits-1)+1;
n = length(x);
adj = 0.5 * ones(size(n));
% adj = 0.5 * ones(1:n);
x_adj = x + adj;
x_qt = round(x_adj);
bigger = find(x_qt >= q_max);
if length(bigger) > 0,
    x_qt(bigger) = (q_max)*(ones(size(bigger)));
    % x_qt(bigger) = (q_max)*(ones (1:length(bigger)));
end
smaller = find(x_qt <= q_min);
if length(smaller) > 0,
    x_qt(smaller) = (q_min)*(ones(size(smaller)));
    % x_qt(smaller) = (q_min)*(ones(1:length(smaller)));
end
x_q = x_qt-.5;
```

APPENDIX 6.C

```
% PEAK detects the highest two peaks
% JT Modified 12 May 1992

function [peak1,ind1,peak2,ind2,dr] = peak(r)
% r = input(' input matrix = ');
rif = r(1);
ril = r(length(r));
th = 1.5; % threshold
[max_r ind_r] = max(r);
if ind_r==1,
```

```
      m1 = [0 r ril-1];
      m2 = [0 0 r];
      m3 = [r ril-1 ril-1];
elseif ind_r==length(r),
      m1 = [rif-1 r 0];
      m2 = [rif-1 rif-1 r];
      m3 = [r 0 0];
else
      m1 = [rif-1 r ril-1];
      m2 = [rif-1 rif-1 r];  % shift right
      m3 = [r ril-1 ril-1];  % shift left
end
m4 = m1-m2 > 0; % compare amp
m5 = m1-m3 > 0; %   " "
m4_th = m1-m2 > th; % threshold right shift
m5_th = m1-m3 > th; %    "      left   "
m6_th = m4_th + m5_th > 0; % Combine threshold ones
m6_zo = m4.*m5;    % combine m4 and m5
m6 = m6_th .* m6_zo;
ind = find(m6);  % find the peaks
m7 = m1(ind);   % form a new matrix of peaks only
[peak1 indm7_1]= max(m7);
ind1 = ind(indm7_1)-1;
m7(indm7_1) = -200;
if length(m7) == 1,
      peak2 = 0;
            ind2 = 0;
      dr = 100;
   else
      [peak2 indm7_2] = max(m7);
      ind2 = ind(indm7_2)-1;
      dr = peak1-peak2;
end
[peak1 ind1 peak2 ind2 dr];
```

CHAPTER 7

Amplifier and Analog-to-Digital Converter Interface

7.1 INTRODUCTION

In order to build a digital receiver, an amplifier chain containing several amplifiers with different gains, noise figures, and third-order intermodulation points is needed in front of the analog-to-digital converter (ADC). To calculate the performance, an amplifier chain can be treated as a single amplifier with a certain gain, noise figure, and third-order intermodulation point. One of the purposes of using amplifiers in front of the ADC is to match the input signal to an ADC. In general, the adding of amplifiers will improve the sensitivity of the receiver.

The main purpose of this chapter is to present an optimum way to match the radio frequency (RF) amplifier with the ADC. The word "optimum" means to obtain a certain sensitivity and dynamic range, desired by the designer, within the limits of the amplifier performance and the ADC. The important parameters for the ADC are the number of bits, maximum sampling frequency, and input power level. It is assumed that the performance of the ADC is ideal. The lower limit of the dynamic range is the noise level rather than the spur levels because the spur levels are difficult to predict, as discussed in the Chapter 6. This same approach can be used to design with nonideal ADCs. For nonideal ADCs, the lower limit of the dynamic range should be considered as limited by the spur's response rather than the noise level.

This chapter will first very briefly present the performance of an analog receiver and point out the difference between it and a digital receiver. Then, the performance of linear amplifiers, their gains, noise figures, and third-order intercept points will be presented. A detailed discussion of analog receivers and amplifiers can be found in [1]. The performance of an ADC related to the amplifier will be discussed next, then the interface between the amplifier and the ADC will be discussed. Finally, a simple program will be provided to produce different combinations of sensitivity and dynamic range of a receiver. The designer can pick the desired performance.

7.2 KEY COMPONENT SELECTION [2–16]

Analog microwave receivers have been built for many years. They include many different types of receivers, (e.g., communication and electronic warfare (EW) receivers). In most of these receivers, microwave components (i.e., amplifiers, attenuators, mixers, local oscillators) are used. At the end of these components a crystal video detector is used to convert the RF into a video signal, which is further processed.

After many years of development, there are many different components available. For example, there is a large selection of microwave amplifiers with different operating frequency ranges, noise figures, gains, and so forth that one can choose from. Even so, it is probably very difficult to choose one amplifier that has the desired performance. However, in receiver design, many different RF amplifiers can be connected in series with proper attenuators added between them to obtain a characteristic close to the desired performance.

On the other hand, the technology of manufacturing ADCs with a high sampling frequency and a high number of bits (i.e., hundreds of megahertz sampling frequency and over 8 bits) is at the stage of infancy. In reality, there are very few choices of ADCs with operating frequencies above 500 MHz that also have 8 bits. Usually, because of the poor choice of available high-speed ADCs, in designing a digital receiver the ADC is the first component to be selected. The RF amplifier chain is then designed to match the selected ADC. The performance of the RF amplifier chain (i.e., the noise figure), the gain, and the third-order intercept point are chosen to optimize the receiver performance. Once the performance is chosen, one can select different microwave components (i.e., amplifiers and attenuators) and connect them in the desired manner to match the calculated performance.

In a receiver, especially a wideband system, one would like both high sensitivity and dynamic range. Unfortunately, the higher the gain, the lower the third-order intercept is in a receiver. As a result, high sensitivity means low dynamic range. Because of this property, one may prefer slightly higher sensitivity in some cases and higher dynamic range in other cases. The design procedure is to provide a list of receiver performances with different sensitivities and dynamic ranges versus different characteristics of amplifier chains. In this way a designer can pick up the desired combination.

7.3 NOTATIONS

It should be noted that the equations used in these chapters are mixed with conventional and logarithmic forms. Equations in logarithmic forms are marked with dB at the end of the equation.

In this discussion, many notations will be used to represent different quantities. Some of the quantities are parameters describing amplifiers, others are related to ADC performance. In addition, sometimes noise levels at various points in the amplifier chain are represented by specific notations. Figure 7.1 shows an amplifier followed by an ADC. The amplifier actually contains a chain of amplifiers. Some of the quantities related to the components are listed below them in the figure.

Many notations will be used in this discussion. It might be difficult to get familiar with them. In order to reduce this difficulty, all the notations are listed below for quick reference in alphabetical order.

b	Total number of bits of the ADC.
B_R	Input RF bandwidth of amplifier.
B_v	Video bandwidth after detector or, in this case, the equivalent video bandwidth after FFT operation.
DR	Overall receiver dynamic range.
F	Noise figure of amplifier chain.
f_s	Sampling frequency.
F_s	Overall receiver noise figure, including ADC.
G	Power gain of amplifier chain.
N	Total number of data points in FFT.
N_1	Noise power at the input of the amplifier per unit bandwidth = kT (−174 dBM), where k is the Boltzman constant (1.38×10^{-16} erg/K) and T (290K) is room temperature.
N_b	Quantization noise power of the ADC.
N_i	Noise power at the input of the amplifier in the bandwidth B_R.
N_o	Noise power at the output of the amplifier in the bandwidth B_R.
N_s	Noise power at the output of the ADC in the bandwidth B_R.
N_v	Noise voltage at the output of the amplifier in the bandwidth B_R.

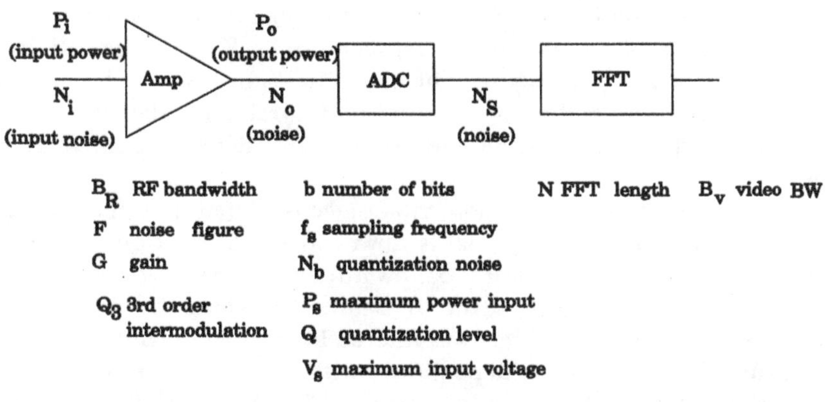

Figure 7.1 Amplifier and ADC connection.

p_3	Third-order intermodulation spur generated at the output of the amplifier.
P_i	Input signal power to the amplifier.
P_I	Input signal power level when p_3 is at a desired noise level.
P_o	Output power level of amplifier.
P_s	Power level at the input of the ADC to generate a full-scale output.
P_{sn}	Power level at the input of the ADC to generate a full-scale output under noisy conditions.
Q	One quantization level in volts.
Q_3	Third-order intercept point of the amplifier chain.
V_n	Voltage reduction caused by noise.
V_s	Maximum input voltage to ADC without causing saturation.

There is a total of 25 quantities that are represented by different notations. However, many of the notations are commonly used by engineers in the receiver area.

7.4 COMPARISON SENSITIVITY OF ANALOG AND DIGITAL RECEIVERS [1]

In an analog receiver, after the RF chain there is a crystal detector to convert the microwave signal into a video signal. For an EW receiver, this video signal is often digitized and further processed through digital signal processing to generate the pulse descriptor word (PDW). Often, the RF chain is designed to have enough gain to amplify the input noise floor to the tangential sensitivity of the detector[1]. Tangential sensitivity is defined for a pulsed signal on a scope display. It is the signal level that the minimum of the noise trace in the pulse region is roughly tangential to the top of the noise trace between pulses. With enough RF gain, the detector itself does not play any role in the noise figure, sensitivity, or dynamic range of the receiver. Only the video bandwidth following the detector and the RF chain (including the RF bandwidth and noise figure) determine the sensitivity of the receiver.

In a digital receiver, there is no crystal detector. The input signal is digitized and processed. Sometimes, the signal is downconverted before digitization. The digital processing determines the RF bandwidth and the video bandwidth. If FFT is used to process the input signal, the RF bandwidth equals the video bandwidth because the signal can be considered as filtered by the FFT operation. As a result, the sensitivity of the receiver is dependent on the length of the FFT operation. Therefore, the ADC should be considered as part of the RF chain. This phenomenon will be further demonstrated in the following sections.

7.5 NOISE FIGURE AND THIRD-ORDER INTERCEPT POINT [17–28]

In this section, the definitions of a noise figure and a third-order intercept point will be presented. In the next section, the noise figure and third-order intercept point of an amplifier chain will be presented.

The gain of an amplifier is defined as the ratio of output power to input power, which can be written as

$$G = \frac{P_o}{P_i} \tag{7.1}$$

This equation can be expressed in decibel form by taking the logarithm on both sides of the equation and multiplying the results by 10. The result is

$$\begin{gathered} 10 \log G = 10 \log\left(\frac{P_o}{P_i}\right) \\ \text{or} \\ G = P_o - P_i \quad \text{dB} \end{gathered} \tag{7.2}$$

Whenever equations are expressed in decibels, this operation applies.

The noise figure of a receiver is defined as

$$\begin{aligned} F &= \frac{\text{actual receiver output noise}}{\text{ideal receiver output noise}} \\ &= \frac{\text{actual receiver output noise}}{\text{receiver input noise} \times G} \end{aligned} \tag{7.3}$$

where G is the gain of the amplifier. This definition will be used to find the overall receiver performance including the ADC.

The third-order intermodulation is a quantity related to the dynamic range of a device (e.g., amplifier and mixer) as well as the receiver. Assume that two signals are of the same amplitude with frequencies of f_1 and f_2. If the two signals are increased to the saturation level of the amplifier, two additional frequency components, $2f_1 - f_2$ and $2f_2 - f_1$, will appear, and this is referred as the third-order intermodulation. Figure 7.2 shows this concept.

The common approach to calculate the third-order intermodulation products is to use the third-order intercept point. The third-order intercept point can be obtained from the input versus output plot and the third-order intermodulation. This result is shown in Figure 7.3. The output versus input of the fundamental frequency is a straight line of unit slope. The third-order intermodulation versus input has a 3:1 slope. The third-order intercept point is the intersection of these two lines.

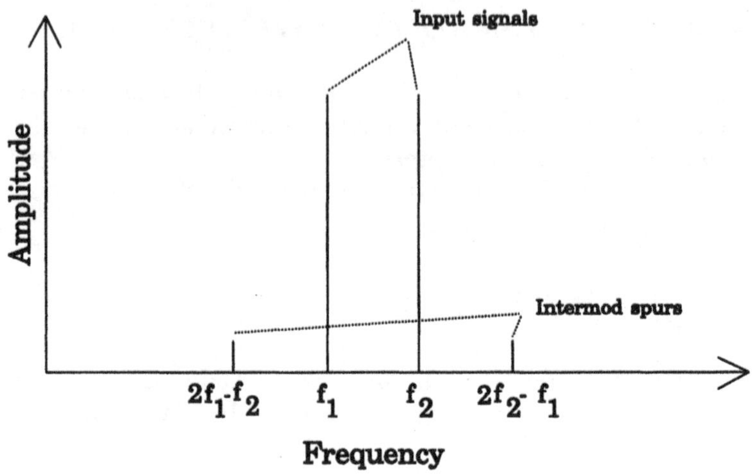

Figure 7.2 Third-order intermodulation products.

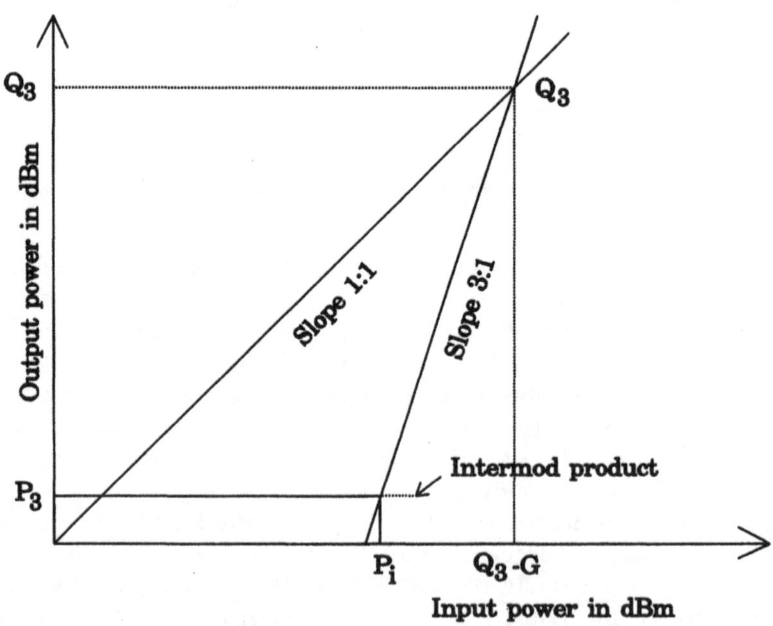

Figure 7.3 Third-order intercept point.

However, it is difficult to obtain the third-order intercept point accurately. Theoretically, when the third-order intermodulation products are at a very low level, the input signals increase by 1 dB and the intermodulation products increase by 3 dB. However, when data are taken experimentally, the 3:1 ratio is seldom realizable. Because of this difficulty, the third-order intercept point sometimes is obtained from one data point. In this approach the third-order intermodulation product is measured at a level near the noise floor. A straight line is drawn through this point with a slope of 3:1 to reach the third-order intercept point.

The third-order intermodulation product P_3, as shown in Figure 7.3, is related to the input signal P_i through the linear relation as

$$\frac{y - Q_3}{x - (Q_3 - G)} = 3 \tag{7.4}$$

Substituting $x = P_i$ and $y = P_3$ into the above equation, the result is

$$P_3 = 3P_i - 2Q_3 + 3G \quad \text{dB} \tag{7.5}$$

However, the input power P_i and the output power P_0 are related by the G of the amplifier, which can be written as

$$P_o = P_i + G \quad \text{dB} \tag{7.6}$$

Using this relation in (7.5), the amplitude of the third-order intermodulation product is

$$P_3 = 3\left(P_o - \frac{2}{3}Q_3\right) \quad \text{dB} \tag{7.7}$$

This relation is often used to determine the two-signal spur-free dynamic range. For example, in many receiver designs the maximum input level is defined as the level when the intermodulation products generated equal the noise level. Under this condition, the input power level is designated as P_I. The corresponding dynamic range is often called the two-signal spur-free dynamic range. This relationship will be used in digital receiver designs to select the performance of the amplifier chain to match a given ADC, and it will be presented in Section 7.9.

7.6 CHARACTERISTICS OF THE AMPLIFIERS IN CASCADE [1, 17–28]

In this section, the parameters of an amplifier chain will be discussed. The amplifier chain is defined as several amplifiers that are connected in cascade. There are three important parameters in an amplifier chain: the noise figure, the gain, and the third-order intercept point. These parameters will affect the performance of the receiver and will be discussed briefly in this section. In the amplifier chain design, if one of the parameters is changed, the other two usually change also. There are many different ways to design an amplifier chain. The general rule will be presented at the end of this section. After this, some examples will be given.

In general, it is desirable to design an amplifier chain to have the lowest possible noise figure and highest possible third-order intercept point. The gain of the amplifier chain is determined by the system designed. This gain value is dependent on the characteristics of the ADC (or the crystal detector in an analog receiver) used at the end of the amplifier chain.

The derivation of the gain, the noise figure, and the third-order intercept point are given in [1]; only the results will be presented here.

If several amplifiers are cascaded in an amplifier chain, the overall gain G can be written as

$$G = G_1 G_2 \ldots G_n \quad \text{or} \quad G = G_1 + G_2 + \ldots + G_n \ \text{dB} \tag{7.8}$$

where $G_1, G_2, \ldots$ are the gains of individual components of the RF chain.

The overall noise figure F of the amplifiers connected in cascade can be written as

$$F = F_1 + \frac{F_2 - 1}{G_1} + \frac{F_3 - 1}{G_1 G_2} + \ldots + \frac{F_n - 1}{G_1 G_2 \ldots G_{n-1}} \tag{7.9}$$

where $F_1, F_2, \ldots$ are the noise figures of individual components of the RF chain. From this equation, one can see that when G_1 is very large, the overall noise figure is determined approximately by F_1. In other words, all the microwave components (i.e., filters and mixers) with insertion losses used before the first amplifier will have an adverse effect on the noise figure. All the components used after a high-gain amplifier have a minor effect on the overall noise figure.

The overall third-order intercept point can be calculated as

$$Q_3 = \frac{G}{\dfrac{G_1}{Q_{3,1}} + \dfrac{G_1 G_2}{Q_{3,2}} + \ldots + \dfrac{G_1 G_2 \ldots G_n}{Q_{3,n}}} \tag{7.10}$$

where $Q_{3,1}, Q_{3,2} \ldots$ are the third-order intercept points of each individual component. The third-order intercept points of the amplifier and the mixer are often provided by the manufacturer. The value is in reference to the output of the device. From this equation, the effect of each component on the overall third-order intercept point is difficult to see. Some examples will be used to demonstrate the effect.

In the above equations, if the component is an amplifier, then G_i, F_i, and $Q_{3,i}$ are all given and the above equations can be used directly. If the component is passive (i.e., an attenuator or a filter), the three quantities are not given, but the insertion loss is. Under this condition, the gain and noise figure can be obtained from the insertion loss; that is, the gain equals the negative value of the insertion loss and the noise figure is equal to the insertion loss. Since a passive component usually does not have a nonlinear region, the third-order intercept point is very high. Thus, a very large number (i.e., 100 dBm) can be assigned to a passive component. In general, such a high value will have little if any effect on the overall third-order intercept point.

A computer program (df7eq1.m) is provided in Appendix 7.A. This program will calculate the overall gain, noise figure, and third-order intercept point. The calculations are based on (7.8–10).

Let us use an example to conclude this section. There are two amplifiers and a 3-dB attenuator, with their characteristics listed in Table 7.1. These three components are connected in different ways as shown in Figure 7.4. Figures 7.4(a, b) show two amplifiers connected in cascade without the attenuator. In these two arrangements, the order of the two amplifiers is reversed. In Figure 7.4(c), the attenuator is placed between the two amplifiers. In Figure 7.4(d), the attenuator is placed at the output of the second amplifier. In Figure 7.4(e), the attenuator is placed at the input of the two amplifiers.

Usually all the given characteristics of amplifiers are given in decibels or decibels referred to 1 mW, but in (7.8–10), all the values represent actual power. Before the equations are used, the given values in decibels and decibels referred to 1 mW must be converted to ratio or power through the following relation:

$$G_{db} = 10\log(G) \quad \text{or} \quad G = 10^{G_{db}/10} \tag{7.11}$$

Table 7.1
Amplifier and Attenuator Characteristics

	Amplifier 1	*Amplifier 2*	*Attenuator*
Gain (dB)	15	15	–3
Noise figure (dB)	3	5	3
Third-order intercept point (dBM)	15	20	100

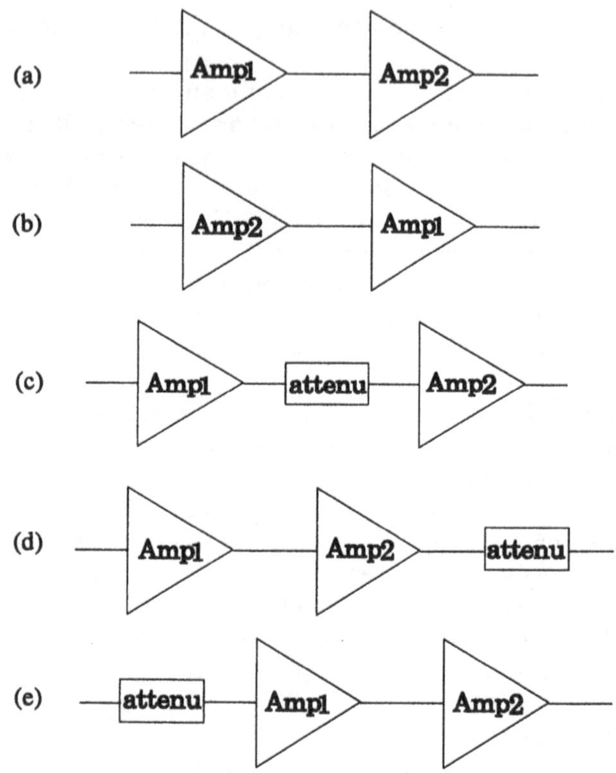

Figure 7.4 Different ways to cascade amplifiers and an attenuator: (a) ampl-amp2, (b) amp2-ampl, (c) ampl-att-amp2, (d) ampl-amp2-att, (e) att-ampl-amp2.

where G_{db} represents the gain in logarithmic scale. This equation can be applied to quantities other than gain, such as the noise figure. The calculated results of the three arrangements are in power ratio or watts. The results are converted back into decibels or decibels referred to 1 mW. All these conversion operations are included in the computer program in Appendix 7.A (df7eq1.m).

The calculated results are listed in Table 7.2. This example is helpful in designing a receiver with given amplifiers, and more discussion will be included in Section 7.12. In order to show the effect of different connections, the results are kept in three decimal points, although this kind of accuracy does not have much meaning in a receiver design because the specifications of components do not have such a high accuracy. The following important factors should be noticed.

1. Comparing the first two cases, one can see that both cases have the same gain. However, the arrangement in Figure 7.4(a) has a lower noise figure and a higher third-order intercept point, which are the desirable results. In gen-

Table 7.2
Results of Different Amplifier Attenuator Connections

	Total Gain (dB)	*Overall Noise Figure (dB)*	*Third-Order Intercept Point (dBM)*
Fig. 7.4(a)	30	3.146	19.586
Fig. 7.4(b)	30	5.043	14.957
Fig. 7.4(c)	27	3.351	19.219
Fig. 7.4(d)	27	3.148	16.586
Fig. 7.4(e)	27	6.146	19.586

eral, RF amplifiers can be briefly divided into two groups: one with low noise and the other with high power. A low-noise amplifier usually has a low third-order intercept point and a high-power one usually has a high noise figure. In a cascade system, the low-noise amplifier should be placed at the beginning of the chain and the high-power one at the end of the chain. From this comparison, one can see that the noise is dominated by the first amplifier and the third-order intercept point is dominated by the last amplifier.

2. When two amplifiers are connected in cascade as shown in Figure 7.4(a), the noise figure is higher than the first amplifier and the intercept point is lower than the second amplifier.
3. When an attenuator is inserted in the amplifier chain, no matter where it is placed, the overall gain will be decreased by the insertion loss of the attenuator.
4. When the attenuator is placed between the two amplifiers as shown in Figure 7.4(c), both the noise figure and the third-order intercept point degrade slightly in comparison with Figure 7.4(a).
5. When the attenuator is placed at the end of the amplifier chain as shown in Figure 7.4(d), the noise figure degrades very little, but the third-order intercept point degrades 3 dB in comparison with Figure 7.4(a).
6. If the attenuator is placed in front of the amplifier chain as shown in Figure 7.4(e), the noise figure suffers 3 dB, but the intercept point does not change.

From this simple example, one can generally say that moving the attenuator to the front of the amplifier chain will degrade the noise figure more, and if the attenuator is moved toward the end, the third-order intercept point will suffer. In general, the attenuator is seldom placed at the input of the amplifier chain because the sensitivity will be reduced by the insertion loss.

This simple example also reveals an important tradeoff factor in receiver design. Assume that the amplifiers and attenuators are connected in the proper order. Under this condition, when the noise figure is low, the third-order intercept is also low. That is why a receiver with high sensitivity usually has low dynamic range. When the dynamic range is high, the sensitivity is low. In actual receiver design, there are usually more than two amplifiers and the attenuator is divided into several separate ones and placed at different locations to obtain the desired results.

One can see that if Figure 7.4(a) is used, it can provide the highest gain, the lowest noise figure, and the highest third-order intercept point. However, this arrangement may not be used because the gain of the amplifier chain must be a specific value. Too high a gain value can produce an adverse effect on the receiver performance. This subject will be discussed in Section 7.9.

7.7 ANALOG-TO-DIGITAL CONVERTER [29]

As mentioned in Section 7.2, an ADC is selected first in a digital receiver. The performance of the ADC was discussed in Chapter 6. In this section the results will be presented again, but these results will be used to determine the specification of the amplifier chain in front of it. The two important parameters of an ADC are the quantization noise and the maximum input power without saturating the device.

The quantization power (N_b) is obtained in (6.13) where the impedance of the system is considered unity. The result is rewritten here as

$$N_b = \frac{Q^2}{12R} \tag{7.12}$$

where Q is the size of the quantization level and R is the input impedance of the ADC. In this equation, the impedance is assumed to be R rather than unity.

The maximum voltage V_s of a sinusoidal wave that can be applied to the input of the ADC without causing saturation is

$$V_s = 2^{(b-1)}Q \tag{7.13}$$

where b is the number of bits. The maximum power P_s is related to the maximum voltage V_s by

$$P_s = \frac{V_s^2}{2R} = \frac{2^{2(b-1)}}{2R}Q^2 \tag{7.14}$$

The signal-to-noise ratio (S/N) can be obtained from (7.12) and (7.14) as

$$\frac{S}{N} = \frac{P_s}{N_b} = \frac{3}{2}2^{2b}$$

or (7.15)

$$\frac{S}{N} = P_s - N_b = 6b + 1.76 \qquad \text{dB}$$

In this case, the only noise considered is the quantization noise.

This ADC is used to collect digitized data. If the sampling frequency is f_s and an N-point FFT is performed on the digitized data, the processing bandwidth B_v can be found as follows. If the input is complex data, the maximum input bandwidth is f_s and there are N output channels. If the input is real data, the maximum input bandwidth is $f_s/2$ and there will be $N/2$ independent output channels. Therefore, in either case the processing bandwidth is

$$B_v = \frac{f_s}{N} \tag{7.16}$$

In a digital receiver, this processing bandwidth is also the RF resolution bandwidth.

7.8 NOISE FIGURE OF AMPLIFIER AND ADC COMBINATION [30]

In this section, the overall noise figure of the amplifier and ADC combination is determined. To find the noise figure of the amplifier and the ADC combination, the ADC can be considered as an additional noise source. As shown in Figure 7.1, the noise at the input of the amplifier is N_i. This noise can be found as

$$N_i = N_1 + B_R \qquad \text{dBm} \tag{7.17}$$

where N_1 (= −174 dBM) is the noise level at room temperature with unity bandwidth and B_R is the RF bandwidth. The output noise of the amplifier is N_o, which can be written as

$$N_o = N_i + F + G \qquad \text{dBm} \tag{7.18}$$

where F and G are the noise figure and the gain of the amplifier, respectively.

The noise N_s at the output of the ADC is the sum of amplifier output noise N_o and quantization noise N_b, assuming the noise is band limited and no noise is folded into the baseband through the ADC. From the definition of noise figure given in (7.3), the overall noise figure F_s can be written as

$$F_s = \frac{N_s}{GB_RN_1} = \frac{N_s}{GN_i} = \frac{N_o + N_b}{GN_i} = F + \frac{N_b}{GN_i}$$

$$\text{or} \tag{7.19}$$

$$F_s = N_s - G - B_R - N_1 \quad \text{dB}$$

where F is the noise figure of the amplifier. This equation is obtained because N_s is the actual noise output of the amplifier-ADC combination and GN_i is the noise output of an ideal system. The noise generated by an ideal system is the input noise multiplied by the gain of the amplifier, and there is no noise contributed by the system. The overall noise figure degrades by the quantity N_b/GN_i.

In order to simplify the operation, the amplifier output noise is measured in terms of quantization noise. Let us define two quantities M and M' as

$$M \equiv \frac{N_o}{N_b} \quad \text{and} \quad M' \equiv M + 1 \tag{7.20}$$

Thus, the noise figure in (7.19) can be written as

$$F_s = \frac{N_o\left(1 + \frac{1}{M}\right)}{GN_i} = \frac{N_o(1 + M)}{GMN_i} = \frac{FM'}{M}$$

$$\text{or} \tag{7.21}$$

$$F_s = F + M' - M \quad \text{dB}$$

It should be noted that in decibel scale, $M' = 10 \log(M + 1)$.

7.9 AMPLIFIER AND ADC INTERFACE [30, 31]

So far, the amplifier and the ADC have been discussed separately. Only the noise figure of the combined system has been calculated. In this section, the output of the amplifier will be made to match the input of the ADC. The meaning of match is twofold. First, at a certain input level the third-order intermodulation equals the noise level. Second, the amplifier amplifies this input signal to the maximum allowable signal level of the ADC. From these relations, the required gain and third-order intercept point of the amplifier can be obtained.

First, let us choose an amplitude of the third-order intermodulation output p_3 from the amplifier to match the noise level. The noise level is defined in the processing band as N_sB_v/B_R. Expressed in decibels and applying the relations in (7.19) and (7.21), the result can be written as

$$P_3 = N_s - B_R + B_v = N_1 + G + F + B_v + M' - M \quad \text{dBM} \tag{7.22}$$

Under this condition, the input power P_i is designed as P_I, which is a special input level. Now, let us find the third-order intercept point Q_3 in terms of this input level. From (7.5), the input P_I can be written as

$$P_I = \frac{P_3 + 2Q_3 - 3G}{3} \tag{7.23}$$

Substituting P_3 from (7.22), this level can be written as

$$P_I = \frac{2Q_3 + N_1 - 2G + F + B_v + M' - M}{3} \quad \text{dBm}$$

or

$$Q_3 = \frac{3P_I - N_1 + 2G - F - B_v - M' + M}{2} \quad \text{dBm} \tag{7.24}$$

This Q_3 is the required overall third-order intercept point of the amplifier with input P_I to produce the third-order intermodulation level to match the noise level in bandwidth B_v.

This input power P_I after the amplifier should equal the maximum power level allowed to the input of the ADC without reaching saturation. The voltage is V_s, as shown in (7.13). However, there is noise at the output of the amplifier. This noise added to the signal can cause the ADC to reach saturation. Thus, in considering the maximum allowable voltage, this noise power should be taken into account. The noise power at the output of the amplifier N_o will reduce the maximum input signal allowable to the ADC. To allow this adjustment, the maximum input power to the ADC is arbitrarily reduced by three standard deviations of N_o. The corresponding voltage is

$$V_n = \sqrt{3N_oR} \tag{7.25}$$

The maximum allowable ADC power considering the noise reduction is

$$P_{sn} = \frac{(V_s - V_n)^2}{2R} = \frac{[2^{(b-1)}Q - V_n]^2}{2R} \tag{7.26}$$

This result represents the maximum power of a single sinusoidal input.

In an evaluation of the third-order intermodulation products, two signals are required. The amplitude of each of the signals must be one-half of the amplitude of the voltage $(V_s - V_n)$. Expressed in terms of power, there is a factor 4, which corresponds to 6 in the logarithmic expression. Therefore, the input power P_I is related to the maximum allowable power as

$$(P_I + 6) + G = P_{sn} \tag{7.27}$$

Combining this equation with the conditions in (7.25) and (7.26) provides the desired gain G.

The dynamic range can be found as

$$DR = P_I + G - P_3 \qquad \text{dB} \tag{7.28}$$

where the third-order intermodulation product P_3 equals the noise floor as described in (7.22).

It should be emphasized here that the third-order intercept point is the minimum required value. If the amplifier chain has a Q_3 greater than the calculated value, it does not cause any adverse effect. The gain calculated is the optimum value. If the gain of the amplifier chain is greater than the calculated value, the amplifier will drive the ADC into saturation and cause generation of spurs.

7.10 THE MEANING OF *M* AND *M′*

It should be noted that M and M' are used to represent both the values and their logarithmic forms. Whenever they are used in logarithmic form, the unit dB is included at the end of the equation.

The result of (7.21) can be rewritten here as

$$F_s = \frac{FM'}{M} = F\frac{M+1}{M} \tag{7.29}$$

If $M = 1$, it means that the quantization noise N_b equals the amplifier output noise. Under this condition, the system noise figure equals 2 times the amplifier noise figure, or the noise figure is degraded by 3 dB. The larger the M value, the less the degradation of the noise figure.

If $M < 1$, from the above equation one can see that the system noise figure will be high, which means the quantization noise dominates the noise figure. This is undesirable because the sensitivity of the receiver will suffer. In order to increase the value of M, the gain of the amplifier must be high.

If $M = 9$, then $M' = 10$ and $M'(\text{dB}) - M(\text{dB}) = 10\log(10) - 10\log(9) = 10 - 9.54 = .46$ dB, which implies the system noise figure will be degraded by 0.46 dB. Under this condition, the amplifier output noise power is 9 times the quantization noise. As M increases to a very large value, the difference $M'(\text{dB}) - M(\text{dB})$ will become very small. Under this condition, the noise figure of the receiver approaches the noise figure of the amplifier and the contribution of the quantization noise is negligible.

7.11 COMPUTER PROGRAM AND RESULTS

In previous sections many equations are used to describe the RF chain design, but it might be difficult to use them in practice. In order to show how these equations are used, a design example will be provided. A computer program (df7eq2.m) is listed in Appendix 7.B. In this computer program, the parameters in the following example are used as input. However, the main purpose of this program is to illustrate the design of other amplifier chains if the input data are available.

In this example two parameters are needed from the amplifier, and they are the noise figure F and the RF bandwidth B_R. The noise figure of the first amplifier can be used as the initial value in the program. If an amplifier chain must be used to obtain the proper gain and third-order intercept point, the noise figure of the amplifier chain is usually higher than the noise figure of the first amplifier. If this situation occurs, the new noise figure of the amplifier chain should be used in the program to make the necessary adjustment.

In an analog receiver, the resolution bandwidth (i.e., the bandwidth of the fine frequency filter) is used to find the noise floor and the sensitivity in the receiver. The overall bandwidth plays no important role in determining the sensitivity. In a digital receiver, the bandwidth is determined by the FFT length and the frequency resolution bandwidth equals the video bandwidth B_v. Under this condition, one may deduce that the RF bandwidth should not play any role in determining the receiver sensitivity. However, this is not true because the digitization noise of the ADC is added to the noise in the bandwidth B_R instead of being added to the noise in processing bandwidth B_v.

Let us use an example to illustrate the application of the computer program listed in Appendix B. Assume the amplifier has the following specifications:

B_R = 30 MHz (RF bandwidth).
F = 3.3 dB (amplifier noise figure).

The ADC has the following parameters:

b = 8 bits (number of bits).
V_s = 270 mV (maximum allowable voltage to ADC).
f_s = 250 MHz (sampling frequency).
R = 50Ω (ADC input impedance).

The FFT is specified as follows:

N = 1,024 points.

Maximum unambiguous input bandwidth = $f_s/2$ = 125 MHz. In this example, the bandwidth is limited by the input amplifier.

$B_v = f_s/N$ = 244 kHz or B_v(dB) = 53.87. This result is obtained from (7.12).

The maximum power P_s can be found from (7.14) where the value of R is used. In using the computer program in the appendix, one must pay attention to the unit. The voltage is given in millivolts and the power is in milliwatts.

The input to the program is a range of M (7.20) values. The outputs are gain, third-order intercept point, noise figure, dynamic range, and the input noise to quantization noise voltage ratio (N_v/Q) with each given M value. The dynamic range discussed here is the two-tone spur-free dynamic range as discussed in Section 2.16. The N_v/Q can provide a general idea of how the quantization levels are filled with noise.

The results are shown in Table 7.3. It should be noted that the P_3 value is also the noise floor after FFT processing. The performance of the receiver is determined by F_s (dB) and DR (dB). In this table the value of M is arbitrarily chosen as .25, .5, 1, 2, 4, 8, 16, 32, 64, 128, 256, and 516. As discussed in Section 7.9, a small value of M will cause a high noise figure. For $M = 1$, the noise figure will degrade about 3 dB. In order to show a trend, some small values of M are chosen in Table 7.3.

A designer should pick the desired receiver performance (i.e., the noise figure and dynamic range). The gain in column 2 of the table is the desired value. This means that in order to obtain the desired performance, the gain must be adjusted to the listed value. On the other hand, the third-order intercept point represents the minimum required values. If Q_3 is less than the listed value, the dynamic range of the receiver will be less than the listed value and the third-order intermodulation products will be at the lower limit of the dynamic range. If Q_3 is larger than the listed value, the dynamic range equals the listed value. Under this condition, the noise floor is the lower limit of the dynamic range. When Q_3 equals the listed value,

Table 7.3
Calculated Performance of an RF Chain

M	G(dB)	Q_3(dBm)	P_3(dBm)	F_3(dB)	DR(dB)	N_3/Q
0.25	38.78	24.41	−71.06	10.29	63.64	0.21
0.5	41.79	24.02	−70.27	8.07	62.86	0.29
1	44.80	23.38	−69.02	6.31	61.60	0.42
2	47.81	22.46	−67.26	5.06	59.81	0.59
4	50.82	21.31	−65.04	4.27	57.57	0.83
8	53.83	19.97	−62.49	3.81	54.97	1.18
16	56.84	18.51	−59.72	3.56	52.15	1.66
32	59.85	16.94	−56.84	3.43	49.19	2.35
64	62.86	15.29	−53.90	3.37	46.13	3.33
128	65.87	13.54	−50.92	3.33	42.98	4.71
256	68.88	11.68	−47.93	3.32	39.74	6.66
512	71.89	9.63	−44.93	3.31	36.37	9.42

the intermodulation products equal the noise floors and both become the lower limit of the dynamic range.

7.12 DESIGN EXAMPLE

Let us use actual hardware to design a receiver with the help of Table 7.3. The ADC used in this design is the Tektronix TKAD20C, which has the performance listed in the last section. Thus, the results listed in Table 7.3 can be used directly. There are two available amplifiers. Their characteristics are listed in Table 7.4.

If these amplifiers are connected in cascade, the overall gain is 71 dB. From the results shown in Table 7.3, when the gain is at about 71 dB and the noise figure of the system is 3.3 dB, the minimum required Q_3 is 9.6 dBm, and the dynamic range is 36 dB. This dynamic range appears low.

The noise figure and dynamic range versus M is plotted in Figure 7.5. It is interesting to note that the noise figure does not change much when $M > 16$, but the dynamic range degrades rapidly with increasing M. For this experiment, the dynamic range of about 52 dB is selected (for $m = 16$). Under this condition, the noise figure will degrade about .26 dB (3.56 – 3.3). The gain of the receiver should be about 57 dB. The minimum required Q_3 is about 18.5 dBm. The dynamic range is 52.1 dB.

In order to achieve this gain, 14 dB of attenuation must be inserted in the amplifier chain to reduce the available gain. In order to keep this design procedure simple, let us use only one 14-dB attenuator rather than splitting them into two attenuators. There are two possible ways to connect the two amplifiers. Case 1 is to place the attenuator in between the two amplifiers and Case 2 is to place the attenuator at the end of the second amplifier. The results can be obtained through the program in Appendix 7.A (df7eq1.m) and are listed in Table 7.5.

From these results, it appears that both cases fulfill the amplifier requirement. Interestingly, both cases have practically the same noise figure. Since the amplifier noise figure does not change from the first amplifier, the calculated system noise figure need not be modified. Otherwise, the new noise figure should be used to repeat the calculation again to generate new performance data.

In general, Case 1 should be used because it provides a large margin in the Q_3 requirement, while in Case 2, the Q_3 barely meets the minimum requirement.

Table 7.4
Amplifier Characteristics

Amplifier	*G (dB)*	*NF (dB)*	Q_3 *(dBm)*
1	42	3.3	12
2	29	4.0	33

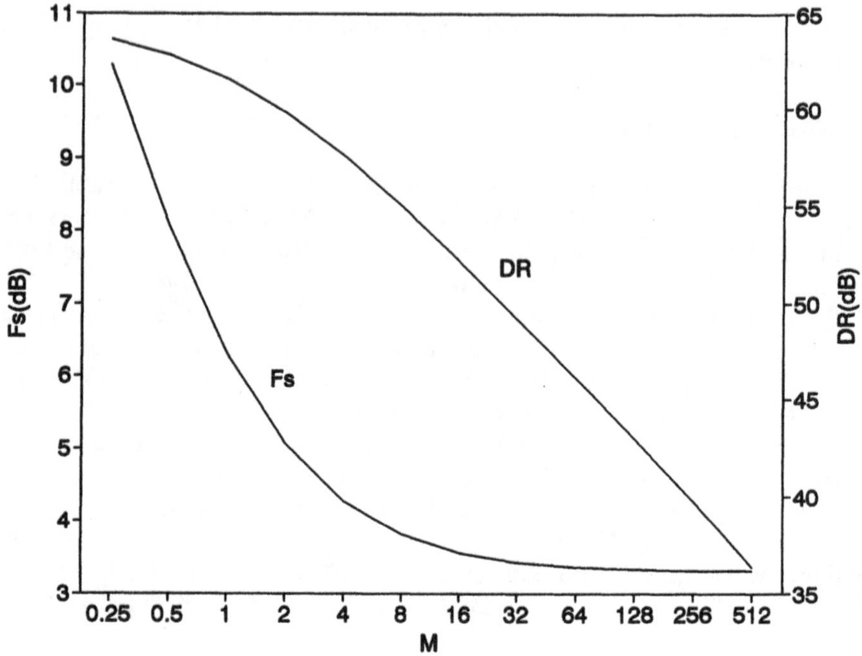

Figure 7.5 Noise figure and dynamic range versus *M.*

Table 7.5
Results of Connecting Amplifiers in Two Different Ways

Connections	*G*	*F*	Q_3
Case 1	57	3.3	26.0
Case 2	57	3.3	18.4

However, in this example Case 2 is used because one can demonstrate that the third-order intermodulation products are very close to the noise floor. This example can demonstrate that with a minimum required Q_3 value, both the noise floor and the third-order intermodulation products can be the lower limit of the dynamic range. If Case 1 is selected, the intermodulation products will be much lower than the noise floor and one will not see the third-order intermodulation products.

7.13 EXPERIMENTAL RESULTS [31]

The experimental setup is shown in Figure 7.6. A 30-MHz (20–50 MHz) bandpass filter is placed in front of the amplifier chain to eliminate possible stray signals in

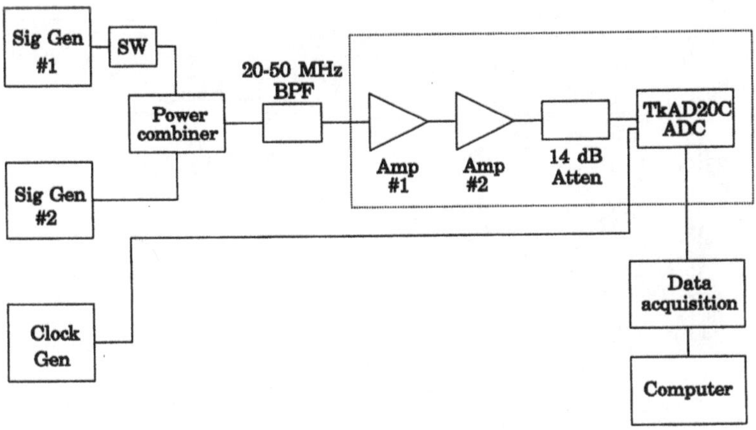

Figure 7.6 Amplifier chain experimental setup.

the environment and limit the noise bandwidth. The noise figure was measured after the filter. The two amplifiers mentioned in the last section were connected in cascade and a variable attenuator was added at the output of the amplifier. The variable attenuator can provide adjustment in the amplifier chain. After the attenuator, the Tektronix TKAD20C ADC was used to collect data. A 1,024-point FFT was used to analyze the data.

7.13.1 Noise Figure Measurement

First, the noise figure of the system was measured. This test was performed using two different techniques. In the first technique, the input of the amplifier chain was terminated with a 50Ω resistor. The noise power was computed in the frequency domain from 20 to 50 MHz by averaging five realizations of a 1,024-point magnitude spectrum.

In the second technique, a full-scale sine wave at 36 MHz was injected into the amplifier and the noise power was again measured in the frequency domain. Once the value of N_s is obtained, the noise figure can be obtained from (7.19). The results are shown in Figure 7.7. This figure contains three sets of data: the theoretically calculated one, the measured one with input terminated, and the measured one with a signal. The noise figure measured with an input signal is very close to the theoretical one.

The noise figure measured with the terminated input matches the theoretical one very well at high gain, but it falls below the theoretical one at low-gain values. This discrepancy can be explained as follows. At low-gain values, the noise from the amplifier fills only a small part of the first quantization level. Since the quantization noise model is based on the premise that the input signal was uniformly

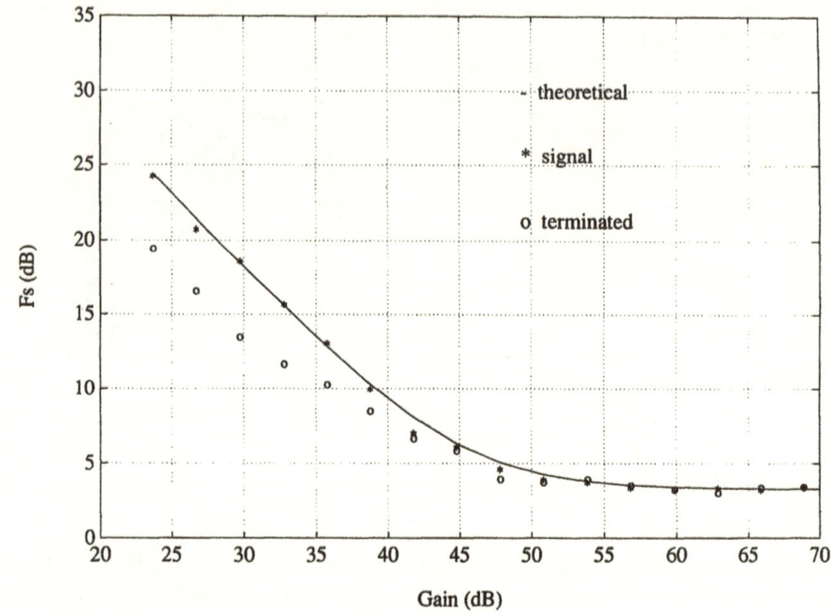

Figure 7.7 Noise figure versus gain.

distributed across a given quantization level, the uniform distribution is no longer valid. When a full-scale signal is applied, the noise model is valid and the result matches the calculated one.

From this experiment, it appears that the noise figure of the receiver is signal-dependent when the gain of the receiver is low. This phenomenon is caused by the nonlinear effect of the ADC. However, in most receiver designs, usually the gain is high enough and the receiver noise figure matches the theoretical value and is signal independent.

7.13.2 Dynamic Range Test

The second test is to find the dynamic range of the receiver. In this test, the goal is to detect the third-order intermodulation products. Two input frequencies at 36 and 41 MHz are kept at the same amplitude. The power of each sine wave was set at –7.3 dBm, which was 6-dB below the full-scale value of the ADC (–1.3 dBm). If the noise effect is considered, the maximum input power should be reduced from –7.3 to –7.5 dBm. Since this difference (0.2 dBm) is within the measurement error, the noise effect is neglected in this experiment. A time domain plot is shown in Figure 7.8. It shows clearly that the signals do not saturate the ADC.

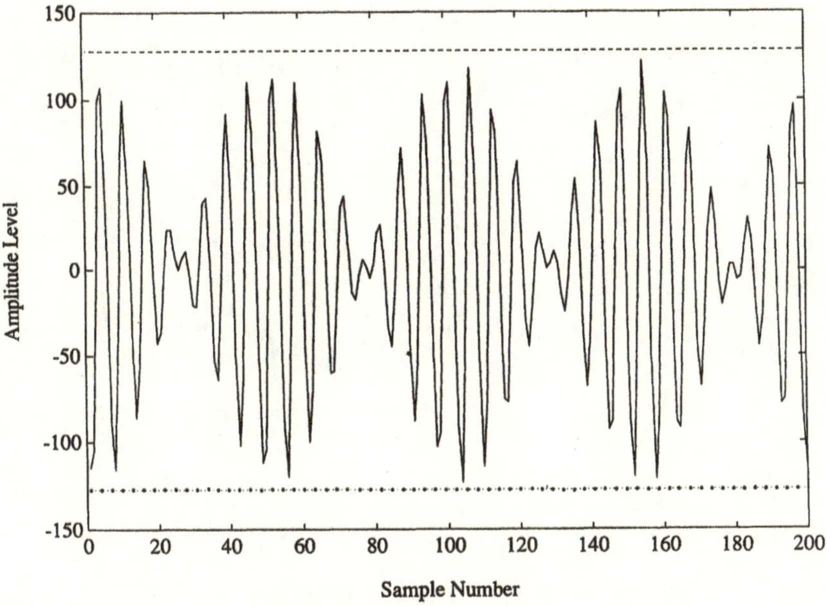

Figure 7.8 Time domain plot of two signals of the same amplitude.

In order to demonstrate the noise floor and the third-order intermodulation products, three attenuation values are selected and they are listed in Table 7.6.

The spectrum plot is shown in Figure 7.9. These results are of five realizations of the magnitude spectrum averaged. In Figure 7.9(a), the noise floor was measured to be –62.2 dBm and the third-order intermodulation products at frequencies are clearly shown. In this case, the third-order intermodulation products are the lower limits of the dynamic range. The third-order intermodulation product at 31 MHz is at –56.4 dBm. Thus, the dynamic range is 49.1 dB (–7.3 + 56.4). This is referred to as the two-tone spur-free dynamic range.

Table 7.6
Gain and Third-Order Intercept Point With Different Attenuation Values

Figure	*Attenuation (dB)*	*Gain (dB)*	Q_3 *(dBm)*
8(a)	17	54	15
8(b)	14	57	18
8(c)	11	60	21

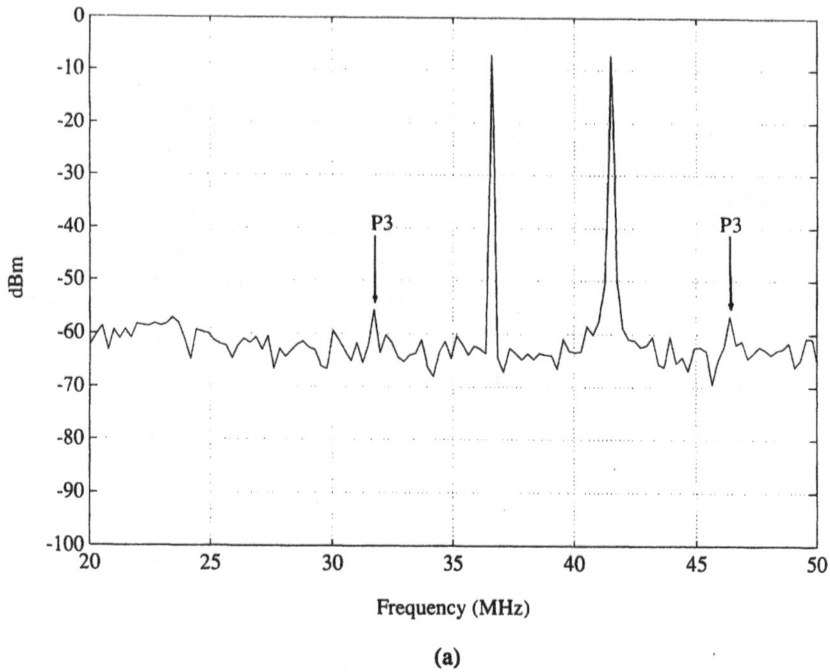

Figure 7.9 Output spectrum: (a) gain = 54 dB, (b) gain = 57 dB, (c) gain = 60 dB.

In Figure 7.9(b), the noise floor is measured at –58.6 dBm. The third-order intermodulation products, which should appear at 31 and 46 MHz, cannot be observed. The noise floor is the lower limit of the dynamic range. The dynamic range in this case is 51.3 dB (–7.3 + 58.6).

In Figure 7.9(c), the noise floor is at –55.2 dBm. The third-order intermodulation products cannot be observed. The noise floor is the lower limit of the dynamic range. The dynamic range is 47.9 dB (–7.3 + 55.2).

These three cases show that Figure 7.8(b) provides the largest dynamic range. The measured value is very close to the designed one. The reason for the two other cases providing lower dynamic range is that the gain is not at optimum value. From these experimental data, one can see that the computer program can predict the performance of the receiver very accurately.

In order to see how close the third-order intermodulation products are to the noise floor (Figure 7.9(b)), the input signals are increased by 0.5 dB. The resulting spectrum is shown in Figure 7.10, where the third-order intermodulation products are clearly visible. Therefore, one can claim that the third-order intermodulation products in Figure 7.9(b) are very close the noise floor, which matches the design goal.

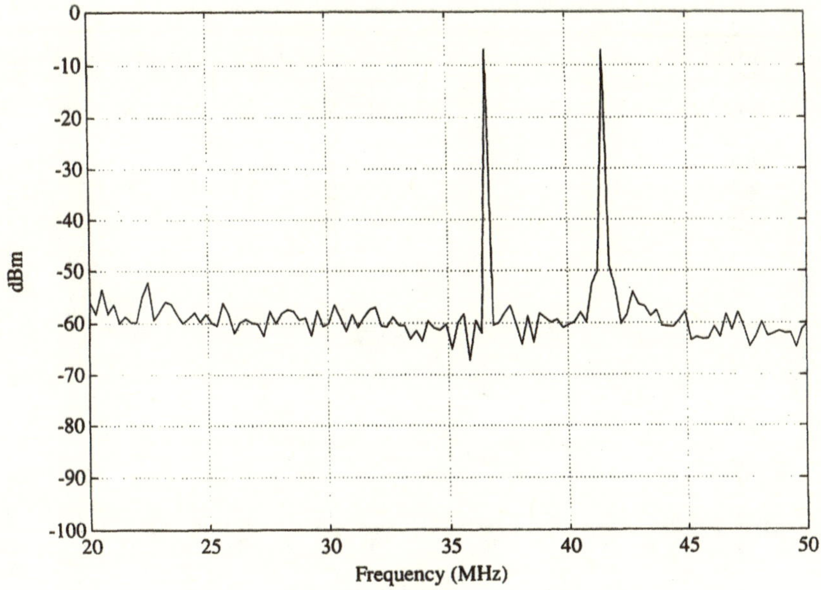

(b)

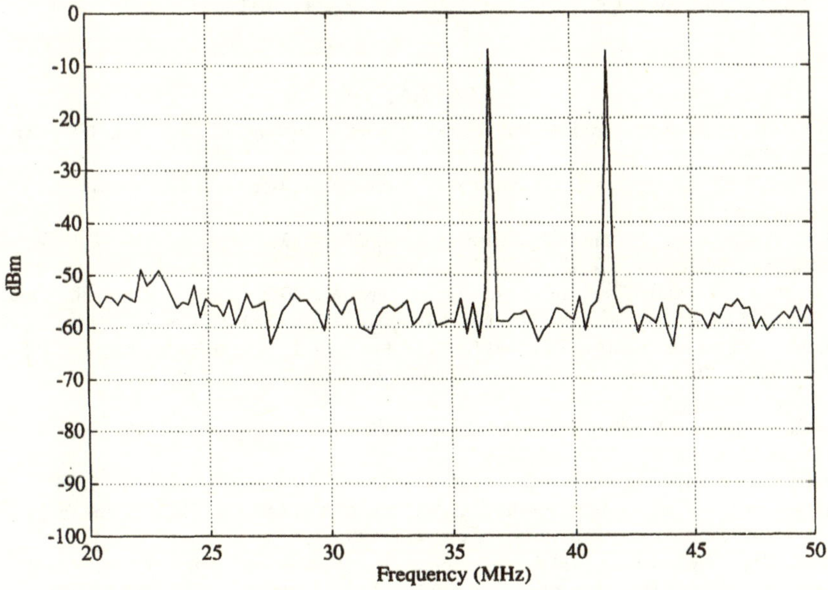

(c)

Figure 7.9 (continued).

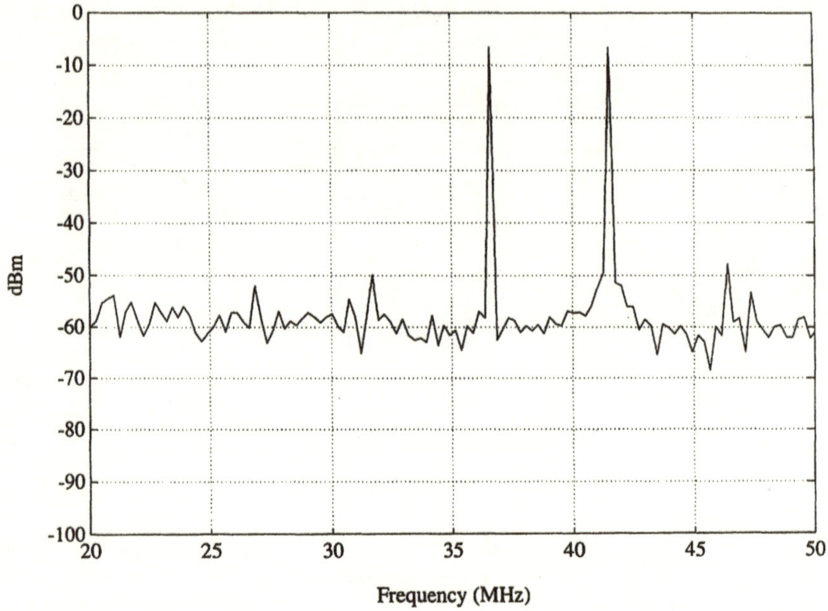

Figure 7.10 Spectrum output with gain = 57 dB and input slightly above the maximum level.

REFERENCES

[1] Tsui, J. B. Y. *Microwave Receivers With Electronic Warfare Applications*, New York, NY: John Wiley & Sons, 1986.

[2] Tserng, H. Q. "Design and Performance of Microwave Power GaAs FET Amplifiers," *Microwave Journal*, June 1979, p. 94.

[3] Ohta, K., Jodai, S., Fukuden, N., Hirano, Y., and Itoh, M. "A Five Watt 4-8 GHz GaAs FET Amplifier," *Microwave Journal*, Nov. 1979, p. 66.

[4] Dilorenzo, J. V., and Wisseman, W. R. "GaAs Power MESFET's: Design, Fabrication, and Performance," *IEEE Trans. Microwave Theory and Techniques*, Vol. MTT-27, May 1979, pp. 367–378.

[5] Peter, G. "Low Noise GaAs FET Dual Channel Front End," *Microwave Journal*, May 1982, p. 153.

[6] Whelehan, J. "Low Noise Amplifiers for Satellite Communications," *Microwave Journal*, Feb. 1984, p. 126.

[7] Ayasli, Y. "Decade Bandwidth Amplification at Microwave Frequencies," *Microwave Journal*, April 1984, p. 71.

[8] Bharj, J. S. "17 GHz Low Noise GaAs FET Amplifier," *Microwave Journal*, Oct. 1984, p. 121.

[9] Sholley, M., Maas, S., Allen, B., Sawires, R., Nichols, A., and Abell, J. "HEMT mm-Wave Amplifiers, Mixers and Oscillators," *Microwave Journal*, Aug. 1985, p. 121.

[10] "6-18 GHz Fully Integrated MMIC Amplifier," *Microwave Journal*, Aug. 1986, p. 121.

[11] Browne, J. "MMIC Chip Amplifier Boosts 0.5 to 5 GHz," *Microwave & RF*, Sept. 1986, p. 157.

[12] Morgan, W. "Minimize IM Distortion in GaAs FET Amplifiers," *Microwave & RF*, Oct. 1986, p. 107.

[13] Franke, E., and Deleon, J. "Broadband Noise Improvement in RF Power Amplifiers," *RF Design*, Nov. 1986, p. 104.

[14] Smith, M. A., Anderson, K. J., and Pavio, A. M. "Decade-Band Mixer Covers 3.5 to 35 GHz," *Microwave Journal*, Feb. 1986, p. 163.
[15] Browne, J. "Microwave Mixer Family Converts 1 to 18 GHz," *Microwave & RF*, Oct. 1986, p. 209.
[16] "Mixer-Amplifier Combination Is Virtually Load-Insensitive," *Microwave Journal*, Dec. 1987, p. 131.
[17] "Solid State Microwave Amplifiers," Aertech Industries, 825 Stewart Dr., Sunnyvale, CA 94086, Catalog No. 5978.
[18] "High Frequency Transistor Primer," Avantek, 2981 Copper Rd., Santa Clara, CA 95051, July 1971.
[19] Designing With GPD Amplifiers," Avantek, 2981 Copper Rd., Santa Clara, CA 95051, June 1972.
[20] Cheadle, D. L. "Cascadable Amplifiers," Watkins-Johnson Co., 3333 Hillview Av., Stanford Industrial Park, Palo Alto, CA 94304, Tech-note, Vol. 6, No. 1, Jan./Feb. 1979.
[21] "Solid State Amplifiers," Watkins Johnson Co., 3333 Hillview Av., Stanford Industrial Park, Palo Alto, CA 94304, June 1979.
[22] Blackham, D., and Hoberg, P. "Minimize Harmonics in Scalar Tests of Amplifiers," *Microwave & RF*, Aug. 1987, p. 143.
[23] Sorger, G. U. "The 1 dB Gain Compression Point for Cascaded Two-Port Networks," *Microwave Journal*, July 1988, p. 136.
[24] "17 Most Asked Questions About Mixers," Mini-Circuits, 2625 E. 14 St., Brooklyn, NY 11235.
[25] Cheadle, D. "Selecting Mixers for Best Intermod Performance," *Microwaves*, Nov. 1973, p. 48, Dec. 1973, p. 58.
[26] Neuf, D., and Brown, D. "What to Look for in Mixer Specs," *Microwaves*, Nov. 1974, p. 48.
[27] Reynolds, J. F., and Rosenzweig, M. R. "Learn the Language of Mixer Specification," *Microwaves*, May 1978, p. 72.
[28] Jacobi, J. H. "IMD: Still Unclear After 20 Years," *Microwave & RF*, Nov. 1986, p. 119.
[29] "A Guide to Waveform Recorder Testing," Prepared by The Waveform Measurement and Analysis Committee of the IEEE Instrumentation and Measurement Society, April 1990. This guide contains 4 papers: 1) Linnenbrink, T. E. "Introduction to Waveform Recorder Testing," 2) Green, P. J. "Effective Waveform Recorder Evaluation Procedures," 3) Souders, T. M., and Flach, D. R. "Step and Frequency Response Testing of Waveform Recorders," 4) Grosby, P. S. "Waveform Recorder Sine Wave Testing-Selecting a Generator."
[30] Steinbrecher D. H. "Broadband High Dynamic Range A/D Converter Limitations" *IEEE International Conference on Analogoue-to-digital and digital-to-analogue Conversion*, Venue University College of Swansea, Wales, pp. 95–99, Sept. 17–19, 1991.
[31] Sharpin, D. L., and Tsui, J. B. Y. "Analysis of the Linear Amplifier/Digital Converter Interface in a Digital Microwave Receiver," *IEEE Trans. Aerospace and Electronic Systems*, Vol. 31, pp. 248–256, January 1995.

APPENDIX 7.A

```
% df7eq1.m :This prog calculates total gain, noise figure and 3rd
order intercept pt
% ******** Input data ********
gc_db = input('gain of all components in dB i.e [15 15 -3] = ');
fc_db = input('noise figure of all components in dB [ ] = ');
qc_db = input('3rd order intercept pt of all components in dB [ ] = ');
len = length(gc_db);
% ******** convert dB into power/ratio ********
gc = 10 .^(gc_db./10);
```

```
fc = 10 .^(fc_db./10);
qc = 10 .^(qc_db./10);
% ******** calculate gain ********
g1 =  cumprod(gc);
g1m = [1 g1(1:len-1)];
gt = g1(len);
gt_db = 10*log10(gt);
% ******** calculate noise figure ********
f1 = [0 ones(1,len-1)];
f2 = fc-f1;
f_div = f2./g1m;
ft = sum(f_div);
ft_db = 10*log10(ft);
% ******** calculate 3rd order intercept pt ********
q1 = g1./qc;
q_den = sum(q1);
qt = gt/q_den;
qt_db = 10*log10(qt);
outp2 = [gt_db ft_db qt_db];
disp(' ')
disp('    Gain       NF        Q3')
disp(outp2)
```

APPENDIX 7.B

```
% df7eq2.m provides the design between an amplifier and ADC.
% JT 24 June 1992

clear
% ******** INPUT ********
     % ** AMP **
n1_db = -174;          % noise at input of amplifer per unit bandwidth
br = 30e6;             % rf bandwidth
br_db = 10*log10(br);
f_db  = 3.3;           % noise figure
     % ** ADC **
b = 8;                 % # of bits
vs = 270;          % saturation voltage in mv
q = vs/(2^(b-1));              % voltage per quantization level
fs = 250e6;            % sampling frequency in Hz
R = 50;                        % input impedance
n = 1024;          % FFT length
m = input('enter the value of m = ');
```

```
m_db = 10*log10(m);
m1 = m+1;                    %Eq 20
m1_db = 10*log10(m1);
md_db = m1_db - m_db;

% ******** GENERATE CONSTANT ********
ps = (vs*vs)*1e-3/(2*R);              %Eq 14 1e-3 changes to mw
ps_db = 10*log10(ps);
nb_db = ps_db - 1.76 - 6*b;           %Eq 15
bv = fs/n;                        %Eq 16
bv_db = 10*log10(bv);
no_db = nb_db + m_db;                 %Eq 20
no = 10 .^(no_db/10);
von = sqrt(no*1e3*2*R);                   %noise voltage
similar as Eq 14
vn = sqrt(3*no*1e5);                  %Eq 25
ns_db = no_db + md_db;                %Eq 17 18 19 21
vsn = vs-vn;                      %Eq 26
psn = (vsn.*vsn)/1e5;                 %Eq 26
psn_db = 10*log10(psn);
g_db = no_db - n1_db - f_db - br_db;      %Eq 17 18
pi_db = psn_db - 6 - g_db;            %Eq 27
p3_db = ns_db - br_db + bv_db;            %Eq 22

% ******** CALCULATION ********
q3_db=(3*pi_db-n1_db+2*g_db-f_db-bv_db-md_db)/2;     %Eq 24
fs_db = f_db + md_db;                     %Eq 21
dr_db = pi_db + g_db - p3_db;             %Eq 28
nqvr = von/q;
disp('     m      Gain      Q3      P3      NF      DR      N/Q')
en = length(m);
dp = [m'  g_db' q3_db'  p3_db'  fs_db'  dr_db'  nqvr'];
disp(dp)
```

CHAPTER 8

Frequency Downconverters

8.1 INTRODUCTION

The frequency range of interest to electronic warfare (EW) applications extends from 100 MHz to 18 GHz. Some long-range searching radars operate under 2 GHz, and one can consider EW operations nominally from 2 to 18 GHz. An EW receiver usually has an instantaneous bandwidth of .5 to 4 GHz, which is limited mainly by hardware constraints. This receiver is often referred to as the intermediate frequency (IF) receiver, which is time-shared to cover the 2- to 18-GHz frequency range. Generally, it is difficult to build an IF receiver with wide instantaneous bandwidth. Even if one could build such an IF receiver, the digital processor following the receiver would not be able to process the data in near real time. As a result, some of the data will be ignored by the processor. In this case, the digital processor limits the bandwidth capability of the IF receiver. In order to increase the bandwidth of the IF receivers, both receiver and processor technologies must be advanced.

To simplify the discussion in this chapter, let us assume that an IF receiver has approximately 1 GHz instantaneous bandwidth, which means any signal in the bandwidth will be intercepted. In order to cover the frequency range of interest, the input frequency range will be broken into many parallel channels (referred as channelization) and each channel will be frequency converted to match the input frequency range of the IF receiver.

In conventional analog receivers, the input frequency of 2 to 18 GHz is divided into 1-GHz bandwidths and each band is converted to some common IF. In a digital receiver, two types of conversion from analog to digital are often considered. One kind is single channel (or real-data conversion) where there is only one output data channel. The other approach is to generate two output channels that are 90-deg out of phase, and this can be referred to as the in-phase and quadrature phase downconversion, or simply called I and Q channels.

This chapter will discuss both schemes of frequency conversion, the one-channel and two-channel conversions. Both analog and digital frequency conversions will be discussed. Several digital approaches to create the *I* and *Q* channels will also be discussed. The impact of imbalance between the *I* and *Q* channels on the receiver performance will be presented. Finally, a correction scheme to rectify the *I* and *Q* channels imbalance will be discussed.

8.2 BASEBAND RECEIVER FREQUENCY SELECTION [1, 2]

In general, the input of an IF receiver should be kept under an octave of bandwidth. An octave of bandwidth means that the high end of the bandwidth is twice the frequency as at the low end of the bandwidth. For example, if the input bandwidth is 1 GHz, the frequency range from 1 to 2 GHz is an octave. If the input frequency range is lower than these values (i.e., 0.5 to 1.5 GHz), the bandwidth is over an octave. If the frequency range is above these values (i.e., 2 to 3 GHz), the bandwidth is under an octave.

If the bandwidth is over an octave, the second harmonic of a low-frequency signal may be in the bandwidth. The second harmonic can limit the dynamic range. For the 0.5 to 1.5-GHz band, if an input frequency is at 600 MHz, then the second harmonic is at 1,200 MHz, which is still in the input bandwidth.

The only known analog receivers that are designed to cover more than an octave of bandwidth are the crystal video and instantaneous frequency measurement (IFM) receivers. However, neither of these receivers can process simultaneous signals. The crystal video receiver only reports the pulse amplitude, pulse width, and time of arrival (TOA). In this case, the second harmonic does not have any adverse effect. The IFM receiver can encode the frequency of one signal as discussed in Chapter 2. The second harmonic amplitude is usually lower than the input signal amplitude and will not affect the frequency encoding circuits. Almost all other types of analog receivers have an input bandwidth less than one octave.

An analog-to-digital converter (ADC) has limited input frequency range. In many cases, due to the limitation of the ADC, it is impractical to have a high-input frequency. As mentioned in Chapter 6, in many ADCs, the dynamic range degrades when the input frequency is high. Thus, there are two choices. One approach is to narrow the input bandwidth and restrict the input frequency range to under one octave. Through technology improvement in ADC, a wider band should be achievable with this approach. A second approach would be to choose a bandwidth over an octave, or even start from a frequency close to dc. In the second approach, many spurs may be contained in the input bandwidth of the receiver, which will be discussed in the next section.

8.3 FREQUENCY CONVERSION [3–7]

The purpose of frequency conversion is to translate the input frequency from one frequency range to a different one at the output of the converter. The common way to achieve frequency conversion is through a mixer, as shown in Figure 8.1. A mixer is a nonlinear device. The output current can be related to the input voltage V as

$$I = a_0 + a_1 V + a_2 V^2 + \ldots \tag{8.1}$$

where a_is are constants. Assume that the input voltage V contains two sine waves as

$$V = V_i \sin(2\pi f_i t) + V_o \sin(2\pi f_o t) \tag{8.2}$$

where V_i and f_i are the amplitude and frequency of the signal, respectively, and V_o and f_o are the amplitude and frequency of the local oscillator, respectively. Substituting this relationship into (8.1) and considering the nonlinear term $a_2 V^2$ in particular,

$$\begin{aligned} a_2 V^2 &= a_2 V_i^2 \sin^2(2\pi f_i) + a_2 V_o^2 \sin^2(2\pi f_o) \\ &\quad + a_2 V_i V_o \{\cos[2\pi(f_o - f_i)t] - \cos[2\pi(f_o + f_i)t]\} \end{aligned} \tag{8.3}$$

The last term in this equation corresponds to the desired output frequency $(f_o - f_i)$ [or $(f_i - f_o)$] and $(f_o + f_i)$. If the output frequency is lower than the input frequency, the process is called frequency downconversion, otherwise, it is called frequency upconversion.

Although a mixer is a nonlinear device, the mixing process is often considered as a linear process in the sense of input/output superposition. The information contained in the input signal does not change, only the frequency is shifted. In an amplifier chain, a mixer can be treated as an amplifier. A mixer has a gain, a noise figure, and a third-order intercept point, as discussed in previous chapter. Its gain is usually less than unity (negative in decibels, or a loss), although in some mixers the gain might be positive. The noise figure may often be taken as equal to the loss of the mixer unless another value is given by the manufacturer. The third-order intercept point is given by the manufacturer.

As (8.3) indicates, there are many frequencies at the output of a mixer besides the desired one. These intermediate output frequencies f_{if} can be written as

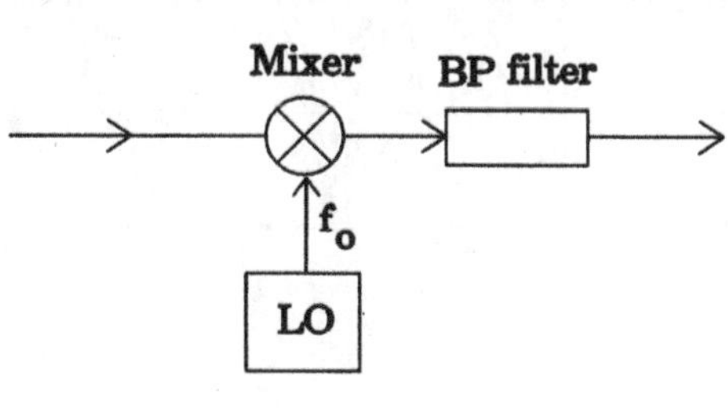

Figure 8.1 A mixer circuit.

$$f_{if} = mf_l + nf_h \tag{8.4}$$

where m and n are positive or negative integers. In this equation, f_l represents the low-input frequency and f_h the high one. Either one can be used to represent the signal or oscillator frequency; thus, f_i and f_o are not used. If both m and n are 1, the output is the sum of the two frequencies. If one of them is +1 and the other is −1, the output is the difference frequency. Other than these two frequencies, all the other frequencies are considered spurious frequencies (or spurs) and should be kept at minimum.

A convenient way to show the spurs is by using a spur chart, which is shown in Figure 8.2. In this mixer chart, the difference frequency is the desired result. To simplify the notation, H is used to represent the higher frequency, which can be either the signal or the local oscillator, and L represents the lower one. The

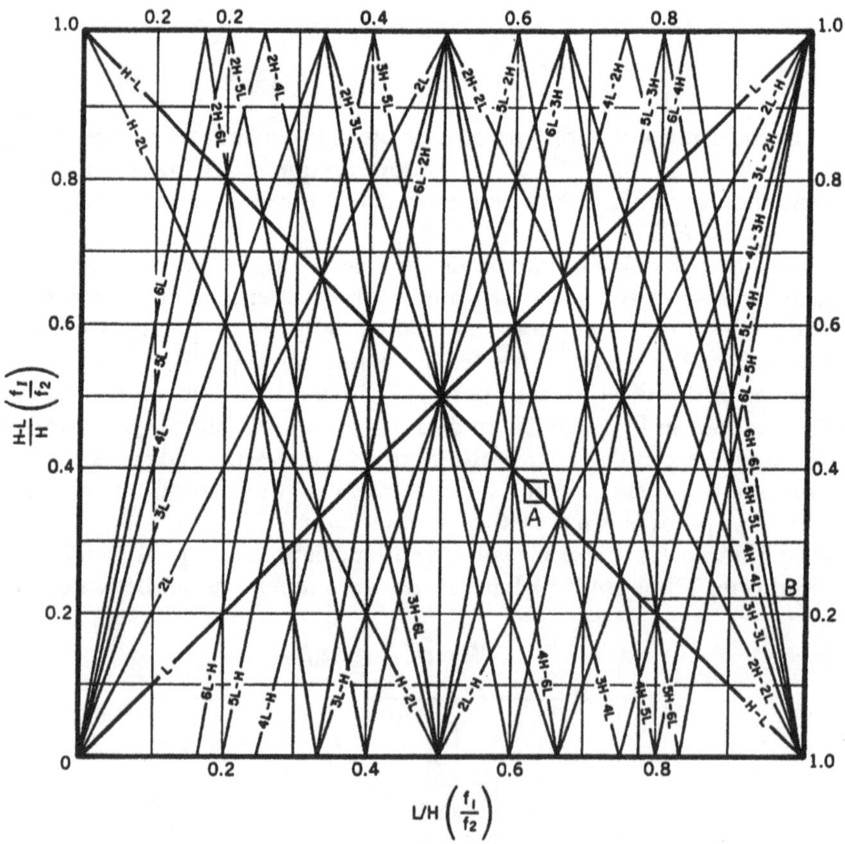

Figure 8.2 Mixer spur chart.

line drawn diagonally and labeled $H - L$ represents the desired output. All the other lines represent spurious outputs. The highest order spur expressed in this figure is 6, which is represented by $6H$ or $6L$. As a general rule, a high-order spur has smaller amplitude.

In Figure 8.2, the square marked A represents an area in which there is no spur. In a narrowband downconverter, this is a desired choice. An IF output starting from zero frequency is represented by a square B on the lower right corner. In the square, there are many spurs (i.e., $2H - 2L$, $3H - 3L$, etc.), therefore the choice of an IF bandwidth starting from zero frequency is not a good one. This is a compromise needed to cover a wide bandwidth. If the technology in ADC and digital processing can be improved to gigahertz operation speed, a wide IF bandwidth of less than an octave might be achievable.

8.4 IN-PHASE (I) AND QUAD-PHASE (Q) CHANNEL CONVERSION

The input of a receiver is always a single channel and the data can be considered as real in contrast to being complex. In order to keep this discussion simple, the following input signal $S(t)$ will be considered:

$$S(t) = A\sin(2\pi f_i t) \tag{8.5}$$

where A and f_i are the amplitude and frequency of the input signal, respectively. For I and Q channel conversion, two outputs are generated and they are 90-deg out of phase with one another. If the frequency is downconverted, the two outputs can be written as

$$\begin{aligned} I(t) &= A\cos[2\pi(f_i - f_o)t] \\ Q(t) &= A\sin[2\pi(f_i - f_o)t] \end{aligned} \tag{8.6}$$

where f_o is the local oscillator frequency. In this equation, the I and Q channels are arbitrarily designated. As long as the two outputs are 90-deg out of phase, one can be called the I channel and the other one the Q channel. Since the two outputs can be combined into a complex form as

$$\begin{aligned} I(t) + jQ(t) &= A\{\cos[2\pi(f_i - f_o)t] + j\sin[2\pi(f_i - f_o)]\} \\ &= Ae^{j2\pi(f_i - f_o)t} \end{aligned} \tag{8.7}$$

the combined outputs are sometimes referenced as complex data.

To achieve the I and Q channel conversion, more hardware is needed, but there are compensating advantages.

1. The bandwidth of the input signal can be doubled if both outputs are digitized. This point can be explained in either the time or the frequency domain. In the time domain, if the sampling frequency is f_s, one must obtain two samples per cycle at the highest input frequency to fulfill the Nyquist sampling rate; thus, the highest frequency is $f_s/2$. If there is a Q channel, two more samples will be collected; thus, the highest frequency can be extended to f_s. In the frequency domain, if the input is real there are positive and negative frequency components, as discussed in Chapter 3, and the highest frequency without ambiguity is $f_s/2$. For complex data, there are no negative frequencies and the unambiguous range extends to f_s.
2. Amplitude information is maintained in an I-Q channel conversion. If real data are processed by an analog receiver, the amplitude information can be recovered from the video detector, which has a lowpass filter to smooth out the RF ripples. Therefore, the amplitude detection has never been considered as a problem in analog microwave receivers. Of course, a similar idea can be extended to digital receivers by taking the absolute value of the real input data. This can be accomplished by passing the square of the real input through a lowpass filter to obtain an approximation of the amplitude information.

   With I and Q channels, an obvious approach is through the relationship

$$A = \sqrt{A^2 \sin^2[2\pi(f_i - f_o)t] + A^2 \cos^2[\pi(f_i - f_o)t]} \tag{8.8}$$

   If there is only one signal, the amplitude A calculated from this equation will be a constant for each sample. If the amplitude changes, one can say that the signal is amplitude modulated or, in the case of an EW receiver, one can conclude that there are simultaneous signals. Thus, from the output of the I-Q channels, either the amplitude of a signal can be found or a simultaneous signal condition can be detected.
3. The I-Q channels can be used to determine instantaneous frequency if only one signal is present. This method can be extended to determine two frequencies, which will be discussed in Chapter 10. Considering that the signals in (8.6) are digitized at time t_i and t_{i+1}, the instantaneous angle at these time instances can be found as

$$\begin{aligned}\theta_i &= \tan^{-1}\left[\frac{A\sin[2\pi(f_i - f_o)t_i]}{A\cos[2\pi(f_i - f_o)t_i]}\right] = 2\pi(f_i - f_o)t_i \\ \theta_{i+1} &= \tan^{-1}\left[\frac{A\sin[2\pi(f_i - f_o)t_{i+1}]}{A\cos[2\pi(f_i - f_o)t_{i+i}]}\right] = 2\pi(f_i - f_o)t_{i+i}\end{aligned} \tag{8.9}$$

The difference in angle θ can be used to find the instantaneous frequency as

$$f_i - f_o = \frac{\theta_{i+1} - \theta_i}{2\pi\Delta t} \tag{8.10}$$

where $\Delta t = t_{i+1} - t_i$.

If the input signal contains a single frequency, this approach can improve the frequency data resolution. By contrast, if a signal is sampled over a 1-μs interval and an FFT is performed on the input, the frequency resolution is 1 MHz. If the above method is used and the delay time is 1 μs, the maximum unambiguous bandwidth is 1 MHz. If one can measure the phase angle with 6 bits of resolution, the resolution will be 5.625 degrees (360 deg/64). Under this condition, the frequency resolution can be measured as 15.625 kHz (1,000 kHz/64), which means the resolution improved 64 times from the FFT approach. However, if there are simultaneous signals, this simple phase approach does not work. However, a slightly different approach can solve two frequencies which will be discussed in Chapter 10.

8.5 IMBALANCE IN *I* AND *Q* CHANNELS [8, 9]

When an *I* and *Q* downconverter is simulated, it is often assumed that the two outputs are perfectly balanced (i.e., the two outputs have equal amplitude and are 90-deg out of phase). However, in actual fabrication of an *I* and *Q* downconverter, this is usually not the case, especially if the converter covers a wide bandwidth. In other words, the two channel outputs may not have the same amplitude and their respective phases are not exactly 90-deg apart. The effect of this imbalance can generate an image signal, which can limit the dynamic range of a receiver.

Let us use a Fourier transform to explain the generation of an image frequency. Instead of finding the Fourier transform of $\exp(j2\pi f_i t)$ directly, let us use the relationship

$$e^{j2\pi f_i t} = \cos(2\pi f_i t) + j\sin(2\pi f_i t) \tag{8.11}$$

The Fourier transforms of $\cos(2\pi f_i t)$ and $\sin(2\pi f_i t)$ can be written as

$$\begin{aligned}\mathcal{F}[\cos(2\pi f_i t)] &= \frac{\delta(f-f_i) + \delta(f+f_i)}{2} \\ \mathcal{F}[\sin(2\pi f_i t)] &= \frac{-j\delta(f-f_i) + j\delta(f+f_i)}{2}\end{aligned} \tag{8.12}$$

The output can be combined as

$$\begin{aligned}\mathcal{F}(e^{j2\pi f_i t}) &= \frac{1}{2}\{\delta(f-f_i) + \delta(f+f_i) + j[-j\delta(f-f_i) + j\delta(f+f_i)]\} \\ &= \delta(f-f_i)\end{aligned} \tag{8.13}$$

The negative frequency components $\delta(f + f_i)$ cancel each other. Only the positive frequency component $\delta(f - f_i)$ is left. These results are shown in Figure 8.3. In this figure, 8.3(a) represents the Fourier transform of the cosine signal, 8.3(b) represents the Fourier transform of the sine signal, and 8.3(c) represents the combined results of (8.13).

The two outputs of I and Q may not have the same amplitude, for example, $\cos(2\pi f_i t)$ versus $.8\sin(2\pi f_i t)$. The negative frequency components of the combined result will not exactly cancel each other. There will be an output in the negative frequency component position, as shown in Figure 8.4, and this component is often referred to as the image of the true signal. In receiver design, the image of a strong signal can limit the dynamic range of the receiver because the weak signal must be higher than the image to be detected.

Now let us find the amplitude of the image as a function of imbalance between the two channels and relate that to the dynamic range of the receiver. This discussion is based on [9]. First, the two outputs are expressed as

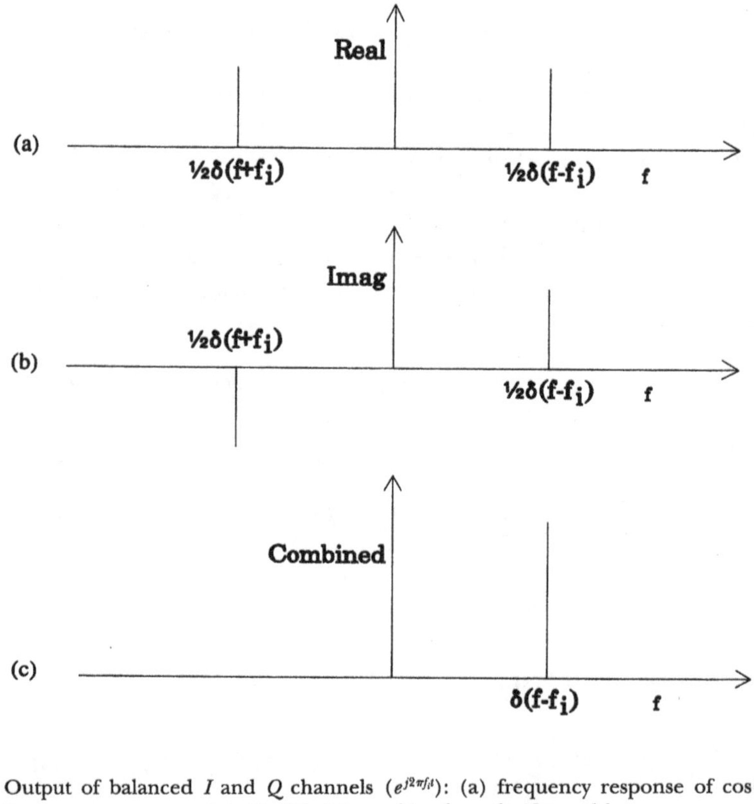

Figure 8.3 Output of balanced I and Q channels ($e^{j2\pi f_i t}$): (a) frequency response of $\cos(2\pi f_o t)$, (b) frequency response of $\sin(2\pi f_o t)$, (c) combined result of a and b.

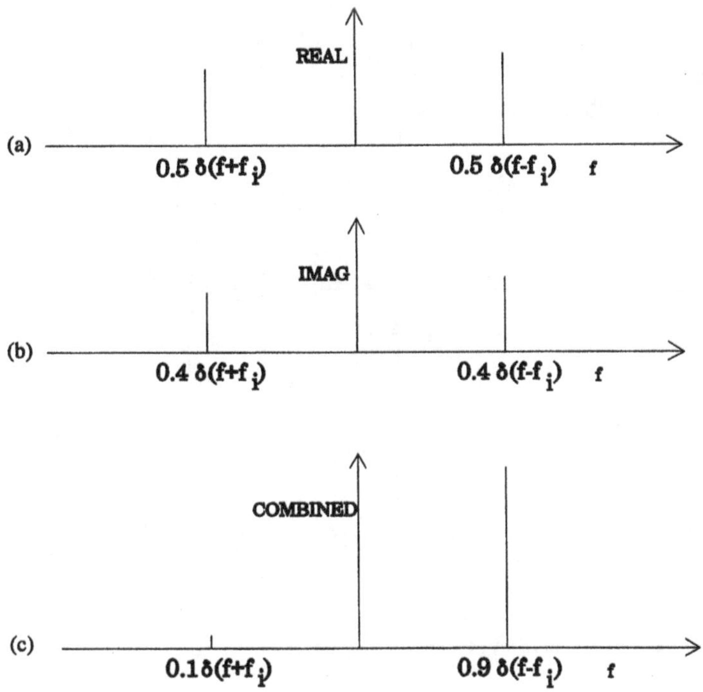

Figure 8.4 Output of imbalanced I and Q channels: (a) frequency response of $\cos(2\pi f_o t)$, (b) frequency response of $\sin(2\pi f_o t)$, (c) combined result of a and b.

$$
\begin{aligned}
s(t) &= \cos(2\pi f_{\text{if}} t) + j\alpha \sin(2\pi f_{\text{if}} t + \epsilon) \\
&= \frac{1}{2}[e^{j2\pi f_{\text{if}} t} + e^{-j2\pi f_{\text{if}} t}] + \frac{\alpha}{2}[e^{j(2\pi f_{\text{if}} t+\epsilon)} - e^{-j(2\pi f_{\text{if}} t+\epsilon)}] \\
&= \frac{1}{2}[e^{j2\pi f_{\text{if}} t}(1 + \alpha e^{j\epsilon})] + \frac{1}{2}[e^{-j2\pi f_{\text{if}} t}(1 - \alpha e^{j\epsilon})]
\end{aligned}
\tag{8.14}
$$

where f_{if} is the IF output angular frequency, α is the amplitude imbalance, and ϵ is the phase imbalance. In this example, the cosine channel is assumed to be perfect and the sine channel has all the imbalance. Since the amplitude and phase errors are relative values between the channels, the error can be included in one channel without loss of generality. From this equation, the desired signal can be considered as $e^{j2\pi f_{\text{if}} t}$ and the image as $e^{-j2\pi f_{\text{if}} t}$. If $\alpha = 1$ and $\epsilon = 0$, the desired term becomes $e^{j2\pi f_{\text{if}} t}$ and the image becomes zero as expected.

In general, the corresponding amplitudes of the signal and image are $1 + \alpha e^{j\epsilon}$ and $1 - \alpha e^{-j\epsilon}$, respectively, which can be represented by phasors, as shown in Figure 8.5. In this figure, A_d is the amplitude of the desired signal and A_i is the amplitude

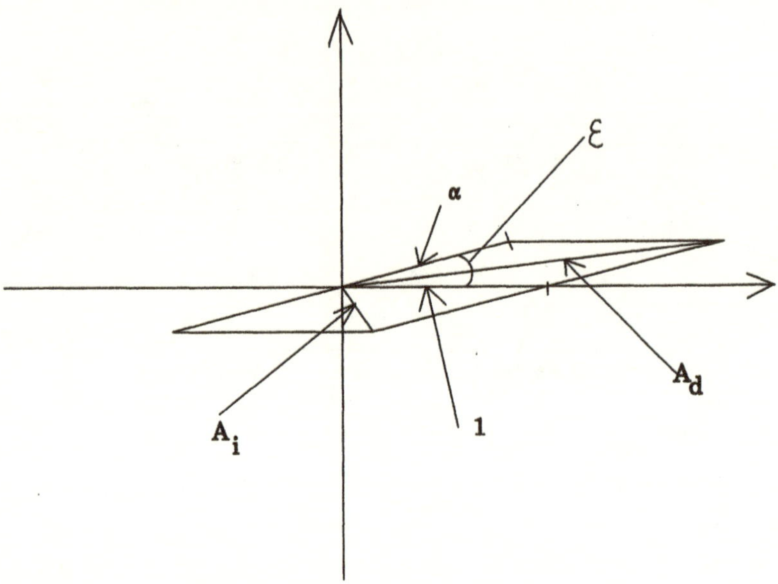

Figure 8.5 Graphical representation of *I* and *Q* outputs.

of the image signal. Through the law of cosine with reference to Figure 8.5, one can see that the desired output A_d and the image A_i can be expressed as

$$\begin{aligned} A_d^2 &= 1 + \alpha^2 + 2\alpha\cos(\epsilon) \\ A_i^2 &= 1 + \alpha^2 - 2\alpha\cos(\epsilon) \end{aligned} \tag{8.15}$$

The image amplitude relative to the desired amplitude can be written in decibels as

$$10\log\left(\frac{A_i}{A_d}\right)^2 = 10\log\frac{1 + \alpha^2 - 2\alpha\cos(\epsilon)}{1 + \alpha^2 + 2\alpha\cos(\epsilon)} \tag{8.16}$$

This value is the dynamic range of the *I-Q* converter as limited by the channel imbalance.

The result of (8.16) is displayed in Figure 8.6. In this figure, the curves are either in a horizontal or a vertical direction. The effect can be explained as follows. Assume that there is a 2-deg imbalance in phase. From the above figure, the image will be 35 dB from the desired signal as long as the amplitude balance is less than approximately 0.15 dB. If the amplitude balance is worse than 0.15 dB, the image amplitude will be dominated by the amplitude imbalance. For example, if there is

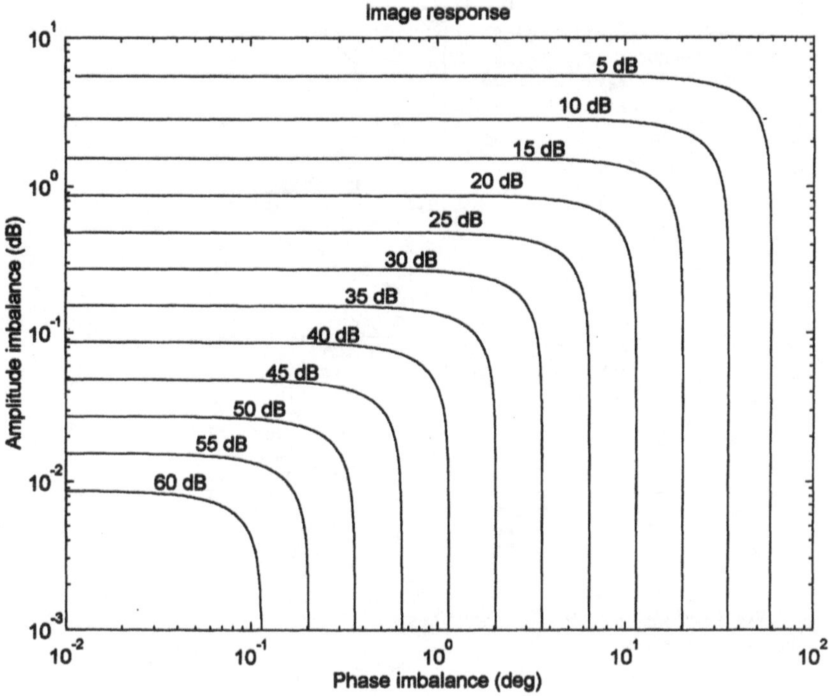

Figure 8.6 Image amplitude as a function of amplitude and phase imbalance.

a 1.5-dB amplitude imbalance, the image will be 15-dB down as long as the phase imbalance is below 20 deg. In other words, the worst performance of the imbalance factor will be the dominant factor in determining the image amplitude. In general, the phase match is more difficult to achieve, especially, in wideband downconverter systems.

8.6 ANALOG *I* AND *Q* DOWNCONVERTERS

Analog *I* and *Q* downconverters are the most commonly used approach in EW receiver design. The main advantage in such a design is the wide bandwidth. It is reasonable to achieve a few gigahertz bandwidth. The major deficiency of this approach is that it is difficult to achieve good balance between the channels.

The main component is a 90-deg phase shifter. There are two different approaches to build analog *I-Q* downconverters and they are shown in Figure 8.7.

In both cases, there are two mixers, but only one local oscillator that feeds both mixers. In Figure 8.7(a), a 90-deg phase shift is introduced in the input of the converter while the local oscillator fed to the mixers is in-phase. In Figure

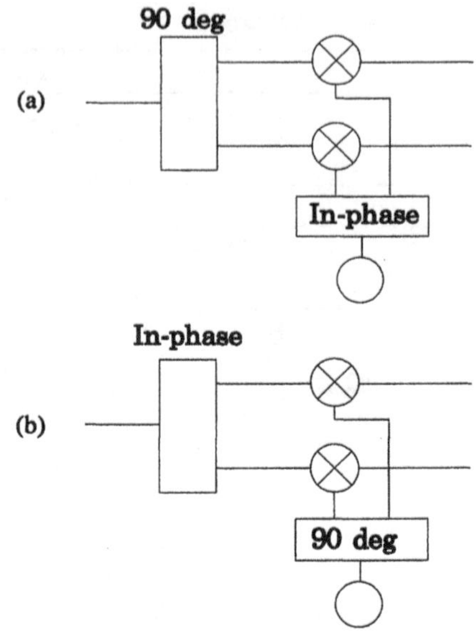

Figure 8.7 Analog I and Q downconverters: (a) 90-deg introduced in input path, (b) 90-deg introduced in local oscillators.

8.7(b), the 90-deg phase shift is introduced in the local oscillator of the two mixers, while the input signals to the mixers are in-phase.

These two approaches provide very similar results. The choice of the approach usually depends on the availability of components. Narrowband components are used in the network between the local oscillator and the mixers. The operating frequency of these components is equal to the local frequency f_o. The components used in the network between the input and the mixers are wideband, which must compromise the bandwidth of the input signal.

There are, in general, three ways to select the local oscillator frequency: 1) f_o is below the input frequency f_i, 2) f_o is above the input frequency, and 3) f_o is in the middle of the input bandwidth. The first two approaches are obvious; thus, only the last approach will be discussed.

Let f_1 and f_2 be the lower and upper bounds of the input frequency band. In the third approach, the local oscillator frequency f_o is often put at the center of f_1 and f_2 or $(f_1 + f_2)/2$. In this approach, the IF bandwidth is half the input bandwidth $(f_2 - f_1)/2$ because both the upper and lower bands fold in the same frequency range. However, whether the input frequency is above or below f_o can be determined from the phase relation between the I and Q channels. Therefore, although the IF bandwidth is half the input bandwidth, the overall bandwidth is equal to the input bandwidth.

Let us use a simple example to demonstrate this idea. Assume the local oscillator signal feeding the two mixers is 90-deg out of phase. Thus, the inputs to the two mixers from the local oscillators can be written as

$$\begin{aligned} v_{o1} &= \sin(2\pi f_o t) \\ v_{o2} &= \cos(2\pi f_o t) \end{aligned} \tag{8.17}$$

and the input signal is

$$v_i = \sin(2\pi f_i t) \tag{8.18}$$

where all the signals are assumed to have unity amplitudes. The output of the mixers can be written as

$$\begin{aligned} v_{if1} &= v_{o1} v_i = \sin(2\pi f_o t)\ \sin(2\pi f_i t) \\ &= \frac{1}{2}\{\cos[2\pi(f_i - f_o)t] - \cos[2\pi(f_i + f_o)t]\} \\ v_{if2} &= v_{o2} v_i = \cos(2\pi f_o t)\ \sin(2\pi f_i t) \\ &= \frac{1}{2}\{\sin[2\pi(f_i - f_o)t] + \sin[2\pi(f_i + f_o)t]\} \end{aligned} \tag{8.19}$$

The high-frequency terms (i.e., the last term in the above two equations) will be neglected in the discussion because in actual design these terms will be filtered out through a lowpass filter.

If the input signal is higher than the local oscillator frequency ($f_i > f_o$) in the above equation, then

$$\begin{aligned} v_{if1} &= \cos[2\pi(f_i - f_o)t] \\ v_{if2} &= \sin[2\pi(f_i - f_o)t] \end{aligned} \tag{8.20}$$

where v_{if1} is leading v_{if2} by 90 deg. If the input signal is lower than the local oscillator frequency ($f_i < f_o$), the results from (8.19) will be

$$\begin{aligned} v_{if1} &= \cos[2\pi(f_i - f_o)t] \\ v_{if2} &= -\sin[2\pi(f_i - f_o)t] \end{aligned} \tag{8.21}$$

where v_{if1} is lagging v_{if2} by 90 deg. Therefore, by measuring the relative phase between the *I* and *Q* channels, the input frequency can be determined, even the IF bandwidth is only half of the input bandwidth.

If the IF bandwidth is wide and conventional operational amplifiers cannot accommodate the bandwidth, an RF amplifier can be used. As discussed before, an RF amplifier usually does not cover dc and very low frequency ranges. As a result, this approach will create a hole in the center of the band, which is highly undesirable.

Two input signals, one above and one below the f_o by the same frequency difference, will occupy the same frequency bin after downconversion. Under this condition, the phase relation will be disturbed and their frequencies may not be identified correctly.

Since two similar analog components are difficult to balance over a wide frequency range, it is very hard to build an analog *I* and *Q* downconverter with balanced outputs. However, highly balanced *I* and *Q* channels can be generated digitally through signal processing. These subjects will be discussed in the following sections.

8.7 DIGITAL APPROACH TO GENERATE *I* AND *Q* CHANNELS

In the previous section, the *I* and *Q* channels are built through analog means. The advantage is wide bandwidth and the disadvantage is the poor balance between the two channels. There are digital approaches that can be used to generate the desired *I-Q* channels. One approach is to use the Hilbert transform, the other is to use a special sampling scheme. In the digital approaches, the data of one channel (say the *I* channel) are obtained from a single-channel downconverter. The data of the *Q* channel have to be generated by processing the *I* channel data. Since the *Q* channel data are generated digitally, the unbalance between the outputs of the *I* and *Q* channels can be kept at a minimum.

One of the major disadvantages of the digital approach is the available operating speed. The bandwidth is limited to tens or a few hundreds of megahertz because of the slow processing speed. For EW applications, such a bandwidth would usually be considered to be too low. However, with an increase in digital operation speed, it might be possible to obtain useful bandwidth for EW applications.

The following sections will discuss the Hilbert transform in the continuous sense. Although the concept of Hilbert transform might be difficult, the mathematic definition appears straightforward. The actual computation is usually difficult to carry out. A popular way to obtain the Hilbert transform is through Fourier transform (as in MATLAB).

8.8 HILBERT TRANSFORM [8, 10–13]

The Hilbert transform of a function $x(t)$ is defined as the convolution of $x(t)$ and a function $h(t)$. This relation can be shown mathematically in the following equation.

$$\begin{aligned} H[x(t)] \equiv x^h(t) &= x(t) \bigstar h(t) = x(t) \bigstar \frac{1}{\pi t} \\ &= \frac{1}{\pi}\int_{-\infty}^{\infty} \frac{x(\tau)}{t-\tau} d\tau \end{aligned} \tag{8.22}$$

where ★ represents convolution and $H[x(t)]$ and $x^h(t)$ represent a Hilbert transform in time domain. The function $h(t)$ is defined as

$$h(t) = \frac{1}{\pi t} \tag{8.23}$$

It is interesting to note that the Hilbert transform of a time domain function stays in the time domain. In the frequency domain, the Hilbert transform $X^h(f)$ can be found as

$$X^h(f) = X(f)H(f) \tag{8.24}$$

The Fourier transform of $h(t)$ can be obtained from (3.92) and (3.94) as

$$\mathscr{F}[h(t)] \equiv H(f) = j\ sgn(f) = -j\begin{cases} 1 & f > 0 \\ -1 & f < 0 \end{cases} \tag{8.25}$$

where *sgn* represents the sign function. Thus, to obtain the Hilbert transform in the frequency domain, the negative frequency of $X(f)$ is multiplied by j and the positive frequency is multiplied by $-j$.

The $X(f)$ can be obtained from (8.24) as

$$X(f) = \frac{X^h(f)}{H(f)} \tag{8.26}$$

However, one can see that $1/H(f) = -H(f)$ from (8.25). Thus,

$$X(f) = -X^h(f)H(f)$$

or

$$x(t) = -x^h(t) \star h(t) = \frac{-1}{\pi}\int_{-\infty}^{\infty}\frac{x^h(\tau)}{t-\tau}d\tau \tag{8.27}$$

Both the Hilbert transform and the inverse transforms are improper integrals because they have infinite discontinuity at $\tau = t$. In order to avoid this discontinuity, the integration should be carried out symmetrically about $\tau = t$. The integral can be written as

$$\int_{-\infty}^{\infty}\frac{g(\tau)}{t-\tau}d\tau = \lim_{\epsilon=0}\left[\int_{-\infty}^{t-\epsilon}\frac{g(\tau)}{t-\tau}d\tau + \int_{t+\epsilon}^{\infty}\frac{g(\tau)}{t-\tau}d\tau\right] \tag{8.28}$$

where ϵ is an infinitesimally small quantity. The Cauchy's principal value of this integral has to be used to calculate this integral. The result of the Hilbert transform can also be obtained from the inverse Fourier transform as

$$x^h(t) = x(t) \star h(t) = \mathscr{F}^{-1}[X(f)H(f)] \tag{8.29}$$

Let us use an example to conclude this section. If the input signal is

$$x(t) = \sin(2\pi f_i t) \tag{8.30}$$

then its Fourier transform is

$$X(f) = \frac{j}{2}[\delta(f + f_i) - \delta(f - f_i)] \tag{8.31}$$

This result is shown in Figure 8.8.

When $f < 0$, it is multiplied by j and the result is $-0.5\delta(f + f_i)$. When $f > 0$, it is multiplied by $-j$ and the result is $-0.5\delta(f - f_i)$. This result is shown in Figure 8.9.

If one is familiar with Fourier transform, it is easy to recognize that the above result is the Fourier transform of $-\cos(2\pi f_i t)$. Thus, one can write the following relation:

$$H[\sin(2\pi f_i t)] = -\cos(2\pi f_i t) \tag{8.32}$$

where $H[x(t)]$ represents the Hilbert transform. This equation shows that the Hilbert transform of a sine function is a negative cosine function. As a result, the phase of the input signal is shifted by $-j$.

If the input signal is a cosine wave, using a similar approach to the one above, the Hilbert transform will change it to a sine wave as

$$H[\cos(2\pi f_i t)] = \sin(2\pi f_i t) \tag{8.33}$$

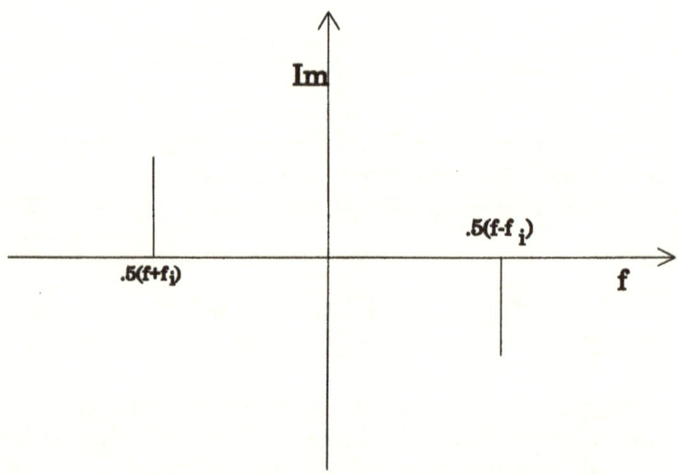

Figure 8.8 Fourier transform of a sine function.

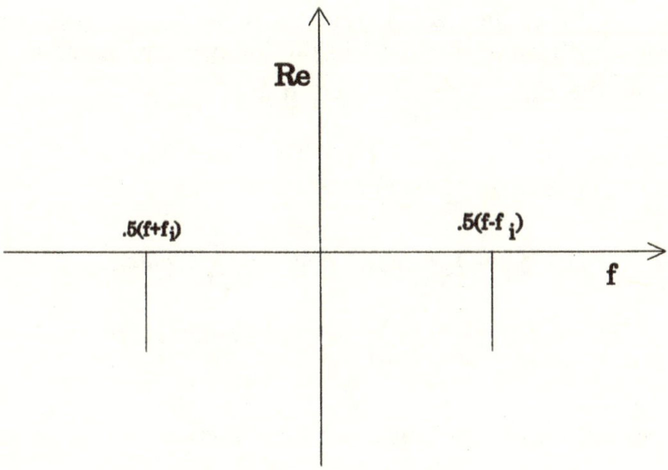

Figure 8.9 $X(f)H(f)$.

This relation also provides a phase shift of $-j$. Thus, a Hilbert transform can provide a 90-deg phase shift without affecting the magnitude of the spectral component.

In this example, the Fourier transform of the input signal can be found easily. In a digital receiver, the input signal $x(t)$ is digitized and a fast Fourier transform (FFT) was to be used to determine $X(f)$. The Hilbert transform in time domain $x^h(t)$ can be obtained from an inverse FFT as shown in (8.29) by using the definition of $H(f)$ in (8.25). However, discrete Fourier transform is periodic in nature, which will affect the discrete Hilbert transform.

8.9 DISCRETE HILBERT TRANSFORM [10, 11, 14, 15]

In this section, the discrete Hilbert transform will be discussed. The function $H(f)$ is extended from $-\infty$ to ∞ in the frequency domain. In the discrete domain, the number of data points must be limited (say, to $2M+1$ points), which is equivalent to adding a rectangular window to the input signal. The z transform of $h(nt_s)$ is

$$H(z) = \sum_{n=-M}^{M} h(nt_s)z^{-n}$$

$$\text{instead of} \tag{8.34}$$

$$H(z) = \sum_{n=-\infty}^{\infty} h(nt_s)z^{-n}$$

By definition, these two equations are not causal because the summation starts from a negative value. In order to implement the filter in practice, the above equations must be made causal.

The finite impulse filter (FIR) design scheme will be used to achieve the discrete Hilbert transform as discussed in the following paragraphs.

1. Let us write the z transform of function $h(nt_s)$ as

$$\begin{aligned} H(z) &= \sum_{n=-\infty}^{\infty} h(nt_s) z^{-n} \\ &= \sum_{n=-\infty}^{-1} h(nt_s) z^{-n} + h(0) + \sum_{n=1}^{\infty} h(nt_s) z^{-n} \\ &= h(0) + \sum_{n=1}^{\infty} [h(-nt_s) z^{n} + h(nt_s) z^{-n}] \end{aligned} \tag{8.35}$$

Substituting $z = \exp(j2\pi f t_s)$, the result can be written as

$$\begin{aligned} H(e^{j2\pi f t_s}) &\equiv H_r(e^{2\pi f t_s}) + jH_i(e^{2\pi f t_s}) \\ &= \sum_{n=-\infty}^{\infty} h(nt_s) e^{-2\pi n f t_s} \\ &= h(0) + \sum_{n=1}^{\infty} [h(-nt_s)\cos(2\pi n f t_s) + jh(-nt_s)\sin(2\pi n f t_s) \\ &\quad + h(nt_s)\cos(2\pi n f t_s) - jh(nt_s)\sin(2\pi n f t_s)] \end{aligned} \tag{8.36}$$

where

$$\begin{aligned} H_r(e^{j2\pi n f t_s}) &= h(0) + \sum_{n=1}^{\infty} [h(-nt_s) + h(nt_s)]\cos(2\pi n f t_s) \\ H_i(e^{j2\pi n f t_s}) &= \sum_{n=1}^{\infty} [h(-nt_s) - h(nt_s)]\sin(2\pi n f t_s) \end{aligned} \tag{8.37}$$

2. When the sampling frequency is f_s, the transfer function $H(f)$ is limited to the bandwidth of $f_s/2$. Due to the periodic property of sampling, the Hilbert transfer function is actually as shown in Figure 8.10. This function can be represented by the Fourier series as

$$H_i(e^{j2\pi f t_s}) = \sum_{n=1}^{\infty} b_n \sin(2\pi n f t_s) \tag{8.38}$$

where b_n can be found as

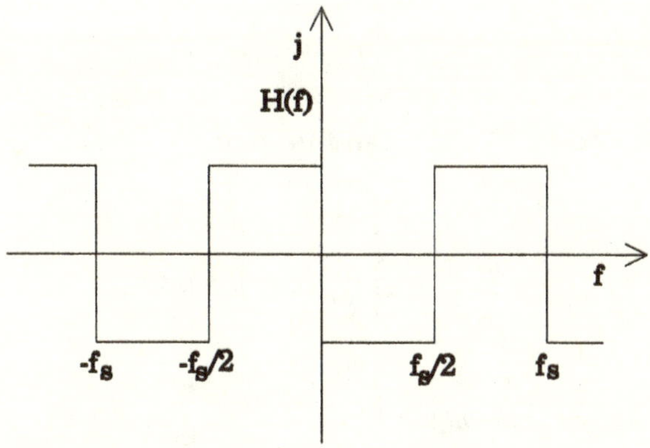

Figure 8.10 Periodical representation of $H(f)$.

$$
\begin{aligned}
b_n &= \frac{2}{f_s}\int_{-f_s/2}^{f_s/2} H_i(e^{j2\pi nft_s})\ \sin(2\pi nft_s)\,df \\
&= \frac{2}{f_s}\left[\int_{-f_s/2}^{0} \sin(2\pi nft_s)\,df + \int_0^{f_s/2} -\sin(2\pi fnt_s)\,df\right] \\
&= \frac{1}{n\pi}[-2 + 2\cos(n\pi)] \\
&= \begin{cases} 0 & n = \text{even} \\ \dfrac{-4}{n\pi} & n = \text{odd} \end{cases}
\end{aligned}
\qquad (8.39)
$$

In the above equation, the relation of $f_st_s = 1$ is used.

3. From the discussion of the Hilbert transform in the continuous sense equation (8.25), it is obvious that the transfer function representing the Hilbert transform has only an imaginary part; therefore, $H_r(f) = 0$, and $H_i(f) \neq 0$. This condition can be fulfilled if

$$
h(0) = 0 \qquad \text{and} \qquad h(-nt_s) = -h(nt_s) \qquad (8.40)
$$

Using this relationship, H_i in (8.37) can be written as

$$
H_i(e^{j2\pi nft_s}) = -2\sum_{n=1}^{\infty} h(nt_s)\ \sin(2\pi nft_s) \qquad (8.41)
$$

Comparing this result and (8.38), one can obtain

$$h(nt_s) = -\frac{b_n}{2} \tag{8.42}$$

Using this relation and (8.39), one can find that

$$h(nt_s) = \begin{cases} 0 & n = \text{even} \\ \dfrac{2}{n\pi} & n = \text{odd} \end{cases}$$
$$h(-nt_s) = \begin{cases} 0 & n = \text{even} \\ \dfrac{-2}{n\pi} & n = \text{odd} \end{cases} \tag{8.43}$$

These are the desired $h(nt_s)$ values.

4. Causality—As mentioned before, the results obtained in (8.43) are not causal. In order to make the results causal, a simple shift in time domain can be used. The value of n in $h(nt_s)$ is windowed from $-M$ to M as shown in (8.34). The shift in time is equivalent to multiplying the first result in (8.34) by z^{-M} and substituting with $k = n + M$. The new result is

$$H(z) = \sum_{n=-M}^{M} h(nT)z^{-(n+M)} = \sum_{k=0}^{2M} h(kT - MT)z^{-k} \tag{8.44}$$

This result is causal because the k value starts from zero instead of a negative value.

5. Windowing—Digitizing and limiting the input signal to a finite number of points is equivalent to providing a rectangular window in the time domain. Its equivalent effect in the frequency domain is convolving with a sinc function (sin x/x), which will disrupt the phase relation and the output. In order to reduce the effect of the sinc function in the frequency domain, special windows $W(nt_s)$ are often used.

8.10 EXAMPLES ON DISCRETE HILBERT TRANSFORM

In this section, two examples will be used to demonstrate the discrete Hilbert transform. The first one uses an FIR filter with a rectangular window. The second one uses an FIR filter with a Hamming window. Let us choose an 11-lag filter to perform the Hilbert transform, which corresponds to $2M + 1 = 11$. The corresponding n value is from -5 to $+5$. The $h(n)$ values obtained from (8.43) are listed in the following table.

It is interesting to note that if $n = -6$ to 6, the same results will be obtained because the two end values are zeros.

Table 8.1
Standard $h(n)$ Values

n	−5	−4	−3	−2	−1	0	1	2	3	4	5
$h(n)$	$-2/5\pi$	0	$-2/3\pi$	0	$-2/\pi$	0	$2/\pi$	0	$2/3\pi$	0	$2/5\pi$
New n	0	1	2	3	4	5	6	7	8	9	10

The second step is to convert $h(n)$ into a causal form. This can be accomplished by rearranging n from 0 to 10 as shown in row 3 of the above table. The direct form of the filter is shown in Figure 8.11. There are only six $h(n)$ values, but each has a delay time of two units. If the input to the filter is a sine wave, the output is obtained from the convolution of the input signal and $h(n)$. Figure 8.12(a) shows the input signal, with 5 data points removed from the beginning and end of the data to match the steady state of the filtered output. Figure 8.12(b) shows the output in the time domain.

Since the filter has a total delay of 10 units, it takes 10 units for the filter output to reach steady state, and the last 10 output data points are not accurate either because a portion of the input data are zeros. These data at the beginning and the end of the output can be considered as in transient states and they are not included in the figure. In general, if the input is a pulse of n data points and the filter has m data points, the output will have $n + m - 1$ data points, and among them there are $n - m$ points in steady state.

The second example is adding a Hamming window to $h(n)$. The Hamming window is given in (4.15) as

$$w(n) = 0.54 + 0.46 \cos\left[\frac{n\pi(n - N/2)}{N}\right] \tag{8.45}$$

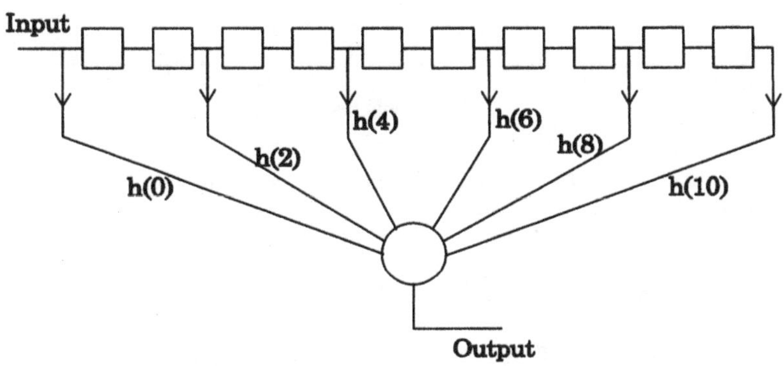

Figure 8.11 Direct form of filter for Hilbert transform.

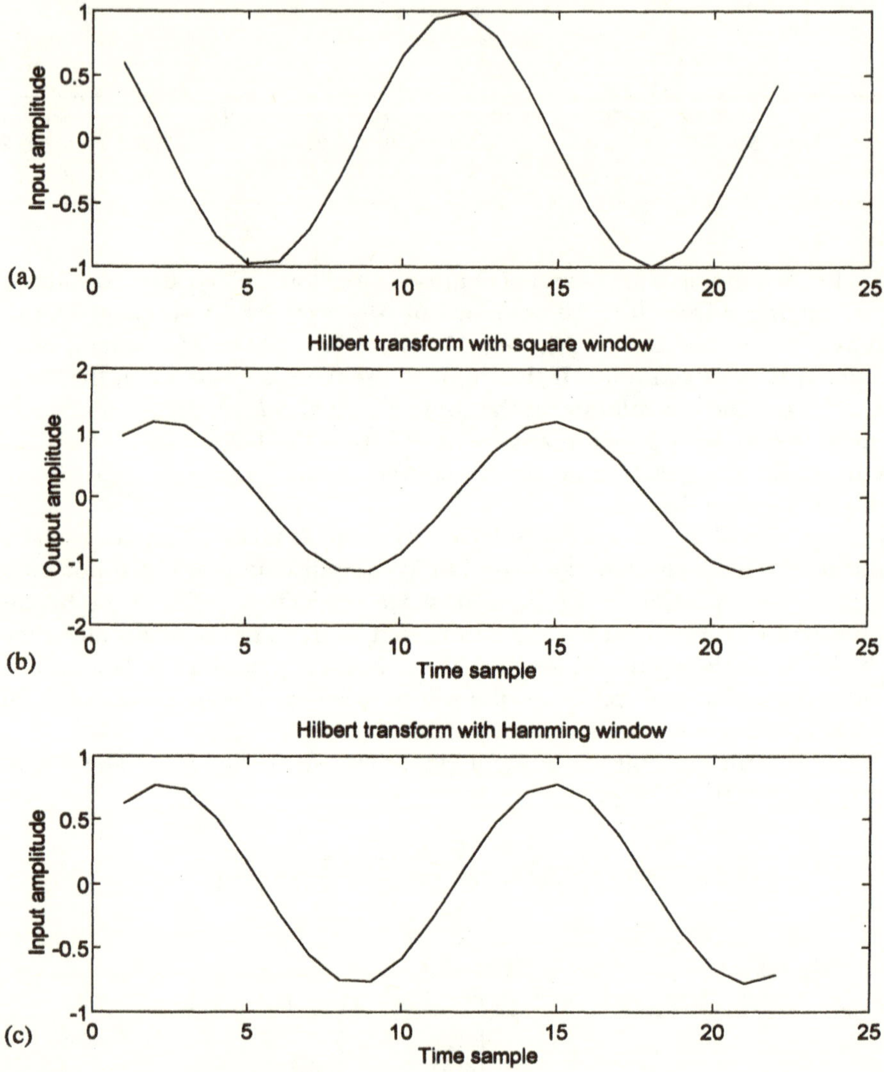

Figure 8.12 Input and output of Hilbert transform filter: (a) input, (b) output (rectangular window), (c) output (Hamming window).

where $N = 10$ and n is from 0 to 10. The filters $w(n)$ and $h(n)w(n)$ are listed in Table 8.2.

The output is shown in Figure 8.12(c). In this particular case, using a Hamming window makes little difference in time domain. In the frequency domain, the Hamming window will limit the highest sidelobe to 43 dB from the mainlobe.

Table 8.2
Filter Values for $w(n)$ and $h(n)w(n)$

n	0	1	2	3	4	5	6	7	8	9	10
$h(n)$	$-2/5\pi$	0	$-2/3\pi$	0	$-2/\pi$	0	$2/\pi$	0	$2/3\pi$	0	$2/5\pi$
$w(n)$	080	.168	.398	.682	.912	1	.912	.682	.398	.168	.080
$h(n)w(n)$	–.010	0	–.084	0	–.581	0	.589	0	.084	0	.010

8.11 NARROWBAND *I* AND *Q* CHANNELS THROUGH A SPECIAL SAMPLING SCHEME [16–22]

Another way to build I and Q channels can be derived from a special sampling scheme. Let us start with a narrowband signal. If the input signal is real and narrowband with a known frequency f_i, it can be written as

$$V_i = A\sin(2\pi f_i t + \theta) \tag{8.46}$$

where A is the amplitude and θ is the initial phase of the input signal.

If the sampling period t_s starts from 0 and operates at twice the Nyquist sampling rate or $4f_i$, which corresponds to $t_s = 1/4f_i$, the outputs starting from $t = 0$ are $V_i = A\sin\theta$, $A\cos\theta$, $-A\sin\theta$, $-A\cos\theta$, $A\sin\theta$. . . . It is obvious that the odd and even terms are 90 deg out of phase. Thus, one can treat the odd terms as the I channel and the even terms as the Q channel. This approach may be applicable to a narrowband system. In an EW system, the frequency of input signal is unknown and hence this approach, in general, is not applicable.

8.12 WIDEBAND *I* AND *Q* CHANNELS THROUGH A SPECIAL SAMPLING SCHEME [22, 23]

The following approach is somewhat similar to the scheme discussed in the last section, but it is for inputting a signal with unknown frequency. This concept can be explained through an analog approach. Let us use Figure 8.7(b) again. The input signal is divided into two in-phase paths and the local oscillator signals applied to the two mixers are 90 deg out of phase. The outputs of the two mixers are

$$\begin{aligned} v_{\text{if1}} &= \sin(2\pi f_i t + \theta)\cos(2\pi f_o t) \\ &= \frac{1}{2}\{\sin[2\pi(f_i - f_o)t + \theta] + \sin[2\pi(f_i + f_o)t + \theta]\} \\ v_{\text{if2}} &= \sin(2\pi f_i t + \theta)\sin(2\pi f_o t) \\ &= \frac{1}{2}\{\cos[2\pi(f_i - f_o)t + \theta] - \cos[2\pi(f_i + f_o)t + \theta]\} \end{aligned} \tag{8.47}$$

where both the low and the high frequencies are 90 deg out of phase. It is interesting to note that basically this is the same result as found in (8.19). In analog design, at each output, a lowpass filter is used to filter out the sum frequency term.

In the above equations, the local oscillator frequency f_o and its initial phase can be considered as known, but the input frequency f_i is unknown. Since the input phase and the sampling frequency of the mixers are known, one can choose the sampling time equal to $1/4f_o$ and start at zero phase. Under this condition, and substituting $t = 1/4f_0$ in (8.47), the two outputs are

$$\begin{aligned} V_{if1} &= \sin(\theta),\ 0,\ -\sin(2\pi f_i t_2 + \theta),\ 0,\ \sin(2\pi f_i t_4 + \theta) \\ V_{if2} &= 0,\ \sin(2\pi f_i t_1 + \theta),\ 0,\ -\sin(2\pi f_i t_3 + \theta),\ 0 \end{aligned} \tag{8.48}$$

In this equation, $t_0 = 0$. According to (8.47), these two outputs are 90 deg out of phase. If these data are combined into one channel and arranged in chronological order in time domain neglecting the zeros, the result is

$$V_{if} = \sin(\theta),\ \sin(2\pi f_i t_1 + \theta),\ -\sin(2\pi f_i t_2 + \theta),\ \sin(2\pi f_i t_3 + \theta),\ \ldots \tag{8.49}$$

These results can be obtained from a real signal. If the input signal is

$$V_{if} = \sin(2\pi f_i t + \theta) \tag{8.50}$$

the result in (8.49) can be obtained by simply making $t = t_0, t_1, t_2, t_3, \ldots$ where $t_0 = 0$. The result is shown in Figure 8.13. In this figure the initial phase is assumed zero.

Now let us look at this problem from a reverse point of view. If a real signal is sampled at any time interval, the result will be the same as in (8.49). Zeros can be added into the sequence, which can then be separated into two channels to match the results in (8.48).

If the sampling frequency is f_s, from the Nyquist sampling theorem, the input bandwidth is limited to $f_s/2$. Based on the condition of obtaining (8.48), let us imagine that the frequency of the local oscillator is related to the sampling frequency as

$$f_o = f_s/4 \tag{8.51}$$

This approach will generate the results in (8.48) which can be converted into two channels that are 90 deg out of phase. The outputs of the two channels are

$$\begin{aligned} v_{if1} &= \frac{1}{2}\{\sin[2\pi(f_i - f_o)t + \theta] + \sin[2\pi(f_i + f_o)t + \theta]\} \\ v_{if2} &= \frac{1}{2}\{\cos[2\pi(f_i - f_o)t + \theta] - \cos[2\pi(f_i + f_o)t + \theta]\} \end{aligned} \tag{8.52}$$

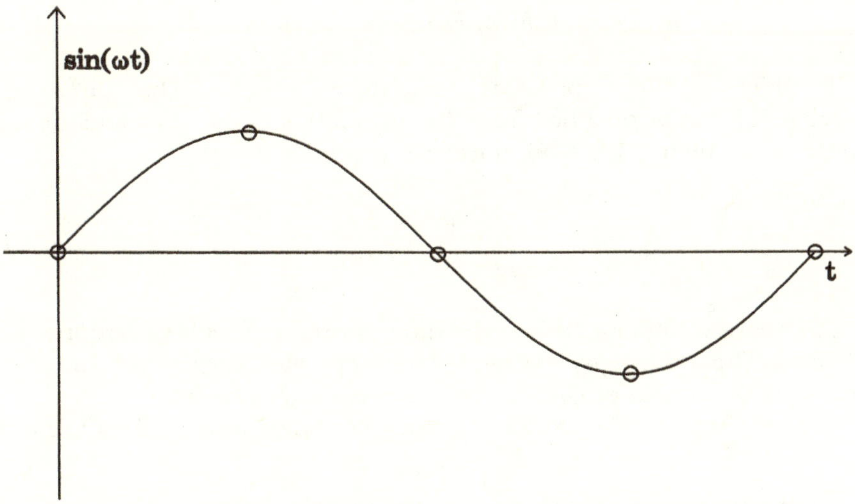

Figure 8.13 A sine wave is sampled at five points.

This is the same result as in (8.47), but it is not the desired result. The desired results are

$$v_{\text{if1}} = \frac{1}{2}\sin[2\pi(f_i - f_o)t + \theta]$$
$$v_{\text{if2}} = \frac{1}{2}\cos[2\pi(f_i - f_o)t + \theta] \tag{8.53}$$

In order to obtain these results, lowpass filters must be used to filter out the high-frequency terms, which are represented by the $f_i + f_o$ terms in (8.52). Since the input frequency is limited to $f_s/2$, the filter cutoff frequency should match this value.

Let us use an example to demonstrate this idea. If an input signal is sampled at 100 MHz, the input bandwidth is limited to 50 MHz. From (8.51), the corresponding local oscillator frequency is 25 MHz. If the outputs are $x(1), x(2), \ldots, x(n)$, according to (8.48), these data can be divided into two groups with zeros added as

$$I(t) = x(1), 0, -x(3), 0, x(5), 0, -x(7)$$
$$Q(t) = 0, x(2), 0, -x(4), 0, x(6), 0 \tag{8.54}$$

Both data must pass through a lowpass filter of $h(t)$ with the cutoff frequency at 50 MHz. Let us use a lowpass FIR filter to filter out the high-frequency components. The impulse response of the filter can be written as

$$h(t) = h(1), h(2), h(3), h(4), h(5), h(6), \ldots \tag{8.55}$$

Two identical filters are needed, one for each output. The outputs of the I and Q channels can be obtained from the convolution of the input and the impulse response of the filter. The relation can be written as

$$\begin{aligned} y_I &= h(t) \star I(t) \\ y_Q &= h(t) \star Q(t) \end{aligned} \tag{8.56}$$

The above equation provides the desired results. This idea is represented in Figure 8.14. The input signal is sampled and separated alternatively into two data channels. At the output of each channel there is a lowpass filter.

The next section will provide hardware considerations on the filter design.

8.13 HARDWARE CONSIDERATIONS ON FILTER DESIGN FOR WIDEBAND DIGITAL *I-Q* CHANNELS [23]

One important design requirement in implementing the approach in Section 8.12 is to reduce the operation speed of the lowpass filter. If the input data are sampled at 100 MHz, the filter must operate at the same speed in order to keep up with the input data rate. However, as shown in (8.54), half of the data are zeros. Taking these zeros into consideration, the filter operation speed can be reduced to $f_s/2$ or 50 MHz. The price one has to pay for such an approach is to double the number of filters. Instead of using one filter per output channel, two filters are required. The approach is discussed as follows.

If the results of (8.54) and (8.55) are substituted into (8.56), and from the definition of convolution, the results are

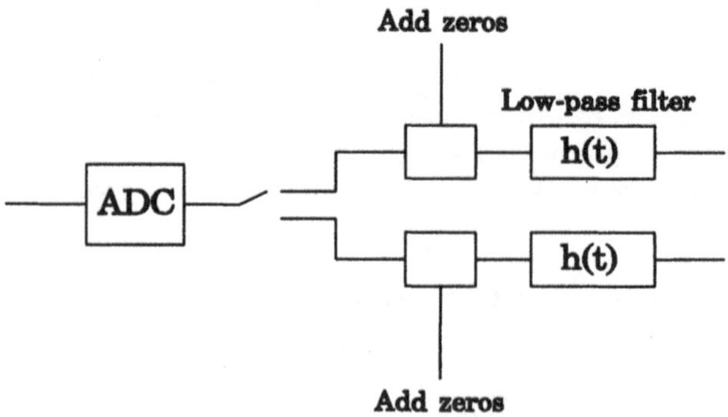

Figure 8.14 Digital mixing and filtering.

$$
\begin{aligned}
y_I(1) &= x(1)h(1) \\
y_I(2) &= x(1)h(2) \\
y_I(3) &= x(1)h(3) - x(3)h(1) \\
y_I(4) &= x(1)h(4) - x(3)h(2) \\
&\cdots \\
y_Q(1) &= 0 \\
y_Q(2) &= x(2)h(2) \\
y_Q(3) &= x(2)h(3) - x(4)h(1) \\
y_Q(4) &= x(2)h(4) - x(4)h(2)
\end{aligned}
\tag{8.57}
$$

In these equations, the outputs y_I and y_Q can be divided into odd and even time terms. The odd time term outputs are $y_I(1)$, $y_I(3)$, . . . , $y_Q(1)$, $y_Q(3)$, . . . , and the even term outputs are $y_I(2)$, $y_I(4)$, . . . , $y_Q(2)$, $y_Q(4)$. . . . It is interesting to note that the odd outputs contain only odd terms of $h(t)$ (i.e. $h(1)$, $h(3)$), and the even terms contain only even terms, of $h(t)$ (i.e., $h(2)$, $h(4)$). Let us use $h_o(t)$ and $h_e(t)$ to represent the odd and even terms of $h(t)$.

If $h(t)$ has N total coefficients, $h_o(t)$ and $h_e(t)$ each will contain $N/2$ of them. The output of I and Q can be subdivided into two channels and each channel will have a lowpass filter. There is a total of four filters as shown in Figure 8.15. Since the filter in the I channel only operates on odd outputs (i.e., $x(1)$, $x(3)$), and the Q channel filter only operates on the even outputs (i.e., $x(2)$, $x(4)$), the operating speed of these filters is only at half of the sampling frequency. In this approach zeros are not added to the output data, as shown in Figure 8.14.

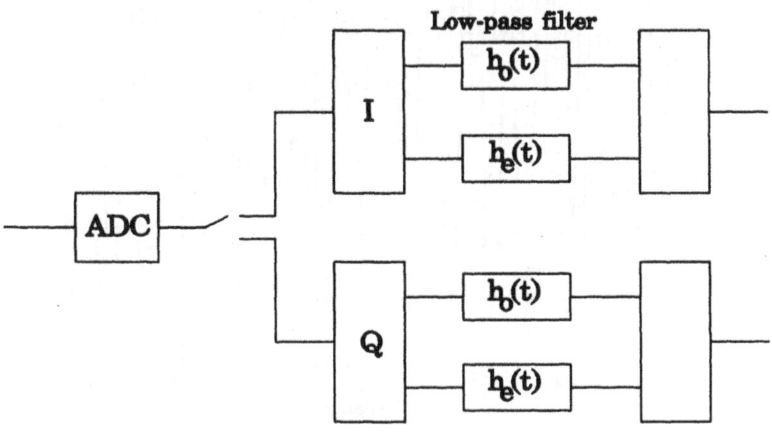

Figure 8.15 Digital mixing with modified filters.

A computer program is listed in Appendix 8.A to demonstrate this approach. The input is a sine wave. It should be noted that in (8.57), the data point $y_Q(1) = 0$. However, from the computer simulation, the first data point generated, $y_Q(1)$, is not zero. In order to make the result correct, a zero is added at the beginning of the $y_Q(t)$ to represent $y_Q(1)$. The outputs are shown in Figure 8.16. One should note that there are transient effects at both the input and the output of the pulse. Since each filter has a length of 16 points, the transient lasts the same length.

8.14 DIGITAL CORRECTION OF *I* AND *Q* CHANNEL IMBALANCE

If the imbalance of two channels can be measured, they can be corrected. The correction scheme is through the Gram-Schmidt procedure, which can be stated as follows for a two-dimensional case[24, 25]. Any two vectors can be represented by two orthonormal vectors. This problem is solved in [26] and the results will be presented here.

In order to simplify this discussion, the amplitude error is included in the *I* channel and the phase error is in the *Q* channel. Usually, there is dc bias on the outputs and this bias must be removed from the outputs. After the dc bias is taken away, the results can be written as

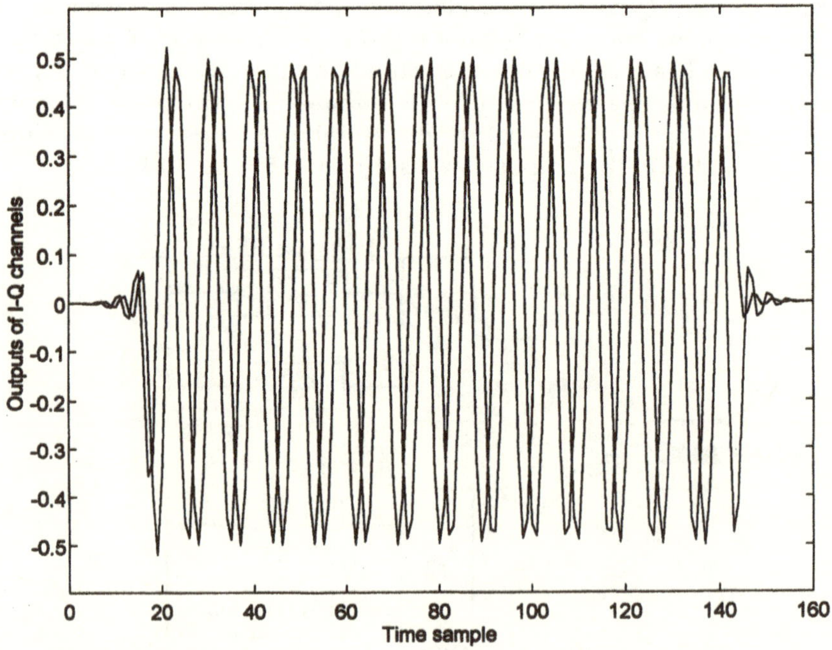

Figure 8.16 *I-Q* outputs from filters.

$$I_1 = (1 + \alpha)A\cos(2\pi f_{if}t)$$
and
$$Q_1 = A\sin(2\pi f_{if}t + \epsilon) \tag{8.58}$$

where α represents amplitude imbalance and ϵ represents phase imbalance. The corrected outputs can be written in matrix form as

$$\begin{bmatrix} I_2 \\ Q_2 \end{bmatrix} = \begin{bmatrix} E_1 & 0 \\ P & 1 \end{bmatrix} \begin{bmatrix} I_1 \\ Q_1 \end{bmatrix} \tag{8.59}$$

where I_2 and Q_2 are orthogonal and balanced. Values of E_1 and P are solved as

$$E_1 = \frac{\cos\epsilon}{1 + \alpha}$$
$$P = \frac{-\sin\epsilon}{1 + \alpha} \tag{8.60}$$

If the values in (8.60) are substituted into (8.59), the results are

$$I_2 = A\cos\epsilon\cos(2\pi f_{if}t)$$
$$Q_2 = A\cos\epsilon\sin(2\pi f_{if}t) \tag{8.61}$$

which indicate that the two outputs have the same amplitudes and they are 90 deg out of phase.

Since ϵ and α are quantities close to zero, E_1 is close to unity. It is conventional for digital arithmetic to make the scaling coefficient E_1 a small number. Thus, a new E is often defined as $E_1 - 1$. The required correction coefficients for gain and phase become

$$E = \frac{\cos\epsilon}{1 + \alpha} - 1$$
$$P = \frac{-\sin\epsilon}{1 + \alpha} \tag{8.62}$$

The following discussion is to find the imbalance and from the collected data to generate the correction coefficients in the above equation. This correction can be accomplished by using a testing signal of frequency f. The output of the I-Q channels can be written in complex form as

$$s_t(t) = (1 + \alpha)A\cos(2\pi ft + \psi) + a + j[A\sin(2\pi ft + \psi + \epsilon) + b] \tag{8.63}$$

where ψ is the initial phase of the input signal, ϵ is the imbalance in phase, and a and b are dc levels. The input frequency is sampled at frequency $f_s = 1/t_s$ where t_s is the unit sampling time. The sampling frequency must be four times the input frequency, which means $f_s = 4f$. Only four samples in time domain are required to calculate the correction coefficients. They are

$$\begin{aligned} s_t(0) &= (1 + \alpha)\ A \cos \psi + a + j[A \sin(\psi + \epsilon) + b] \\ s_t(t_s) &= -(1 + \alpha)\ A \sin \psi + a + j[A \cos(\psi + \epsilon) + b] \\ s_t(2t_s) &= -(1 + \alpha)\ A \cos \psi + a + j[-A \sin(\psi + \epsilon) + b] \\ s_t(3t_s) &= (1 + \alpha)\ A \sin \psi + a + j[-A \cos(\psi + \epsilon) + b] \end{aligned} \tag{8.64}$$

The corresponding FFT can be written as

$$S_t(k) = \sum_{n=0}^{N-1} s_t(t)\ e^{-j2\pi nk/N} \tag{8.65}$$

Since the input frequency is f and the input is sampled at $4f$ and only four samples are obtained, the frequency domain and $S_t(k)$ has only four outputs. Three of the four frequency components are as follows: $S_t(0)$ is the dc component, $S_t(1)$ is the input frequency, and $S_t(3)$ is the image.

Substituting the results from (8.64) into this equation, the frequency domain output is

$$S_t(0) = 4(a + jb) \tag{8.66}$$

This information can be used to correct the dc bias.

The coefficients E and P for correction of gain and phase errors can be obtained from the frequency domain outputs at test signal frequency $S_t(1)$ and at the image frequency $S_t(3)$. The output at the test signal frequency is

$$S_t(1) = 2A[(1 + \alpha) + \cos \epsilon + j \sin \epsilon]e^{j\psi} \tag{8.67}$$

The output of the filter at the image of the test signal frequency is

$$S_t(3) = 2A[(1 + \alpha) - \cos \epsilon + j \sin \epsilon]e^{-j\psi} \tag{8.68}$$

It is easily shown that

$$S_t^*(1) + S_t(3) = 4A(1 + \alpha)\ e^{-j\psi} \tag{8.69}$$

where $S_t^*(t)$ represents the conjugate of $S_t(t)$. From (8.68) and (8.69), one can find

$$\frac{2S_t(3)}{[S_t^*(1) + S_t(3)]} = 1 - \frac{\cos \epsilon}{1 + \alpha} + \frac{j \sin \epsilon}{1 + \alpha} \tag{8.70}$$

From (8.62) and (8.69), the following results can be obtained:

$$\begin{aligned} E &= -Re\left[\frac{2S_t(3)}{S_t^*(1) + S_t(3)}\right] \\ P &= -Im\left[\frac{2S_t(3)}{S_t^*(1) + S_t(3)}\right] \end{aligned} \tag{8.71}$$

These relations are used to estimate the correction factors. This correction method should be tested at different frequencies.

This correction can be applied at one frequency at a time. When the I and Q channels cover a wide bandwidth, the imbalance is a function of frequency. The simple calibration as discussed in this section might be tedious to apply.

REFERENCES

[1] East, P. W. "Design Techniques and Performance of Digital IFM," *IEEE Proc.*, Vol. 129, 1982, pp. 154–163.

[2] Tsui, J. B. Y. *Microwave Receivers With Electronic Warfare Applications*, New York, NY: John Wiley & Sons, 1986.

[3] Brown, T. T. *Mixer Harmonic Chart*, Electronic Buyers' Guide, June 1953, pp. R58–59.

[4] Smith, T., and Wright, J. "Spurious Performance of Microwave Double-Balanced Mixers," Watkins-Johnson Application note, Aug. 1975.

[5] Huany, M. Y., Buskirk, R. L., and Carlile, D. E. "Select Mixer Frequencies Painlessly," *Electronic Design 8*, April 12, 1976, pp. 104–109.

[6] Henderson, B. C. "Mixers: Part 1 Characteristics and Performance," Watkins-Johnson Tech-notes, March/April 1981.

[7] Henderson, B. C. "Predicting Intermodulation Suppression in Double-Balanced Mixers," Watkins-Johnson Tech-notes, July/August 1983.

[8] Papoulis, A. *The Fourier Integral and its Applications*, New York, NY: McGraw Hill Book Co., 1962.

[9] Engler, H. Georgia Tech Research Institute, Atlanta, GA, Private communication.

[10] Stremler, F. G. *Introduction to Communication Systems*, 2nd Edition, Reading, MA: Addison-Wesley Publishing Co., 1982.

[11] Ziemer, R. E., Tranter, W. H., and Fannin, E. R. *Signals and Systems Continuous and Discrete*, New York, NY: Macmillan Publishing Co., 1983.

[12] Urkowitz, H. "Hilbert Transforms of Bandpass Functions," *Proc. IRE.*, Vol. 50, Oct. 1962, p. 2143.

[13] Bedrosian, E. "A Product Theorem for Hilbert Transforms," *IEEE Proc.*, Vol. 51, May 1963, pp. 868–869.

[14] Rabiner, L. R., and Gold B. *Theory and Application of Digital Signal Processing*, Englewood Cliffs, NJ: Prentice-Hall, Inc., 1975.

[15] Oppenheim, A. V., and Schafer, R. W. *Digital Signal Processing*, Englewood Cliffs, NJ: Prentice-Hall, Inc., 1975.

[16] Waters, W. M., and Jarrett, B. R. "Bandpass Signal Sampling and Coherent Detection," *IEEE Trans. on Aerospace and Electronic Systems*, Vol. AES-18, Nov. 1982, pp. 731–736.

[17] Rice, D. W., and Compton, R. T. "Quadrature Sampling With High Dynamic Range," *IEEE Trans. on Aerospace and Electronic Systems*, Vol. AES-18, Nov. 1982, pp. 736–739.
[18] Rader, C. M. "A Simple Method for Sampling In-Phase and Quadrature Components," *IEEE Trans. on Aerospace and Electronic Systems*, Vol. AES-20, Nov. 1984, pp. 821–824.
[19] Waters, W. M., and Jarrett, B. R. "Tests of Direct-Sampling Coherent Detection With a Laboratory Analog-to-Digital Converter," *IEEE Trans. on Aerospace and Electronic Systems*, Vol. AES-21, May 1985, pp. 430–432.
[20] Mitchell, R. I. "Creating Complex Signal Samples From a Band-Limited Real Signal," *IEEE Trans. on Aerospace and Electronic Systems*, Vol. AES-25, May 1989, pp. 426–427.
[21] Liu H. L., Ghafoor, A., and Stockmann, P. H. "A New Quadrature Sampling and Processing Approach," *IEEE Trans. Aerospace and Electronics Systems*, Vol. 25, Sept. 1989, pp. 733–748.
[22] Pellon, L. E. "A Double Nyquist Digital Product Detector for Quadrature Sampling," *IEEE Trans. Signal Processing*, Vol. 40, July 1992, pp. 1670–1681.
[23] Miniuk, J., and Shoemaker, M. Naval Surface Warfare Center, Dahlgren, VA, Private communication.
[24] Van Trees, H. L. *Detection, Estimation, and Modulation Theory, Part I*, New York, NY: John Wiley & Sons, 1968.
[25] Carlson, A. B. *Communication Systems*, 2nd Edition, New York, NY: McGraw Hill Book Co., 1968.
[26] Churchill, F. E., Ogar, G. W., and Thompson, B. J. "The Correction of *I* and *Q* Errors in a Coherent Processor," *IEEE Trans. on Aerospace and Electronic Systems*, Vol. AES-17, Jan. 1981, pp. 131–137.

APPENDIX 8.A

```
% df8_16.m IQBK generates I and Q channels from fs/2
% clear
n1 = 128;
n=[0:n1-1];
% f = input('enter input frequency in MHz from 1 to 49 = ');
f=14*10^6;
x = sin(2*pi*f*1e-8*n+.01);  % sampled at 100 MHz

% ******** gnerate I and Q channels ********
xx = reshape(x,4,n1/4);
xrm = [xx(1,1:n1/4); -xx(3,1:n1/4)];
xim = [xx(2,1:n1/4); -xx(4,1:n1/4)];
xr = reshape(xrm,1,n1/2);
xi = -1*reshape(xim,1,n1/2);

% ******** create a low pass filter ********
dbst = 70;
betat = .1102*(dbst-8.7);
windowt = kaiser(32,betat);
hkt = fir1(31,0.5, windowt);
filtab = reshape(hkt,2,16);
filtat = filtab(1,1:16);
filtbt = filtab(2,1:16);
```

```
% ******** filtering ********
cnvlra = conv(filtat,xr);
cnvlrb = conv(filtbt,xr);
cnvlia = conv(filtat,xi);
cnvlib = conv(filtbt,xi);
I1 = [cnvlra; cnvlrb];
It = reshape(I1,1,2*length(cnvlra));
Q1 = [cnvlia; cnvlib];
Qt1 = reshape(Q1,1,2*length(cnvlra));
Qt = [0 Qt1(1:n1+30-1)]; % add 0 as the first point

% ******** plot the results ********
plot(1:n1+30,Qt, 1:n1+30,It)
axis([0 160 -.6 .6])
xlabel('Time sample')
ylabel('Outputs of I Q channels')
```

CHAPTER 9

Sensitivity and Detection Problems

9.1 INTRODUCTION

In this chapter, the sensitivity of a receiver will be discussed. The sensitivity of a receiver is defined as its capability to measure the weakest signal. The output of an analog-to-digital converter (ADC) is digital. This output will impact the measurement approach of input signals and thus the receiver sensitivity. The conventional way to calculate sensitivity is through mathematical modeling. However, due to the nonlinear property of the ADC, it is difficult to manipulate the calculations mathematically. As a result, the basic analysis is still based on an analog approach. If the ADC has a large number of bits, the result will be close to the analog approach. The threshold for detecting false alarms and signals in a digital receiver cannot be set arbitrarily in a continuous sense. The threshold can only be chosen from a finite number of levels.

This chapter discusses different types of detection schemes. The difference between the detection schemes of analog and digital receivers will be emphasized. The possible advantages of a digital receiver from a detection point of view will be discussed. The detection based on one data sample will be discussed first. The one sample case will be extended to multiple samples. Finally, frequency domain detection, which can be applied to digital receivers, will be examined. Examples with computer programs are used to illustrate some of the detection schemes.

Before the general discussion, let us clarify the meaning of the word detection. This word has two distinct applications in a discussion of receivers. In the first sense, it refers to the process of converting a radio frequency (RF) signal or noise into a video signal. In the second sense, it refers to the process that determines whether a signal is present. In spite of this ambiguity, the meaning should be clear from the context.

9.2 ELECTRONIC WARFARE RECEIVER DETECTION APPROACH

In this section, the detection procedure in an analog receiver will be discussed. The detection is based on pulsed signals. Although this may not be true for all conventional receivers, such as communication receivers, it represents most electronic warfare (EW) receivers. In an analog receiver, the input RF signal is converted into a video signal by removing the RF through a crystal video detector. Sometimes the signal is converted into an intermediate frequency (IF), then converted to a video signal through a crystal video detector. A simple threshold comparator is used at the output of the video detector. If the video signal is higher than the threshold, the receiver will detect the signal as shown in Figure 9.1. This type of detection is achieved in the time domain because the video signal is in the time domain.

In a radar receiver, since the input signal shape is known, it is possible to design the receiver with a matched filter to match the input signal and maximize the receiver sensitivity. The receiver can also integrate many pulses to further improve the receiver sensitivity.

In an EW receiver, since the input signal is unpredictable, it is impossible to design a receiver with a matched filter to obtain an optimum sensitivity. Usually, an EW receiver is designed to match the shortest pulse anticipated. For example, if an EW receiver is designed to intercept pulses with 100-ns minimum pulse width, the minimum filter bandwidth that determines the frequency data resolution is about 10 MHz (or 1/100 ns). The sensitivity of the receiver is determined by the

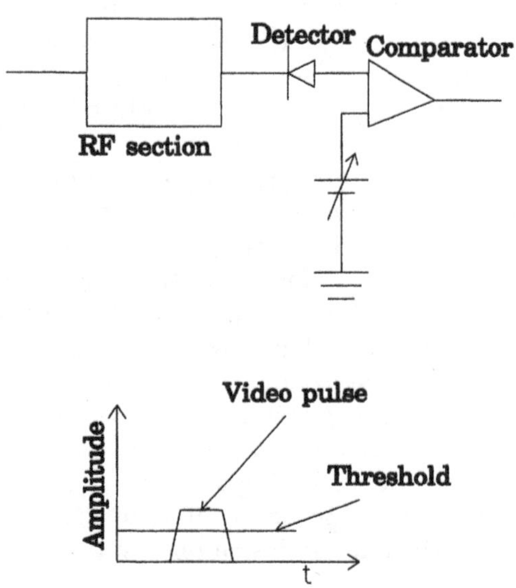

Figure 9.1 Detection circuit in an EW receiver.

noise floor, the noise figure (discussed in Chapter 7), and the threshold of the system. This minimum filter bandwidth will determine the noise floor, thus determining the sensitivity of the receiver. Any signal with a pulse width of more than 100 ns will be detected with this sensitivity. Theoretically, a receiver can be designed to intercept a longer pulse with narrow bandwidth to produce higher sensitivity. It is desirable to design an EW receiver with variable sensitivity (i.e., for longer input pulse, the receiver will have higher sensitivity).

An EW receiver is usually designed to receive the input signal on a pulse-by-pulse basis and convert the information on a pulse into a pulse descriptor word (PDW). This subject has been discussed in Chapter 2. The PDWs of the received pulses are compared through a digital signal processor to deinterleave the pulses into a pulse train emitted by a certain radar. Thus, it is unusual to design an EW receiver with multiple pulse integration to improve sensitivity.

9.3 POTENTIAL DETECTION ADVANTAGES IN A DIGITAL EW RECEIVER

In a digital receiver, the IF output is digitized at very high speed. Within one pulse, many data points can be collected. A fast Fourier transform (FFT) can be applied to the data to convert the ADC output to the frequency domain. Therefore, the output can be detected either in the time domain or in the frequency domain. The potential detection in both the time and the frequency domains will be discussed in the following sections.

9.3.1 Frequency Domain Detection

In this type of detection, FFT operation is performed on all the input data. In order to perform the FFT, the length of the input data points must be predetermined (e.g., 64 points chosen as a group). The FFT can be performed on the data points in a nonoverlapping mode as shown in Figure 9.2(a) or in a 50% overlapping mode as shown in Figure 9.2(b). This operation is sometimes called the short time fast Fourier transform (STFT), and will be discussed in detail in Chapter 11.

The frequency domain detection can be defined as the detection performed on each individual FFT output. Strictly speaking, if this detection is performed on a sequence of FFT outputs, one can consider that the detection is in the time domain. However, in this chapter let us limit the frequency domain detection to detection from one FFT output.

More overlapping in consecutive FFT outputs requires more operations, but less overlapping may lose information. For example, if a pulsed signal is 64 data points long and divided into two nonoverlapping consecutive data sets, the FFT outputs will be spread into several frequency bins and the amplitude will be lower than the expected value. Thus, the sensitivity of the receiver will be lower. The highest sensitivity can be obtained by sliding the processing window across the data

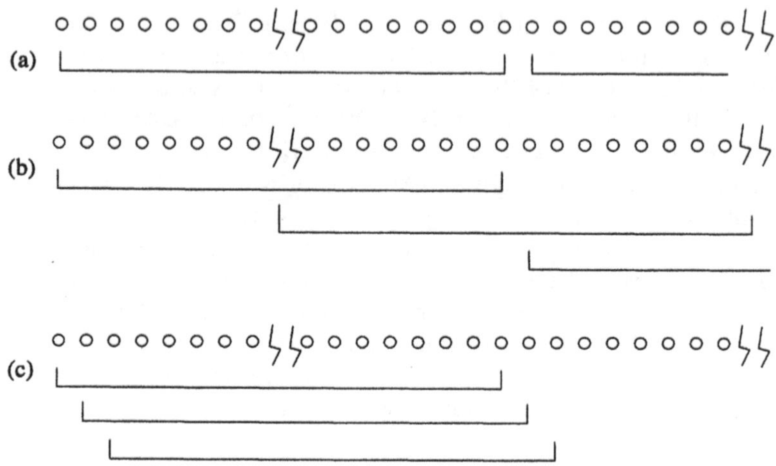

Figure 9.2 Data points grouping: (a) nonoverlapping, (b) 50% overlapping, (c) one point sliding.

one point at a time, as shown in Figure 9.2(c). This approach can be called sliding FFT, which is discussed in Section 4.7. The FFT outputs are compared with a certain threshold to determine whether there are signals, and the number of signals is determined. In this approach, sidelobes must be neglected or eliminated.

The major advantage of frequency domain detection is that the FFT operating on a large number of data points will pull the signal out of the noise. The disadvantages are that many FFT operations are required and the FFT length is predetermined. In other words, the FFT speed must match the sampling speed. Even if the data points contain only noise, the FFT must be performed. Since the length of the FFT is predetermined, when the signal length does not match the FFT, the receiver cannot achieve optimum sensitivity. It is theoretically possible to perform the FFT with many different data lengths. However, this approach may increase the complexity of signal processing excessively.

9.3.2 Time Domain Detection

A simple time domain detection is relatively easy to perform. The amplitude of the signal is compared with a fixed threshold to determine whether there are signals in the data. Assume that the signal is collected in complex form. The amplitude of the signal can be obtained as

$$
\begin{aligned}
x_r &= A\cos(2\pi ft) \\
x_i &= A\sin(2\pi ft) \\
A &= \sqrt{x_r^2 + x_i^2}
\end{aligned}
\tag{9.1}
$$

where x_r and x_i are the real and imaginary parts of the input signal, and A and f are the amplitude and frequency of the input signal, respectively.

Obviously, the time domain detection is easier than the frequency domain detection mentioned above. If the output data from the ADC can be detected first in the time domain, the FFT can be applied only to data strings with signals to obtain frequency information. When there is only noise in the data, FFT processing on the data can be avoided. If this scheme can be effectively implemented, the FFT speed may not be required to match the ADC sampling speed. However, the sensitivity of the receiver is determined by the time domain detection scheme.

In a digital receiver, the number of data points that can be obtained on one pulse depends on the pulse width (*PW*) and the sampling rate (f_s). Each individual data point can be compared with a threshold to obtain a sensitivity. However, the sensitivity can be improved by processing many data points together. The number of data points to be processed together must also be predetermined. However, since comparison in the time domain is rather simple, many different data lengths may be selected. If the input PW is close to one of these predetermined data lengths, the detection should be close to the matched filter for rectangular pulse shapes and high sensitivity can be achieved. Although this approach also increases the complexity of signal processing, it is more likely to be implemented than implementing FFT with various lengths.

Before the detection scheme can be meaningfully discussed, the threshold must be determined. The threshold setting determines the false alarm rate, which must be specified by the user of the receiver. First, the false alarm of the receiver is discussed.

9.4 FALSE ALARM TIME AND PROBABILITY OF FALSE ALARM FOR ONE DATA SAMPLE

In an EW receiver, the false alarm time is often specified. The false alarm time is the average time the receiver takes to produce a false alarm when there is no input signal. For example, if the false alarm time is 100 sec, it means on the average the receiver will produce one false alarm every 100 sec. If the ADC sampling rate is f_s and the probability of false alarm on one single sample is P_{fas}, the false alarm time T_f can be written as

$$T_f = \frac{1}{P_{\text{fas}} f_s} \quad \text{or} \quad P_{\text{fas}} = \frac{1}{T_f f_s} \tag{9.2}$$

In the above equation, it is assumed that every output sample of the ADC is compared to the threshold. If any output crosses the threshold, it is considered a false alarm. If the sampling frequency $f_s = 100$ MHz, in order to generate one false alarm every 100 sec, the probability of false alarm $P_{\text{fas}} = 1 \times 10^{-10}$ is obtained from

(9.2). Once the allowable probability of false alarm has been determined, the next step is to set the threshold at the output to generate such a probability false alarm.

9.5 THRESHOLD SETTING FOR ONE DATA SAMPLE [1–5]

In this section, only one data point is compared with a fixed threshold. The probability density of noise $p(x)$ is assumed Gaussian; thus,

$$p(x) = \frac{1}{\sqrt{2\pi}\sigma} e^{-\frac{x^2}{2\sigma^2}} \tag{9.3}$$

where σ^2 is the variance of the noise. In order to generate an envelope for the input signal, I and Q channels are required. If (9.3) represents the output noise of the I channel, the Q channel can be similarly represented by

$$p(y) = \frac{1}{\sqrt{2\pi}\sigma} e^{-\frac{y^2}{2\sigma^2}} \tag{9.4}$$

The probability density of the envelope is the product of $p(x)p(y)$, which can be written as a function of r as

$$p(r) = \int_0^{2\pi} rp(x)p(y)\, d\phi = \int_0^{2\pi} \frac{r}{2\pi\sigma^2} e^{\frac{-(x^2+y^2)}{2\sigma^2}}\, d\phi = \frac{r}{\sigma^2} e^{\frac{-r^2}{2\sigma^2}} \tag{9.5}$$

where $r^2 = x^2 + y^2$ and $\phi = \tan^{-1}(y/x)$. This probability density function is referred to as the Rayleigh distribution.

The probability of false alarm can be written as

$$P_{fas} = \int_{r_1}^{\infty} p(r)\, dr = \int_{r_1}^{\infty} \frac{r}{\sigma^2} e^{-\frac{r^2}{2\sigma^2}}\, dr = e^{-\frac{r_1^2}{2\sigma^2}} \tag{9.6}$$

where r_1 is the threshold. The above equation can be written in a slightly different form as

$$-\frac{r_1^2}{2\sigma^2} = \ln(P_{fas}) \qquad \text{or} \qquad r_1 = \sqrt{-2\sigma^2 \ln(P_{fas})} \tag{9.7}$$

The value obtained from this equation, in general, will not coincide with one of the quantization levels of an ADC. One should choose a quantization level close to r_1. A higher value will reduce the probability of false alarm and sensitivity, and a lower value will have the opposite effect.

In the above equation, if r_1 is chosen to match one of the quantization levels of the ADC and P_{fas} is determined by (9.2), the only variable that can be changed is σ. As discussed in Chapter 7, the noise is the combination of the quantization noise from the ADC and the noise from the RF amplifier in front of the ADC. Thus, from (7.12) and (7.20), the variance of the noise is

$$\sigma^2 = (1 + M)\frac{Q^2}{12R} \tag{9.8}$$

In this equation, M is the ratio of the amplifier output noise to the quantization noise. From (7.29), one can see that the sensitivity of the receiver is degraded by

$$F_s - F = 10 \log\left(\frac{1 + M}{M}\right) \quad \text{dB} \tag{9.9}$$

where F_s and F are the noise figures of the amplifier chain without and with the ADC, respectively. If the M value is small, the sensitivity of the receiver will suffer.

If (9.8) is substituted into (9.7), the result is

$$\frac{r_1^2}{2\sigma^2} = \frac{(nQ)^2}{2(1 + M)\frac{Q^2}{12}} = -\ln(P_{fas}) \tag{9.10}$$

In this equation, r_1 is replaced by nQ where n is an integer number representing the threshold level.

The M value can be obtained as

$$M = -\left[\frac{6n^2}{\ln(P_{fas})} + 1\right] \tag{9.11}$$

The value of M can be adjusted by varying the amplifier gain in front of the ADC.

9.6 PROBABILITY OF DETECTION FOR SINGLE-SAMPLE DETECTION [1–5]

When there is a signal in the data, the probability density function from the outputs of the I and Q channels can be written as

$$\begin{aligned} p(x) &= \frac{1}{\sqrt{2\pi}\sigma} e^{\frac{-(x-\mu_x)^2}{2\sigma^2}} \\ p(y) &= \frac{1}{\sqrt{2\pi}\sigma} e^{\frac{-(y-\mu_y)^2}{2\sigma^2}} \end{aligned} \tag{9.12}$$

where μ_x, μ_y are the means of the Gaussian distribution and they are related to the input signal as

$$\mu_x = A \cos \alpha \qquad \mu_y = A \sin \alpha \tag{9.13}$$

where α is the initial phase of the signal. Similarly, the x and y are related to r as

$$x = r \cos \phi \qquad y = r \sin \phi \tag{9.14}$$

The joint density function equals

$$\begin{aligned} p(r, \phi/\alpha) &= rp(x)p(y) \\ &= \frac{r}{2\pi\sigma^2} e^{\frac{-[r^2+A^2-2rA(\cos\alpha\cos\phi+\sin\alpha\sin\phi)]}{2\sigma^2}} \\ &= \frac{r}{2\pi\sigma^2} e^{\frac{-[r^2+A^2-2rA\cos(\alpha-\phi)]}{2\sigma^2}} \end{aligned} \tag{9.15}$$

Integrating over ϕ, the probability density is

$$\begin{aligned} p(r|\alpha) &= \int_0^{2\pi} p(r, \phi|\alpha)\, d\phi \\ &= \frac{r}{\sigma^2} e^{-\frac{r^2+A^2}{2\sigma^2}} I_0\left(\frac{rA}{\sigma^2}\right) = p(r) \end{aligned} \tag{9.16}$$

where $I_0(x)$ is the modified Bessel function of zero order. It should be noted that this equation is independent of α, thus it is represented by $p(r)$. This distribution is referred to as the Rician distribution.

The probability of detection P_{ds} for one single sample can be calculated from

$$P_{ds} = \int_{r_1}^{\infty} p(r)\, dr = 1 - \int_0^{r_1} p(r)\, dr \tag{9.17}$$

where r_1 is the threshold. The second portion of the integral is often calculated because the limit ∞ can be avoided and a better accuracy can be expected. Unfortunately the above integration cannot be evaluated analytically. Only numerical integration can be used to obtain the result.

This result is for a single digitized data point. Sometimes the false alarm can be based on multiple data points; thus, the detection problem must also be analyzed for multiple data points. The following sections will discuss the multiple sample detection problem.

9.7 DETECTION BASED ON MULTIPLE DATA SAMPLES

The sampling speed is usually in the hundreds of megahertz and the corresponding sampling time is a few nanoseconds. It is not necessary to make a decision based on one data sample. A more reasonable way is to detect the input signal through the entire PW, which contains multiple data samples. In order to detect the input signal based on multiple data samples, a new set of equations must be derived. They are the probability of false alarm and probability of detection, based on multiple samples. These relations can be obtained from equations of single data point (discussed in previous sections). Several detection approaches will be discussed in this section.

The relation between the false alarm time and the probability of false alarm given in (9.2) can be modified to accommodate multiple samples. Suppose a total of N samples are used to determine the false alarm and the probability of detection. In order to keep the desired false time, the probability of false alarm can be increased by the factor N because instead of making a decision every sampling time $t_s = 1/f_s$, it makes a decision every Nt_s time. Thus, the probability of false alarm can be written as

$$P_{\text{fam}} = \frac{N}{T_f f_s} \tag{9.18}$$

where P_{fam} is used to represent the probability of false alarm for multiple samples. Use of multiple samples to set the threshold is similar to narrowing the video bandwidth in an analog microwave receiver. The equivalent video bandwidth can be considered as f_s/N. In an analog receiver, the video bandwidth must match the minimum PW anticipated. In a digital receiver, the equivalent video bandwidth should match the minimum PW also. However, if one can design a receiver with several different N values, the receiver can match many different PWs. This approach will improve the sensitivity of the receiver on various PWs.

The next question to be answered is how to make a decision on every N sample. Two approaches will be presented:

1. Use the result derived from the single detection scheme to form an L-out-of-N approach. Under this condition, a certain quantization level will be used as the threshold. By selecting an L value that is smaller than N, if within the N samples L or more samples cross the threshold, an input signal is detected.
2. The second approach is based on the sum of N samples. This approach is more akin to traditional radar detection where the sum over N pulses is used to make a decision. Both approaches will be discussed in the following sections.

9.8 DETECTION SCHEME FOR MULTIPLE SAMPLES (*L*-OUT-OF-*N*) [6–8]

In this section, the *L*-out-of-*N* approach will be discussed. First, a given false alarm time is used to generate the probability of false alarm P_{fa}. From the value of P_{fa}, a threshold will be selected.

This false alarm can be treated as a binomial distribution problem. In order to generate the probability of false alarm from N consecutive samples, a total number of L or more samples must cross the threshold. If the probability of one sample crossing the threshold (a certain quantization level) is p, the probability of not crossing the threshold is $(1 - p)$. The probability of exactly L samples crossing the threshold is [6–8]

$$p(L|N) = \frac{N!}{L!(N-L)!}p^L(1-p)^{N-L} \tag{9.19}$$

where $p(L|N)$ represents the probability that L samples are above the threshold.

The $p(L|N)$ value in the above equation can be used to either represent the false alarm rate or to represent the probability of detection depending on the p value. If p_{fas} is used to replace p, $p(L|N)$ represents probability of false alarm. If P_{ds} is used to replace p, $p(L|N)$ represents the probability of detection.

Let us use $p(L+|N)$ to represent the probability that at least L samples cross the threshold (i.e., from L to N samples crossing the threshold). The probability $p(L+|N)$ can be written as

$$p(L+|N) = \sum_{i=L}^{N} p(i|N) \tag{9.20}$$

This equation can be used to determine the desired probabilities, either false alarm or detection.

The calculation of the probability of detection from multiple samples is quite similar to the approach mentioned in the previous section. First, the desired P_{fam} must be determined. Equations (9.19) and (9.20) will be used to match the desired P_{fam}. However, there are many combinations of L and N values, and it is difficult to solve for unique values of L, N, and P_{fas}. As a result, it is suggested to first select an N value and a value of P_{fas}. From (9.19) and (9.20), find a value of L that will generate a P_{fam} value close to the desired value. In this approach, the exact value of P_{fam} cannot be achieved, but a value close to it can be obtained.

Examples will be shown in Section 9.13 to illustrate the usage of these equations.

9.9 PROBABILITY DENSITY FUNCTION AND CHARACTERISTIC FUNCTION [8–12]

The next detection approach is to sum the output of N data points and compare the result with one threshold. In order to perform this analysis, the probability density functions of these N data points must be formulated. These probability functions will be used to produce the false alarm and detection probabilities. Finally, the detection scheme will be discussed. In this section, the relation of probability density function and characteristic function will be presented. This relation will be used to produce the probability density functions.

In order to determine the probability density functions of a sum of random variables, it is convenient to define a characteristic function. The characteristic function is defined as

$$C(\omega) \equiv \int_{-\infty}^{\infty} p(x)\, e^{j\omega x}\, dx \tag{9.21}$$

where $p(x)$ is the probability density function and ω is an arbitrarily chosen variable. The function is identical to the inverse Fourier transform. Thus, the properties of Fourier transform can be used in the following discussion.

The probability density function of a sum of n mutually independent random variables is given by the $n-1$ fold convolution of their individual probability density function $p_i(x)$, which can be written mathematically as [8]

$$p_n(x) = \int_{-\infty}^{\infty} \cdot\cdot \int_{-\infty}^{\infty}\int_{-\infty}^{\infty} p_{n-1}(x - x_{n-1}) \cdot\cdot\, p_2(x - x_2)\, p_1(x_1)\, dx_1 dx_2 \cdot\cdot\, dx_{n-1} \tag{9.22}$$

where x_i are dummy variables. From Section 3.7, one can realize that the above equation is the convolution of p_1, p_2, and so forth in the x domain. Convolution in the x domain is equivalent to the product of the their characteristic functions in the ω domain as

$$C_n(\omega) = C_{n-1}(\omega) \cdot\cdot\, C_2(\omega)\, C_1(\omega) \tag{9.23}$$

because $C(\omega)$ is the inverse Fourier transform of $p(x)$ as shown in (9.21). The inverse transform of this equation will give the desired combined probability density function as

$$p_n(x) = \frac{1}{2\pi}\int_{-\infty}^{\infty} C_n(\omega)\, e^{-j\omega x}\, d\omega \tag{9.24}$$

It is often easier to use the characteristic function to obtain the desired probability density function than carry out the integral shown in (9.22).

9.10 PROBABILITY DENSITY FUNCTION OF SUM SAMPLES WITH A SQUARE LAW DETECTOR [1, 8–12]

It appears in [1] that the mathematical analysis for an envelope detector is more complicated than that for a square law detector. However, the results from these two detectors are very close; the worst-case difference is less than .2 dB. In an actual receiver measurement, if one counts all the error in the test equipment, .2-dB error can be ignored. In a square law detector, the output samples are obtained from a squaring operation of the digitized data. The square law detector will be presented here and the result should be applicable to both envelope and square law detectors.

This discussion is again based on [1]. When there is only noise, the probability density function in (9.5) will be used here. For a square law detector, $z = r^2$ can be used to replace r, and $dz/dr = 2r$. This relation between $p(z)$ and $p(r)$ can be written as

$$p(z)\,dz = p(r)\,dr \qquad \text{or} \qquad p(z) = p(r)\frac{dr}{dz} = \frac{p(r)}{dz/dr} \tag{9.25}$$

Thus, the corresponding probability density function $p_{1n}(z)$ is

$$p_{1n}(z) = \frac{p(r)}{2r} = \frac{1}{2\sigma^2}e^{-\frac{r^2}{2\sigma^2}} = \frac{1}{2\sigma^2}e^{-\frac{z}{2\sigma^2}} \tag{9.26}$$

This is the probability density of a square law detector. The subscript $1n$ of $p(z)$ represents one data point with only noise as input. In order to find the probability density of N samples, the characteristic function will be used.

It is important to note that the integration presented in the following sections is difficult to perform. However, all the integrals can be found in [11]. In order to match the formulas in the reference, the following changes are made. In the conventional sense, the characteristic equation is defined as the inverse Fourier transform as shown in (9.22). The probability density function is obtained through the Fourier transform. In the following sections, the characteristic function is redefined as the Fourier transform and the probability density function is obtained through the inverse Fourier transform. Since the operations are performed in pairs, the result will be the same.

The characteristic function of (9.26) can be written as

$$C_{1n}(\omega) = \int_0^\infty \frac{1}{2\sigma^2}e^{-\frac{z}{2\sigma^2}}e^{-j\omega z}\,dz = \frac{1}{j2\omega\sigma^2 + 1} \tag{9.27}$$

If N samples are used to determine the overall probability density function, the characteristic function can be written as

$$C_{Nn} = [C_{1n}(\omega)]^N = \left[\frac{1}{j2\omega\sigma^2 + 1}\right]^N = \left(\frac{1}{2\sigma^2}\right)^N \left[\frac{1}{j\omega + \frac{1}{2\omega^2}}\right] \tag{9.28}$$

The last expression can be used directly from the inverse Fourier transform formula No. 431 (p. 44 in [11]). The result is

$$P_{Nn}(z) = \frac{y^{N-1} e^{-\frac{y}{2\sigma^2}}}{(2\sigma^2)^N (N-1)!} \tag{9.29}$$

The subscript Nn represents the probability density of N data points of data with only noise. This equation will be used to find the false alarm rate.

If there is a signal present, the probability density function can be obtained from (9.16) in a similar manner as from (9.25) and (9.26). The result is

$$p_{1s}(z) = \frac{1}{2\sigma^2} e^{-\frac{z+A^2}{2\sigma^2}} I_0\left(\frac{\sqrt{z}A}{\sigma^2}\right) \tag{9.30}$$

This is the probability density of one sample from a square law detector with a signal. The characteristic function $C_{1s}(\omega)$ of this equation can be written from [11] (No. 655.1, p. 79). The Fourier transform result is

$$\begin{aligned} C_{1s}(\omega) &= \frac{1}{2\sigma^2} e^{-\frac{A^2}{2\sigma^2}} \int_0^\infty e^{-\frac{z}{2\sigma^2}} I_0\left(\frac{A\sqrt{z}}{\sigma^2}\right) e^{-j\omega z}\, dz \\ &= \frac{1}{j2\omega\sigma^2 + 1} e^{-\frac{jA^2\omega}{j2\omega\sigma^2 + 1}} \end{aligned} \tag{9.31}$$

where the subscript $1s$ is a single data point with a signal. For the N sample case, the characteristic function can be written as

$$\begin{aligned} C_{Ns}(\omega) &= [C_{1s}(\omega)]^N = \frac{1}{(j2\omega\sigma^2 + 1)^N} e^{-\frac{jN\omega A^2}{j2\omega\sigma^2 + 1}} \\ &= \left(\frac{1}{2\sigma^2}\right)^N e^{-\frac{NA^2}{2\sigma^2}} \frac{1}{\left(j\omega + \frac{1}{2\sigma^2}\right)^N} e^{-\left[\frac{1}{\frac{4\sigma^4}{NA^2}\left(j\omega + \frac{1}{2\sigma^2}\right)}\right]} \end{aligned} \tag{9.32}$$

The last expression matches the formula in [11] (No. 650.0, p. 77). The probability density function can be obtained from the inverse Fourier transform as

$$p_{Ns}(z) = \frac{1}{2\sigma^2}\left(\frac{z}{NA^2}\right) e^{-\frac{NA^2+z}{2\sigma^2}} I_{N-1}\left(\frac{A\sqrt{Nz}}{\sigma^2}\right) \tag{9.33}$$

where I_{N-1} is the modified Bessel function of the first kind. The results in (9.29) and (9.33) are used to find the probability of false alarm and the probability of detection.

9.11 DETECTION OF MULTIPLE SAMPLES BASED ON SUMMATION [1, 8–19]

In this approach, N samples will be used to determine the probability of false alarm. The difference between this method and the L-out-of-N method is that the output from each sample will be added together. After N samples of data are collected, the result will be compared with one threshold. This case is similar to the pulsed radar detection problem, which has been studied in [1]. A brief discussion will be presented here. The probability of false alarm and probability of detection can be determined through the same procedure as discussed for the single data point case.

The first step is to find the probability of false alarm P_{fam} from (9.18). This value will be used to find the threshold. Once the threshold is found, the probability of detection can be determined. The threshold can be found in a similar way as that shown in (9.6) with (9.29) used as the probability density function. The result is

$$P_{fam} = \int_{r_1}^{\infty} p_{Nn}(z)\, dz \tag{9.34}$$

In this equation, P_{fam} is given; thus, threshold r_1 can be found.

The probability of detection can be obtained in a similar manner as that shown in (9.18). The result is

$$P_{dm} = \int_{r_1}^{\infty} p_{Ns}(z)\, dz = 1 - \int_{0}^{r_1} p_{Ns}(z)\, dz \tag{9.35}$$

There are no closed form solutions to these two equations. Numerical integrals will be used to find the results. Several examples will be presented in the following sections to demonstrate the procedure to find the false alarm and probability of detection.

In actual application, N can have different values with different thresholds. Detection can be declared as soon as one or more thresholds are crossed. The value of N can even start from one.

9.12 AN EXAMPLE OF SINGLE-SAMPLE DETECTION

The following example is used to demonstrate the application of the equations to find the probability of false alarm and the probability of detection of a digital receiver. The design examples in Sections 7.11 and 7.12 will be used here. Let us assume that the input bandwidth is 125 MHz and the desired false alarm time is about 100 sec, which means on the average the receiver is allowed to generate one false every 100 sec. The sampling rate is 250 MHz and the ADC is 8 bits; thus,

$T_f = 100$ sec
$f_s = 250$ MHz
$b = 8$ bits

In this example, let us use only one sample to detect the existence of the input signal. From (9.2), the probability of false alarm is

$$P_{fas} = \frac{1}{T_f f_s} = 4 \times 10^{-11} \tag{9.36}$$

From (9.7), the threshold should be set as

$$-\frac{r_1^2}{2\sigma^2} = -23.94 \qquad \text{or} \qquad \frac{r_1^2}{\sigma^2} = 47.88 \tag{9.37}$$

It should be noted that the threshold r_1 is an integer number of Q (or $r_1 = nQ$) where n is an integer and Q is the quantized unit. If the noise from (9.8) is substituted into the above equation, the result is

$$\frac{(nQ)^2}{\frac{(1+M)Q^2}{12}} = \frac{12n^2}{1+M} = 47.88 \tag{9.38}$$

where M is the amplifier output noise measured in units of quantization noise (see (7.20)). The M value is approximately equal to 16 (from Section 7.12). The corresponding n value is 8.24. There are three ways to choose the M and n values.

1. Change the M value to make either $n = 8$ or 9 to fulfill the result of this equation. This approach will change the gain of the front end slightly.
2. Keep $M = 16$ and choose $n = 8$.
3. Keep $M = 16$ and choose $n = 9$.

Since the mathematical manipulation of all the approaches are exactly the same, keeping $M = 16$ and choosing $n = 8$ and $n = 9$ will be used in the following discussion.

With $M = 16$, these choices of n will reflect

$$\frac{12n^2}{1+M} = \begin{cases} 45.18 & \text{for} \quad n = 8 \\ 57.18 & \text{for} \quad n = 9 \end{cases} \tag{9.39}$$

The corresponding probability of false alarm is

$$P_{\text{fas}} = \begin{cases} e^{-22.59} = 1.55 \times 10^{-10} & n = 8 \\ e^{-28.59} = 3.84 \times 10^{-13} & n = 9 \end{cases}$$
$$T_f = \frac{1}{P_{\text{fas}} f_s} = \begin{cases} 25.8 \text{ sec} & n = 8 \\ 10{,}416 \text{ sec} & n = 9 \end{cases} \tag{9.40}$$

One can see that if $n = 8$ is chosen, the false alarm time is worse than the desired value, although it might still be acceptable. If $n = 9$ is chosen, the false alarm time is much better than the desired value. In this discussion, both cases will be carried out. The last step is to find the probability of detection from (9.16) and (9.17). The corresponding r_1 can be found from (9.7).

The Rayleigh and Rician distributions are shown in Figure 9.3. The Rayleigh distribution is labeled noise only and the Rician distribution is at a signal-to-noise

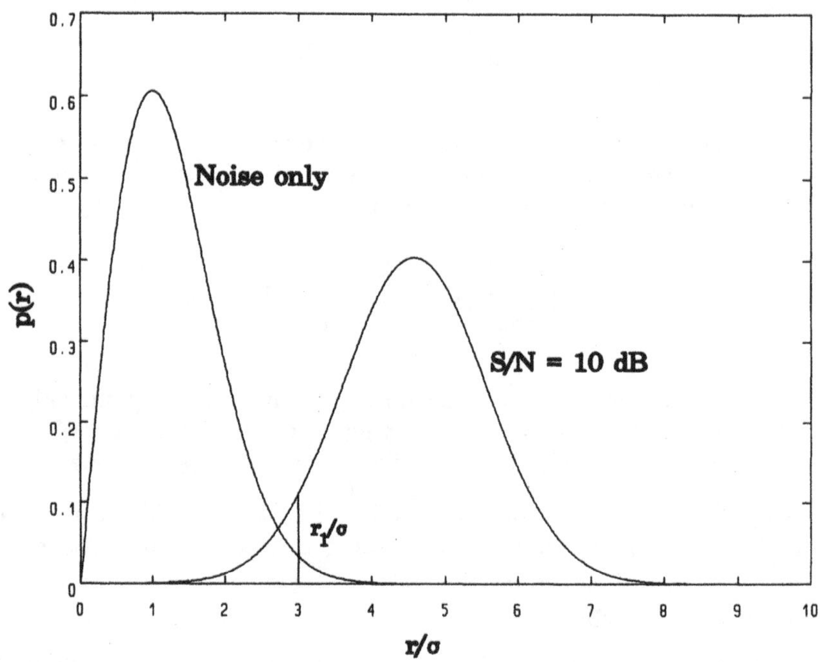

Figure 9.3 Rayleigh and Rician distributions.

ratio (S/N) of 10 dB. An r_1 is arbitrarily chosen. The area on the right side of the noise-only curve represents the probability of false alarm. The area on the right side of the S/N = 10-dB curve represents the probability of detection.

A combination of several programs are used to generate Figure 9.4, and they are listed in the Appendix. Figure 9.4(a, b) shows the probability of detection versus S/N for n = 8 and 9, respectively. In order to produce a 90% probability of detection, the S/N required is approximately 14.95 dB for T_f = 25.8 sec and 15.8 dB for T_f = 10,416 sec. This is the tradeoff between sensitivity and false alarm rate. Accepting a higher false alarm rate implies the receiver can have higher sensitivity.

9.13 AN EXAMPLE OF MULTIPLE-SAMPLE (L-OUT-OF-N) DETECTION

In this example, the requirement is the same as the example in the last section. The only difference is to use 64 points (N = 64) to determine the false alarm rate and probability of detection. The amplifier in front of the ADC is kept the same. In other words, the amplification factor M = 16 is used.

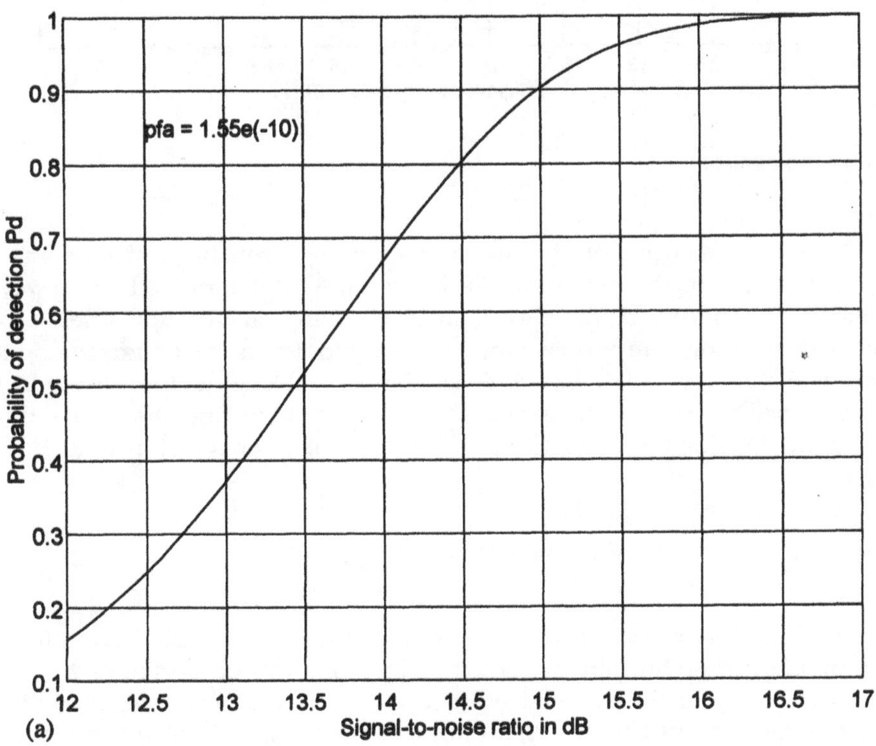

Figure 9.4 Probability of detection versus S/N: (a) n = 8, (b) n = 9.

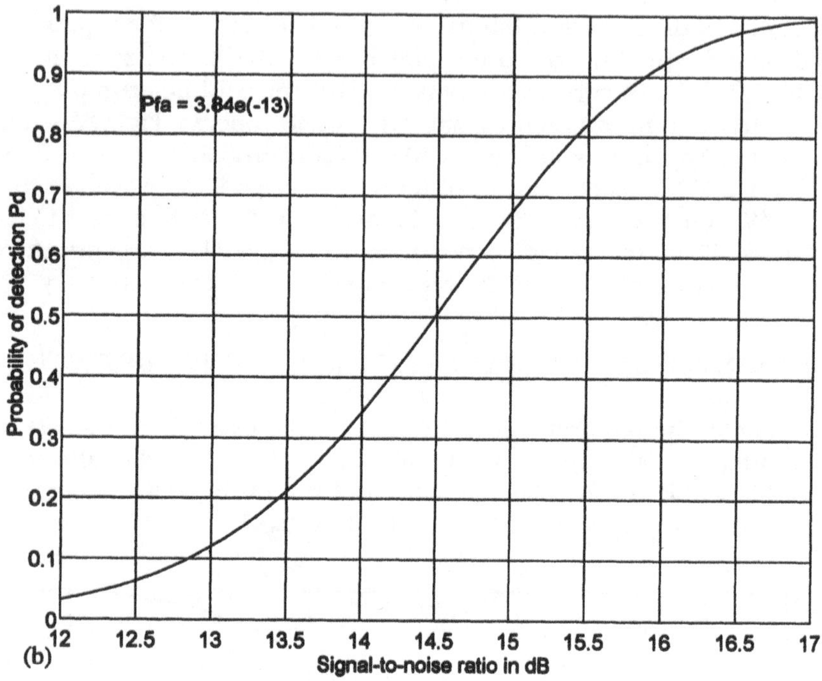

Figure 9.4 (continued).

Since $N = 64$, from (9.18), the corresponding probability of false alarm is $P_{fam} = 2.56 \times 10^{-9}$. In the next step, the L value will be determined. However, the L value is determined by the threshold level. Many quantization levels can be selected as thresholds. In this example, let us try to use the first quantization level as the detection threshold. If a different level is selected as the threshold, the procedure will be exactly the same. The first step is to find the probability of crossing the threshold by noise alone. The result can be obtained from (9.10) with $n = 1$ as

$$P_{fas}(1) = e^{-\frac{6}{1+M}} = e^{-\frac{6}{1+16}} = 0.7026 \tag{9.41}$$

where $P_{fas}(1)$ is the probability of the noise crossing the first quantization level on a one-sample basis. Thus, the probability p in (9.19) can be written as $p = P_{fas}(1)$. In order to determine how many samples are needed to cross the threshold in the total of 64 samples, (9.19) and (9.20) are to be used. In this calculation, because the threshold must be one of the quantization levels, the result obtained will be close to the desired value and the exact false alarm time T_f cannot be achieved.

The calculation will be illustrated as follows. Using the result from (9.41) as the p value in (9.20), the results shown in Table 9.1 are obtained from Program binomial.m.

For $L = 63$, the probability is 4.351×10^{-9}, which is greater than the desired value of 2.56×10^{-10}. These results can be obtained from a trial and error approach by using Program df9_4.m in the Appendix. This means the number of false alarms generated is greater than the design goal. If only $L = 64$ is included, the overall probability of false alarm is $P_{fam} = 1.549 \times 10^{-10}$, which is less than the desired value of 2.56×10^{-9}. Therefore, if the first quantization level is chosen as threshold, all 64 samples should cross the threshold for it to be considered that a signal is detected. Thus, this example turns out to be a very special case because $L = N$. In general, L will be less than N.

Now let us consider the probability of detection, and use a 90% probability of detection as a criterion. A slightly different procedure will be taken here. We want to find the $p(i|N)$ value in (9.20) that can generate $p(L + |N) = P_{dm} = 0.9$. However, there is no direct solution either. A plot of $p = p(i|N)$ versus $P(L + |N) = P_{dm}$ is shown in Figure 9.5(a). In this special case of $L = N$, this curve is generated through $P_{dm} = p^{64}$. If $L < N$, this curve can be plotted by the program df9_5.m listed in the Appendix. By further refining the value of $p(i|N)$, it is found that $p(i|N) = (0.90)^{1/64} = .998355$ will generate the desired P_{dm}.

The last step is to find the S/N of a single sample crossing that will generate a P_{dm} of 0.998355, with the probability of false alarm of 0.7026 given by (9.41). The procedure for finding the probability of detection is exactly the same as that discussed in the previous section. The result is plotted in Figure 9.5(b). The S/N required is about 8 dB, which is much less than the 15 dB required by the single sample detection scheme. This indicates that the sensitivity of the receiver can be improved. It should be noted that the false alarm time is also improved because the P_{fam} obtained is 1.549×10^{-10}, which corresponds to a false alarm time of 1,653 sec.

This improvement is obtained by integrating the incoming data points. If the signal is longer than L, this approach will detect the input with a high sensitivity. If the signal length is less than L data points, the chance of detecting the signal is extremely small because the chance of crossing the first threshold by noise alone

Table 9.1
Results From (9.41) and (9.19)

L	$P(L + N)$
63	4.351×10^{-9}
64	1.549×10^{-10}

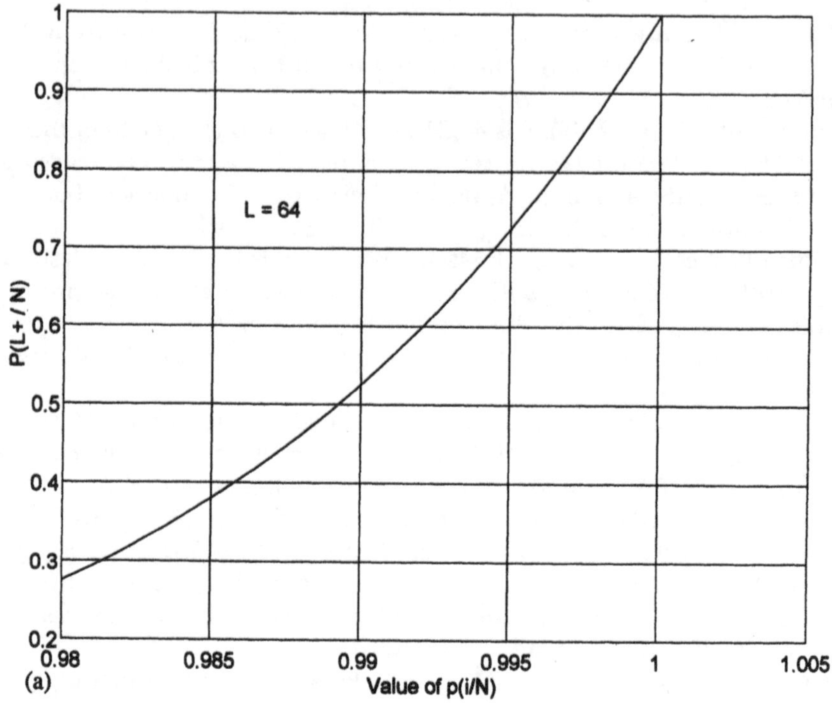

Figure 9.5 Use 1st quantization level as threshold: (a) P_{dm} versus p, (b) P_{dm} versus S/N.

is small. The selection of N should be based on the minimum PW the receiver is designed to intercept.

9.14 SELECTION OF THRESHOLD LEVEL

In the last section, the first level is arbitrarily chosen as the threshold. Other levels can also be used as the threshold. The approach is the same as discussed in the previous section. Since the first quantization level used as the threshold turns out to be a special case, the detailed procedure will be presented again.

With the second quantization level as the threshold, the probability of crossing the second level by noise alone is

$$p = e^{-\frac{6n^2}{1+M}} = e^{-\frac{24}{17}} = .2437 \tag{9.42}$$

where $n = 2$ and M is still equal to 16. In order to generate a probability of false alarm less than 2.56×10^{-9}, but close to it, $L = 39$ is chosen. The value is obtained

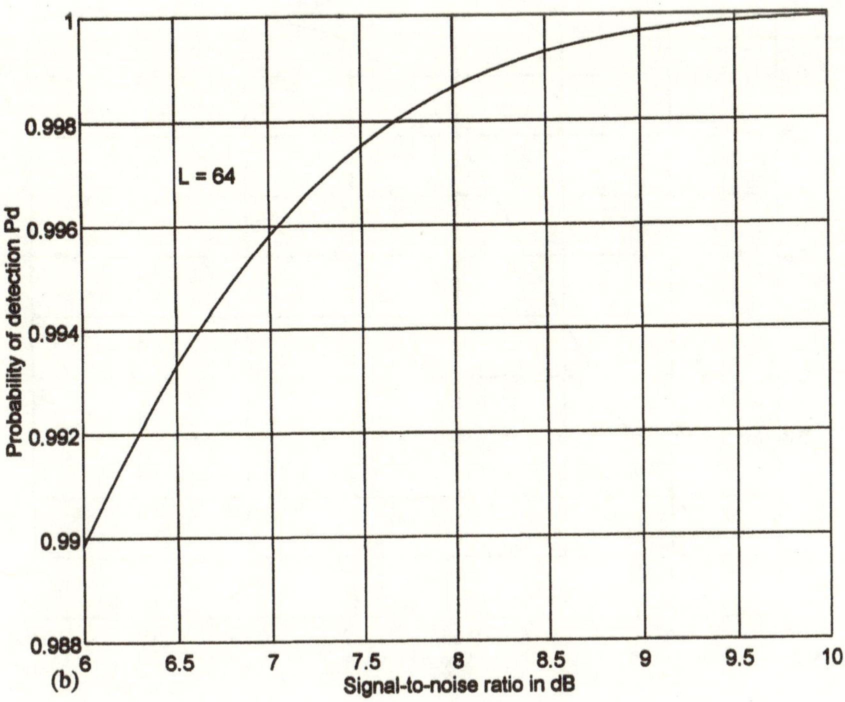

Figure 9.5 (continued).

numerically through trial and error from (9.19) and (9.20) or from Program binomial.m. This means 39 samples out of 64 must cross the threshold to be detected as a signal. The corresponding probability of false alarm from (9.20) is

$$p_{\mathrm{fam}} = p(L + |N) = \sum_{i=39}^{64} p(i|64) = 5.6835 \times 10^{-10} \tag{9.43}$$

Next, the probability of detection will be found. In order to generate $P_{\mathrm{dm}} = .90$, $p = .67698$ is required. This result is obtained from Figure 9.6(a), generated from Program df9_5.m. Finally the required S/N for one sample to cross the second threshold ($n = 2$) with a probability of false alarm of .2437 is obtained. The result is shown in Figure 9.6(b), from Program 9_4.m. The required S/N value to obtain $p = .676$ is about 2.3 dB, which is much less than the 8 dB required before.

In a very similar manner, the third, fourth, fifth, and sixth levels are used for threshold. The seventh level will be too high for the threshold because the false alarm generated is below the desired value. The results are listed in Table 9.2, including the results from single sample detection.

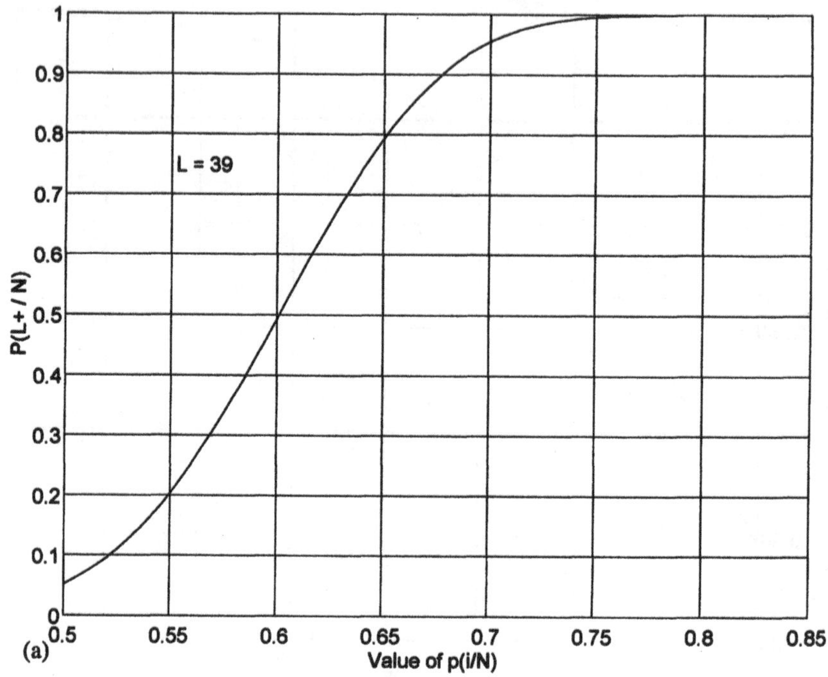

Figure 9.6 Use 2nd quantization level as threshold: (a) P_{dm} versus P, (b) P_{dm} versus S/N.

These results are obtained from Program ta9_1.m. In this program, the N, L, and P_{fam} values are given and it will find the correct S/N. From these results, it is obvious that multiple-sampled cases provide better sensitivity, which is well known in radar detection. However, among the multiple-sampled cases, it is difficult to make a very accurate comparison because the probability of false alarm in each case is different. However, one can see that the best result appears at $n = 2$. Under this condition, $L = 39$.

The last row in the table is the result from the summation method. The required S/N is less than the L-out-of-N method. The calculation will be presented in Section 9.16.

In general, when a certain quantization level is chosen as the threshold, if the corresponding L is close to $N/2$, the best sensitivity can be expected. This point will be further discussed in the next section.

9.15 OPTIMIZING THE SELECTION OF THRESHOLD [9]

In this section, it is intended to demonstrate that the optimum threshold selected is close to $L = N/2$ in the L-out-of-N scheme. Since the mathematic operations

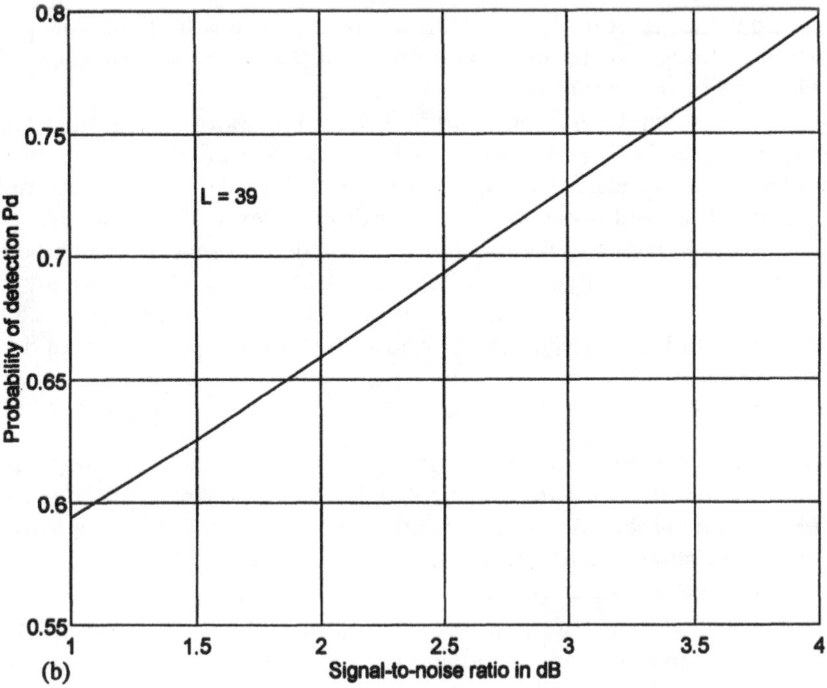

Figure 9.6 (continued).

Table 9.2
Results From Different Detection Schemes

N	n	L	P_{fas}	P_{fam}	T(sec)	S/N (dB)
	8			1.55×10^{-10}	25.8	15.00
1	9			3.84×10^{-13}	10,417	16.86
	1	64	0.7026	1.55×10^{-10}	1,652	7.84
	2	39	0.2437	5.68×10^{-10}	451	2.26
	3	17	0.0417	7.41×10^{-10}	345	2.32
64	4	8	3.52×10^{-3}	8.91×10^{-11}	2,873	4.03
	5	4	1.47×10^{-4}	2.96×10^{-10}	865	5.87
	6	3	3.03×10^{-6}	1.16×10^{-12}	220,690	7.93
		Summation		2.295×10^{-9}	111.5	0.81

involved are rather tedious, it is difficult to prove analytically. Thus, this point will be illustrated though a numerical approach. Hopefully, this illustration will also make the steps in the previous section clearer.

In order to compare the sensitivity at different threshold levels, the probability of false alarm must be kept the same. It is shown in previous sections that if the threshold levels are selected in a discrete manner, this goal cannot be accomplished. Thus, the threshold will be selected in a continuous sense. The basic procedure is to find the L value from 1 to 64. At each L value, the required S/N will be obtained to generate the same probability of detection with the same probability of false alarm. Let us still use $N = 64$ samples, the same sampling frequency of 250 MHz, the overall probability of false alarm of 2.56×10^{-9}, and the probability of detection of 90%.

The procedure can be divided into three steps for each given L value.

1. The first step is from the overall given probability of false alarm (from all 64 samples) to find each individual probability of false alarm required to generate the desired value. The approach uses (9.20) and (9.21) by adjusting the p value to generate the desired $P(L + |N) = 2.56 \times 10^{-9}$. A ±0.1% error in $P(L + |N)$ is used in this calculation.
2. The second step is from the overall given probability of detection (from all 64 samples) to find each individual probability of detection required to generate the desired value. The procedure is exactly the same as in step 1. The only difference is that in the case $P(L + |N) = 0.900$, the same ±0.1% error in $P(L + |N)$ is used in this calculation.
3. The last step is to find the required S/N for each individual sample. In this approach, (9.17) and (9.18) are used. The S/N is adjusted to match the desired probability of detection. The minimum step size in S/N is 0.01 dB. Program ta9_1.m is used to perform these calculations. The result is shown in Figure 9.7. The minimum S/N required in this calculation is 1.85 dB, which occurs at both $N = 27$ and 28. The curve at the minimum is rather flat; thus, it is not very critical to determine the minimum S/N.

From this illustration, one can realize the rule to choose the proper quantization level as the threshold. In the L-out-of-N method, if the threshold can be chosen such that the L value is about half of N, it is a reasonable approach. Similar results are shown in [9].

9.16 AN EXAMPLE OF N SAMPLE DETECTION (SUMMATION METHOD)

This example illustrates the summation method discussed in Section 9.11. As discussed before, a threshold that will generate the desired probability of false alarm must be found first. The threshold can be found from the numerical integration equation (9.29). In order to obtain the desired probability of false alarm of 2.56

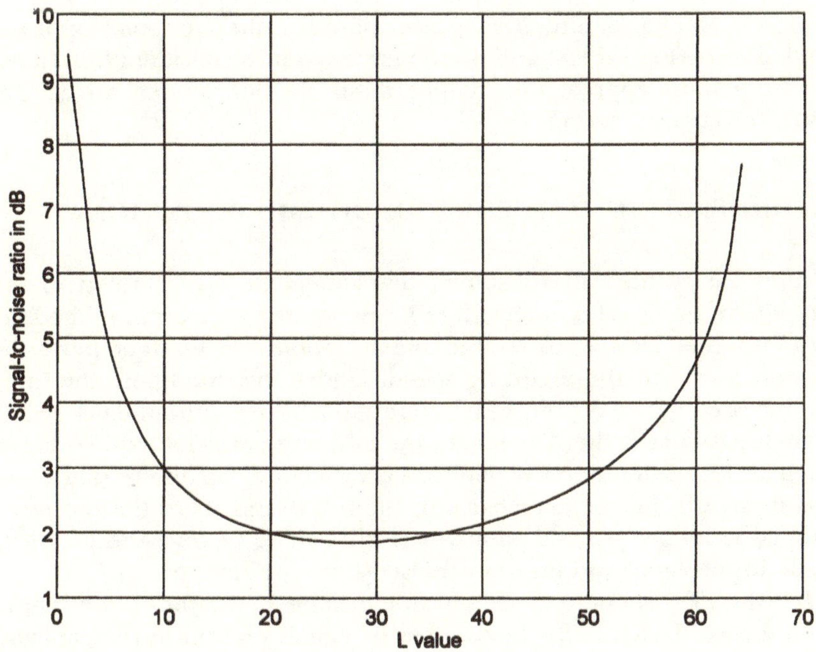

Figure 9.7 *S/N* versus *L* value.

$\times 10^{-9}$, the threshold is at 244.5562. However, because the data are quantized, 245 is used as the threshold. Under this condition, the probability is 2.295×10^{-9}. The probability of detection can be found from (9.33) and (9.34). In the integral, the modified Bessel function $I_n(x)$ is approximated by [1]

$$I_n(x) \underset{x \gg n}{\approx} \frac{e^x}{\sqrt{2\pi x}} \tag{9.44}$$

Program ta9_1a can be used to calculate the result. The required inputs to the program are the values of the probability of false alarm. The result, listed in the last row of Table 9.2, is that a *S/N* of 0.81 dB is required.

It should be noted that the summation is made on the outputs of a square law detector and digitization is not taken into consideration.

Comparing the summation and the *L*-out-of-*N* methods, two advantages can be found for the summation method.

1. The summation method has higher sensitivity.
2. If the signal is shorter than 64 data points and the signal is strong, the summation method can detect the signal more effectively than the *L*-out-of-*N* approach.

Once a sampled datum crosses a threshold in the *L*-out-of-*N* approach, the amplitude information is lost and no longer taken into consideration. In contrast, with the summation approach the amplitudes of the entire signal are accumulated and taken into consideration.

9.17 INTRODUCTION TO FREQUENCY DOMAIN DETECTION

In the previous section, it is assumed that detection is performed in the time domain. When the signal is detected, FFT can be used to determine the frequency of the input signal. Now let us assume that an *N*-points FFT can be performed just fast enough to match the digitizing speed. Under this condition, the input data will be processed by the *N*-points FFT continuously, but without data overlapping. It is desirable to match the *N* points to the minimum anticipated PW. However, a pulse signal with a minimum PW might be divided into two different FFT windows, and this signal will not be detected with the full sensitivity of the receiver. If the input signal is longer than *N* points, it is difficult to change the FFT length to match the input signal and improve the sensitivity on long pulses.

After the FFT operation, one must determine 1) whether there is any signal in the windowed data, and 2) the number of signals present in the data and their frequencies. An important technical challenge is to avoid sidelobes generated by a strong signal and detect weak signals in the presence of strong ones. Of course, this is also a challenge to other types of EW receivers with simultaneous signal detection capability.

It appears that the problem in frequency domain detection is much more complicated than for time domain detection. One of the major problems is that one does not know when the signal appears in the data points in the time domain. For example, if the total length of the FFT is 64 points, 64 points of time domain data will be processed. If all the data points in this time frame contain signals, the output of the FFT should be very well behaved and detection should be relatively easy.

If only a small portion of the time frame data contain signals, the mainlobe will be very wide (reciprocal of signal length). One extreme case is that the time window collects only one point of data from the leading edge of a strong pulse. The results are shown in Figure 9.8.

Figure 9.8(a) shows noise alone in time domain and Figure 9.8(b) shows the corresponding power spectrum from the FFT output. Figure 9.8(c) shows there is one data point containing a signal and Figure 9.8(d) shows its corresponding output in the frequency domain. From these figures, it is obvious that when there is even one signal data in the time domain, the amplitude of the output of the frequency domain increases. If a fixed threshold is used to determine whether there is a signal, the threshold might be crossed at many frequency components. In addition, the peaks in the frequency domain do not necessarily correspond to the correct

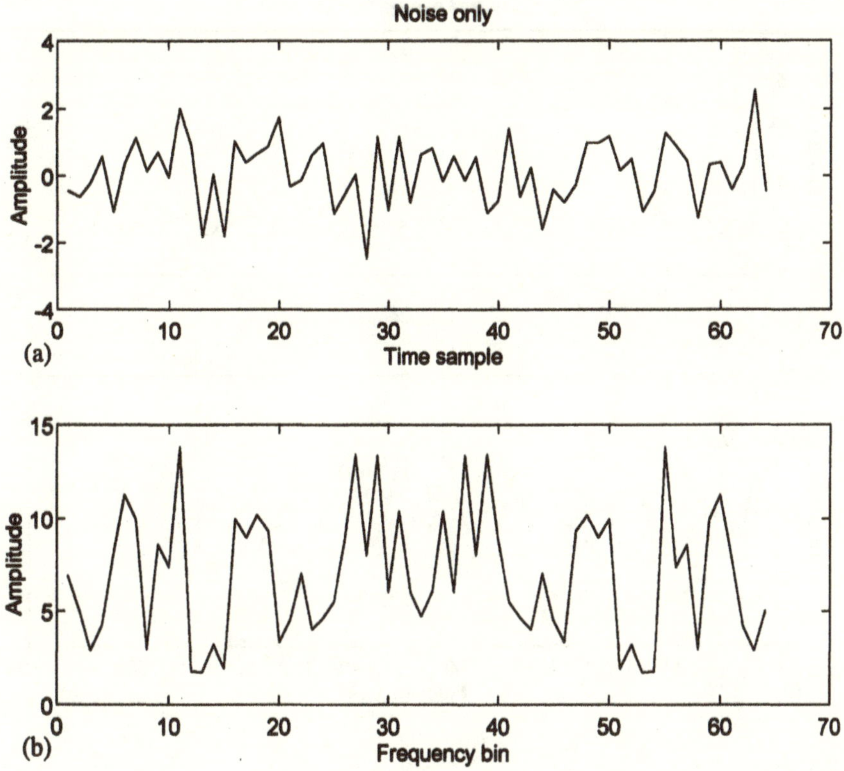

Figure 9.8 Time and frequency domain responses: (a) noise alone in time domain, (b) noise alone in frequency domain, (c) noise and one datum containing signal (time), (d) noise and one datum containing signal (frequency).

frequencies of the input signals because the one sample can create a very wide spectrum.

If most of the data in the time domain contain signals, the peaks in the frequency domain will represent the frequency of the input signals. The following sections will discuss the detection of signals in the frequency domain.

9.18 A SUGGESTED APPROACH TO FREQUENCY DOMAIN DETECTION

In this discussion, it is assumed that the FFT length is shorter than the shortest anticipated pulse. The FFT is performed consecutively without data overlap. In other words, a signal can be divided into several (at least two) time frames.

One possible solution to the frequency domain detection problem is to take a similar procedure to that used for a microscan receiver [13]. The input signal of

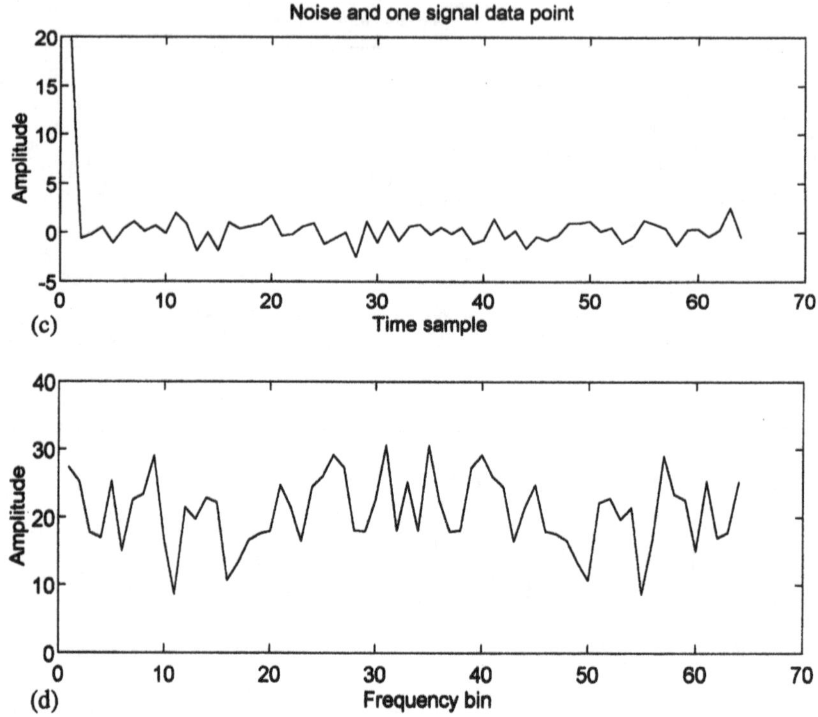

Figure 9.8 (continued).

a microwave receiver is divided into many segments (scan time) in the time domain. Each time segment contains many serial outputs, representing the input frequency. The receiver is usually designed in such a way that the scan time is equal to or less than the minimum PW to be processed. If one compares a microscan receiver with digital receiver with fixed FFT length operation, the results will be quite similar.

One way to obtain the frequency information in a microscan receiver is to generate frequency information from two consecutive scan times. There are five possibilities in which the data can be divided into two consecutive time segments.

1. The pulse is equal to or longer than twice the minimum PW. In a digital receiver, this situation corresponds to a PW ($PW > 2N$). Under this condition, the first time frame is likely to be filled with data partially containing signals. But all the data in the second time frame will contain signals. Thus, the second time frame should be used to determine the input frequency.
2. In the following four cases, it is assumed that the PW is equal to the window time or ($PW = N$). If the signal fills one entire time frame, the frequency calculated from this time window will be well defined.

3. The signal can be divided equally between two time frames. Under this condition, the power spectrum obtained from these two time windows are the same. Either the first or the second time segment can be used to determine the frequency.
4. The input signal is split into two consecutive time windows and the first time frame contains more signal data.
5. The input signal is split into two consecutive time windows and the second time frame contains more signal data.

In a microscan receiver, the output information is generated on a pulse-by-pulse basis rather than a scan-by-scan basis. Usually, the frequency components from two consecutive time frames are compared. If the two time windows contain the same frequency (or very close in frequency), it is considered that they belong to the same signal. The frame with the stronger signal will be used to determine the signal frequency. In all the five cases mentioned above, this approach produces reasonable frequency readings. An identical approach can be used in a digital receiver.

9.19 PROBABILITY OF FALSE ALARM IN FREQUENCY DOMAIN

The calculation of probability of false alarm is quite similar to the time domain approach. The probability of false alarm is calculated as shown in Section 9.5. The only difference is that the amplitude of the frequency component is used instead of the sampled amplitude in the time domain. The first step is to find the noise distribution in the frequency domain. Since the FFT is a linear operation, the noise distribution in the frequency domain is similar to the time domain distribution. This can be proven mathematically from the noise power spectrum. The noise is assumed uncorrelated in the time domain and the noise power in the time domain can be written as

$$E[x_n x_n] = \sigma^2 \tag{9.45}$$

where $E[\ ]$ represents expectation value. The corresponding noise power in frequency domain σ_f can be obtained from the expectation value of the spectrum components X_k and X_l as follows:

$$\begin{aligned}
\sigma_f^2 &= E\left[\left(\sum_m^{N-1} x_m e^{-\frac{j2\pi mk}{N}}\right)\left(\sum_n^{N-1} x_n e^{-\frac{j2\pi nl}{N}}\right)^*\right] \\
&= \sum_m \sum_n [x_m x_n^*] e^{-\frac{j2\pi(mk-nl)}{N}} = \sum_n [x_n x_n^*] e^{-\frac{j2\pi n(k-l)}{N}} \\
&= \sigma^2 \sum_n^{N-1} e^{-\frac{2\pi n(k-l)}{N}} = \begin{cases} 0 & \text{for} \quad k \neq l \\ N\sigma^2 & \text{for} \quad k = l \end{cases}
\end{aligned} \tag{9.46}$$

where * presents the complex conjugate. In the above equation, the noise spectrum is represented by its Fourier transform. This equation shows that the noise has the same distribution, but the variance is increased by N. This is reasonable because in order to calculate the noise power spectrum, all the N points in the time domain are used.

This result implies the distribution of the power spectrum is Rayleigh. The probability density function can be written in the same form as (9.5) by replacing σ^2 with $N\sigma^2$. This result can be obtained also from the real and imaginary parts of the FFT. They can be expressed as

$$\begin{aligned} p_f(x) &= \frac{1}{\sqrt{2\pi N\sigma}} e^{-\frac{x^2}{2N\sigma^2}} \\ p_f(y) &= \frac{1}{\sqrt{2\pi N\sigma}} e^{-\frac{y^2}{2N\sigma^2}} \\ p_f(r) &= \int_0^{2\pi} r p_f(x) p_y \, d\phi = \frac{r}{N\sigma^2} e^{-\frac{r^2}{2N\sigma^2}} \end{aligned} \tag{9.47}$$

where the subscript f represents the probability density in the frequency domain. The probability of false alarm is

$$P_{\text{faf}} = \int_{r_1}^{\infty} p_f(r) \, dr = e^{-\frac{r_1^2}{2N\sigma^2}} \tag{9.48}$$

For a given probability of false alarm, this equation can be used to set the threshold r_1.

If the false alarm time T_f is given, the corresponding probability of false alarm in frequency domain P_{faf} is

$$P_{\text{faf}} = \frac{N}{T_f f_s} \tag{9.49}$$

where N is the total number of points in the FFT and f_s is the sampling frequency. Since a decision is made every N samples, this result is identical to (9.18).

9.20 INPUT SIGNAL CONDITIONS IN FREQUENCY DOMAIN DETECTION

The probability of detection in the frequency domain depends on the input signal conditions. For example, if the input power is constant and the frequency matches one of the frequency bins after the FFT, its power spectrum will be high in comparison with the input frequency just between two frequency bins. Four signal conditions will be discussed here. In the first two conditions, the signals fill up the time window, with the input frequency on a frequency bin and a frequency at the middle of two bins. In the last two conditions, the signal only fills half the time window, with the input frequency on a frequency bin and at the middle of two bins. The input signal is assumed to be a complex sinusoidal wave, which can be written as

$$x(n) = Ae^{\frac{j2\pi nk_o}{N}} \tag{9.50}$$

where A is the amplitude of the signal, k_o is the frequency, and N is the length of the FFT used to process this signal.

The four signal conditions can be written as follows.

1. Signals fill up the time domain window and the input frequency matches one of the spectrum lines. This condition will produce the highest spectrum output and thus the highest probability of detection. Under this condition, $k_o = k_i$ where k_i is a certain spectrum component. The highest output of the FFT is

$$\begin{aligned} X(k_i) &= \sum_{n=0}^{N-1} Ae^{\frac{j2\pi nk_o}{N}} e^{\frac{-j2\pi nk_i}{N}} \\ &= A\sum_{n=0}^{N-1} e^{\frac{j2\pi n(k_o-k_i)}{N}} = NA \end{aligned} \tag{9.51}$$

2. Signals fill up the time domain window and the input frequency is exactly at the center of two frequency lines. Under this condition, $k_o = k_i + 0.5$ and there are two highest outputs with the same amplitude. The spectrum output $X(k_i)$ is

$$\begin{aligned} X(k_i) &= A\sum_{n=0}^{N-1} e^{\frac{j2\pi n(k_o-k_i)}{N}} \\ &= A\sum_{n=0}^{N-1} e^{\frac{j\pi n}{N}} = A\frac{1 - e^{j\pi}}{1 - e^{j\pi/N}} \end{aligned} \tag{9.52}$$

Its amplitude is

$$|X(k_i)| = \frac{2A}{\sqrt{2 - 2\cos\frac{\pi}{N}}} \tag{9.53}$$

It should be noted that the next frequency component $X(k_{i+1})$ has the same amplitude. However, if noise is present, these two amplitudes may be different. To determine the probability of detection, both frequency components will be considered.

3. Signals fill only half of the time domain window and the input frequency is one of the spectrum lines. Under this condition, $k_o = k_i$ and the highest FFT output is

$$X(k_i) = A\sum_{n=0}^{\frac{N}{2}-1} e^{\frac{j2\pi n(k_o-k_i)}{N}} = \frac{NA}{2} \tag{9.54}$$

4. Signals fill only half of the time domain window and the input frequency is exactly at the center of two frequency lines. This condition produces the worst detection probability. Under this condition, $k_o = k_i + .5$ and the $X(k_i)$ component is

$$\begin{aligned} X(k_i) &= A\sum_{n=0}^{\frac{N}{2}-1} e^{\frac{2\pi n(k_o-k_i)}{N}} \\ &= A\sum_{n=0}^{\frac{N}{2}-1} e^{\frac{j\pi n}{N}} = A\frac{1 - e^{j\pi/2}}{1 - e^{j\pi/N}} \end{aligned} \tag{9.55}$$

Its amplitude is

$$|X(k_i)| = \frac{A}{\sqrt{1 - \cos\left(\frac{\pi}{N}\right)}} \tag{9.56}$$

and $X(k_{i+1})$ has the same amplitude.

Cases 1 and 4 provide the highest and lowest probability of detection, respectively. The general probability of detection should fall between the highest and lowest values. The following sections will calculate the probability of detection of the four cases.

9.21 PROBABILITY OF DETECTION IN FREQUENCY DOMAIN

The probability of detection will be discussed in this section. The approach to solving this problem is similar to the solution for that in the time domain. If there are signals in the output data, the probability density function has the same form of (9.16) by replacing σ^2 by $N\sigma^2$

$$p(r) = \frac{r}{N\sigma^2} e^{-\frac{r^2+x_m^2}{2N\sigma^2}} I_0\left(\frac{rX_m}{N\sigma^2}\right) \tag{9.57}$$

This result will be used to find the probability of detection.

The probability of detection for signal conditions 1 and 3, where the input frequency is one of the spectrum lines, can be found as

$$P_{df} = \int_{r_1}^{\infty} p(r)\, dr = 1 - \int_{0}^{r_1} p(r)\, dr \tag{9.58}$$

where P_{df} represents the probability of detection in the frequency domain and $p(r)$ is the probability density function from (9.57).

For signal conditions 2 and 4, where the input frequency is at the center of two adjacent spectrum lines, the probability of detection should be considered slightly different. If any one of the two adjacent spectrum lines crosses the threshold or both the spectrum lines cross the threshold, it is considered that there is an input signal. The probability of detection of two adjacent spectrum lines can be found by summing the probability of all three cases: 1) $X(k_i)$ crosses the threshold, 2) $X(k_{i+1})$ crosses the threshold, and 3) both of them cross the threshold.

Here, a slightly different approach is used. First, the chance of the signal not crossing the threshold is evaluated, which can be written as

$$P_{uf} = 1 - P_{df} = 1 - \int_{r_1}^{\infty} p(r)\, dr = \int_{0}^{r_1} p(r)\, dr \tag{9.59}$$

where P_{uf} is the probability that one of the outputs does not cross the threshold. The probability that both the power spectrum lines do not cross the threshold is P_{uf}^2. Thus, the probability of either one or both the spectrum lines crossing the threshold is

$$\begin{aligned} P_{df2} &= 1 - P_{uf}^2 = 1 - [\int_{0}^{r_1} p(r)\, dr]^2 \\ &= 1 - (1 - P_{df})^2 = P_{df}(2 - P_{df}) \\ &\text{or} \\ P_{df} &= 1 - \sqrt{1 - P_{df2}} \end{aligned} \tag{9.60}$$

In the above equation, P_{df2} is used to represent the probability of detection from two adjacent spectrum lines to avoid the confusion with P_{df}.

The last step is to relate the frequency outputs to the time domain information. Since the S/N in the frequency domain is used to find the probability of false alarm and the probability of detection, this information should be related back to the time domain S/N. This calculation is also considered in four different signal conditions as mentioned in the last section.

1. For case 1, the highest spectrum line is $X_m = X(k_i) = NA$ and the S/N in the time and frequency domains can be related by

$$\left(\frac{S}{N}\right)_f = \frac{X_m^2}{2\sigma_f^2} = \frac{(NA)^2}{2N\sigma^2} = N\frac{A^2}{2\sigma^2} = N\left(\frac{S}{N}\right)$$

or

$$\left(\frac{S}{N}\right) = \frac{1}{N}\left(\frac{S}{N}\right)_f \tag{9.61}$$

In this equation, it should be noted that N is used to represent both the total points of the FFT operation and the noise. When it represents noise, it associates with signal S. The subscript f is used to represent notations in frequency domain, and no subscript is associated with notations in the time domain. The noise in the frequency domain σ_f is obtained from (9.46).

2. For case 2, $X_m = 2A[2 - 2\cos(\pi/N)]^{-1/2}$, the S/N in the time and frequency domains is related by

$$\left(\frac{S}{N}\right)_f = \frac{X_m^2}{2\sigma_f^2} = \frac{2}{N\left[1 - \cos\left(\frac{\pi}{N}\right)\right]}\frac{A^2}{2\sigma^2} = \frac{2}{N\left[1 - \cos\left(\frac{\pi}{N}\right)\right]}\left(\frac{S}{N}\right)$$

or

$$\left(\frac{S}{N}\right) = \frac{N\left[1 - \cos\left(\frac{\pi}{N}\right)\right]}{2}\left(\frac{S}{N}\right)_f \tag{9.62}$$

3. For case 3, $X_m = NA/2$, the relation is similar to case 1, and the result is

$$\left(\frac{S}{N}\right)_f = \frac{N}{2}\left(\frac{S}{N}\right)$$

or

$$\left(\frac{S}{N}\right) = \frac{2}{N}\left(\frac{S}{N}\right)_f \tag{9.63}$$

4. For case 4, $X_m = A[1 - \cos(\pi/N)]^{-1/2}$, the result is similar to case 2, and that result is

$$\left(\frac{S}{N}\right)_f = \frac{1}{N\left[1 - \cos\left(\frac{\pi}{N}\right)\right]}\left(\frac{S}{N}\right)$$

or

$$\left(\frac{S}{N}\right) = N\left[1 - \cos\left(\frac{\pi}{N}\right)\right]\left(\frac{S}{N}\right)_f \tag{9.64}$$

These four equations can be used to find the S/N in the time domain once the S/N in the frequency domain is obtained. The following two examples will be used to demonstrate the calculation.

9.22 EXAMPLES ON FREQUENCY DOMAIN DETECTION

In this section, the probability of false alarm and detection in the frequency domain will be illustrated by two examples. In the first example, the best detection case is considered when all the data contain signal and the frequency is coincident with one of the output frequency bins. In the second example, the worst case will be considered, which means the data are half-filled with signal and the frequency is between two frequency bins. For simplicity, the quantization effect will be neglected in this discussion. The result will be compared with the result of time domain detection. Thus, the example in the time domain will be used. The information is given as

$T_f = 100$ sec
$f_s = 250$ MHz
$N = 64$
$b = 8$ bits

The probability of false alarm is calculated previously from (9.18) as

$$P_{fa} = 2.56 \times 10^{-9}$$

The noise from the amplifier in front of the ADC is determined as $M = 16$ from the time domain calculation. The corresponding variance of the noise in the frequency domain is $N\sigma^2$.

First, a threshold must be found to provide the desired probability of false alarm. The threshold is set at

$$e^{-\frac{r_1^2}{2\sigma_f^2}} = 2.56 \times 10^{-9} \quad \text{or} \quad \frac{r_1}{\sigma_f^2} = 6.2902 \tag{9.65}$$

This result is obtained from (9.6).

1. In the first example, the input signal will fill up the window and the frequency matches one of the output frequency bins. Under this condition, using (9.58) and Program 9_4.m with $P_{fa} = 2.56e - 9$, one can find that $(S/N)_f = 14.5$ dB is required to generate a 90% probability of detection. The corresponding (S/N) in the time domain can be obtained from (9.61) in decibel form as

$$\left(\frac{S}{N}\right) = -10\log(N) + \left(\frac{S}{N}\right)_f = -10\log(64) + 14.5 \tag{9.66}$$
$$= -18.1 + 14.5 = -3.6 \text{ dB}$$

This result is even better than the summation method in time domain detection, which requires about 0.81 dB. The reason was that in the time domain detection, the signal can be considered summed incoherently, while the FFT integrates the signal coherently.

2. In the second example, the input signal only fills half the window and the frequency is at the center of two frequency bins. In other words, this is the worst signal condition according to the previous discussion.

For the $P_{fa} = 2.56 \times 10^{-9}$, the threshold $r_1/\sigma_f = 6.2902$, which is the same as the previous example. For a 90% probability of detection $P_{df2} = 0.9$, the corresponding P_{df} can be found from (9.60) as

$$P_{df} = 1 - \sqrt{1 - P_{df2}} = 0.6838 \tag{9.67}$$

To achieve this probability of detection, a $(S/N)_f = 13.5$ dB is needed. The (S/N) in time domain can be found from (9.64), written in decibel form as

$$\left(\frac{S}{N}\right) = 10\log(N) + 10\log\left[1 - \cos\left(\frac{\pi}{N}\right)\right] + \left(\frac{S}{N}\right)_f \tag{9.68}$$
$$= 18.1 - 29.2 + 13.5 = 2.4 \text{ dB}$$

This result is very close to the S/N required for the L-out-of-N method in time domain detection. In general, the S/N required should be between the most optimum signal conditions and the worst case. In other words, the S/N required is between −3.6 and 2.4 dB.

9.23 COMMENTS ON FREQUENCY DOMAIN DETECTION

The discussion on frequency domain detection is limited to determine the sensitivity of a receiver. It is difficult to generalize this approach to process strong signals, and especially to detect time-coincident (simultaneous) signals. In general, the detection of simultaneous signals through a simple threshold crossing is not satisfactory. The main peaks of the spectrum will be detected, but the sidelobes should be neglected. Special effort and special algorithms are needed to process simultaneous signals. The major effort in designing analog EW receivers with simultaneous signals detection capability is concentrated on picking the true input signals and avoiding generation of false information from the sidelobes of the signals. It is anticipated that in a digital EW receiver, a similar problem will be encountered.

REFERENCES

[1] Marcum, J. I. "A Statistical Theory of Target Detection by Pulsed Radar, Mathematical Appendix," *IRE Trans. Information Theory,* Vol. IT-6, April 1960, pp. 145–267.

[2] Robertson, G. H. "Operating Characteristics for a Linear Detector of CW Signals in Narrowband Gaussian Noise," *The Bell System Technical Journal,* April 1967, pp. 755–774.

[3] Van Trees, H. L. *Detection, Estimation and Modulation Theory,* Part I, New York, NY: John Wiley & Sons, 1968.

[4] DiFranco, J. V., and Rubin, W. L. *Radar Detection,* Englewood Cliffs, NJ: Prentice Hall, 1968.

[5] Whalen, A. D. *Detection of Signals in Noise,* New York, NY: Academic Press, 1971.

[6] Papoulis, A. *Probability, Random Variables, and Stochastic Process,* New York, NY: McGraw-Hill Book Co., 1965.

[7] Drake, A. W. *Fundamentals of Applied Probability Theory,* New York, NY: McGraw-Hill Book Co., 1967.

[8] Davenport, W. B., Jr. *Probability and Random Process,* New York, NY: McGraw-Hill Book Co., 1970.

[9] Schwartz, M., and Shaw, L. *Signal Processing: Discrete Spectral Analysis, Detection and Estimation,* New York, NY: McGraw-Hill Book Co., 1975.

[10] Scharf, L. L. *Statistical Signal Processing, Detection, Estimation, and Time Series Analysis,* Reading, MA: Addison-Wesley Publishing Co., 1991.

[11] Campbell, G. A., and Foster, R. M. *Fourier Integrals for Practical Applications,* Princeton, NJ: Van Nostrand, 1948.

[12] Shaw, A., and Xia, W. Wright State University, Dayton, OH, Private communication.

[13] Tsui, J. B. Y. *Microwave Receivers With Electronic Warfare Applications,* New York, NY: John Wiley & Sons, 1986.

[14] Press, W. H., Flannery, B. P., Teukolsky, S. A., and Vetterling, W. T. *Numerical Recipes,* Cambridge, NY: Cambridge University Press, 1986.

[15] Hansen, V. G. "Optimization and Performance of Multilevel Quantization in Automatic Detectors," *IEEE Trans. on Aerospace and Electronic Systems,* Vol. AES-10, March 1974, pp. 274–280.

[16] Knight, W. C., Pridham, R. G., and Kay, S. M. "Digital Signal Processing for Sonar," *IEEE Proc.,* Vol. 69, pp. 1451–1506.

[17] Rohling, H. "Radar CFAR Thresholding in Clutter and Multiple Target Situations," *IEEE Trans. on Aerospace and Electronic Systems,* Vol. AES-19, July 1983, pp. 608–621.

[18] Gandhi, P. P., and Kassam, S. A. "Analysis of CFAR Processors in Nonhomogeneous Background," *IEEE Trans. on Aerospace and Electronic Systems,* Vol. AES-34, July 1988, pp. 427–445.

[19] Polydoros, A., and Nikias, C. L. "Detection of Unknown Frequency Sinusoids in Noise Spectral Versus Correlation Domain," *IEEE Trans. on Acoustics Speech and Signal Processing,* Vol. ASSP-35, June 1987, pp. 897–900.

APPENDIX 9.A

```
% ta9_1.m for table calculation
%find SNR using Tsui (binomial) method

clear;

r=[8:.1:20]';
global snr var st_dev

ns=input(' How Many Samples N ?  ');
%q=input(' Q = ?  ');
q=1;
%m=input(' M = ?  ');
m=16;
pfa=input(' Probability of False Alarm = ?  ');
%pd=input(' Pd = ?  ');
pd=0.9;
l=input(' L = ?  ');
var=(1+m)*(q.^2)/12;

if ns==1,
n=sqrt(-1*(1+m)*log(pfa)/6)
else
pk=inverse(ns,round(l),pfa);
n=sqrt(-1*(1+m)*log(pk)/6)
end

st_dev=1;
snr=100;
for ii=1:2:5,
for jj=1:10,
snr=snr-(10.^(2-ii))
if ns==1,
p_DET=1-quad8('pd1',0,sqrt(abs(2*var*log(pfa))));
else
det=1-quad8('pd1',0,sqrt(abs(2*var*log(pk))));
p_DET=binomial(ns,round(l),det);
end
if (p_DET<=pd),break,end
end
for j=1:10,
snr=snr+(10.^(1-ii))
```

```
if ns==1,
p_DET=1-quad8('pd1',0,sqrt(abs(2*var*log(pfa))));
else
det=1-quad8('pd1',0,sqrt(abs(2*var*log(pk))));
p_DET=binomial(ns,round(l),det);
end
if (p_DET>=pd),break,end
end
end
snr
end
```

APPENDIX 9.B

```
%ta9_1a.m the summing method
%Find SNR using Square-Law detection from known ns,pfa and pd

clear;

r=[8:.1:20]';
global snr var st_dev ns pfa

ns=input(' How Many Samples N ? ');
pfa=input(' Probability of False Alarm = ?  ');
%pd=input(' Pd = ?  ');
pd=0.9;
var=1;
d_square=msr1
st_dev=1;
%st_dev=0.5;

snr=1;
%snr=10;
for ii=1:2:5,
for jj=1:20,
snr=snr-(10.^(1-ii))
if ns==1,
p_DET=1-quad8('pd1',0,sqrt(abs(2*var*log(pfa))))
else
p_DET=quad8('pd3',d_square,2*d_square)
end
if (p_DET<=pd),break,end
```

```
end
for jj=1:20,
snr=snr+(10.^(-ii))
if ns==1,
p_DET=1-quad8('pd1',0,sqrt(abs(2*var*log(pfa))))
else
p_DET=quad8('pd3',d_square,2*d_square)
end
if (p_DET>=pd),break,end
end
end
snr
end
```

APPENDIX 9.C

```
% df9_3.m
clg
r = [0:.01:10];
out1 = rayleigh(r);
plot(r,out1)
hold on
out2 = ricianns(r,10);
plot(r, out2)
plot([3, 3], [0, .11])
xlabel('r')
ylabel('P(r)')
text(1.5, .55, 'noise only')
text(5.5, .35, 'S/N = 10 dB')
text(3.1, .08, 'r1')
```

APPENDIX 9.D

```
% df9_4.m RICI_INT This program generates prob of detection given the prob of
% false alarm rate and signal-to-noise ratio. Plot Pd vs. snr
clear
clg

snrends = input('enter starting and ending snr values [a b] = ');
r = [8:.1: 20]';
global snr
```

```
p_fa = input('enter probability of false alarm p_fa =' );
%1.55e-10;3.84e13
threshold = sqrt(abs(2*log(p_fa)));
  l=0;
for snr = snrends(1,1):.1:snrends(1,2);
     l = l+1;
     p_d(l) = 1- quad8('rician', 0, threshold);
end
plot([snrends(1,1) : .1 : snrends(1,2)], p_d)
grid
xlabel('Signal-to-noise ratio in dB')
ylabel('Probability of detection P_d')
text(12.5,.85,'P_fa=3.84e(-13)');
```

APPENDIX 9.E

```
% df9_5.m BINOPD_P This program calculates the probability of binomial
distribution
% and plot P_dm versus p.
clear

xx = input('enter that starting pt of L (from 1 to 64) = ');
x = [xx:64];

p = input('ent a range of p(i/N) values [.7:.01:.99] = ');
% exp(-12/(1+M));
v = length(p);
b = length(x);
ex = zeros(v,b);
p1 = zeros(v,b);

coff = gamma(65)./(gamma(65-x).*gamma(x+1));

n = 1;
while n <= v,
     ex(n,1:b) = (p(n) .^x) .* ((1-p(n)) .^(64-x));
     coff(n,1:b) = coff(1,1:b);
     p1(n,1:b) = coff(n,1:b) .* ex(n,1:b);
     n = n+1;
end
if b == 1
     pd = p1';
```

```
  else
    pd = sum(p1');
end
plot(p, pd)
grid
xlabel('value of p(i/N)')
ylabel('P(L+/N)')
```

APPENDIX 9.F

```
% ricianns.m This function calculates the rician pdf given r,
snr, sigma, and A

function y = rician(r,snr);
sigma = 1;
A = (10^(snr/20))*sqrt(2);
arg = (r.*A)./(sigma^2);
[J, digits] = bessela(0, i*arg);
I = real(J);
y = (r.*exp(- (r.^2 + A^2) ./ (2*(sigma^2))) .* I) ./(sigma^2);
```

APPENDIX 9.G

```
% rayleith.m  This function generates the rayleith distribution

function y = ray(r);
sigma = 1;
y = r .*exp(-(r.^2) ./(2*(sigma^2))) ./(sigma^2);
```

APPENDIX 9.H

```
% rician.m  This function calculates the rician pdf given r,
sigma, and A

function y = rician(r);
global snr
sigma = 1;
%snr = 10 % this line must be commended out if work with Rici_int
A = (10^(snr/20))*sqrt(2);
arg = (r.*A)./(sigma^2);
[J, digits] = bessela(0, i*arg);
```

```
I = real(J);
y = (r.*exp(- (r.^2 + A^2) ./ (2*(sigma^2))) .* I) ./(sigma^2);
```

APPENDIX 9.I

```
% binomial.m
function y=binomial(n,l,p),
y=0;
for a=l:n,
pl(a)=prod(linspace(n-a+1,n,a))*((1-p).^(n-a))/prod(linspace(1,a,a))*(p.^a);
y=y+pl(a);
end
end
```

APPENDIX 9.J

```
function d=inverse(n,l,p);

pfas=1;

%for i=1:2:13,
for i=1:2:9,
for j=1:10,
pfas=pfas-(0.1.^(i));
%pp=pfas/1000000;
y=binomial(n,l,pfas);
if (y<=p),break,end
end
for j=1:10
pfas=pfas+(0.1.^(1+i));
%pp=pfas/1000000;
y=binomial(n,l,pfas);
if (y>=p),break,end
end
end

d=pfas;

end
```

APPENDIX 9.K

```
%file name: msr1.m
%Find the threshold d for square-law detection

function d=msr1;

global ns pfa

if ns==1,

d=-2*1*log(pfa);

else

x=1000;

for i=1:ns-1,
h(i)=ns-i;
end
g=1;
for i=1:ns-1,
g=g*h(i);
end

ga=gamma(ns);

for i=1:2:5,
for j=1:10,
x=x-(10.^(3-i));
y=1-ga*gamma(ns,x)/g;
if (y>=pfa),break,end
end
for j=1:10
x=x+(10.^(2-i));
y=1-ga*gamma(ns,x)/g;
if (y<=pfa),break,end
end
end

d=2*1*x;

end
```

APPENDIX 9.L

```
%file name: pd1.m

function y=pd1(r);

global snr st_dev var

a=sqrt(2)*st_dev*(10^(snr/20))*sqrt(var);
arg=(r.*a)./var;
[j, digits]=bessela(0,i*arg);
b=real(j);
y=(r.*exp(-(r.^2+a^2)./(2*var)).*b)./var;
end
```

APPENDIX 9.M

```
%file name: pd3.m
%ideal one or multiple samples square_law
%probability density function of detection

function y=pd3(r);

global st_dev snr var ns

a=sqrt(2)*st_dev*(10^(snr/20))*sqrt(var);
arg=a.*sqrt(ns.*r)./var;
[j,digits]=bessela((ns-1),i.*arg);
b=real(j.*((-i).^(ns-1)));
y=((r./ns./(a.^2)).^((ns-1)./2)).*exp(-((ns.*(a.^2)+r)./var./2)).*b./var./2;
end
```

CHAPTER 10

Phase Measurements and Zero Crossings

10.1 INTRODUCTION

In this chapter, a phase measurement method and a zero crossing method will be discussed. These methods are relatively simple, but they can operate only under limited conditions. If the input contains only one sinusoidal wave, these methods can provide very accurate frequency measurement. Theoretically, these methods can detect the existence of simultaneous signals. The phase measurement approach uses analog in-phase and quadrature (*I-Q*) channels as the front end. If the *I-Q* channels are perfectly balanced, the phase measurement method can measure the frequencies of two input signals. The zero crossing method can measure the frequencies of multiple input signals. However, when multiple signals are present, the frequency accuracy measured by both these two methods will suffer compared with one input signal. Some of these methods need more investigation to determine their performances.

If the input signal is frequency modulated (FM), these methods can find the instantaneous frequency of the signal. In other words, the frequency versus time can be obtained. Such information is difficult to obtain from the Fourier transform. If the signal modulation is phase shift keying (PSK), the phase-sensitive methods can detect phase reversal in the signal. Therefore, it can detect the chip rate (or the clock frequency) of the PSK. These capabilities are also limited to a single signal. If there are simultaneous signals, these methods may generate erroneous information.

The instantaneous frequency measurement (IFM) receiver, an analog receiver discussed in Chapter 2, measures the phase between an input signal and its delayed version to obtain frequency information. The receiver can process only one signal and will be disturbed by simultaneous signals. The phase measurement method is quite similar to this type of receiver. The first element of an IFM receiver is a limiting amplifier, which is a nonlinear device. This nonlinear device makes the analysis on multiple signals very difficult, if not impossible. In the phase measurement approach, no limiting amplifier is used in front of the analog-to-digital converter (ADC); therefore, it is possible to measure two signals.

Compared with an IFM receiver, the digital phase measurement approach has another advantage (i.e., in its adaptability). In an analog IFM receiver, the delay time is fixed, and is selected based on the minimum anticipated pulse width (PW). As a result, the frequency resolution is fixed. In the digital version, the delay time can be adaptive, or changed in units of sampling time. In other words, the delay time can be PW-dependent. As a result, the frequency resolution can be PW-dependent, which is a very desirable property for an EW receiver. Since this method can generate very fine frequency resolution, they may have special EW applications. Some of the possible applications will be stated later in this chapter.

The major disadvantage of the digital phase measurement method is that the bandwidth of the receiver is narrow. An analog IFM receiver can cover an instantaneous bandwidth of 16 GHz (from 2 to 18 GHz). The digital approach is limited by the speed of the ADC and the digital signal processor following it. The positive aspect of the narrow input bandwidth is its low probability of having simultaneous signals. With a single input signal, this method can measure the frequency very accurately.

10.2 DIGITAL PHASE MEASUREMENT [1–8]

The simplest digital phase measurement is shown in Figure 10.1. This figure is identical to Figure 8.7. The input signal is divided into two paths and downconverted. The two output channels are 90 deg out of phase and referred to as the I and Q outputs. In order to keep this discussion simple, let us assume that the two channels are perfectly balanced.

From (8.20), the intermediate frequency (IF) outputs are

$$\begin{aligned} V_{\text{if1}} &= A\cos[2\pi(f_i - f_o)\,t + \phi] \\ V_{\text{if2}} &= A\sin[2\pi(f_i - f_o)\,t + \phi] \end{aligned} \tag{10.1}$$

where A and f_i are the amplitude and the frequency of the input signal, respectively, f_o is the frequency of the oscillator, and ϕ is the initial phase of the input signal. Let $I(n)$ and $Q(n)$ represent the digitized signal where n is an integer. They can be written as

$$\begin{aligned} I(n) &= A\cos[2\pi(f_i - f_o)\,nt_s + \phi] \\ Q(n) &= A\sin[2\pi(f_i - f_o)\,nt_s + \phi] \end{aligned} \tag{10.2}$$

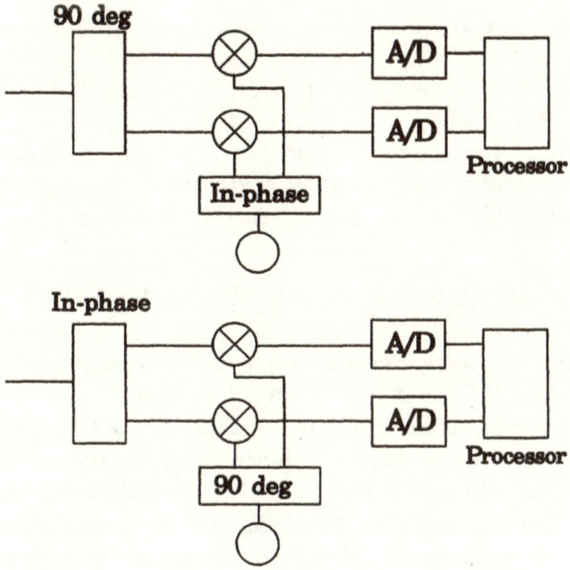

Figure 10.1 Digitizing I and Q channels.

where t_s is the digitization time interval.

In order to measure the phase of the signal, a processor is needed to perform the following operations: 1) take the arctangent of $Q(n)/I(n)$ to generate the phase $\theta(n)$, and 2) take the difference of the phase θ at different time intervals. These two steps can be written as

$$\theta(n) = \tan^{-1}\left\{\frac{Q(n)}{I(n)}\right\} = 2\pi(f_i - f_o)\, nt_s + \phi \tag{10.3}$$

$$\Delta\theta(i) = \theta(n+i) - \theta(n) = 2\pi(f_i - f_o)\, it_s$$

where $\theta(n)$ represents the phase angle at time n and $\Delta\theta(i)$ is the phase difference between two phase angles separated by time i unit of time t_s. If the $\Delta\theta(i)$ can be obtained, the frequency of the input signal can be found as

$$(f_i - f_o) = \frac{\Delta\theta(i)}{2\pi it_s} \tag{10.4}$$

In this equation, there is one important point; that is, the phase angle is limited from 0 to 2π (or $-\pi$ to π) radians. If the phase change is over 2π (or 360 deg), the phase will be aliased to less than 2π. Under this condition, an ambiguity problem of reporting an erroneous frequency will occur. In order to keep the input signal unambiguous, the angle difference should be kept less than 2π by selecting a small value of i.

Another important issue is the phase discontinuity. The phase difference calculated from (10.3) is obtained from two phase values. If one value is below 2π and the other above, there is phase discontinuity between these two values. The phase discontinuity must be taken into consideration in order to avoid generating an erroneous result. This procedure is often referred to as phase unwrapping.

10.3 ANGLE RESOLUTION AND QUANTIZATION LEVELS

In this section, the relationship between angle data resolution and the ADC number of bits will be discussed. Instead of deriving the relation analytically, a few simple examples are shown graphically in Figures 10.2(a) to 10.2(d). In these figures, the amplitude of the signal always matches the full scale of the ADC.

In Figure 10.2(a), the ADCs have only 1 bit. One bit is along the x-axis (voltage output of the I channel) and one bit along the y-axis (voltage output of the Q channel). The circle represents angle change in an analog sense. When the circle is in a certain square, the center of the square represents the digitized angle. In

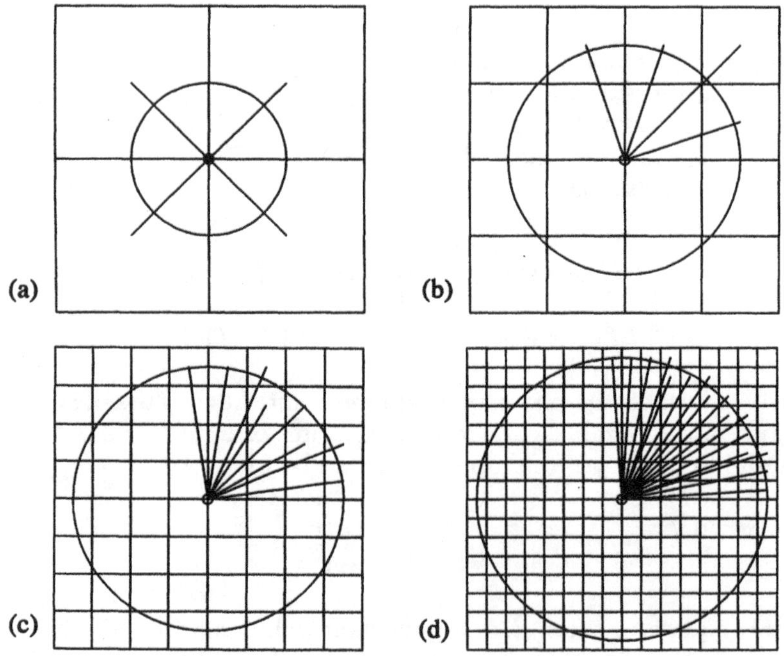

Figure 10.2 Angle resolution versus number of ADC bits: (a) 1 bit, (b) 2 bits, (c) 3 bits, (d) 4 bits.

each quadrant, there is only one angle quantization cell. There are a total of four quantization cells and each cell has the same size. In Figure 10.2(b), the ADCs have two bits. Each quadrant has three quantization cells. The cell size is determined by the length of the arc in the square. One can see that the cell sizes are not uniform. There is a total of 12 cells. In Figure 10.2(c), the ADCs have three bits and each quadrant has seven cells. The nonuniformity of the cell size still exists. The total number of cells is 28. In Figure 10.2(d), the ADCs have four bits and each quadrant has 15 nonuniform cells. The total number of cells is 60.

From this illustration, one can conclude that when the signal amplitude matches the full scale of the ADC, the angle quantization cell per quadrant A_q is

$$A_q = 2^b - 1 \tag{10.5}$$

where b is the number of bits of the ADC. Since a circle contains four quadrants, the total number of quantization cells A_r is

$$A_r = 4(2^b - 1) \tag{10.6}$$

This quantity represents the maximum number of angle quantization cells. In general, these cell sizes are nonuniform. If the amplitude of the input signal is less than the full scale, the number of angle quantization cells will decrease. For example, if the amplitude of the signal in Figure 10.2(b) reaches only the first bit level, the angle quantization will drop to four cells, as shown in Figure 10.2(a). If the amplitude of the input signal is very strong and drives the ADC into saturation, the number of quantization cells will decrease. In the limit, when the signal is very strong, the sine wave will be digitized more or less like a square wave. The result will be similar to the one-bit situation. In order to simplify the discussion, in later discussion the cell size is assumed uniform and the input signal is matched to the maximum of the ADC.

10.4 COMPARISON OF PHASE MEASUREMENT AND FFT RESULTS

In this section, the frequency data resolution generated by phase measurement and by an FFT will be compared. Let us use an example in which the ADC has eight bits with a sampling frequency f_s of 250 MHz (or $t_s = 1/f_s = 4 \times 10^{-9}$ sec = 4 ns). If a 64-point complex FFT is performed, the total time T is 256 ns (4 × 64). In this calculation, the period should be used rather than the actual time between samples, as discussed in Section 5.7. The frequency data resolution between adjacent frequency bins is $1/T$ = 3.906 MHz, which is independent of the number of bits of the ADC.

If the amplitude of the signal matches the highest level of the ADC, the angle data resolution obtained from (10.6) is 512 units. If the phase calculation is obtained from the first and the last data points, the actual time difference is $T' = 252$ ns and the corresponding unambiguous bandwidth of the phase measurement method is 3.968 MHz. Another way to look at the unambiguous bandwidth is that whenever the input frequency changes (increase or decrease) by 3.968 MHz, the phase will change by 2π. Thus, the phase angle can be measured, theoretically, to one of 512 cells and each cell has a frequency data resolution of 7.750 kHz (3.968 MHz/512).

From this example, one can see that from the same input data, if the input is a simple sinusoidal wave, the phase measurement method can provide a frequency data resolution much better than the FFT can provide. In addition, the processing scheme is rather simple in comparison with the FFT processing. However, if there are simultaneous signals, this approach may produce an erroneous frequency, as in an IFM receiver, and this is the major drawback.

10.5 APPLICATION OF THE PHASE MEASUREMENT SCHEME

Equations (10.3) and (10.4) are used to find the frequency of an input signal. In general, using the phase measurement method to obtain fine frequency has two applications. The first one is to measure the instantaneous frequency on an FM input signal. The second application is to find the highest possible fine-frequency data resolution of a pure sinusoidal wave.

For the first application, the phase difference between every data point is taken. If the data are sampled at a 4-ns (250 MHz) time interval and a total of 64 data points are considered, the phase angle must be calculated at every data point and the phase between each pair of consecutive data points must be obtained. These calculations can be written as

$$\begin{aligned} \Delta\theta(1) &= \theta(2) - \theta(1) \\ \Delta\theta(1) &= \theta(3) - \theta(2) \\ \Delta\theta(1) &= \theta(4) - \theta(3) \\ &\cdots \\ \Delta\theta(1) &= \theta(64) - \theta(63) \end{aligned} \tag{10.7}$$

where $\Delta\theta(1)$ represents the phase difference of two adjacent phases in time. A total of 63 calculations are required. If the frequency of input signal changes with time, this method can find the frequency change every 4 ns. However, the frequency data resolution is rather poor because of the short delay time. The unambiguous bandwidth is 250 MHz (1/4 ns). If the ADC used has 8 bits, the best frequency data resolution that can be achieved is 488 kHz (250 MHz/512). If better frequency data resolution is desirable, the phase difference between every other data point can be calculated. In this case, the best frequency data resolution that can be achieved is 244 kHz. In general, if a certain delay time is selected, the frequency within the delay period is considered a constant.

If the input signal has a constant frequency, one can use the phase measurement method to obtain very high frequency precision. In this case, it is not necessary to calculate the frequency every 4 ns and a simpler way can be used to obtain the frequency. A binary approach can save time and calculations. For example, the first, second, third, fifth, and so forth data points can be used to obtain the difference phase angle as

$$\begin{aligned}\Delta\theta(1) &= \theta(2) - \theta(1)\\ \Delta\theta(2) &= \theta(3) - \theta(1)\\ \Delta\theta(4) &= \theta(5) - \theta(1)\\ &\cdots\\ \Delta\theta(63) &= \theta(64) - \theta(1)\end{aligned} \tag{10.8}$$

The difference phase angles can be written as $\theta(2^n)$ except for the last data point, where n is an integer and 2^n is the delay time between two data points. The last delay time is $2^n - 1$.

In the above equations, there are 64 (2^6) data points, and the difference phases are calculated only 7 (6 + 1) times with delays of 1, 2, 4, 8, 16, 32, and 63. This number of calculations is much less than the number of calculations (63) required in the first approach. In general, if the data have 2^b points, a total of $b + 1$ calculations are needed. In these calculations, the short delay time is used to resolve frequency ambiguity and the long delay is used to provide frequency data resolution. In this example, the shortest delay is 4 ns, which corresponds to an unambiguous bandwidth of 250 MHz. The frequency data resolution generated from the short delay does not need to be very high. It is only required to resolve the ambiguity of the next level (i.e., a minimum of two frequency data resolution cells are needed). For example, $\theta(2)$ has a delay time of 8 ns, which corresponds to an unambiguous bandwidth of 125 MHz. This approach is quite similar to the design of an analog IFM receiver.

10.6 ANALYSIS OF TWO SIMULTANEOUS SIGNALS

As mentioned in previous sections, the major deficiency in the phase measurement approach is that it is susceptible to simultaneous signals. In this section, the theoretical analysis of another phase measurement approach that can measure the frequencies of two simultaneous signals is presented. This approach is limited to two simultaneous signals, and is based on [8].

The simultaneous signal detection scheme is rather simple. If there is only one sinusoidal signal, the I and Q channel outputs can be written as in (10.2). The amplitude can be found as

$$\sqrt{I(n)^2 + Q(n)^2} = A \tag{10.9}$$

which is a constant and independent of time.

If there are two signals of angular frequency f_1 and f_2, the combined input signal can be written in complex form as

$$s(t) = e^{-j2\pi f_o t}(e^{j2\pi f_1 t} + Re^{j2\pi f_2 t}) \tag{10.10}$$

where f_o is the frequency of the local oscillator and $0 \le R \le 1$ represents the amplitude of the second signal. This assumes arbitrarily that the first signal has a unity amplitude and the second signal is weaker than the first one, and both signals have zero phase at time $t = 0$. This assumption does not restrict generality. This equation can be written as

$$\begin{aligned} s(t) &= e^{-j2\pi f_o t} e^{j2\pi f_1 t}(1 + Re^{j2\pi \Delta f_2 t}) \\ &= e^{-j2\pi f_o t} e^{j2\pi f_1 t}[p(t) + jq(t)] \end{aligned} \tag{10.11}$$

where $\Delta f = f_2 - f_1$ is the difference frequency, and

$$p(t) = 1 + R\cos(2\pi\Delta f t) \qquad q(t) = R\sin(2\pi\Delta f t) \tag{10.12}$$

This equation can be written in terms of amplitude and phase, and the result is

$$s(t) = e^{-2\pi f_o t} E(t)\, e^{j\theta(t)} \tag{10.13}$$

where $E(t)$ and $\theta(t)$ are the time-dependent envelope and the instantaneous phase of the input signals, which can be written as

$$E(t) = \sqrt{p(t)^2 + q(t)^2} = \sqrt{2 + 2R\cos(2\pi\Delta f t)} \tag{10.14}$$

and

$$\theta(t) = 2\pi f_1 t + \tan^{-1}\frac{q(t)}{p(t)} \tag{10.15}$$

From (10.14), it is obvious that if there is more than one signal, then the amplitude is time dependent. By measuring the amplitude at different times, one can determine whether there is one, or more than one, signal.

The instantaneous frequency $f(t)$ can be defined as the derivative of the phase $\theta(t)$ [3], and the result is

$$\begin{aligned} f(t) &= \frac{1}{2\pi}\frac{d[\theta(t)]}{dt} = f_1 + \frac{1}{2\pi}\left(\frac{p\dot{q} - q\dot{p}}{p^2 + q^2}\right) \\ &= f_1 + R\Delta f\left[\frac{R + \cos(2\pi\Delta f t)}{1 + R^2 + 2R\cos(2\pi\Delta f t)}\right] \\ &\equiv f_1 + R\Delta f k_r(t) \end{aligned} \tag{10.16}$$

where $\dot{p}$, $\dot{q}$ representing derivatives with respect to t, and

$$k_r(t) = \frac{R + \cos(2\pi\Delta f t)}{1 + R^2 + 2R\cos(2\pi\Delta f t)} \tag{10.17}$$

These two quantities—the amplitude of the signal and the instantaneous frequency—can be measured to determine the existence of simultaneous signals and to obtain the frequencies of them. Since the frequency of the local oscillator is known, the instantaneous frequency of the signal can be determined.

The amplitude of the envelope, given from (10.14), versus time is shown in Figure 10.3. In this special case, $R = 0.8$.

From (10.17), the maxima of $k_r(t)$ appear at $2\pi f t = 0, 2\pi, 4\pi, 6\pi$, and so forth where $\cos(2\pi\Delta f t) = 1$ and $k_r(t) = 1/(1 + R)$. The minima of $k_r(t)$ appear at $\pi, 3\pi, 5\pi$, and so forth where $\cos(2\pi\Delta f t) = -1$ and $k_r(t) = -1/(1 - R)$. The instantaneous frequency depends on the sign of Δf.

If Δf is positive (or $f_1 < f_2$), the maximum value of the instantaneous frequency is

$$f_{max}(t) = f_1 + \frac{R\Delta f}{1 + R} \tag{10.18}$$

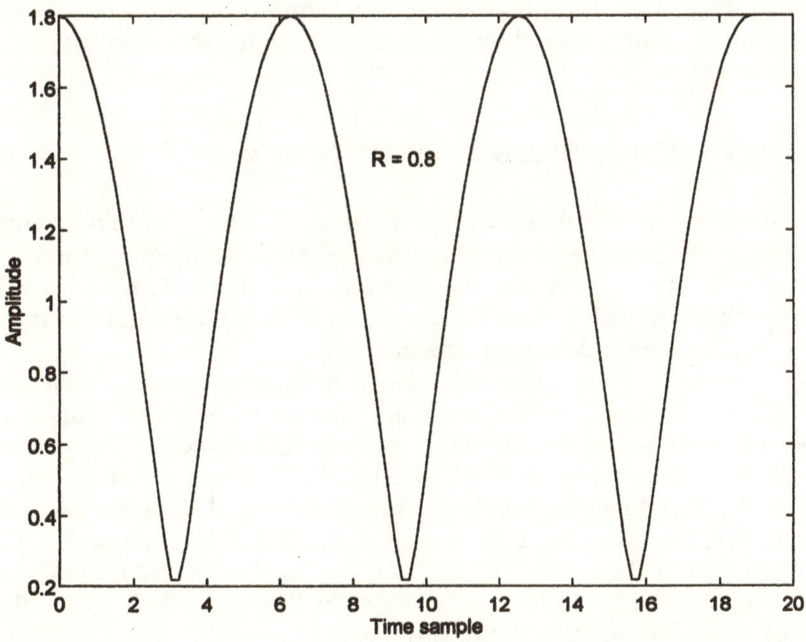

Figure 10.3 Signal envelope versus time.

and the minimum is

$$f_{\min}(t) = f_1 - \frac{R\Delta f}{1 - R} \tag{10.19}$$

The minimum instantaneous frequency can become negative. This condition can be found from (10.19) as

$$f_1(1 - R) < R\Delta f \equiv R(f_2 - f_1) \qquad \text{or} \qquad f_1 < Rf_2 \tag{10.20}$$

Since $0 < R < 1$ as defined in (10.10), in order to obtain a negative frequency, instantaneous frequency f_2 must be higher than f_1 and the condition in the above equation must be fulfilled. Negative instantaneous frequency can be easily understood through the phase angle defined in (10.15). When the phase rotates in the negative direction the corresponding instantaneous frequency is negative.

If Δf is negative (or $f_1 > f_2$), (10.19) represents the maximum value of the instantaneous frequency and (10.18) represents the minimum value. Under this condition, (10.20) cannot be fulfilled; thus, the instantaneous frequency is always positive.

This discussion provides the basic equations to measure the frequencies of two input signals. This discussion is based on continuous functions in the time domain rather than the digitized version. The next section will provide an approach to measure these frequencies.

10.7 FREQUENCY MEASUREMENT ON TWO SIGNALS [8]

This section will use the result obtained from last section to measure the frequencies of two signals. The frequency information will be obtained from the instantaneous frequency of (10.16). First, let us take a look at the variable $k_r(t)$ of (10.17). This is a periodic equation with period T such that $\Delta fT = 1$. If the period T is measured, the difference frequency Δf can be determined.

Figure 10.3 shows that the amplitude of the envelope has the same period T as $k_r(t)$. Therefore, one can measure either from the amplitude change of the envelope or $k_r(t)$ to obtain the difference frequency Δf. However, $k_r(t)$ is not a term that can be measured explicitly (but instantaneous frequency $f(t)$ in (10.16) can be measured). The function $f(t)$ is plotted in Figure 10.4 under two different conditions, and they are 1) in Figure 10.4(a) $f_1 = 100$, $f_2 = 150$, and $R = 0.8$ so Δf is positive, and 2) in Figure 10.4(b) $f_1 = 100$, $f_2 = 50$, $R = 0.8$ so Δf is negative. The two frequencies f_1 and f_2 and the amplitude ratio R are arbitrarily chosen because their values bear little significance in this discussion.

In Figure 10.4(a), portions of the instantaneous frequency are negative. The width of the minimum is very sharp in comparison with the maximum values. This

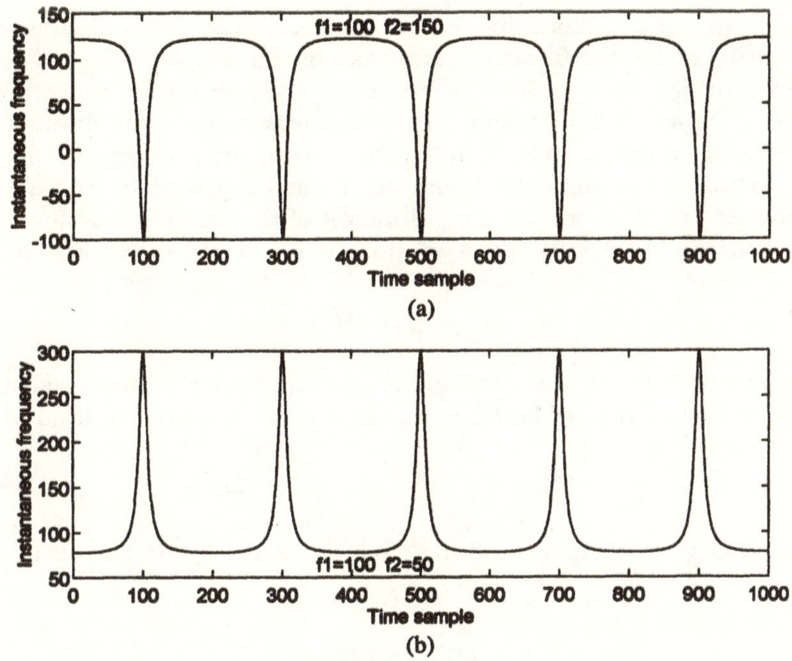

Figure 10.4 Instantaneous frequency versus time: (a) Δf is positive, (b) Δf is negative.

shape can indicate Δf is positive ($f_1 < f_2$), while in Figure 10.4(b) the opposite is true and the instantaneous frequency stays in the positive range. This shape can indicate Δf is negative ($f_1 > f_2$). The frequency variation is periodic, and this period T can be measured either from the minima in Figure 10.4(a) or from the maxima in Figure 10.4(b). In either case, sharp points in $f(t)$ should be used to determine the period T because a more accurate result can be obtained. After the value and sign of the difference frequency Δf are determined, both the frequencies of the strong signal and the weak signal can be found.

First, let us find the frequency of the strong signal. In order to obtain this frequency, one can take the average of the instantaneous frequency over a period of nT. The result obtained from (10.15) and (10.16) will be

$$\hat{f} = \frac{1}{2\pi T} \int_{f_1}^{t_1+nT} \frac{d[\theta(t)]}{dt} dt = f_1 + \tan^{-1}\left\{\frac{q(t_1 + nT)}{p(t_1 + nT)}\right\} - \tan^{-1}\left\{\frac{q(t_1)}{p(t_1)}\right\} = f_1 \quad (10.21)$$

where t_1 is any starting time, n is an integer. Since $p(t)$ and $q(t)$ from (10.12) are periodical of T, the two arctangent terms cancel. This result shows that the average of $f(t)$ is f_1. In this equation, the period T is not exactly known; however, if one averages many cycles, the frequency error can be made small.

Next, the sign of Δf will be determined. The sign can be determined with the help of Figure 10.4. The approach is to measure the time duration of the maximum and minimum of $f(t)$. In Figure 10.4(a), the average of $f(t)$ (or f_1) is close to the maximum value of $f(t)$ and the minima of $f(t)$ occur for a short time, which implies $\Delta f > 0$. If one finds the duration of the maximum is short, then $\Delta f < 0$. Once the sign of Δf is determined, the second frequency can be obtained as

$$f_2 = f_1 \pm \Delta f \tag{10.22}$$

If the amplitude of the envelope is measured, such as the maximum and minimum, the amplitude of both signals can also be measured from (10.14) as

$$E_{max}^2 = 2 + 2R \qquad E_{min}^2 = 2 - 2R \tag{10.23}$$

Thus, the strong signal A_{st} is 1 and the weak one A_{wk} is R, and they can be found as

$$A_{st} = \frac{E(t)_{max}^2 + E(t)_{min}^2}{4} = 1 \tag{10.24}$$

$$A_{wk} = \frac{E(t)_{max}^2 - E(t)_{min}^2}{4} = R$$

This discussion outlines how to obtain two frequencies from the phase measurement scheme.

This discussion is based on the assumption that the I and Q channels are perfectly balanced. If the I and Q channels are not perfectly balanced, even a single signal with constant amplitude and frequency can cause the amplitude output to vary as a function of time. Thus, in a real receiver design, the unbalance of the I and Q channels limits the applications of measuring two simultaneous signals.

10.8 SINGLE-FREQUENCY MEASUREMENT FROM ZERO CROSSING

In the phase measurement method, the input signal must be divided into two parallel outputs: the I and Q channels. If the input is a simple sinusoidal, the frequency can be measured from the period of the signal. The period can be measured from the maxima, the minima, or the zero amplitude points of the digitized data. The zero amplitude points are usually referred to as the zero crossings because they are usually obtained from two adjacent data points, one positive and one negative.

In this section, the zero crossings are used to find the frequency. This method has the same limitation as the IFM receiver measurement method. Simultaneous signals may cause erroneous results. In the phase measurement system, a minimum of two samples per cycle are required to measure the input frequency. The zero crossing method discussed here uses a real signal; therefore, a minimum of four samples per cycle are required.

If the input signal is digitized at sampling frequency $f_s = 1/t_s$, it is not likely that zero amplitude points will be sampled. Thus, the first attempt is to find the time of zero crossing points from the digitized data. This can be accomplished from either of the following trigonometric relations [9].

$$\cos[(n+2)2\pi f t_s + \alpha] = 2\cos(2\pi t_s)\cos[(n+1)2\pi f t_s + \alpha] - \cos(n2\pi f t_s + \alpha) \quad (10.25)$$
$$\sin[(n+2)2\pi f t_s + \alpha] = 2\cos(2\pi f t_s)\sin[(n+1)2\pi f t_s + \alpha] - \sin(n2\pi f t_s + \alpha)$$

where n is an integer, t_s is the sampling period, f is the frequency of the input signal, and α is any initial angle. Now, using the second equation, three consecutive sampled data can be written as

$$\begin{aligned} x(n) &= A\sin(n2\pi f t_s + \alpha) \\ x(n+1) &= A\sin[(n+1)2\pi f t_s + \alpha] \\ x(n+2) &= A\sin[(n+2)2\pi f t_s + \alpha] \end{aligned} \quad (10.26)$$

Substituting these data into (10.25), the result is

$$x(n+2) = 2x(n+1)\cos(2\pi f t_s) - x(n) \quad (10.27)$$

From this equation, the frequency can be obtained as

$$2\pi f t_s = \cos^{-1}\left[\frac{x(n+2) + x(n)}{2x(n+1)}\right] \quad (10.28)$$

The frequency of the input signal can be found from this equation. This method will be further discussed in Chapter 14.

Figure 10.5(a) shows an example of the sampled data. In this figure, three points are chosen and there is a sign change between $x(n+1)$ and $x(n+2)$. The point before the sign change is used as the $t = 0$ reference. In this case, the reference is $x(n+1)$, which will be written as $\sin(\alpha)$. In order to create this condition, it is assumed $n = -1$ in (10.26) and the results are

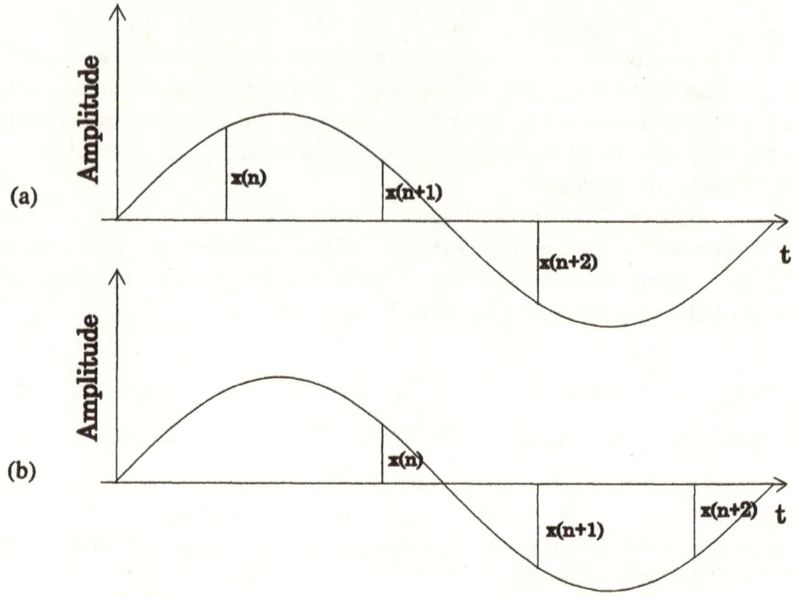

Figure 10.5 Three points selected to calculate zero crossing: (a) between 2nd and 3rd samples, (b) between 1st and 2nd samples.

$$
\begin{aligned}
x(-1) \equiv x_1 &= A\sin(-2\pi f t_s + \alpha) \\
&= A\sin(\alpha)\cos(2\pi f t_s) - A\cos(\alpha)\sin(2\pi f t_s) \\
x(0) \equiv x_2 &= A\sin(\alpha) \\
x(1) \equiv x_3 &= A\sin(2\pi f t_s + \alpha) \\
&= A\sin(\alpha)\cos(2\pi f t_s) + A\cos(\alpha)\sin(2\pi f t_s)
\end{aligned}
\tag{10.29}
$$

In this equation, x_1, x_2, and x_3 are just new notations used in the following derivations for simplicity of the result. From this equation, it is easy to see that

$$
\cos(2\pi f t_s) = \frac{x_1 + x_3}{2x_2} \tag{10.30}
$$

which is the same result as (10.28). It is also obvious that in order to find the time from x_2 to the zero crossing point, the angle α must be found. Therefore, the next step is to obtain $\sin(\alpha)$ from (10.29) and (10.30), which can be written as

$$
\frac{x_2}{x_3} = \frac{\sin\alpha}{\sin\alpha\cos(2\pi fT) + \cos\alpha\sin(2\pi fT)}
$$

$$x_3 \sin\alpha - x_2 \sin\alpha \cos(2\pi fT) = x_2\sqrt{1-\sin^2\alpha}\,\sin(2\pi fT) \tag{10.31}$$

$$\sin^2\alpha x_3^2 - 2x_2x_3\cos(2\pi fT) + x_2^2\cos^2(2\pi fT) = x_2^2(1-\sin^2\alpha)\sin^2(2\pi fT)$$

$$\sin^2\alpha = \frac{x_2^2[1-\cos^2(2\pi fT)]}{x_3^2 - 2x_2x_3\cos(2\pi fT) + x_2^2}$$

From this equation, the time difference between x_2 and the zero crossing can be found as

$$\alpha = \sin^{-1}\left\{\sqrt{\frac{x_2^2[1-\cos^2(2\pi fT)]}{x_2^2 - 2x_2x_3\cos(2\pi fT) + x_3^2}}\right\} \tag{10.32}$$

From the second equation of (10.29), x_2 (or $x(0)$) is $\sin\alpha$ away from zero. The corresponding time δt from x_2 to the zero crossing point can be found as

$$\delta t = \frac{\alpha}{2\pi f} \tag{10.33}$$

because the phase angle changes 2π, covering one cycle with a period of $1/f$, and f is obtained from (10.28). Once this time is determined, the true zero crossing can be obtained by adding δt to $n+1$, the second data point in (10.26).

10.9 ILL CONDITION IN ZERO CROSSING FOR SINGLE-SIGNAL AND REMEDY

There is one problem in the above approach. That is the case where the value of $x(n+1)$ in (10.26) (or x_2 in (10.30)) is very small. Under this condition, the error in calculating f will be large. To avoid this problem, one can select three different points. For example, if $x(n+1)$ in Figure 10.5(a) is close to zero, one can choose the next three points, as shown in Figure 10.5(b). Under this condition, the zero crossing is between the first and the second samples.

The data point before the zero crossing is still used as $t = 0$. Three values can be written as

$$\begin{aligned}
x(0) &\equiv x_1 = A\sin(\alpha) \\
x(1) &\equiv x_2 = A\sin(2\pi ft_s + \alpha) = A\sin(\alpha)\cos(2\pi ft_s) + A\cos(\alpha)\sin(2\pi ft_s) \\
x(2) &\equiv x_3 = A\sin(4\pi ft_s + \alpha) = A\sin(\alpha)\cos(4\pi ft_s) + A\cos(\alpha)\sin(4\pi ft_s)
\end{aligned} \tag{10.34}$$

From these relations, it is easy to show that

$$\cos(2\pi f t_s) = \frac{x_1 + x_3}{2x_2} \tag{10.35}$$

and

$$\sin^2 \alpha = \frac{x_1^2[1 - \cos^2(2\pi f t_s)]}{x_2^2 - 2x_1x_2\cos(2\pi f t_s) + x_1^2} \tag{10.36}$$

The same approach as before can be used to find the time delay from x_1 to the zero crossing points.

It should be noted that in both cases the time calculated is referenced to the data point before the zero crossing. From simulated data it is demonstrated that when the signal-to-noise (S/N) ratio is high, the calculated result is very accurate. If the S/N = 1,000 dB, the error is less than 10^{-8} Hz. When S/N = 100 dB, the error is less than 10 Hz, while for S/N = 10 dB, the error is about 100 kHz.

Although the zero crossing time calculated from the above equations is based on the exact solution, the calculation itself is rather tedious. It involves multiplication, square rooting, and an inverse sine calculation. Besides, it may have ill-conditioned cases. Thus a simplified approach will be discussed in the next section.

10.10 SIMPLIFIED ZERO CROSSING CALCULATION FOR SINGLE-SIGNAL

A simpler way to estimate a zero crossing is to connect a straight line between two points on either side of a zero crossing. This is an approximate approach. Figure 10.6 shows such an arrangement. In this approach, two consecutive data points are located, one above zero and one below, or vice versa. A straight line is drawn between the two points and the point where this line crosses the x-axis is considered as the approximate zero crossing point.

The mathematics to calculate this value are as follows. Let these two points be x_1 and x_2 at t_1 and t_2, respectively. The straight line passing through these two points is

$$\frac{x - x_1}{t - t_1} = \frac{x_1 - x_2}{t_1 - t_2} \tag{10.37}$$

This line intercepting with $x = 0$ provides

$$t = t_1 + \frac{x_1(t_2 - t_1)}{x_1 - x_2} = t_1 + \frac{x_1 t_s}{x_1 - x_2} \tag{10.38}$$

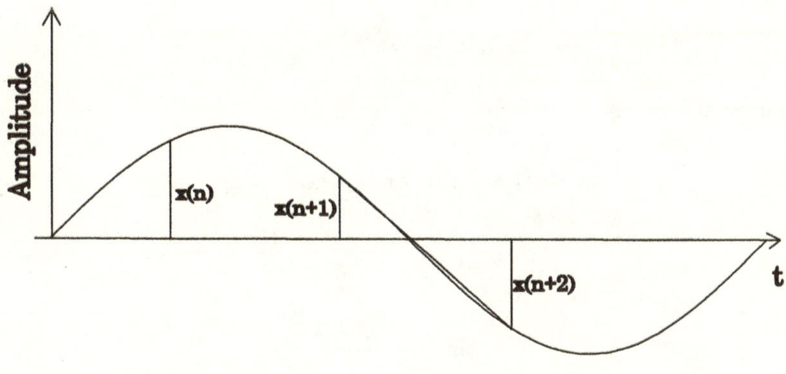

Figure 10.6 A straight line to approximate a zero crossing.

where the relation of $t_2 - t_1 = t_s$ is used in the above derivation.

Since (10.37) does not generate the true zero crossing, it is desirable to find the error. In this discussion, the maximum error will be derived. The maximum error occurs at the highest input frequency, where four samples are obtained per cycle. Under this condition, $t_s = \pi/2$. Assume that $x_1 = \cos(t)$, then $x_2 = \cos(t + \pi/2)$. The error as a function of t can be found from the above equation by substituting these values for x_1 and x_2 as

$$\begin{aligned} e(t) &= \left\{ t + \frac{t_s \cos(t)}{\cos(t) - \cos(t + \pi/2)} \right\} - t_z \\ &= \left\{ t + \frac{\frac{\pi}{2}\cos(t)}{\cos(t) + \sin(t)} \right\} - t_z \end{aligned} \tag{10.39}$$

where t_z is the true zero crossing time. To find the worst error, one takes the derivative of $e(t)$ with respect to t. The result is

$$\begin{aligned} \frac{de(t)}{dt} &= 1 + \frac{\pi}{2}\left[\frac{(\sin t + \cos t)(-\sin t) - \cos t(\cos t - \sin t)}{(\sin t + \cos t)^2}\right] \\ &= 1 - \frac{\pi}{2}\left[\frac{\sin^2 t + \cos^2 t}{(\sin t + \cos t)^2}\right] \\ &= 1 - \frac{\pi}{2}\left[\frac{1}{(\sin t + \cos t)^2}\right] \end{aligned} \tag{10.40}$$

Setting this result to zero yields

$$(\sin t + \cos t)^2 = \frac{2}{\pi} \tag{10.41}$$

which can be solved as

$$\sin^2 t + 2\sin t \cos t + \cos^2 t = \frac{\pi}{2}$$

$$2\sin t \cos t = \frac{\pi}{2} - 1$$

$$\sin 2t = \frac{\pi - 2}{2} \tag{10.42}$$

or

$$t = \frac{1}{2}\sin^{-1}\left(\frac{\pi - 2}{2}\right) = \begin{cases} 0.304 \\ \pi - 0.304 = 2.838 \end{cases}$$

This result is shown in Figure 10.7. The zero crossing produced under this condition can be found through (10.38) as follows:

$$\begin{aligned} t_1 &= 2.838 \\ x_1 &= \sin(t_1) = \sin(2.838) = 0.299 \\ x_2 &= \sin\left(t_1 + \frac{\pi}{2}\right) = \sin(4.409) = -0.954 \\ t_{zs} &= t_1 + \frac{0.299 x \frac{\pi}{2}}{0.299 + 0.9541} = 3.213 \\ t_z &= \pi \\ e(t_1) &= t_{zs} - t_z = 0.071 \end{aligned} \tag{10.43}$$

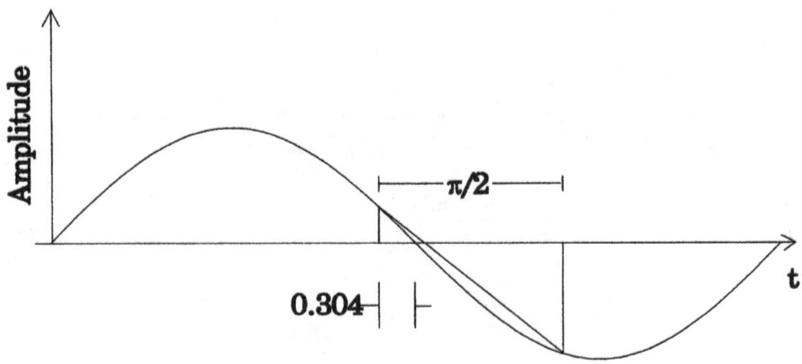

Figure 10.7 Worst error in straight-line approximating zero crossing.

where t_1 is the result obtained from (10.42), t_{zs} is the zero crossing calculated from this straight-line approximation, and t_z is the true zero crossing time. Thus, the worst error in percentage per sampling period is $e(t_1)/t_z = 4.5\%$.

Besides the simplicity in frequency calculation, this approach does not have ill-conditioned cases. When a data point is close to zero, it does not create an inaccurate result, as discussed in Section 10.9.

10.11 EXPERIMENTAL RESULTS FROM SINGLE-FREQUENCY ZERO CROSSING METHODS

In this section, the zero crossing will be applied to data collected from a sine wave. The data were collected by a Hewlett Packard digital scope. The input signal is 200 MHz. The output of the scope has 8 bits. The sampling time was a nominal 1 ns; however, the actual sampling time was 9.99401×10^{-10} sec. The digitized data are shown in Figure 10.8(a). In this demonstration, only 200 samples were used. The frequency was calculated from the nth zero crossing with respect to the first one. The relation can be written as

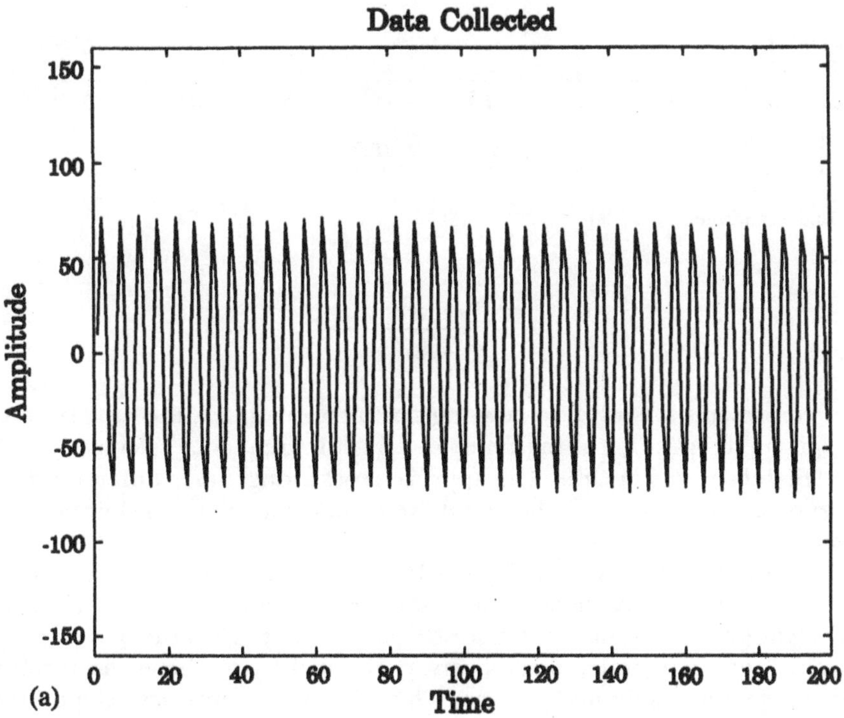

Figure 10.8 Error frequency calculated from zero crossing: (a) input signal, (b) without fine zero crossing, (c) with three-point fine zero crossing, (d) with straight-line fine zero crossing.

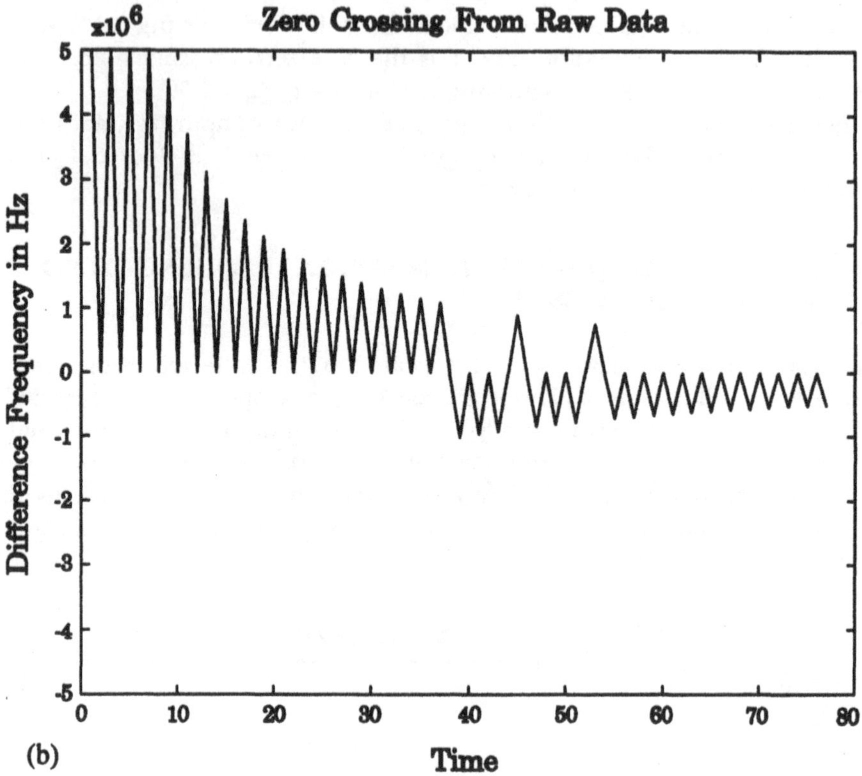

Figure 10.8 (continued).

$$f_n = \frac{n-1}{2(t_{zn} - t_{z1})} \tag{10.44}$$

where n represents the nth zero crossing starting from 2, t_{zn} is the nth zero crossing time, t_{z1} is the first zero crossing time, and f_n is the frequency calculated from the nth and first zero crossing time. This approach uses a longer time difference between many zero crossings to find the input frequency, and better resolution can be obtained.

The results are shown in Figures 10.8(b–d). These figures show the error frequency (measured frequency minus the input frequency) versus time. Let us only compare the variation of the error frequency and neglect the bias error, which may result from experimental setup error. Figure 10.8(b) shows the result when the time resolution is limited to 1 ns. When two data points have opposite signs, the first data point is assumed to be the zero crossing. The variations of the error frequency converge to about ±200 kHz. Figure 10.8(c) shows the results obtained

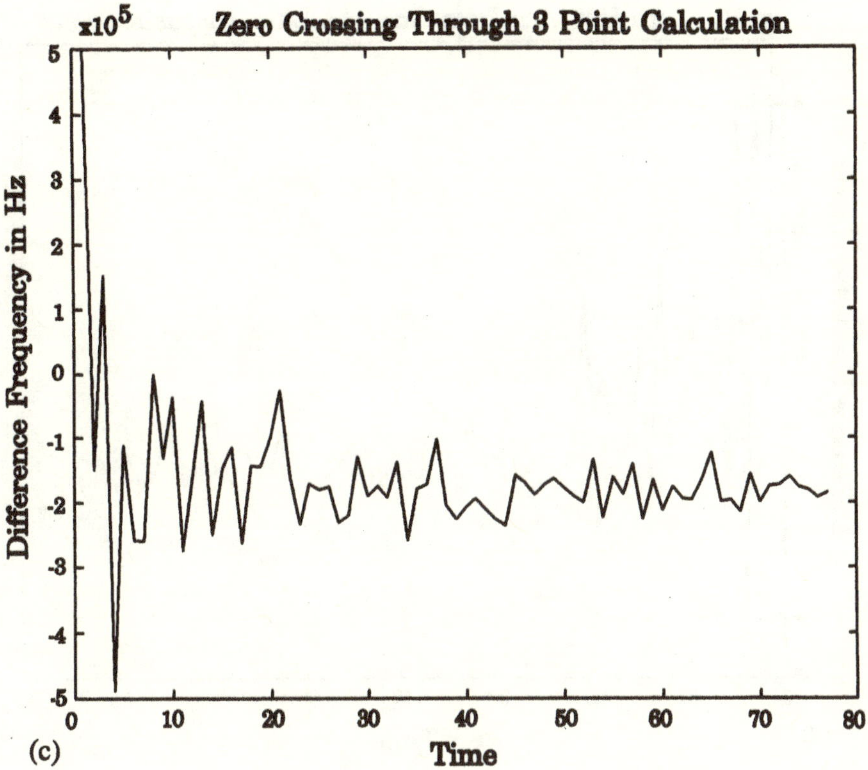

(c)

Figure 10.8 (continued).

from the zero crossings calculated from the discussion in Section 10.8. In this calculation, the ill-conditioned cases were detected and corrected as discussed in Section 10.9. If the ill-conditioned cases were not properly corrected, some large errors appeared. However, those results are not presented here. The maximum variation of the error frequency converges to about ±20 kHz. The result obtained from the straight-line approach presented in Section 10.10 is shown in Figure 10.8(d). The result is similar to Figure 10.8(c).

10.12 APPLICATION TO COHERENT DOPPLER RADAR FREQUENCY MEASUREMENT [10]

Both the phase measurement and the zero crossing methods can measure the frequency of one input signal very accurately. The previous discussion is limited to continuous wave (CW) signals. However, these approaches can also be used to measure the frequency of coherent Doppler radars.

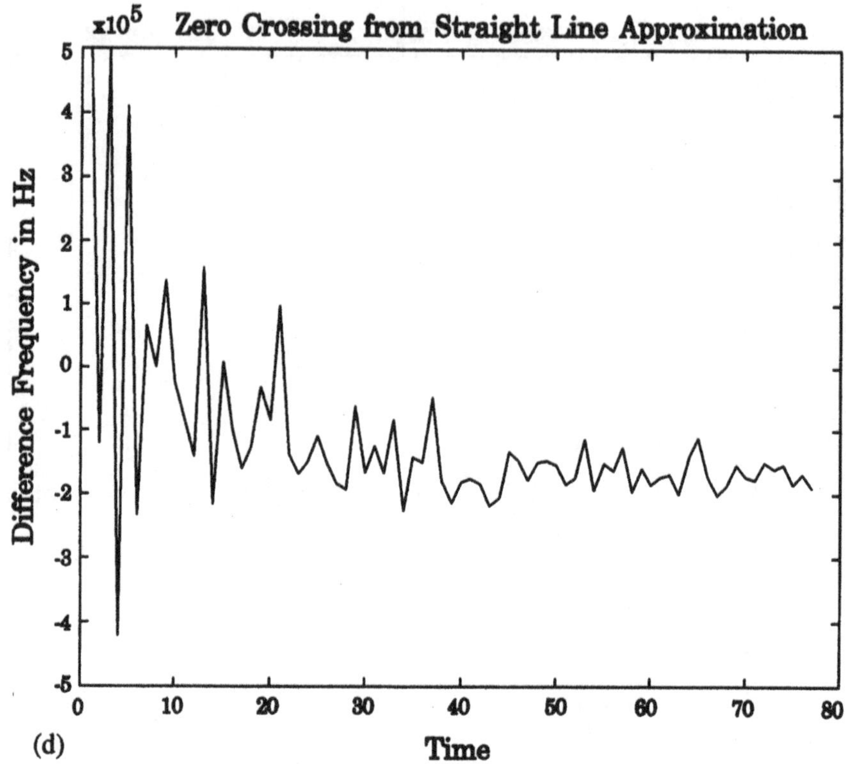

(d)

Figure 10.8 (continued).

A coherent radar is a pulsed radar in which the pulses are obtained from gating on and off a CW signal, as shown in Figure 10.9.

In this figure, the CW signal is gated by the rectangular windows and they represent the radar pulses transmitted. There is a certain phase relation between the pulses because they are part of the same CW signal.

A Doppler radar measures the velocity of the target from the Doppler effect. In order to measure the Doppler frequency over a certain range, the spectrum lines must be separated far apart to avoid ambiguity. To accomplish wide spectrum line separation, the pulses must be close together in time. This effect has been discussed in Section 3.9.

The two critical requirements that the phase method and zero crossing approach have are: 1) the pulsed signal must be part of the same sine wave (coherent), and 2) the pulses must be close together. Figure 10.10 shows the four pulses with a PW of τ and pulse repetition interval (PRI) of T. If the carrier frequency of the radar is f_i, the corresponding time per cycle is $t_i = 1/f_i$. The frequency accuracy Δf_m measured in time τ can be written as

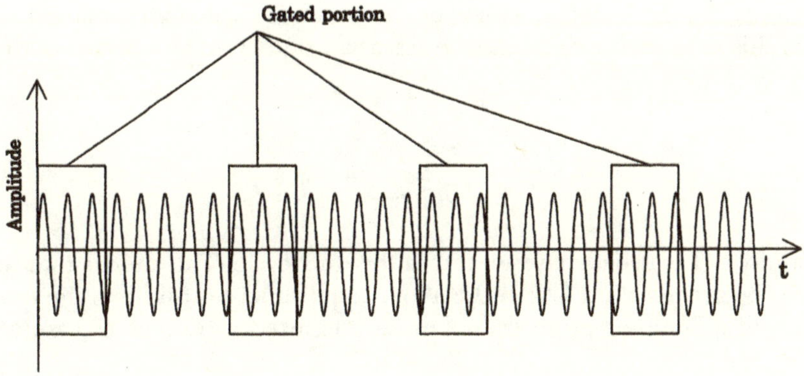

Figure 10.9 Coherent pulse train.

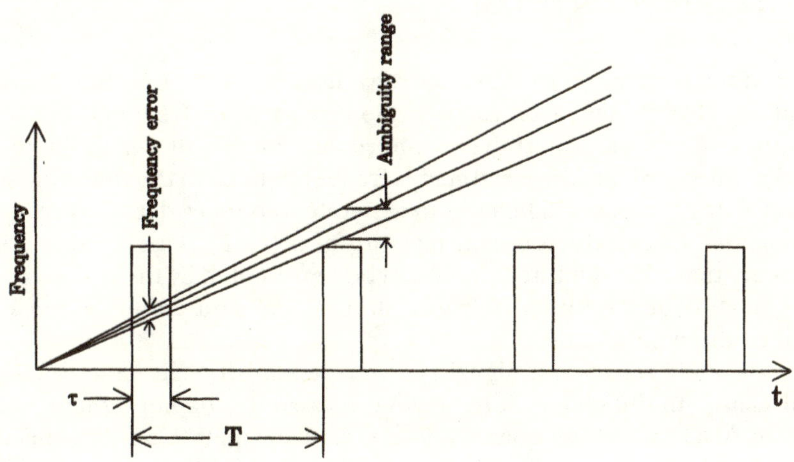

Figure 10.10 Phase error in coherent radar measurement.

$$\Delta f_m = \frac{1}{k\tau} \tag{10.45}$$

where k is a constant. In general, $k > 1$ and it depends on the measurement method and the amount of noise. The percentage error can be written as

$$\frac{\Delta f_m}{f_i} = \frac{1}{k\tau f_i} \tag{10.46}$$

In order to use this information to continue the measurement on the second pulse, the error extended to the next pulse must be evaluated. The error is amplified by

T, as shown in Figure 10.10. However, in order for this extrapolation to work properly, this extended error must be less than one cycle of f_i; thus, the relation can be written as

$$\frac{T\Delta f_m}{f_i} = \frac{T}{k\tau f_i} \le \frac{1}{f_i} \quad \text{or} \quad \frac{\tau}{T} \ge \frac{1}{k} \tag{10.47}$$

The quantity τ/T is usually considered as the duty cycle. This means that the duty cycle must be greater than $1/k$. If the above requirement is fulfilled, the frequency resolution of a Doppler pulse train can be measured as well as in case of a CW signal.

10.13 ZERO CROSSING USED FOR GENERAL FREQUENCY DETERMINATION [11–25]

In the previous sections, zero crossings were used to determine the frequency of the input signal. This approach can provide very accurate frequency information, but only for one input signal. If there is more than one input signal, the previously discussed methods will produce erroneous frequency data. In the following sections, the zero crossing method will be used to generate frequency data that are the same as that obtained from a discrete Fourier transform (DFT). This operation is based on the theory that a band-limited signal can be represented by the real and complex zeros in the function or by a polynomial. The real and complex zeros will be discussed in the next section.

The dynamic range of a digital receiver depends on the quantization levels of the ADC used in the system. It has been discussed in Chapter 6 that it is difficult to make an ADC with many quantization levels that operate at high speeds. One of the potential advantages of zero crossing spectrum analysis is that an ADC with many quantization levels is no longer needed, provided high clock speed and accurate zero crossing detection can be implemented. However, if an ADC is to be used to measure zero crossing, the number of bits is important because the higher the number of bits, the more accurate the time of zero crossing can be measured. One possible disadvantage of using zero crossing times for multiple signals spectral analysis is that the required signal processing may be relatively complicated. The numerical operations required may actually be comparable to those of the DFT.

The discussion in the next few sections will be kept relatively simple, and some of the theoretical analysis will be omitted. The discussion is based on [23] and a thorough theoretical discussion can be found in the reference. This discussion will concentrate on the mechanism used to solve the problem. The presentation will be divided into three parts: 1) the basic definition of the problem, 2) the proper generation of zero crossings, and 3) the spectrum estimation.

10.14 BASIC DEFINITION OF THE ZERO CROSSING SPECTRUM ANALYSIS [11–25]

The input signal $x(t)$ must be bandwidth-limited, which means the signal (or all the signals with the narrowband frequency combined) must have a bandwidth $\leq B$, where B is the single-sided bandwidth of the system. The observation time t is from $-T/2 \leq t \leq T/2$. The signal will repeat itself outside the time window T. This assumption comes from basic idea of digital signal processing, and this phenomenon is explained in Chapter 3. The time bandwidth of the system is BT. The signal $x(t)$ can be represented through the exponential Fourier series as [23]

$$x(t) = \sum_{n=-N}^{N} C_n e^{\frac{j2\pi nt}{T}} \tag{10.48}$$

where $N = BT$ and C_n are complex constants, $C_n = C_{-n}^*$, and * represents a complex conjugate. Now let us assume the argument that t has complex values, say $t \rightarrow \xi = t + j\sigma$ and $Z = e^{j2\pi t/T}$. Then, $x(t)$ can also be written in terms of Z as

$$x(Z) = \sum_{n=-N}^{N} C_n Z^n \tag{10.49}$$

In this equation, there are $2N$ zeros and the zeros of $x(Z)$ are given by

$$Z_i = e^{\frac{j2\pi\xi_i}{T}} = e^{\frac{j2\pi(t_i+j\sigma_i)}{T}} \tag{10.50}$$

where $i = 1, 2, \ldots, 2N$. The actual axis crossing zeros of $s(t)$ are the real zeros for which $\xi_i = t_i$ (or $Z_i = e^{j2\pi ti/T}$, $\sigma = 0$). The complex zeros of $x(t)$ are the ones for which $\xi_i = t_i + j\sigma_i$ for $\sigma_i \neq 0$.

Assuming that $x(t)$ has only real zeros, the function $x(t)$ can be written in terms of sine functions as

$$x(t) = 2^{2N}|C_N| \prod_{i=1}^{2N} \sin\left[\frac{\pi}{T}(t - t_i)\right] \tag{10.51}$$

where t_i is the zero crossing time. Whenever $t = t_i$, $x(t) = 0$. It should be noted that the total number of real zeros is $2N$. The function of $\sin[\pi(t - t_i)/T]$ can be written as

$$\sin\left[\frac{\pi(t - t_i)}{T}\right] = \frac{e^{\frac{j\pi(t-t_i)}{T}} - e^{\frac{-j\pi(t-t_i)}{T}}}{2j} \tag{10.52}$$

$$= \frac{-j}{2}\left[\sqrt{\frac{Z}{Z_i}} - \sqrt{\frac{Z_i}{Z}}\right] = \frac{-j}{2\sqrt{Z_i}} Z^{-1/2}(Z - Z_i)$$

If $|C_N|$ is unknown, this function $x(t)$ may be reconstructed to within a scale factor. Using the relation in this equation, the polynomial $x(Z)$ in (10.51) can be written as

$$\frac{x(Z)}{C_N} = Z^{-N}\prod_{i=1}^{2N}(Z - Z_i) = Z^{-N}\prod_{i=1}^{2N}(Z - e^{\frac{j2\pi t_i}{T}}) \tag{10.53}$$

where C_N is a constant and $x(Z)$ is given in (10.49). Once $x(Z)$ is written in the form shown in (10.49) with all the coefficients C_n known, the amplitude of each frequency component can be found as

$$X(k) = (2BT + 1)\,C_k \tag{10.54}$$

where $k = -N, -N + 1, \ldots, 0, \ldots, N$. Therefore, one can start from (10.52); that is, from the product of $(Z - Z_i)$ to achieve the form of (10.49). The magnitude of C_n represents the amplitude of that frequency component.

In this zero crossing approach, there must be enough real zeros; otherwise, the above approach cannot be applied. In the next section, the minimum required zero crossings and one approach to generate them will be discussed.

10.15 GENERATING REAL ZERO CROSSINGS [23]

In order to obtain information from zero crossings, the number of real zero crossings must equal to $2BT$ (or $2N$). If there are not enough real zero crossings, they have to be created to fulfill the required condition.

One way to create all the necessary zeros is by adding a high-frequency signal to the input. The high-frequency signal can be written as

$$x_h(t) = A_h \cos(2\pi f_h t) \qquad \frac{-T}{2} \le t \le \frac{T}{2} \tag{10.55}$$

where

$$A_h > \max|x(t)| \qquad f_h = B + \frac{1}{T} \tag{10.56}$$

This frequency f_h is slightly higher than the bandwidth B. The amplitude A_h of the this signal must be slightly greater than the sum of all the input signals combined. If the added high-frequency signal is too strong, it will dominate all the zero crossings and the signals to be measured will only add little effect on the zero crossings. As a result, the accuracy of the zero crossing measurement will suffer. In a practical receiver design, a strong high-frequency signal can be used. The relation in (10.56) can be considered as the upper limit of the dynamic range.

The new signal with the high-frequency signal added is

$$y(t) = x(t) + x_h(t) \tag{10.57}$$

After the frequency analysis, this high frequency should be subtracted from the input signal in the frequency domain. With this modification, the total zero crossings are $2(BT + 1)$, which satisfies the requirement.

Figure 10.11 is used to demonstrate the generation of zero crossings. Figure 10.11(a) shows the sum of two sinusoidal waves: the low-frequency one has unit amplitude, while the second signal has an amplitude of 0.25 with a frequency 5.5 times the first one. Since the first and last points do not cross the zero axis, they are not counted as zero crossings. In this figure, there are five zero crossings. Intuitively, one can determine that there are not enough zero crossings because many fine changes do not cross the real axis. Under this condition, one can consider that there are complex zero crossings, which are difficult to realize from this figure. If another signal has an amplitude of 1.3, which is greater than 1.25 (or 1+0.25), with a frequency 10 times the first one, the result is shown in Figure 10.11(b). In

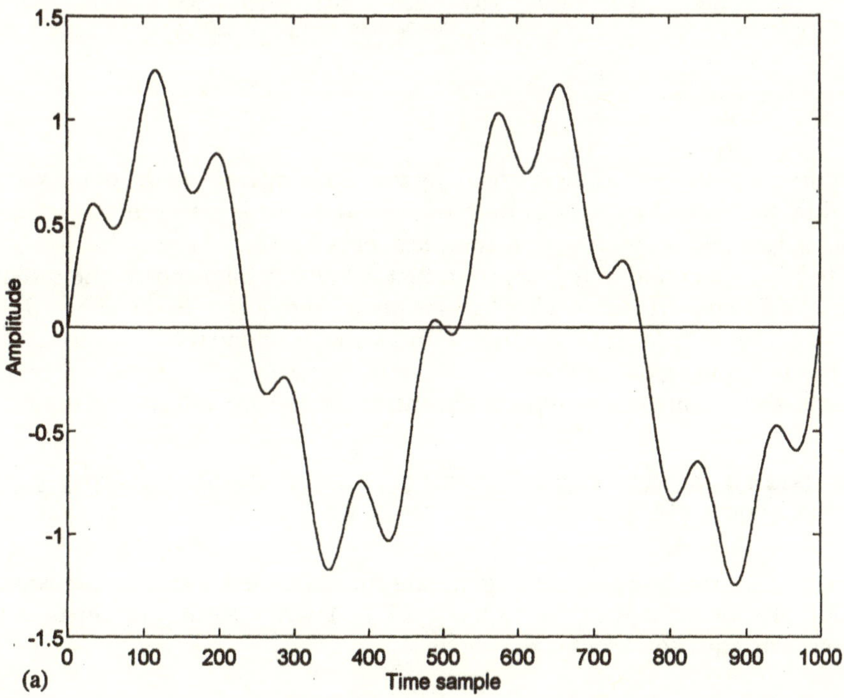

Figure 10.11 Zero crossings of multiple signals: (a) two signals, (b) with added high frequency signal.

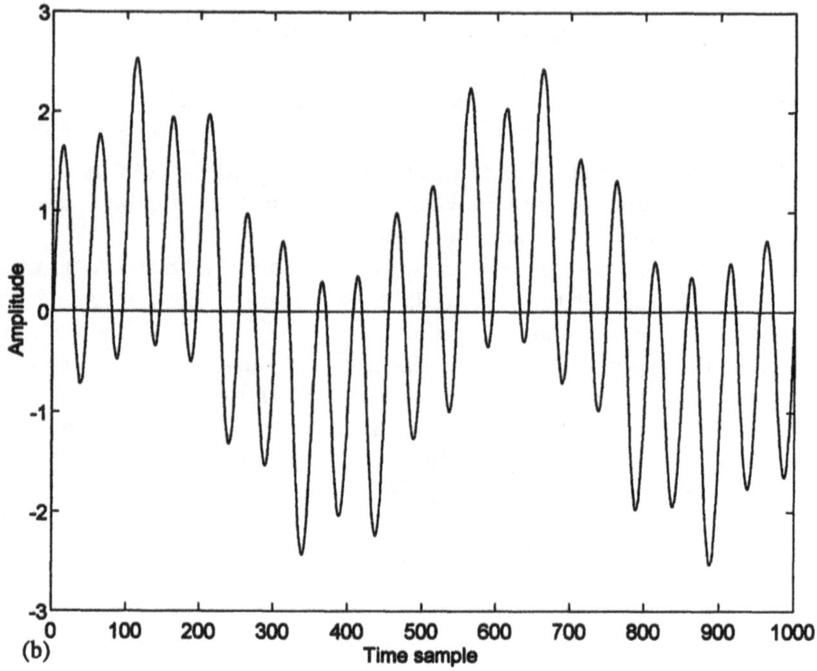

Figure 10.11 (continued).

this figure one can see that there are many zero crossings and all the detail variations caused by the second signal can be represented by these zero crossings. This processing is referred to as changing complex zeros into real zeros.

If B is 1 GHz and T is 1 μs, then $2BT = 2{,}000$, which means the system will require 2,000 zero crossings. The high-frequency signal required will be $(B + 1/T)$ or 1,001 (1,000 + 1) MHz. Therefore, if this signal is added to the input signal, the overall zero crossings should be 2,002. Obviously, with this large number of zero crossings, the calculation of the coefficients from (10.53) will not be easy.

10.16 CALCULATING COEFFICIENTS FOR ZERO CROSSING SPECTRUM ANALYSIS [23, 26]

In this section, the approaches to generate the coefficients will be discussed. The function of concern is $y(Z)$, which has $2(BT + 1)$ zero crossings rather than $2BT = N$. Thus, this equation can be written as

$$y(t) = \prod_{n=-N-1}^{N+1} C_n e^{\frac{j2\pi nt}{T}} \tag{10.58}$$

In this equation, $C_{N+1} = A_h/2$ since the amplitude and the frequency of the highest frequency are known signals. From (10.53) $y(Z)$ can be written as

$$\frac{y(Z)Z^{N+1}}{C_{N+1}} = \prod_{i=1}^{2(N+1)} (Z - Z_i) = a_0 Z^{2N+2} + a_1 Z^{2N+1} + \ldots + a_{2N+2} \tag{10.59}$$

$$\text{and} \qquad Z_i = e^{\frac{j2\pi t_i}{T}}$$

where t_i is the zero crossing time. In receiver applications, it is important to measure the relative amplitude of all the frequency components, which are equivalent to the amplitudes of the coefficient C_n.

One approach to find the coefficients of (10.59) is referred to as the direct calculation. In this approach, all the terms $(Z - Z_i)$ in (10.59) are multiplied together to obtain the coefficients of Z^i. When the number of zeros is small this method may be used.

If the number of zeros is large, the direct multiplication method may become cumbersome. Under this condition, a recursive method might be used. The recursive method can be started from direct multiplication as follows:

$$\begin{aligned}
y^1(Z) &= Z - Z_1 \\
y^2(Z) &= y^1(Z)(Z - Z_2) = (Z - Z_1)(Z - Z_2) = Z^2 - (Z_1 + Z_2)Z + Z_1 Z_2 \\
y^3(Z) &= y^2(Z)(Z - Z_3) = (Z - Z_1)(Z - Z_2)(Z - Z_3) \\
&= Z^3 - (Z_1 + Z_2 + Z_3)Z^2 + (Z_1 Z_2 + Z_2 Z_3 + Z_3 Z_1)Z - Z_1 Z_2 Z_3 \\
&\cdots\cdots \\
y^{k+1}(Z) &= y^k(Z)(Z - Z_{k+1})
\end{aligned} \tag{10.60}$$

where $y^k(Z)$ represents the product of k zeros. This relation can be extended to a more general case as

$$\begin{aligned}
y^k(Z) &= \prod_{i=1}^{k} (Z - Z_i) = Z^k + \sum_{i=1}^{k} a_{i,k} Z^{k-1} \\
&= Z^k + a_{1,k} Z^{k-1} + a_{2,k} Z^{k-2} + \ldots + a_{k-1,k} Z + a_{k,k} \\
y^{k+1}(Z) &= (Z - Z_{k+1}) y^k(Z) \\
&= Z^{k+1} + (a_{1,k} - Z_{k+1}) Z^k + (a_{2,k} - Z_{k+1} a_{1,k}) Z^{k-1} + \ldots \\
&\quad + (a_{k,k} - Z_{k+1} a_{k-1,k}) Z - a_{k,k} Z_{k+1} \\
&= Z^{k+1} + \sum_{i=1}^{k} (a_{i,k} - Z_{k+1} a_{i-1,k}) Z^{k+1-i} - a_{k,k} Z_{k+1}
\end{aligned} \tag{10.61}$$

In this equation, each coefficient has two subscripts. The first subscript represents the numerical order of the coefficient and the second subscript represents recursion order. This equation reveals that the coefficients a_{k+1} can be obtained from the zero Z_{k+1} and the two coefficients of a_k. The recursion relations are

$$\begin{aligned} a_{0,k} &= 1 \\ a_{i,k+1} &= a_{i,k} - Z_{k+1}\, a_{i-1,k} \qquad i = 1, 2, \ldots, k \\ &= -Z_{k+1}\, a_{k,k} \qquad i = k+1 \end{aligned} \tag{10.62}$$

Let us use these relations to obtain the results in (10.62).

$$\begin{aligned} a_{0,0} &= 1 \\ a_{1,1} &= -a_{0,1}\, Z_1 = -Z_1 \\ a_{1,2} &= a_{1,1} - Z_2\, a_{0,1} = -Z_1 - Z_2 \\ a_{2,2} &= -Z_2\, a_{1,1} = Z_1 Z_2 \\ a_{1,3} &= a_{1,2} - Z_3\, a_{0,1} = -Z_1 - Z_2 - Z_3 = -(Z_1 + Z_2 + Z_3) \\ a_{2,3} &= a_{2,2} - Z_3\, a_{1,2} = Z_1 Z_2 + Z_2 Z_3 + Z_3 Z_1 \\ a_{3,3} &= -Z_3\, a_{2,2} = -Z_1 Z_2 Z_3 \end{aligned} \tag{10.63}$$

The $a_{i,3}$ values are the final result, and they are the same results obtained from $y^3(Z)$ in 10.60.

Obviously, both direct and recursive methods to calculate the coefficients are much more complicated than the calculation in the single-frequency case, but this method can process simultaneous signals.

If these calculated coefficients represent the frequency components of the FFT, only half of them carry information because the frequency components of the FFT have this property. From (10.59), one can see that there are $2N$ ($N = BT$) roots and $2N + 1$ coefficients from a_0 to a_{2N}. Mathematically, these coefficients should have the following relations:

$$a_i = a^*_{2N-i} \qquad \text{where } i = 0, 1, \ldots, N \tag{10.64}$$

But one can see that these relations do not hold. For example,

$$a_0 = 1 \qquad a_{2N} = \prod_{i=1}^{2N} Z_i \tag{10.65}$$

In general, they are not the complex conjugate of each other.

However the following relation does hold [26]:

$$|a_i| = |a_{2N-i}| \qquad \text{where } i = 0, 1, \ldots, N \tag{10.66}$$

Thus, if the amplitude of the frequency components is of interest, half of the coefficients can provide all the needed information. This result agrees with the FFT result.

The coefficient a_i can be modified by a factor to fulfill the relation in (10.64) [26]. This factor is

$$Z_f = \frac{-1}{2}\sum_{i=1}^{2N}\theta_i \qquad \text{where } \theta_i = \tan^{-1}\left(\frac{Im\ Z_i}{Re\ Z_i}\right) \tag{10.67}$$

where *Im* and *Re* represent the imaginary and real parts of a quantity.

The coefficient a_i can be modified as

$$b_i = a_i Z_f \tag{10.68}$$

These b_i values can fulfill the relation

$$b_i = b^*_{2N-i} \qquad \text{where } i = 0, 1, \ldots, N \tag{10.69}$$

where b_N has real value.

Therefore, strictly speaking, the coefficients of (10.59) can be made to have the property shown in (10.64) if they are multiplied by a constant phase term.

10.17 POSSIBLE CONFIGURATION OF ZERO CROSSING SPECTRUM ANALYZER

Summarizing the previous discussion on zero crossing spectrum analysis, one can conclude that only two simple steps are needed. First, enough real zeros must be generated. Second, these real zeros are used to calculate the coefficients of the function created by the product of $(Z - Z_i)$.

In using the zero crossing for spectrum analysis, a strong signal with a known high frequency must be added to the input signal to create the required real zeros. It might be bothersome for receiver designers that a high frequency must be injected into the input signal. One must perform spectrum analysis on the entire spectrum, and at the same time the injected signal with the highest frequency must be neglected. If there is an input signal with a frequency close to the injected one, the results could be confusing.

One way to separate the injected signal is shown in Figure 10.12. In this figure, the input signal bandwidth is limited to f_2, which can be much less than f_h. Thus, the frequencies of the input signal will be far away from the frequency of the injected signal. Since the input signal is limited to f_2, only the coefficients related to frequencies up to f_2 need to be calculated.

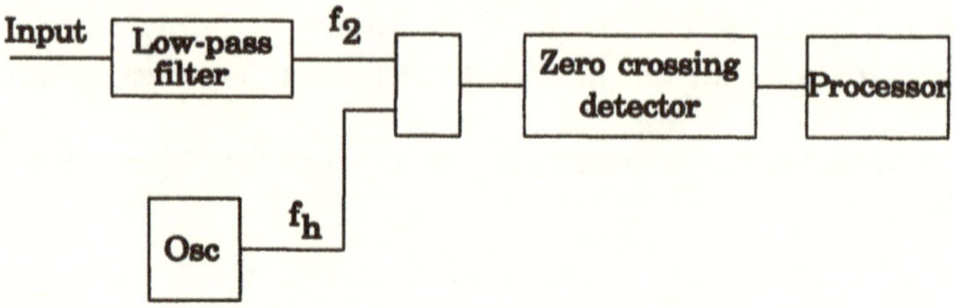

Figure 10.12 Configuration of zero crossing spectrum analysis.

Suppose that the signal bandwidth is B and the bandwidth including f_h is B_h and $B_h > B$. There will be $2B_hT$ zeros. Equation (10.59) can be written as

$$\frac{y(Z)\,Z^{B_hT}}{C_{B_hT}} = \prod_{i=1}^{2B_hT} (Z - Z_i) \tag{10.70}$$

In this equation, only BT coefficients need to be evaluated. A processor to calculate these coefficients at real time must be built.

These outputs are the same as the frequency components of an FFT operation. Further signal processing is required to determine the number of input signals and their center frequencies.

REFERENCES

[1] Earp, C. W. "Frequency Indicating Cathode Ray Oscilloscope," US Patent 2434914, Jan. 27, 1948.

[2] Wilkens, M. W., and Kincheloe, W. R., Jr. "Microwave Realization of Broadband Phase and Frequency Discriminators," Technical Report No. 1962/1966-2, Stanford Electronics Laboratories, SU-SEL-68-057, Nov. 1968.

[3] Myers, G. A., and Cumming, R. C. "Theoretical Response of a Polar Display Instantaneous Frequency Meter," IEEE *Trans. Instrumentation Measurement,* Vol. IM-20, Feb. 1971, PP. 38–48.

[4] Lang, S. W., and Musicus, B. R. "Frequency Estimation From Phase Differences," *presented at ICASSP,* May 23–26, 1989, pp. 2140–2143.

[5] Kay, S. M. "A Fast and Accurate Single Frequency Estimator," *IEEE Trans. Acoustics, Speech, Signal Processing,* Vol. ASSP-37, Dec. 1990, pp. 1987–1990.

[6] Panter, P. F. *Modulation, Noise, and Spectral Analysis,* New York, NY: McGraw Hill Book Co., 1965.

[7] Chu, D. "Phase Digitizing Sharpens Timing Measurements," *IEEE Spectrum,* Vol. 25, No. 7, July 1988, pp. 28–32.

[8] McCormick, W. S., and Lansford, J. L. "Time Domain Algorithm for the Estimation of Two Sinusoidal Frequencies," *IEE Proc. Vision, Image and Signal Processing,* Vol. 141, No. 1, Feb. 1994, pp. 33–38.

[9] Kay, S. M., and Marple, S. L. "Spectrum Analysis-A Modern Perspective," *IEEE Proc.,* Vol. 69, Nov. 1981, pp. 1380–1419.

[10] Paciorek, L. J. Anaren Microwave Inc., Private communication.

[11] Bond, F. E., and Cahn, C. R. "On Sampling the Zeros of Bandwidth Limited Signals," *IRE Trans. Information Theory,* Vol. IT-4, Sept. 1958, pp. 110–113.

[12] Voelcker H. B. "Toward a Unified Theory of Modulation Part I: Phase-Envelope Relationships," *IEEE Proc.*, Vol. 54, March 1966, pp. 340–353.

[13] Voelcker, H. B. "Toward a Unified Theory of Modulation Part II: Zero Manipulation," *IEEE Proc.*, Vol. 54, May 1966, pp. 735–755.

[14] Seeky, A. "A Computer Simulation Study of Real-Zero Interpolation," *IEEE Trans. Audio and Electroacoustics*, Vol. AU-18, March 1970, pp. 43–54.

[15] Voelcker, H. B. "Zero-Crossing Properties of Angle-Modulated Signals," *IEEE Trans. Communications*, Vol. COM-20, June 1972, pp. 307–315.

[16] Voelcker, H. B., and Requicha, A. A. G. "Clipping and Signal Determinism: Two Algorithms Requiring Validation," *IEEE Trans. Communications*, June 1973, pp. 738–744.

[17] Voelcker, H. B., and Requicha, A. A. G. "Band-Limited Random-Real-Zero Signals," *IEEE Trans. Communications*, Vol. COM-21, Aug. 1973, pp. 933–936.

[18] Logan, B. F., Jr. "Information in the Zero Crossings of Bandpass Signals," *Bell System Technical Journal*, Vol. 56, April 1977, pp. 487–510.

[19] Papoulis, A. *Signal Analysis*, New York, NY: McGraw Hill Book Co., 1977.

[20] Requicha, A. A. G. "The Zeros of Entire Functions: Theory and Engineering Applications," *IEEE Proc.*, Vol. 68, March 1980, pp. 308–328.

[21] Requicha, A. A. G. "The Zeros of Entire Functions: Theory and Engineering Applications," *IEEE Proc.*, Vol. 68, March 1980, pp. 308–328.

[22] Higgins, R. C. "The Utilization of Zero-Crossing Statistics for Signal Detection," *J. Acoust. Soc. Am.*, Vol. 67, May 1980, pp. 1818–1820.

[23] Kay, S. M., and Sudhaker, R. "A Zero Crossing-Based Spectrum Analyzer," *IEEE Trans. Acoustics, Speech, and Signal Processing*, Vol. ASSP-34, Feb. 1986, pp. 96–104.

[24] Keden, B., "Spectral Analysis and Discrimination by Zero-Crossings," *IEEE Proc.*, Vol. 74, Nov. 1986, pp. 1477–1493.

[25] Marvasti, F. A. "A Unified Approach to Zero-Crossings and Nonuniform Sampling of Single and Multidimensional Signals and Systems," Dept. Electrical Engineering, Illinois Institute of Technology, 1987.

[26] Marden, M., *Geometry of Polynomials*, 2nd Edition, American Mathematical Society, Providence, RI, 1985.

CHAPTER 11

Frequency Channelization

11.1 INTRODUCTION

Channelization is one of the most important operations in building digital electronic warfare (EW) receivers. The equivalent analog operation is the filter bank. Therefore, digital channelization can be considered a digital filter bank. It can also be considered as an N-port network with one input and $N-1$ outputs. An input signal will appear at a certain output according to its frequency. By measuring the outputs from the filter bank, the frequency of the input signal can be determined.

The only practical approach to building a wideband digital EW receiver with today's technology is through channelization. A common method of performing channelization is by employing the fast Fourier transform (FFT). To build a receiver using FFT, the length and the overlap of the FFT are very important parameters. These parameters are related to the minimum pulse width and the frequency resolution, which determines the sensitivity of the receiver. The frequency information can be obtained from the outputs of the digital filters. In order to obtain the input frequency, the filter outputs must be further processed. The main objectives of a receiver are to determine the number of input signals and their frequencies. The circuit used to accomplish these goals is referred to as the encoder.

The encoding circuit is the most difficult subsystem to design in an EW receiver. Most research effort is spent on the encoder design. This is true for both digital and analog receivers. The main problems are to avoid the generation of false signals and the detection of weak signals. In an analog filter bank, the shape of the filter is difficult to control, and it is difficult to build filters with uniform performance such as the bandwidth and the ripple factor; therefore, the encoder must accommodate this problem. The shape of each individual filter in a digital filter bank can be better controlled. As a result, the encoder should be slightly easier to design because it does not need to compensate for the filter differences. Because of the complexity of the encoder, its design will not be discussed in detail.

The design of a specific digital filter bank will be presented. This specific example is used to illustrate the design procedure while avoiding the unnecessary mathematical complexity of a general design. In this example the concepts of polyphase filter and multirate operation will be introduced. In order to understand these concepts, decimation and interpolation are discussed first.

11.2 FILTER BANKS

The straightforward approach for building a filter bank is to build individual filters, each one with a specific center frequency and bandwidth. Figure 11.1 shows such an arrangement. Each digital filter can be either a finite impulse response (FIR) or infinite impulse response (IIR) type. Theoretically, each filter can be designed

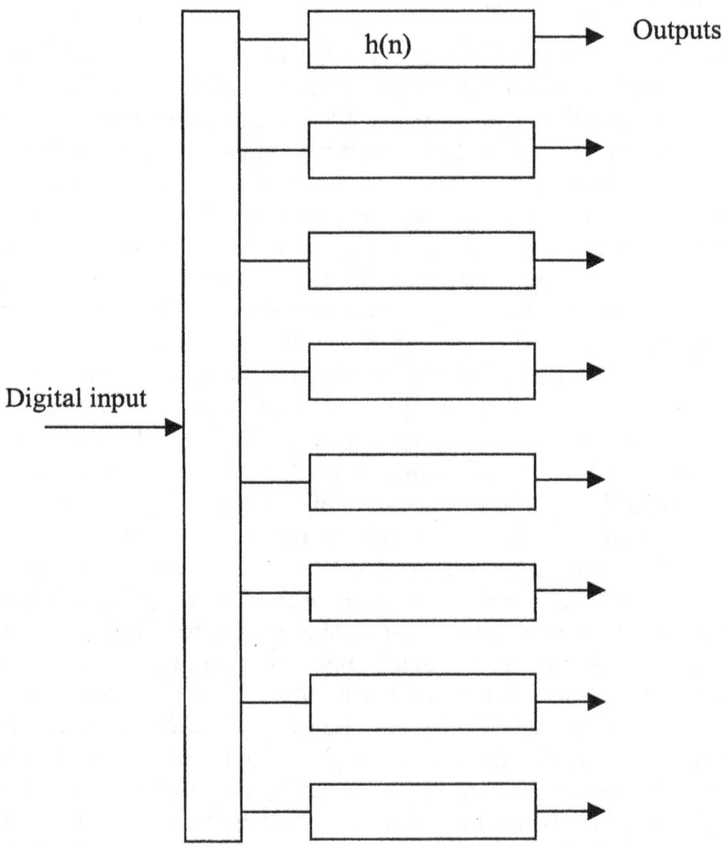

Figure 11.1 A filter bank.

independently with a different bandwidth or shape. In this arrangement, if the input data are real (as opposed to being complex) the output data are also real. The output is obtained through convolving the input signal $x(n)$ and the impulse response of the filter $h(n)$. One of the disadvantages of this approach is that the operation of the filter bank is computationally complex.

It is desirable to build a receiver with uniform frequency resolution; that is, the filters have the same shape and bandwidth. It is easier to build such a filter bank through FFT techniques than by using individual filter design because there is less computation.

11.3 FFT AND CONVOLUTION OPERATIONS [1, 2]

In the previous section, it is stated that the outputs of a filter bank can be obtained from convolution and also from the FFT operation. In this section the similarity between the FFT and convolution operation will be illustrated. In the FFT operation, one set of data in the time domain can be used to find one set of data in the frequency domain. In order to process the input data in a continuous manner, the FFT must also operate continuously. This subject will be further discussed in Sections 11.4 and 11.5.

This discussion is similar to that in [1]. Let us assume that one frequency component from the FFT output is equivalent to one filter output at a specific point in time from a filter bank. The output of the k component $X(k)$ from an N point FFT can be written as

$$X(k) = \sum_{n=0}^{N-1} x(n) e^{\frac{-j2\pi kn}{N}} \tag{11.1}$$

The k_0 component of $X(k_0)$ can be written as

$$X(k_0) = \sum_{n=0}^{N-1} x(n) e^{\frac{-j2\pi k_0 n}{N}} \tag{11.2}$$

In order to relate this output to convolution, let us define an impulse function as

$$h(k) = e^{\frac{j2\pi k k_0}{N}} \tag{11.3}$$

where $k = -(N-1), -(N-2), \ldots, -1, 0$; then

$$h(k-n) = e^{\frac{j2\pi(k-n)k_0}{N}} \tag{11.4}$$

If $k = 0$, then

$$h(k-n)|_{k=0} = e^{\frac{-j2\pi nk_0}{N}} \tag{11.5}$$

thus,

$$X(k_0) = \sum_{n=0}^{N-1} x(n)h(k-n)|_{k=0} \tag{11.6}$$

One can see that this expression is a discrete convolution as shown in (5.6). It represents the input signal $x(n)$ convolving with $h(k-n)$.

This operation illustrates that a certain frequency bin from the FFT operation can be treated as an input signal convolved with a certain impulse function. Therefore, one can consider that each individual FFT output can be represented by a filter impulse function convolved with the input signal, which is the concept introduced in Section 11.2. Because the FFT operation is rather simple compared with an individual filter design, the FFT will be used for filter bank design in the rest of this chapter. A similar discussion showing that one frequency bin of the FFT outputs is equivalent to a convolution of the input signal with a filter impulse can be found in reference [2].

11.4 OVERLAPPING INPUT DATA IN THE FFT OPERATION [2–4]

In the previous section it was demonstrated that each FFT output can be considered as a filter output. In order to operate on a continuous input signal, the FFT must operate on different intervals of data at different times. Usually, the initial data point is labeled $n = 0$, and the data interval can slide M points and be represented by $n = M$. The corresponding FFT can be written as

$$X(k) = \sum_{n=M}^{N+M-1} x(n)e^{\frac{-j2\pi kn}{N}} \tag{11.7}$$

The M value must be changed continuously with the input signal. This operation is sometimes referred to as the short time Fourier transform (STFT).

Figure 11.2 is used to illustrate the input data overlapping condition. In this figure the FFT uses only 8 data points. When $M = 0, 1, 2, \ldots$, as shown in Figure 11.2(a), the input data slide one point every time, which is referred to as the sliding DFT and is discussed in Section 4.8. For this case the data can be considered 100% overlapping. If the minimum pulse width is 8 data points long, this approach can always fill one of the FFT windows with the shortest pulse. The time of arrival (TOA) resolution mentioned in Section 2.6 is equal to the t_s, where $t_s = 1/f_s$ is the sampling time and f_s the sampling frequency. With 100% overlapping, the FFT must be performed every t_s, thus, the computation load is very high.

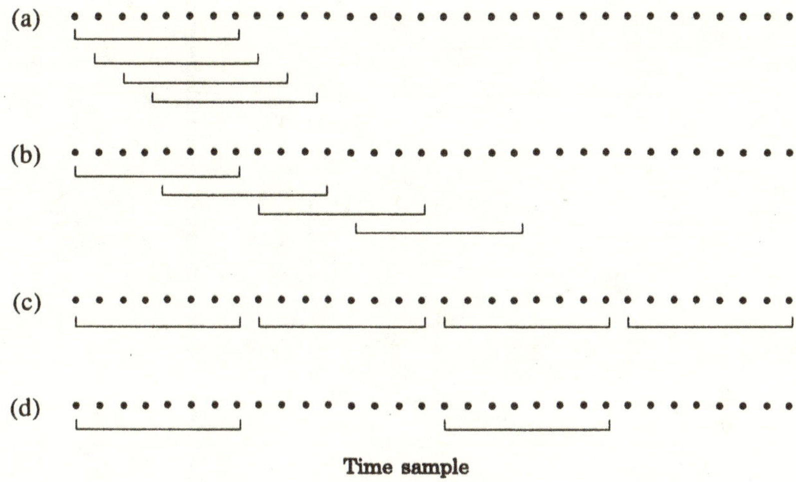

Figure 11.2 Data string in time domain: (a) 100% overlap, (b) 50% overlap, (c) zero overlap, and (d) missing 50% of data.

Figure 11.2(b) shows a 50% data overlapping, which corresponds to $M = 0$, 4, 8, ... The shortest pulse under the worst-case condition can fill 75% of the window. Under this condition the sensitivity of the receiver will degrade. The TOA resolution will be reduced to $4t_s$. The FFT, however, only operates every $4t_s$, which means less computation is required.

Figure 11.2(c) shows zero overlapping, but no missing data either. The corresponding $M = 0, 8, 16, \ldots$. The shortest pulse under the worst-case condition can fill 50% of the window. The TOA resolution and the FFT operation rate are both decreased to $8t_s$. Usually in receiver design this situation provides the lowest acceptable FFT operation rate. This phenomenon will be explained later in this section.

Figure 11.2(d) shows that some data are missing. The $M = 0, 16, 32, \ldots$, are selected for this illustration. In general this is not an acceptable choice because the receiver will miss pulses.

The degradation of filling half a window is illustrated in Figure 11.3 with 64 input data points. Figure 11.3(a) shows that the data fill a rectangular window in time domain; Figure 11.3(b) shows that the data fill only half of the window. The corresponding FFT outputs are shown in Figure 11.3(c) and (d). In order to smooth the outputs in the frequency domain, zero padding is used in performing the FFT. The spectrum in Figure 11.3(d) not only decreases in amplitude but also spreads in frequency. This frequency spreading causes difficulty in designing the parameter encoder following the FFT outputs. Even for a long pulse, the signal usually fills the window only partially at the leading and trailing edges. These partially filled

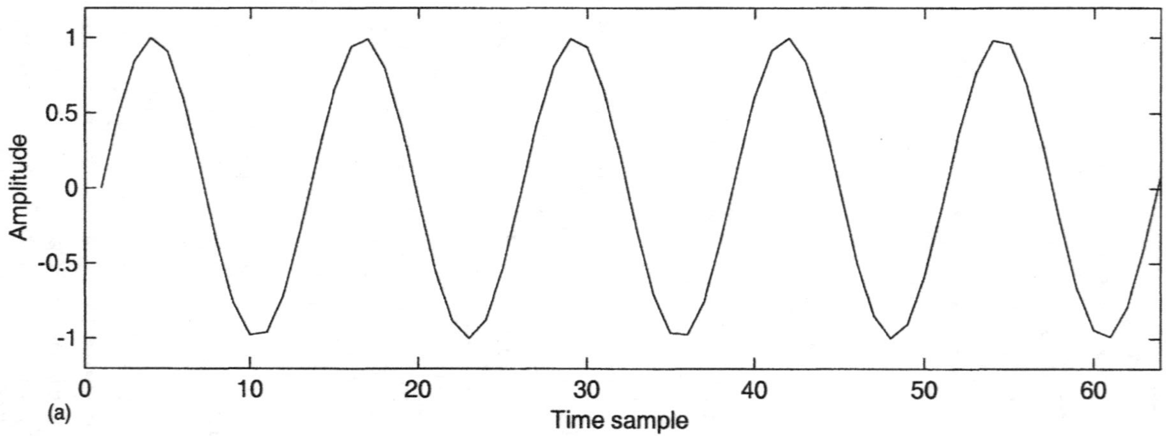

Figure 11.3 FFT on filled and partially filled windows: (a) signal filled window, (b) partially filled window (time domain), (c) signal filled window, and (d) partially filled window (frequency domain).

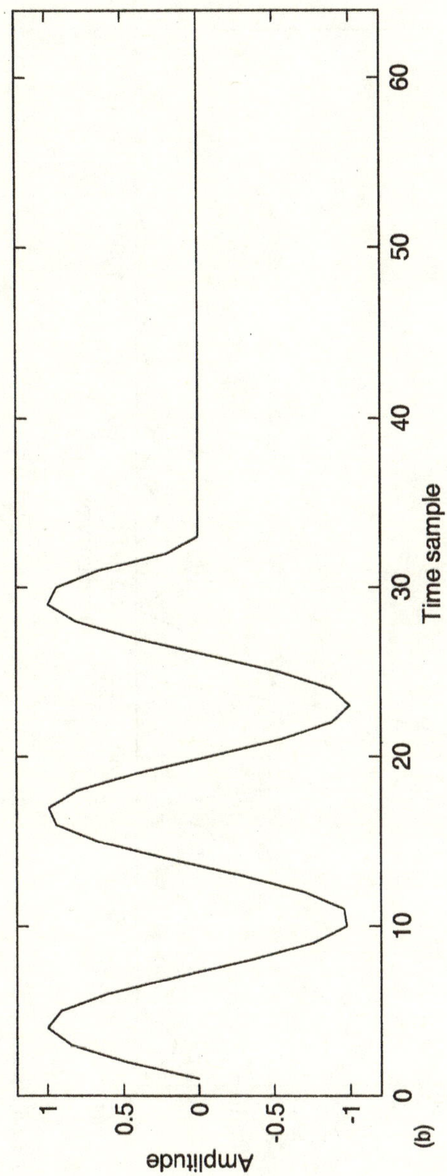

Figure 11.3 (continued).

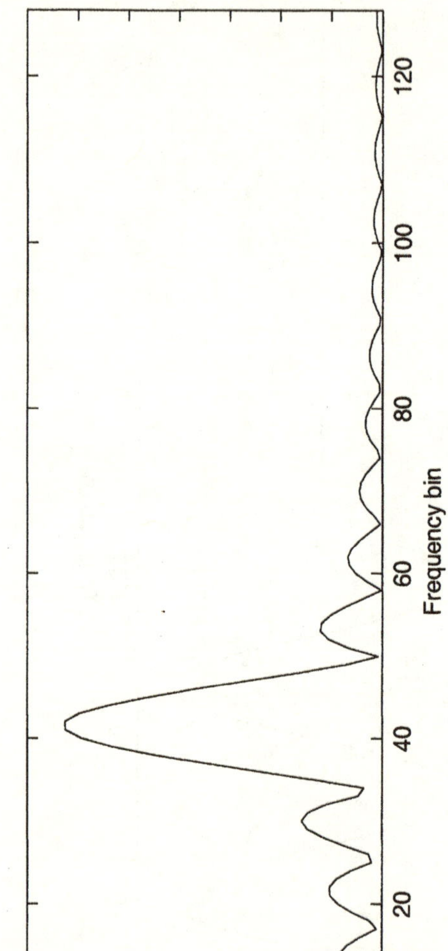

Figure 11.3 (continued).

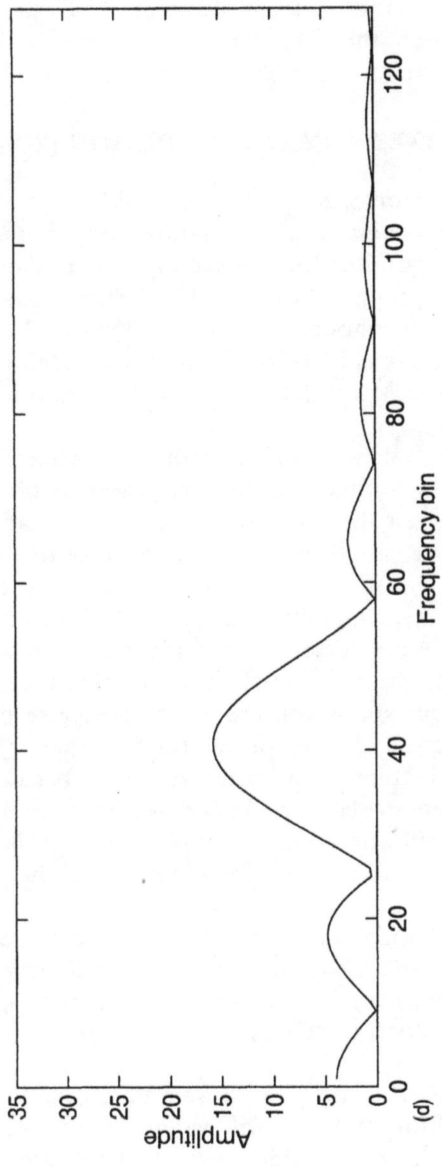

Figure 11.3 (continued).

windows will spread the energy in frequency domain to adjacent channels and may cause detection problems.

The selection of the FFT operation rate depends on the technology development of the FFT chip. With today's technology, the FFT operation speed is much less than the sampling rate of the ADC. This point is one key factor in determining the receiver design approach.

11.5 OUTPUT DATA RATE FROM FFT OPERATION [2–6]

If the FFT operation is performed on N points of data, the resulting N frequency outputs will occur at the same time. If the sampling frequency is f_s, which is the input data rate, the corresponding input sampling time is $t_s = 1/f_s$. The output rate depends on the data overlapping rate discussed in the previous section. For example, if the input data is 100% overlapped as shown in Figure 11.2(a), the output rate is also f_s. If the input data are shifted by M samples between every FFT operation, the output sampling time is Mt_s, which corresponds to an output sampling rate of f_s/M.

The output sampling rate is very important in building a receiver because the outputs from the FFT are usually further processed to obtain finer frequency resolution. For this processor the input rate is the output sampling rate from the FFT. This output sampling rate determines the bandwidth of this processor. If a rectangular window containing N points is used to process the input signal with zero overlapping, the output sampling rate is f_s/N. The corresponding output bandwidth is about f_s/N. For a rectangular window in the time domain the corresponding output in the frequency domain is a sinc function and the response is shown in Figure 11.4. This shape represents the response of one filter output. Figure 11.4(a) shows the detailed shape of this filter and Figure 11.4(b) shows three adjacent filters. This filter shape is not desirable because the sidelobes are very high and the first sidelobe is only 13 dB down. If a signal of frequency f_i falls in channels B and C, it will also enter channel A through its first sidelobe. This phenomenon limits the instantaneous dynamic range of the receiver to less than 13 dB.

Because the output bandwidth is limited to f_s/N, channel A cannot properly process a signal at f_i because it is outside of its bandwidth. However, if channel A processes this signal, it may assign an erroneous frequency. A higher output sampling rate will help this situation. This problem will be further discussed in the next chapter.

To lower the sidelobes of the filter, a weighting function can be applied in the time domain. A weighting function will widen the main lobe but suppress the sidelobes. Because the main lobe is wide, fewer channels are needed to cover the desired input bandwidth. In order to further process the signal, the output sampling rate must be increased to match the bandwidth. Figure 11.5 shows the FFT output of a Hamming window (Section 4.6). The need of increasing the output sampling

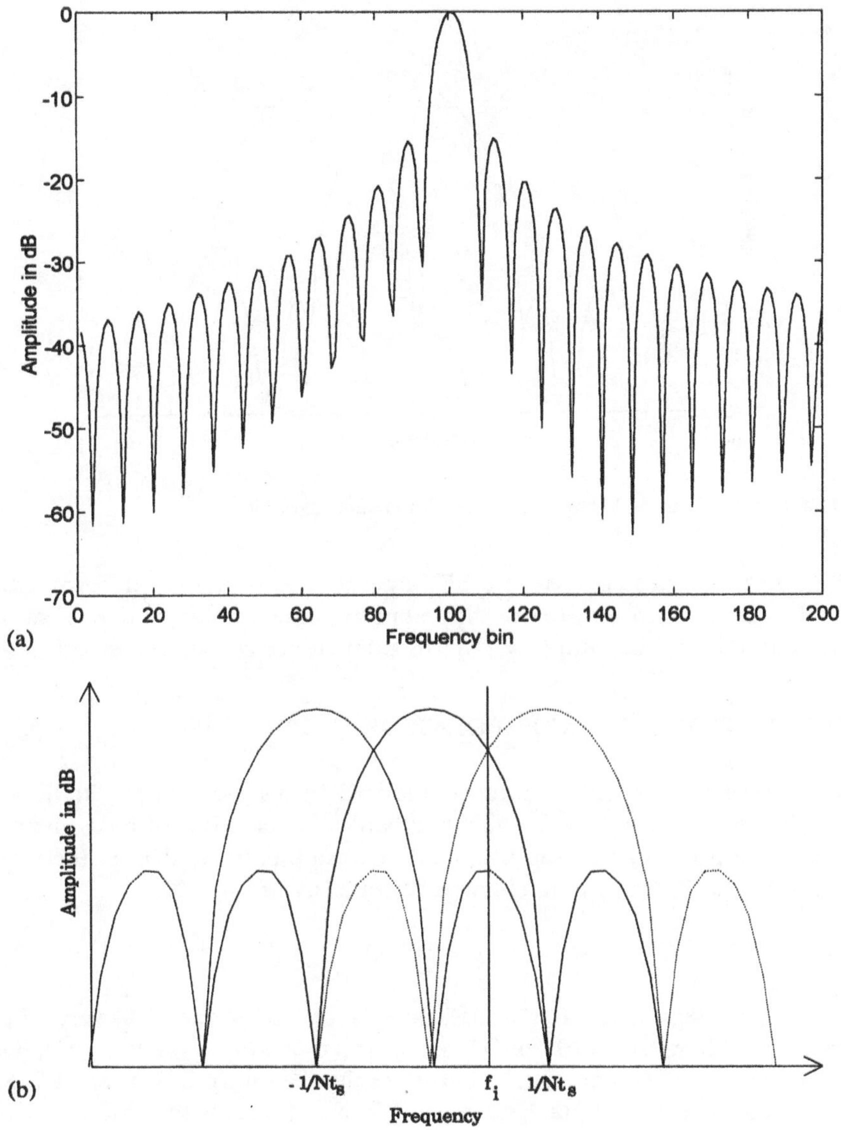

Figure 11.4 FFT of a rectangular window: (a) detail filter response and (b) three adjacent filters.

rate can be explained in the time domain also. Figure 11.6 shows the time domain of a Hamming function. One can see that only the data near the center of the weighting function are given weights near unity. The data close to the edges of the window are heavily attenuated. If a zero overlapping approach is used, the

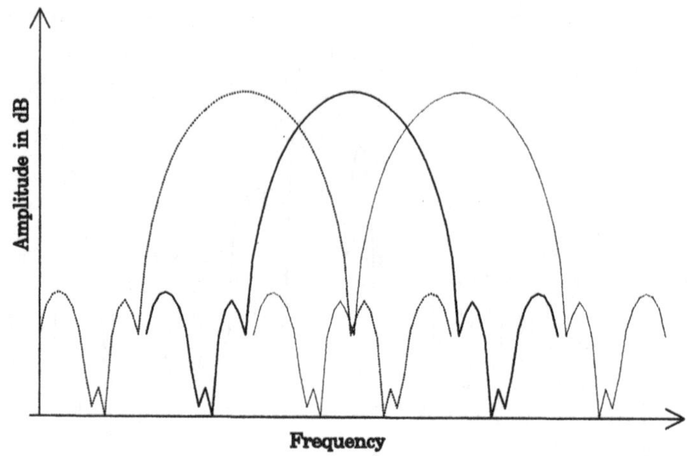

Figure 11.5 Outputs from three adjacent filters with Hamming window.

contribution from these data in the FFT operation will be non-uniform, which is undesirable. If the windows are overlapped in the time domain, which is equivalent to increasing the output sampling rate, this deficiency can be remedied.

11.6 DECIMATION AND INTERPOLATION [7–10]

The concepts of decimation and interpolation will be discussed briefly in this section because they will be used to design filter banks. Decimation of data means only using one data point in a group of data. For example, if the data points are $x(n)$ where $n = 0, 1, 2, 3, \ldots$, the decimated result x_d by M is

$$x_c = x(Mn) \tag{11.8}$$

where M is an integer. In other words, one data point is selected every M points of data. If $M = 2$, $x_d(0) = x(0)$, $x_d(1) = x(2)$, $x_d(2) = x(4)$, ... If $M = 3$, $x_d(0) = x(0)$, $x_d(1) = x(3)$, $x_d(2) = x(6)$, ... These results are shown in Figure 11.7. Figure 11.7(a) shows the original data; Figures 11.7(b) and (c) show the results decimated by 2 and 3, respectively. Decimation by M can be represented by $M\downarrow$.

It is obvious that decimation loses information. If a signal is sampled at 3 GHz, decimation by 2 is equivalent to sampling the data at 1.5 GHz. Decimation by 3 is equivalent to sampling the data at 1 GHz.

In interpolation, additional data points are added to the original input data. Because it is difficult to add information to the data, only zeros will be added. Mathematically, if the input data are interpolated by L, the result $x_l(n)$ can be related to the input signal as

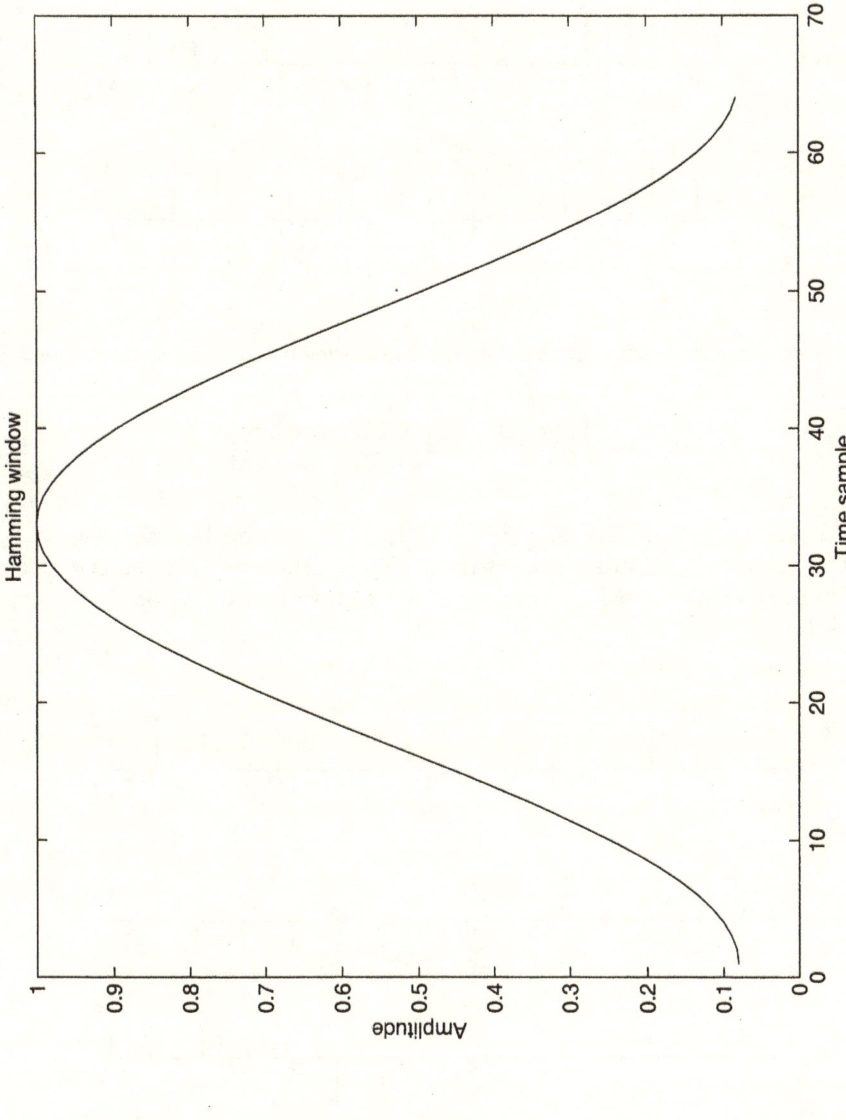

Figure 11.6 Hamming window in time domain.

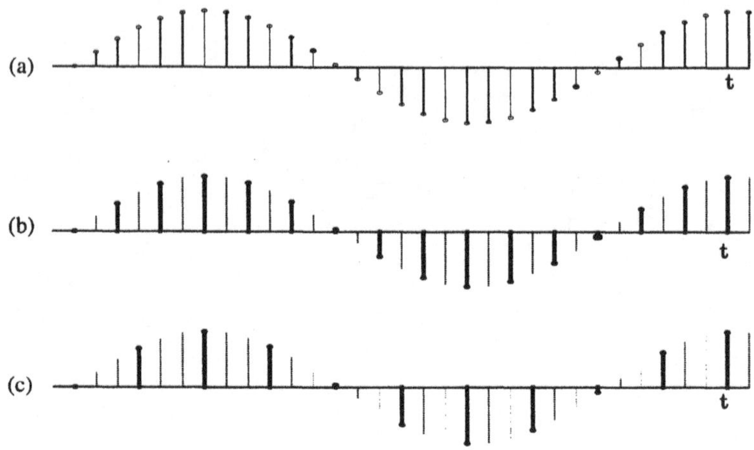

Figure 11.7 Example of decimation: (a) input data, (b) decimated by 2, and (c) decimated by 3.

$$x_l(n) = \begin{cases} x(x/L) & \text{if } n/L = \text{integer} \\ 0 & \text{if } n/L \neq \text{integer} \end{cases} \tag{11.9}$$

where L is an integer. If $L = 3$, $x_l(0) = x(0)$, $x_l(1) = 0$, $x_l(2) = 0$, $x_l(3) = x(1)$, $x_l(4) = 0, \ldots$. Figure 11.8 shows the results of interpolation by 2 and 3. As expected, interpolation does not add information to the input data. Interpolation is represented by $L\uparrow$.

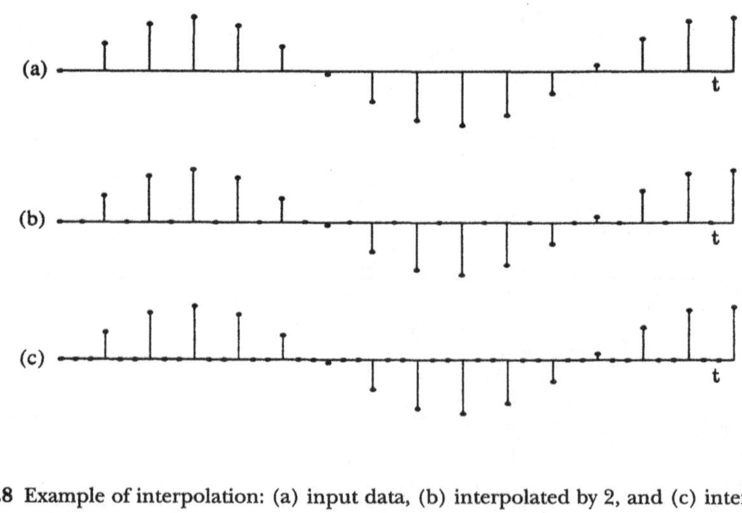

Figure 11.8 Example of interpolation: (a) input data, (b) interpolated by 2, and (c) interpolated by 3.

If the input data are interpolated by L and decimated by L, the original data will be obtained. This is obvious because interpolation by L adds $L - 1$ zeros per data point and decimation by L takes the added zeros away. If, however, the data are decimated by M first and interpolated by M, the process will not produce the original data. The decimation process loses data. The interpolation process cannot recover lost data.

11.7 DECIMATION AND INTERPOLATION EFFECTS ON THE DISCRETE FOURIER TRANSFORM [7–10]

In this section, the decimation and interpolation effects on DFT will be presented. The decimation effect will be easy to understand. As mentioned in the previous section decimation slows down the sampling rate by a factor of M. If the input data are real (as opposed to being complex) and sampled at f_s, the input bandwidth is $f_s/2$. If an N-point FFT is performed on the input data, the frequency resolution is f_s/N. If the N-point FFT is performed on the data decimated by M, the input bandwidth is reduced to $f_s/(2M)$ and the resolution is $f_s/(MN)$. Thus, the input bandwidth is reduced by M, but the width of the frequency resolution cell is increased by M.

The effect of the FFT with interpolated data is slightly more complicated. The effect of the FFT can be written as follows. If the input data $x(n)$ are interpolated by L, the resulting data $x_l(n)$ have NL points. The FFT of $x_l(n)$ can be written as

$$X_l(k) = \sum_{n=0}^{NL-1} x_l(n)\, e^{\frac{-j2\pi nk}{NL}} \tag{11.10}$$

From the previous equation, whenever $n/L \neq$ integer, $x_l(n) = 0$ and $n/L =$ integer, $x_l(n) = x(n)$. Therefore, this equation can be written as

$$X_l(k) = \sum_{n=0}^{NL-1} x_l(n)\, e^{\frac{-j2\pi nk}{NL}} = \sum_{n=0\ stepL}^{NL-1} x(n/L)\, e^{\frac{-j2\pi nLk}{NL}} \tag{11.11}$$

For the frequency component of $k < N$ the above function can be written as

$$X_l(k) = \sum_{n=0}^{N-1} x(n)\, e^{\frac{-j2\pi nk}{N}} \tag{11.12}$$

which is the same result obtained from a conventional FFT with the original data. Because there are NL data points in the interpolated data, the FFT will have NL frequency components. The value of $k > N$ can be found by replacing $k = mN + k'$ in the above equation where m is an integer. The result is

$$X(k) = X(mN + k') = \sum_{n=0}^{N-1} x(n)\, e^{\frac{-j2\pi n(mN+k')}{N}} = \sum x(n)\, e^{\frac{-j2\pi nk'}{N}} \tag{11.13}$$

This equation shows that the output data are periodic with a period of N.

Figure 11.9 shows an example of this operation. Figure 11.9(a) shows the amplitude of the FFT outputs of a sine wave, and the output has 16 frequency components. Figure 11.9(b) shows the same data interpolated by 3. The output has 48 frequency components consisting of three cycles. In general, if the input data are interpolated by L, the FFT outputs will have L cycles.

From this example one can see that an interpolator does not introduce any additional information in the frequency domain. The additional data points result in many repeated cycles. This effect is different from the zero padding operation discussed in Section 4.2. Although zero padding does not add more information to the input data, it interpolates the FFT outputs and a better frequency estimate can be obtained. Detailed information on zero padding can be found in Section 4.2.

11.8 FILTER BANK DESIGN METHODOLOGY

From this section to the end of the chapter, a filter bank design will be presented. The general filter design method used today is based on polyphase filters and

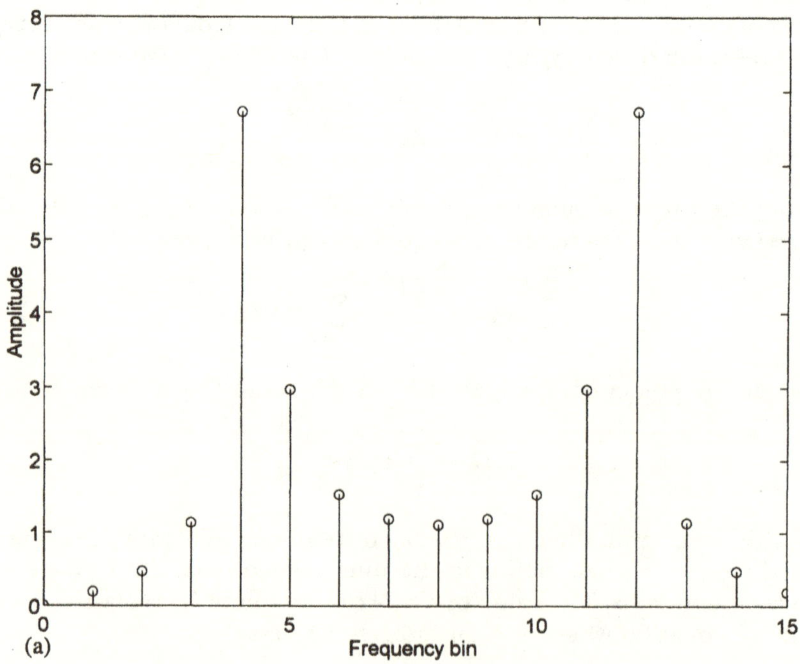

Figure 11.9 FFT of interpolated outputs: (a) 16-point FFT and (b) interpolated by 3.

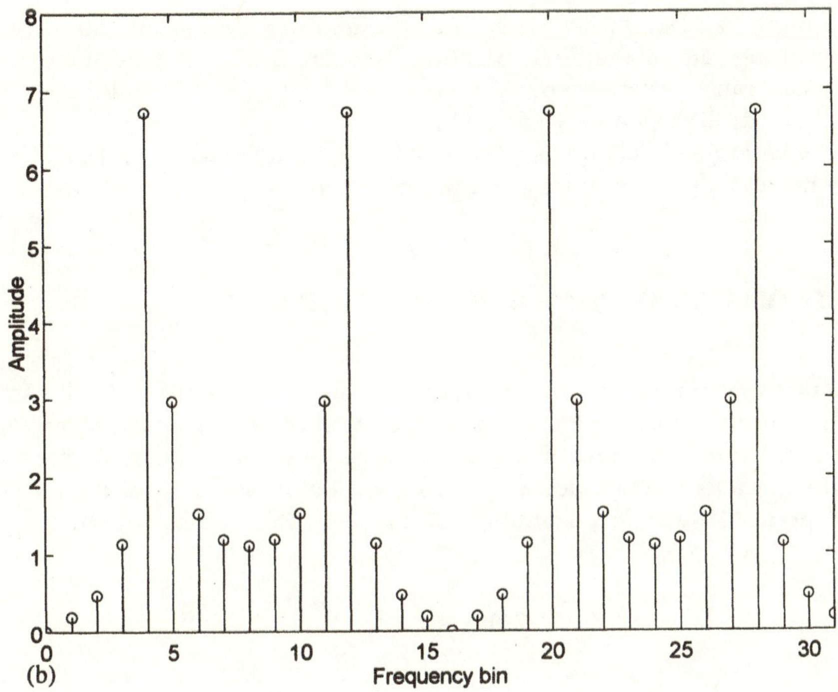

Figure 11.9 (continued).

multirate processing. This design concept has been adopted for digital channelized receivers. An example will be used to illustrate the idea. Actually, the number of bits and the sampling frequency of the ADC are not very important for this illustration. An example of an existing ADC as used in a receiver, however, provides a feeling of realism. Let us assume that the ADC is the one measured in Section 6.16. It has 8 bits and operates at 3 GHz. Under this sampling rate, a pulse width of 100 ns contains 300 data points. A base-2 number such as 256 can be chosen as the FFT length, which is slightly less than 100 ns. This value is slightly less than the minimum pulse width required in Section 2.19.

Performing a 256-point FFT with no overlapping data, the FFT must operate at about 11.72 MHz (3,000 MHz/256). From Section 11.3, one can see that using a rectangular window will not result in a receiver with high enough instantaneous dynamic range due to the presence of high sidelobes. In order to suppress the sidelobes, a weighting function (or window) must be applied. As mentioned in Section 11.4, a window in the time domain increases the main lobe width in the frequency domain. This main lobe width is the bandwidth of each individual filter. Increasing the filter bandwidth reduces the number of filters required to cover the desired input bandwidth. A window function, however, necessitates an increase in the processing rate of the FFT operation. No data overlapping operation with a window can cause different contributions from data in the time domain. This is undesirable, as discussed in Section 11.5.

The basic goal in channelized receiver design is to increase the FFT operating rate by decreasing the number of output channels.

11.9 DECIMATION IN THE FREQUENCY DOMAIN

Decimation also can be used in the frequency domain processing. In this section, the FFT outputs will be decimated. This operation can decrease the complexity of the FFT operation. Instead of presenting a general case, a special case will be presented because the notation will be simpler. Let us assume that the outputs of the 256-point FFT are decimated by 8. A 256-point FFT can be written as

$$X(k) = \sum_{n=0}^{N-1} x(n) e^{\frac{-j2\pi nk}{N}} \tag{11.14}$$

where $N = 256$. There are 256 outputs in the frequency domain. If every eighth output is kept and the other outputs are discarded, the resulting outputs are $k = 0, 8, 16, \ldots, 248$. There are a total of 32 (256/8) outputs. These outputs can be written as

$$\begin{aligned} X(0) &= \sum_{n=0}^{255} x(n) \\ X(8) &= \sum_{n=0}^{255} x(n) e^{\frac{-j2\pi 8n}{256}} \\ X(16) &= \sum_{n=0}^{255} x(n) e^{\frac{-j2\pi 16n}{256}} \\ &\cdots \\ X(248) &= \sum_{n=0}^{255} x(n) e^{\frac{-j2\pi 248n}{256}} \end{aligned} \tag{11.15}$$

First let us arbitrarily choose two frequency components $k = 16$ and $k = 248$ and rewrite in slightly different form. The results are

$$
\begin{aligned}
X(16) &= \sum_{n=0}^{255} x(n) e^{\frac{-j2\pi 16n}{256}} = \sum_{n=0}^{255} x(n) e^{\frac{-j2\pi 2n}{32}} \\
&= [x(0) + x(32) + x(64) + \ldots + x(224)] \\
&\quad + [x(1) + x(33) + x(65) + \ldots + x(225)]e^{\frac{-j2\pi 2}{32}} \\
&\quad + [x(2) + x(34) + x(66) + \ldots + x(226)]e^{\frac{-j2\pi 2\times 2}{32}} \\
&\quad + \ldots \\
&\quad + [x(31) + x(63) + x(95) + \ldots + x(255)]e^{\frac{-j2\pi 2\times 31}{32}}
\end{aligned}
\tag{11.16}
$$

and

$$
\begin{aligned}
X(248) &= \sum_{n=0}^{255} x(n) e^{\frac{-j2\pi 248n}{256}} = \sum_{n=0}^{255} x(n) e^{\frac{-j2\pi 8n}{32}} \\
&= [x(0) + x(32) + x(64) + \ldots + x(224)] \\
&\quad + [x(1) + x(33) + x(65) + \ldots + x(225)]e^{\frac{-j2\pi 31}{32}} \\
&\quad + [x(2) + x(34) + x(66) + \ldots + x(226)]e^{\frac{-j2\pi 31\times 2}{32}} \\
&\quad + \ldots \\
&\quad + [x(31) + x(63) + x(95) + \ldots + x(255)]e^{\frac{-j2\pi 31\times 31}{32}}
\end{aligned}
\tag{11.17}
$$

In the above equations the relation of $e^{-j2\pi n} = 1$ when n = integer is used. Now let us define a new quantity $y(n)$ as

$$
\begin{aligned}
y(n) &= x(n) + x(n+32) + x(n+64) + \ldots + x(n+224) \\
&= \sum_{m=0}^{7} x(n+32m)
\end{aligned}
\tag{11.18}
$$

where $n = 0$ to 31. This $y(n)$ represents the values in the bracket of (11.16) and (11.17). Each $y(n)$ value contains a total of 8 data points. This operation can be graphically represented in Figure 11.10. In this figure, the 256 input data points are divided into eight 32-point sections. The beginning data point of each section is shown. These eight sections are stacked and summed vertically as shown in the figure. The results are the 32 $y(n)$ values.

Using these $y(n)$ values, the FFT results from (11.15) can be rewritten as

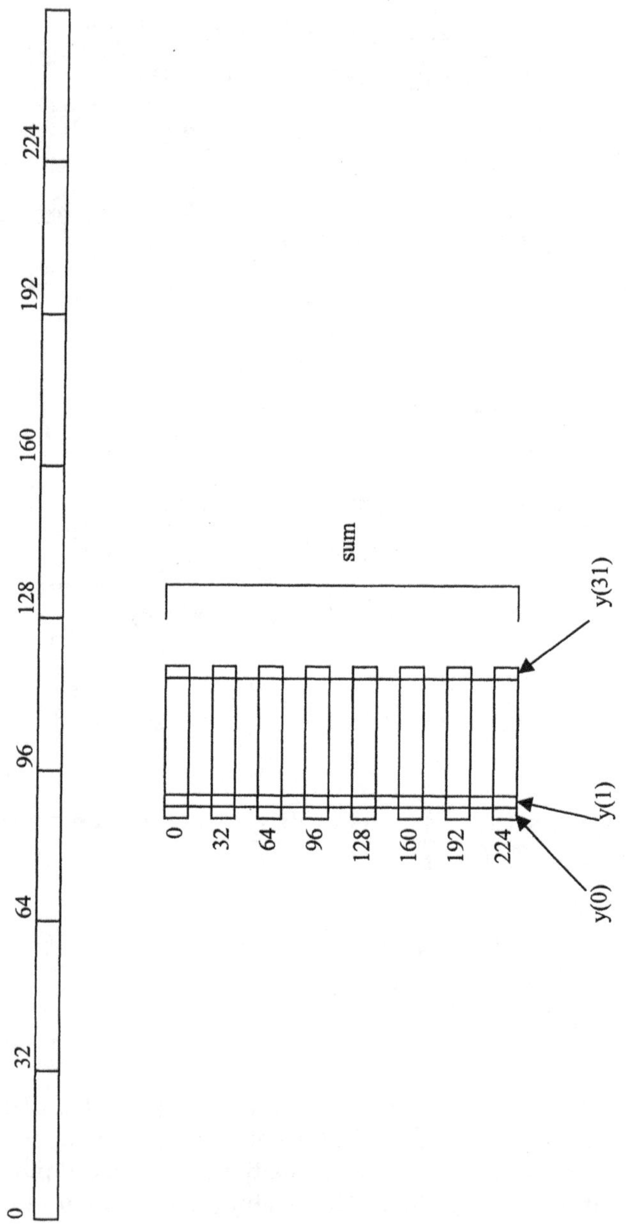

Figure 11.10 Graphic representation of obtaining the y values.

$$
\begin{aligned}
X(0) &= \sum_{n=0}^{31} y(n) \\
X(8) &= y(0) + y(1)e^{\frac{-j2\pi}{32}} + y(2)e^{\frac{-j2\pi 2}{32}} + \ldots + y(31)e^{\frac{-j2\pi 31}{32}} \\
&= \sum_{n=0}^{31} y(n)e^{\frac{-j2\pi n}{32}} \\
X(16) &= y(0) + y(1)e^{\frac{-j2\pi 2}{32}} + y(2)e^{\frac{-j2\pi 2\times 2}{32}} + \ldots + y(31)e^{\frac{-j2\pi 2\times 31}{32}} \\
&= \sum_{n=0}^{31} y(n)e^{\frac{-j2\pi 2n}{32}} \\
&\ldots \\
X(248) &= y(0) + y(1)e^{\frac{-j2\pi 31}{32}} + y(2)e^{\frac{-j2\pi 31\times 2}{32}} + \ldots + y(31)e^{\frac{-j2\pi 31\times 31}{32}} \\
&= \sum_{n=0}^{31} y(n)e^{\frac{-j2\pi 31 n}{32}}
\end{aligned}
\tag{11.19}
$$

All these equations can be written into one as

$$X(8k) = \sum_{n=0}^{31} y(n)e^{\frac{-j2\pi kn}{32}} \tag{11.20}$$

where $k = 0, 1, 2, \ldots, 31$; and $n = 0, 1, 2, \ldots, 31$.

The output $X(8k)$ can be relabeled as $Y(k)$; thus, the above equation can be written as

$$Y(k) = \sum_{n=0}^{31} y(n)e^{\frac{-j2\pi kn}{32}} \tag{11.21}$$

This equation represents a 32-point FFT. In order to obtain the outputs of a 256-point FFT decimated by 8, a 32-point FFT can achieve the goal. Thus, the design of the FFT can be simplified. The input signal must be manipulated, however, in order to obtain the desired result.

A general statement without further proof will be presented here. If one wants to perform an N-point FFT and the outputs in the frequency domain are decimated by M, one can achieve the goal by performing an N/M-point FFT. A new input format $y(n)$ must be built first. The generalization of the $y(n)$ can be written as

$$y(n) = \sum_{m=0}^{M-1} x(n + mN/M) \tag{11.22}$$

where $n = 0, 1, 2, \ldots, (N/M) - 1$. The outputs in the frequency domain can be obtained as

$$X(n) = \sum_{n=0}^{(N/M)-1} y(n) e^{\frac{-j2\pi kn}{N/M}} \tag{11.23}$$

This section illustrates that when the FFT outputs are decimated by M, the outputs can be obtained from N/M-point FFT. The design of the FFT chip can be simplified tremendously. Using the number of computations $(N/(2\log_2 N))$ discussed in Section 4.7, the saving can be estimated.

11.10 OUTPUT FILTER SHAPE FROM A DECIMATED FFT

In this section the outputs from the FFT will be discussed. If 256 data points are used for FFT and every output is kept, there are 128 independent outputs in the frequency domain. The filters are overlapped at about –3.9 dB $\{20 \times \log[\sin(\pi/2)/(\pi/2)]\}$ down, as shown in Figure 11.4. If 32 points are used for the FFT operation and every output is kept, there are 16 independent outputs in the frequency domain. The filter shape is independent of the length of the FFT, but the bandwidth is inversely proportional to the data length used in the FFT.

If the FFT uses 256 data points but only one out of eight of the outputs are kept, there are a total of 32 outputs. Among these outputs 16 of them carry redundant information. Therefore, only 16 outputs are displayed in Figure 11.11. Each filter output is represented by a sinc function. Only a few side lobes are shown in this figure and the highest two are only 13 dB below the main lobe. This filter bank has many holes (high insertion loss region). If an input signal falls in one of the holes, the receiver will miss it entirely. The shape of this filter is definitely unacceptable. In the following section this problem will be fixed.

11.11 USING WEIGHTING FUNCTION TO WIDEN THE OUTPUT FILTER

To widen the individual filters and at the same time suppress the sidelobes, a window (or weighting) function can be applied to the input data. There are many different window functions. The one used here is the Parks-McClellan window because it can provide the desired frequency response. The coefficients of the window can be obtained from the "remez" function of the MATLAB. The window function is shown in Figure 11.12. Figure 11.12(a) shows the time domain response as obtained from the MATLAB "remez" program. Only the relative amplitude of the window function is of interest. Figure 11.12(b) shows the corresponding frequency domain response with very low pass band ripples and side lobes lower than 70 dB. This is a desirable filter shape. The frequency response is obtained from the "freqz" function of the MATLAB. One can see from the time domain response that the window function passes with moderate attenuation on fewer than 50 of the 256 samples. The rest of the input data are highly attenuated. The corresponding effect

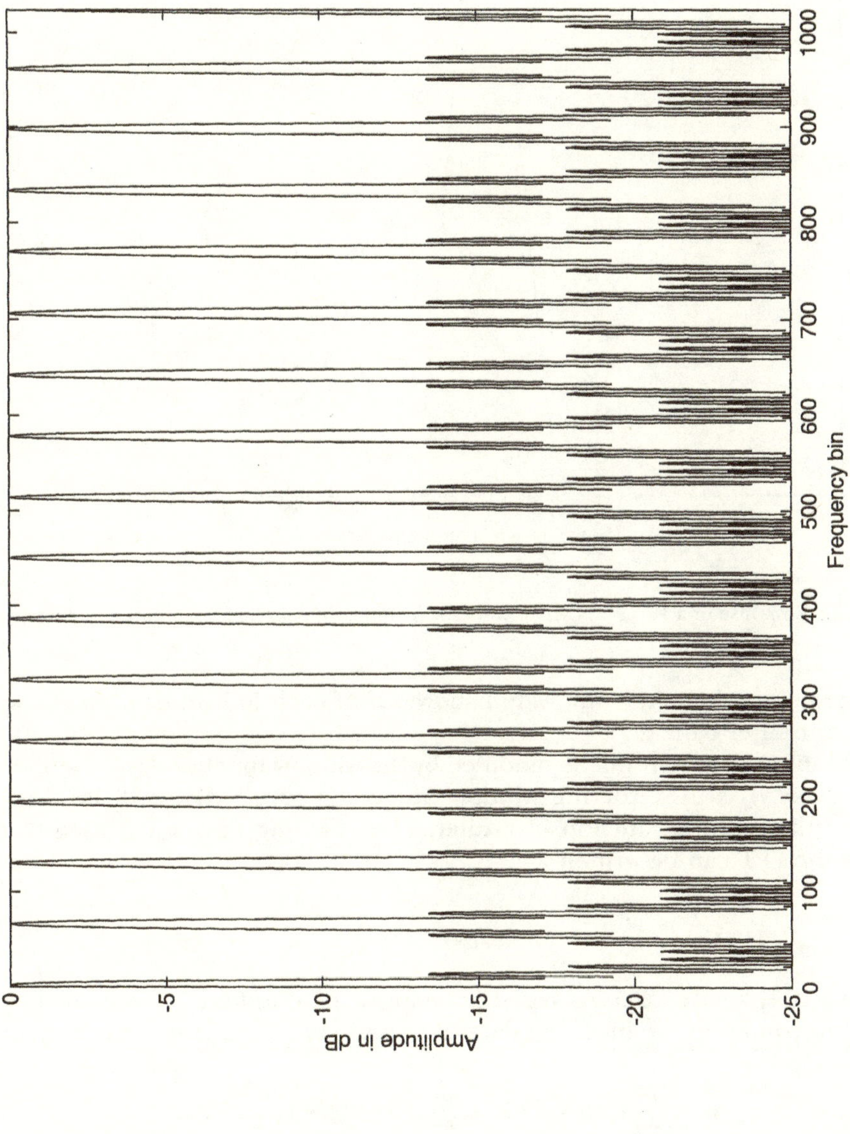

Figure 11.11 Filter outputs from decimated outputs with a rectangular window.

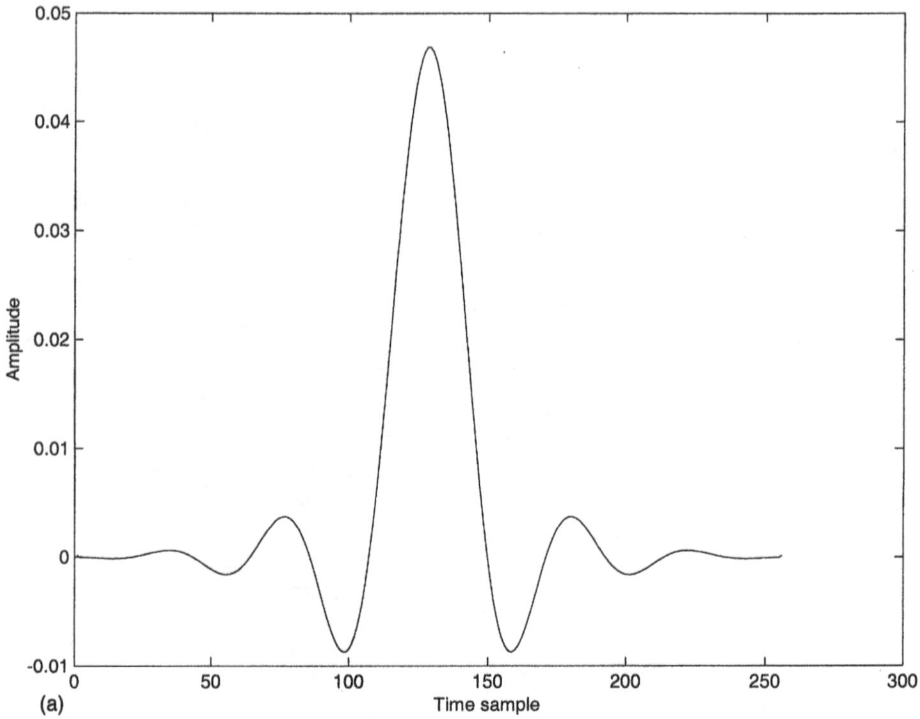

Figure 11.12 Response of a Parks-McClellan window: (a) time domain and (b) frequency domain.

in the frequency domain is the wide bandwidth of each individual filter shape in the uniform filter bank.

The input data $x(n)$ will be modified by the window function $h(n)$. Here, $h(n)$ instead of $w(n)$ is used for the window function because $h(n)$ will be used to represent the impulse function of a filter. The resulting data $x_m(n)$ used as the input of the FFT can be written as

$$x_m(n) = x(n)h(n) \tag{11.24}$$

where $n = 0, 1, 2, \ldots, 255$. As stated previously, the outputs are decimated by 8. Under this condition the modified data can be used in (11.18) to find the $y(n)$ as

$$y(n) = \sum_{m=0}^{7} x_m(n + 32m) = \sum_{m=0}^{7} x(n + 32m)h(n + 32m) \tag{11.25}$$

where $n = 0, 1, 2, \ldots, (N/M) - 1$. A few $y(n)$ terms are written as

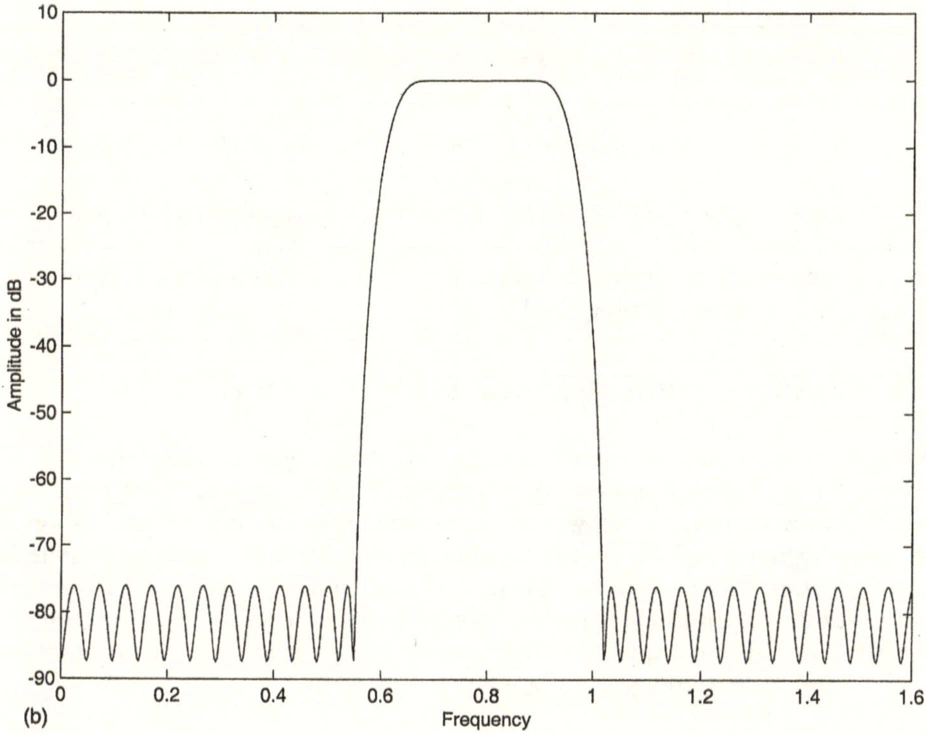

Figure 11.12 (continued).

$$
\begin{aligned}
y(0) &= x(0)h(0) + x(32)h(32) + \ldots + x(224)h(224) \\
y(1) &= x(1)h(1) + x(33)h(33) + \ldots + x(225)h(225) \\
&\ldots \\
y(31) &= x(31)h(31) + x(63)h(63) + \ldots + x(255)h(255)
\end{aligned}
\tag{11.26}
$$

If a 32-point FFT is performed on these $y(n)$ values, 16 individual filters will be generated. Each filter shape is as shown in Figure 11.12(b).

11.12 CHANGING OUTPUT SAMPLING RATE

The operation mentioned in the previous section can be considered as a software approach because the value in (11.16) can be calculated. With this approach it is easy to change the output sampling rate. If one would like to shift the input data by M points, all that is required is to calculate the results from (11.16) as

$$
\begin{aligned}
y(0) &= x(M)h(0) + x(M+32)h(32) + \ldots + x(M+224)h(224) \\
y(1) &= x(M+1)h(1) + x(M+33)h(33) + \ldots + x(M+225)h(225) \qquad (11.27) \\
&\ldots \\
y(31) &= x(M+31)h(31) + x(M+63)h(63) + \ldots + x(M+255)h(255)
\end{aligned}
$$

In this equation the only change is the input data points, which determines the output sampling rate. If $M = 1$, the output sampling rate equals to the input sampling rate, which corresponds to the 100% data overlapping case in Figure 11.2. This software approach is very flexible.

11.13 CHANNELIZATION THROUGH POLYPHASE FILTER [7–10]

Although the approach discussed in the two previous sections is very flexible, it is not suitable for high-speed operation because of the limitation of the calculation speed. The same operation, however, can be accomplished in hardware with much higher operation speed. Now let us consider in more detail the process to generate the $y(n)$ values. The $y(n)$ values listed in (11.26) must be generated from input data shifting with time. One can see that each of these values can be generated from the convolution output of a filter with the input signal. The 256-point window function in the time domain can be written as

$$
h(n) = h(255)\delta(n) + h(254)\delta(n-1) + h(253)\delta(n-2) + \ldots + h(0)\delta(n-255) \qquad (11.28)
$$

where the δ function indicates the $h(n)$ value occurs at time n. The impulse sequence of the filter is written in an inverse way. This impulse function can generate the results from (11.26) through convolution with the input signal. Because the window function shown in Figure 11.12(a) is symmetric in the time domain, this inverse is just a subscript change. This function can be decimated by 32, which results in 32 individual filters, each having eight taps. This filter decimation is often referred to as the polyphase filter. Each of the 32 filters has the response as indicated below:

$$
\begin{aligned}
h_0(n) &= h(224)\delta(n) + h(192)\delta(n-1) + h(160)\delta(n-2) + \ldots + h(0)\delta(n-7) \\
h_1(n) &= h(225)\delta(n) + h(193)\delta(n-1) + h(161)\delta(n-2) + \ldots + h(1)\delta(n-7) \\
&\ldots \\
h_{31}(n) &= h(255)\delta(n) + h(223)\delta(n-1) + h(191)\delta(n-2) + \ldots + h(31)\delta(n-7)
\end{aligned} \qquad (11.29)
$$

These filters must be convolved with the proper input data to obtain the result in (11.26). In order to obtain the correct data format, the input data must be decimated by 32 also. When the decimated input signal and the decimated filter are convolved and reach steady state, the output is equal to the result of (11.26).

The next time the 32-point FFT is performed, the input $y(n)$ values to the FFT are

$$
\begin{aligned}
y(0) &= x(32)h(0) + x(64)h(32) + \ldots + x(256)h(224) \\
y(1) &= x(33)h(1) + x(65)h(33) + \ldots + x(257)h(225) \\
&\ldots \\
y(31) &= x(63)h(31) + x(95)h(63) + \ldots + x(287)h(255)
\end{aligned}
\tag{11.30}
$$

In this equation, the first data point is $x(32)$; thus, the input is shifted 32 points. The hardware to accomplish this goal is shown in Figure 11.13. In this figure, there are 32 filters and each filter has eight taps. Two cycles of input data are shown and each cycle contains 32 data points. The outputs are shown as $y(n)$ and they are used as the input of the FFT. The final results in the frequency domain are represented by $Y(k)$. In this case the inputs are decimated by 32, and the final frequency domain also has 32 outputs. The input data are shifted 32 points, which is also the output frequency bin number. This case is referred to as the critically sampled case. A critically sampled case is one where the number of output frequency bins equals to the input data shift. This means that the output sampling rate is $1/M$ times the input sampling rate where M is the number of input data points shifted. If one wants to increase the output sampling rate, the hardware must be modified, which is not as flexible as the software approach discussed in the previous

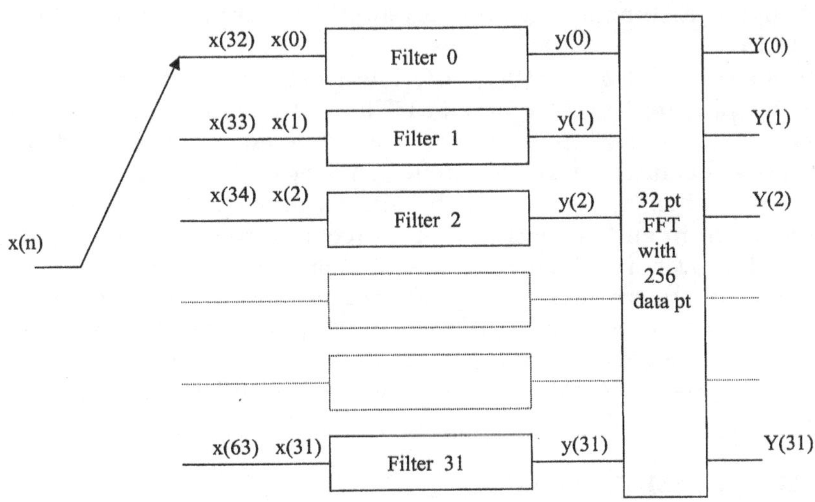

Figure 11.13 Channelization approach.

section. The detailed approach of doubling the output sampling rate with the same number of output channels will be discussed in Section 13.9.

A finite impulse response (FIR) filter design is used this discussion. Filter number 0 with $y(0)$ as output is shown in Figure 11.14. In this figure the decimated input data points are shown. When the input signal reaches steady state, the output of the filter contains eight terms. The first and second consecutive filter outputs after steady state are also listed. The lower line represents the first output, which matches the $y(0)$ output of (11.26). The upper line represents the second time output where the input data are shifted by 32 points. This result matches the $y(0)$ output from (11.30). The rest of the filter outputs can be obtained in a similar manner.

11.14 OPERATION OF THE POLYPHASE FILTER [7–10]

In this section the detailed operation of the polyphase filter will be discussed. First the speed of operation will be considered. The input data are sampled at 3,000 MHz, which is the input data coming out the analog-to-digital converter (ADC) at about every 0.33 ns per sample. If the input data to the polyphase filter are decimated by 32, the input rate will be 93.75 MHz (3,000/32) and the filter operates at this speed. In order to process all the data, there are 32 parallel channels. It is relatively easy to operate at this lower rate. Because this system has two operation rates, it is often referred to as a multirate system. The 32-point FFT following the filters also operates at this low rate. Progress in FFT signal processing technology may allow one to implement new techniques.

From the output of Figure 11.12 one can see that the input data to the filter move 32 points each time. The progression of the data processing is illustrated in Figure 11.15.

As shown in Figure 11.10(a), the input data under the main lobe of the window function are less than 50 points. This channelization approach shifts the 256-point window only 32 points. There should be enough data overlapping that all the data will contribute to the output in a near uniform manner.

Another advantage of shifting the window function of 32 points is the fine time resolution. In an EW receiver, time is required to generate two parameters, as discussed in Section 2.6. One is the time of arrival (TOA) and the other is the pulse width. In modern signal sorting algorithms, it is desirable to have fine time resolution. The time resolution provided by the polyphase filter is about 10.7 ns, which is suitable for most of the operations. This output rate is also influenced by the signal processing following the filter, which will be further discussed in Chapter 13.

11.15 FILTER DESIGN [11–12]

Finally, the filter design will be presented. The window function is the same as the one displayed in Figure 11.12(a). The approach to obtain this filter is as follows.

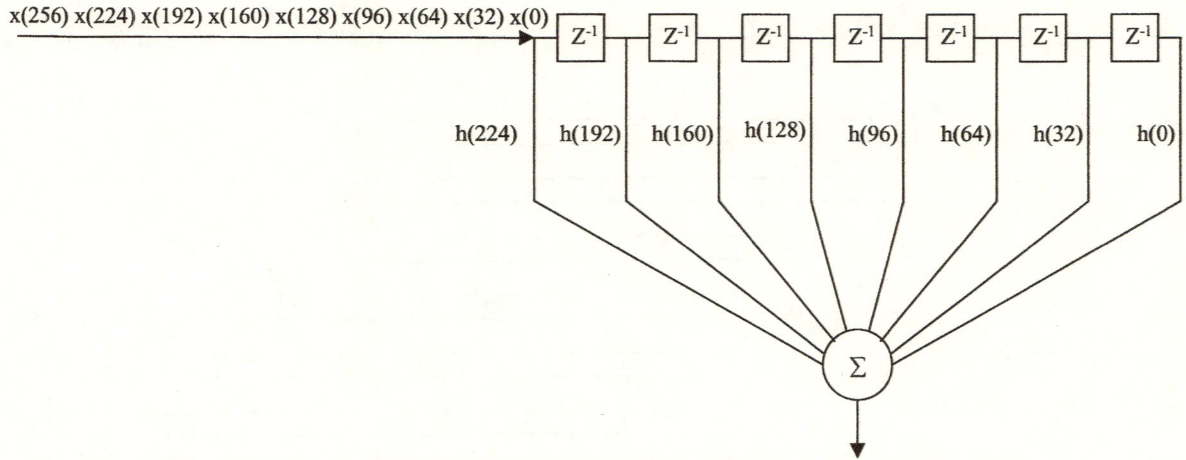

x(256)h(224)+x(224)h(192)+x(192)h(160)+x(160)h(128)+x(128)h(96)+x(96)h(64)+x(64)h(32)+x(32)h(0)

x(224)h(224)+x(192)h(192)+x(160)h(160)+x(128)h(128)+x(96)h(96)+x(64)h(64)+x(32)h(32)+x(0)h(0)

Figure 11.14 An individual polyphase filter.

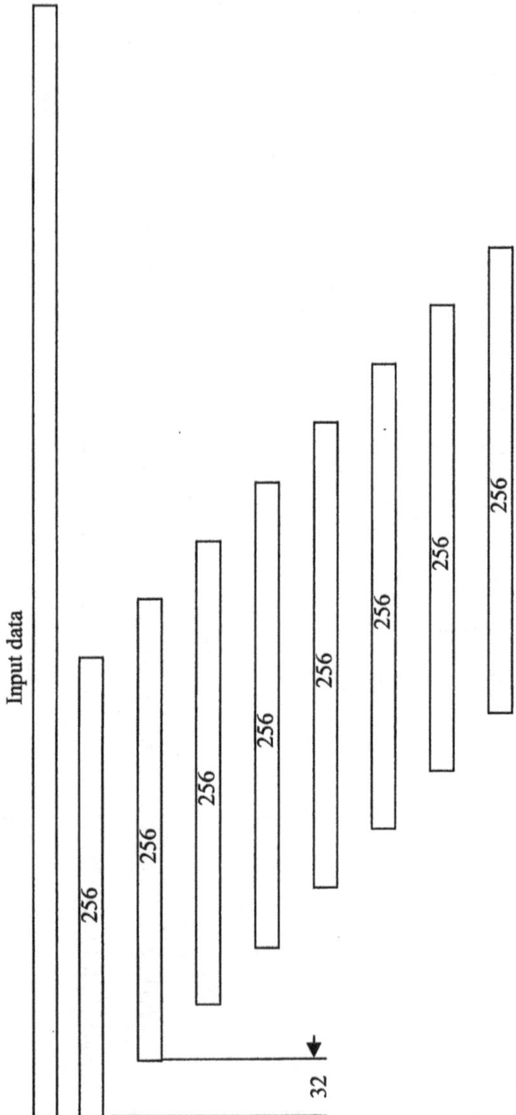

Figure 11.15 Input data rate.

The sampling frequency is 3,000 MHz; thus, the unambiguous bandwidth is 1,500 MHz. Because only 16 independent channels can be obtained from the 32-point FFT, the equivalent filter bank has 16 outputs. The bandwidth of each channel is 93.75 MHz (1,500/16), which can be considered as the 3-dB bandwidth. It is desirable to have at least 60-dB attenuation at a bandwidth of 187.5 MHz (93.75×2), which is double the 3-dB bandwidth. This required filter shape is shown in Figure 11.16. Only three adjacent filters are shown.

In order to realize this filter response, one must determine how long the window function (or filter taps) should be. This can be determined from [11]

$$T_p = \frac{-10\log(R_p R_s) - 13}{2.324 B_{tr}} + 1 \tag{11.31}$$

where R_p and R_s are related to the passband ripple factor and the insertion loss of the stopband, respectively, and B_{tr} represents the transition period in radians, which is $2\pi/64$. The passband ripple in decibels related to R_p as $20\log(1 + R_p)$ and the stopband insertion loss is related to R_s as $20\log(R_s)$. With $R_p = 0.01$ [R_p(dB) = 0.086 dB] and $R_s = 0.001$ [R_s(dB) = −60 dB], $T_p = 163$. Because the total window function contains 256 points, the required filter response should be achieved or exceeded.

Using the MATLAB program and adjusting some of the parameters, such as the 3-dB bandwidth and the 60-dB bandwidth, one can select a window function to result in the filter bank shown in Figure 11.17. This filter bank has a dynamic range of about 75 dB, which is higher than the design goal of 60 dB because of the high number of taps. The filter shape is quite uniform. This is superior to the performance of an analog filter bank. Because the 3-dB bandwidth of the filter is about 93.75 MHz, its capability to separate two signals close in frequency is quite limited. If two signals fall into one filter, without further signal processing, the receiver cannot separate them. The performance of the MATLAB example shown here is quite close to what can be done in an actual filter implementation in the laboratory.

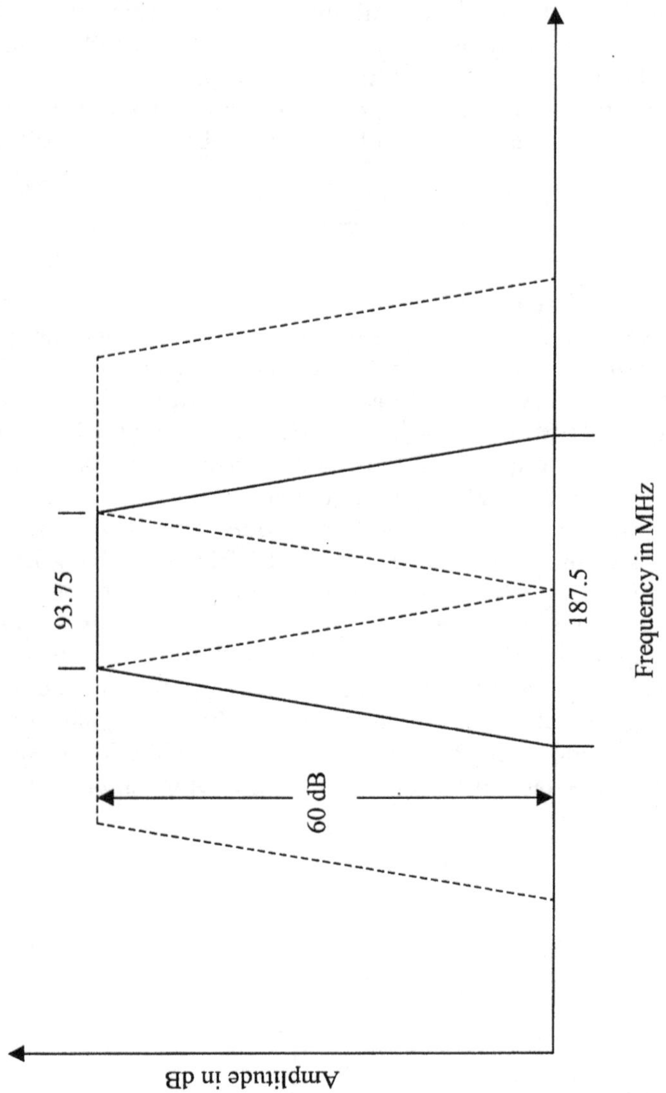

Figure 11.16 Desired filter response.

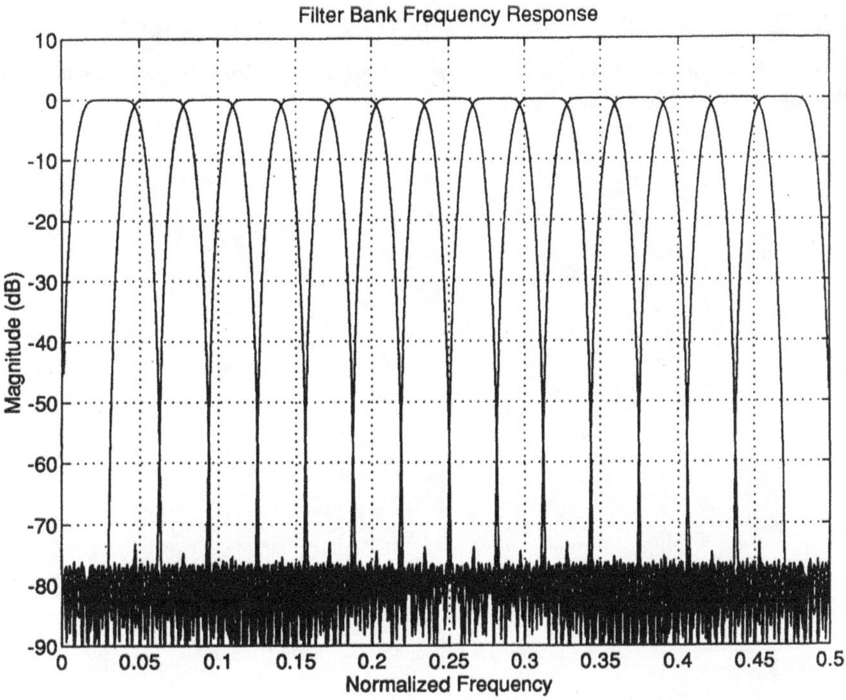

Figure 11.17 Designed filter bank.

REFERENCES

[1] Kay, S. M., *Modern Spectral Estimation, Theory and Application*, Englewood Cliffs, NJ: Prentice Hall, 1987.

[2] Harris, F. J., "Time Domain Signal Processing with the DFT," Ch. 8 of Elliot, D. F., Editor, *Handbook of Digital Signal Engineering Applications*, San Diego, CA: Academic Press, Inc., 1987.

[3] Allen, J. B. "Short Term Spectral Analysis, Synthesis and Modification by Discrete Fourier Transform," *IEEE Trans. on Acoustics, Speech and Signal Processing*, Vol. ASSP-25, June 1977, pp. 235–238.

[4] Allen, J. B., and Rabiner, L. R., "A Unified Approach to Short Time Fourier Analysis and Synthesis," *IEEE Proc.*, Vol. 65, Nov. 1977, pp. 1558–1564.

[5] Harris, F. J., "On the Use of Windows for Harmonic Analysis with the Discrete Fourier Transform," *IEEE Proc.*, Vol. 66, Jan. 1987, pp. 51–83.

[6] Thong, T., "Practical Consideration for a Continuous Time Digital Spectrum Analyzer," *ISCAS*, Vol. 2, 1989, pp. 1047–1050.

[7] Crochiere, R. E., and Rabiner, L. R., *Multirate Digital Signal Processing*, Englewood Cliffs, NJ: Prentice Hall, 1983.

[8] Vaidyanathan, P. P., *Multirate Systems and Filter Banks*, Englewood Cliffs, NJ: Prentice Hall, 1993.

[9] Vaidyanathan, P. P., "Multirate Digital Filters, Filter Banks, Polyphase Networks, and Applications: A Tutorial," *Proc. of IEEE*, Vol. 78, No. 1, January 1990, pp. 55–93.

[10] Ansari, R., and Liu, B., "Multirate Signal Processing," Ch. 14 of Mitra, S. K., and Kaiser, J. F., Editors, *Handbook for Digital Signal Processing*, New York, NY: John Wiley & Sons, 1993.

[11] Oppenheim, A. V., and Schafer, R. W., *Discrete-Time Signal Processing,* Englewood Cliffs, NJ: Prentice Hall, 1989.

[12] Rabiner, L. R., and Gold, B., *Theory and Application of Digital Signal Processing,* Englewood Cliffs, NJ: Prentice Hall, 1975.

CHAPTER 12

Monobit Receiver

12.1 INTRODUCTION

In this chapter the concept of the monobit receiver will be introduced. This technique can be considered as a digital channelized approach. The fast Fourier transform (FFT) is very simple and can be built on one chip. A simple frequency encoder is used after the FFT outputs to determine the number of input signals and their frequencies. The design of a candidate encoder will be presented. The encoder and the FFT can be built on one chip. The chip has been fabricated and the monobit receiver concept has been validated successfully in the laboratory.

The monobit receiver concept is inspired by commercial Global Positioning System (GPS) receiver designs. Usually, the analog-to-digital converter (ADC) in a commercial GPS receiver has only 1 or 2 bits, and the GPS signal is more complicated than a pulsed radio frequency (RF). This idea is adopted in wideband receiver applications and the name monobit receiver is used. This technique can be used to reduce to a minimum the hardware required for a given receiver function, with only a slight reduction in performance.

Because the ADC used in the monobit receiver has very few bits, the system is basically nonlinear. A nonlinear system is difficult to analyze theoretically. Thus, in designing the receiver one uses data collected from a data collection system. The collected data are processed in a computer to evaluate the performance. In order to determine the number of input signals, thresholds must be incorporated in the chip design. These threshold values are based on computer simulation results.

The major advantage of the monobit receiver is its simplicity. The price for simplicity is reduced performance in certain configurations. The monobit concept should not be considered as direct replacement of or compared with the digital channelized approach discussed in the previous chapter. The monobit receiver may be used for some special applications or to enhance or supplement other receivers. The current demonstration chip only implements the FFT and the frequency encoder. Because the receiver is nonlinear, the RF front-end design can be very simple. In the future, the RF chain and ADC could be included on the same chip. In other words, the entire receiver can be fabricated on one chip. For electronic warfare applications, size is a very important factor, especially for an airborne system. The ability to easily reconfigure and minimize costs of integration is very important to fielded system applications.

12.2 ORIGINAL CONCEPT OF THE MONOBIT RECEIVER

The original idea of the monobit receiver is to reduce the FFT complexity by eliminating the multiplication in the FFT operation. One simple way to eliminate multiplication is to use a 1-bit ADC. One bit generates ±1 as output, and this is the input to the FFT operation.

The discrete Fourier transform (DFT) can be written as

$$X(k) = \sum_{n=0}^{N-1} x(n) e^{\frac{-j2\pi kn}{N}} \tag{12.1}$$

where $x(n)$ is the input data and $e^{-j2\pi kn/N}$ is the Kernel function. If the input $x(n)$ is ±1, no multiplication is needed with the Kernel function. Because the FFT only requires addition and subtraction, the chip design can be very simple.

The next step is to manipulate the Kernel function. In the computer a large number of bits are used to represent the value of the Kernel function. Because the operation will be built in hardware, the number of bits is important and should be minimized. Because the input only has 1 bit, a low number of bits in the Kernel function do not degrade the output in the frequency domain significantly. An optimum combination of signal bits and Kernel function bits can be determined experimentally by observing the FFT outputs in the frequency domain.

12.3 MONOBIT RECEIVER IDEA [1]

Another way to avoid multiplication in the FFT operation is to reduce the number of bits of the Kernel function to 1 bit. This idea can be illustrated by using (12.1). The Kernel function, however, is a complex function, and therefore, it cannot be represented by a 1-bit real number. The minimum to represent the Kernel function by 1 bit is with 1 bit in the real and 1 bit in the imaginary domain. Mathematically this idea can be written as

$$e^{\frac{-j2\pi kn}{N}} \Rightarrow +1, -1, +j, -j \tag{12.2}$$

The Kernel function can be equal to one of these four values. Under this condition, no multiplication is required in performing the FFT.

Graphically, the values of the Kernel function are equally spaced around the unit circle in the complex plane as shown in Figure 12.1. The value starts from the real value of 1 and spaced by angle $2\pi/N$, where N is the total number of FFT points. In this figure $N = 8$ is shown. In Figure 12.2 the values of the Kernel function are digitized by 1 bit on the real axis and 1 bit on the imaginary axis. One can consider that all the Kernel function values in the range of $7\pi/4 \le \theta < \pi/4$ are quantized to 1; $\pi/4 \le \theta < 3\pi/4$ are quantized to j; $3\pi/4 \le \theta < 5\pi/4$ are quantized to -1; and $5\pi/4 \le \theta < 7\pi/4$ are quantized to $-j$ as illustrated in this figure.

Once the Kernel function is digitized into 1 bit, the input signal can be multiple bits and there is still no multiplication needed. Using simulation, the number of input data bits is increased to find the effect on the output. There is improvement from 1 to 2 bits, but from 2 to 3 bits the improvement is very small. Increasing beyond 3 bits shows no noticeable improvement. In order to keep the chip design simple, a 2-bit ADC is used.

12.4 DESIGN CRITERIA

Once the basic approach is determined, some design criteria can be chosen. In order to achieve 1-GHz input bandwidth, the Nyquist sampling frequency required is 2 GHz. To take the finite slope of the input filter into consideration, however, 2.5 times the input bandwidth is often used. Thus, the ADC should operate at

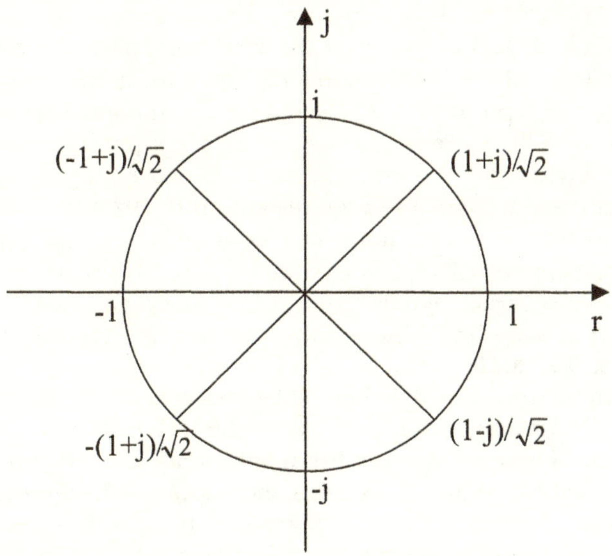

Figure 12.1 Kernel function of DFT with $N = 8$.

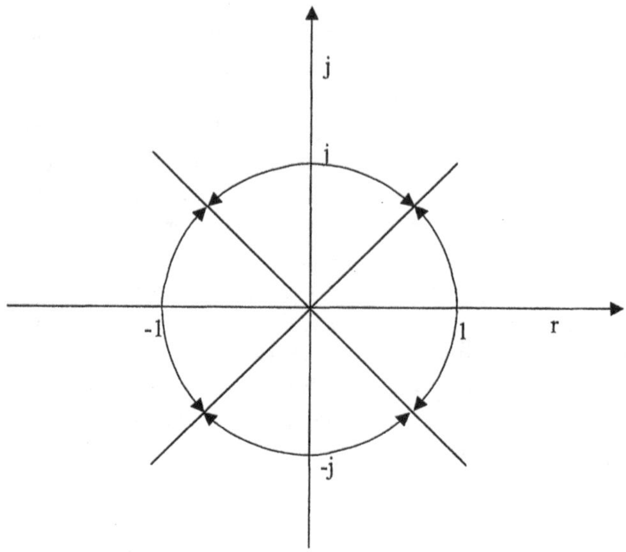

Figure 12.2 Kernel function is digitized 1 bit on real and imaginary axis.

2.5 GHz with only 2 bits. The corresponding sampling time is 0.4 ns. As stated in Section 6.17, the ADC should be able to take an input of 2.5 GHz. This kind of ADC is readily available because it only requires 2 bits.

If 256 points of data are used in the FFT operation, the equivalent time is 102.4 ns (256 × 0.4), which is approximately equal to the desired minimum pulse of 100 ns. Thus, a 256-point FFT is selected for the design. A rectangular window will be used for the FFT operation, which means the input data are not attenuated. In order to simplify the chip design, there is no data overlapping between adjacent FFTs. This arrangement limits the time resolution to 102.4 ns. If a faster FFT chip can be designed, the time resolution can be improved. The bandwidth of each channel is approximately 9.77 MHz (1/102.4 ns). Thus, the receiver has good sensitivity and can separate signals close in frequency, such as 10 MHz. With this design, the output sampling time is 102.4 ns, which corresponds to an output sampling rate of 9.77 MHz.

The rectangular window has high sidelobes and limits the receiver dynamic to less than 10 dB. Because the input data to the FFT have only 2 bits, the ADC is equivalent to a hard limiter. The amplitude information will be lost anyway. A hard limiter, such as a 1-bit digitizer, exhibits the capture effect under simultaneous signal conditions. The capture effect means that the strong signal suppresses the weak one. This effect also generates many harmonics in the frequency domain.

Figure 12.3 shows the capture effect of a 1-bit digitizer. Figure 12.3(a) shows the spectrum of two sinusoidal waves separated by 3 dB in amplitude. Figure 12.3(b)

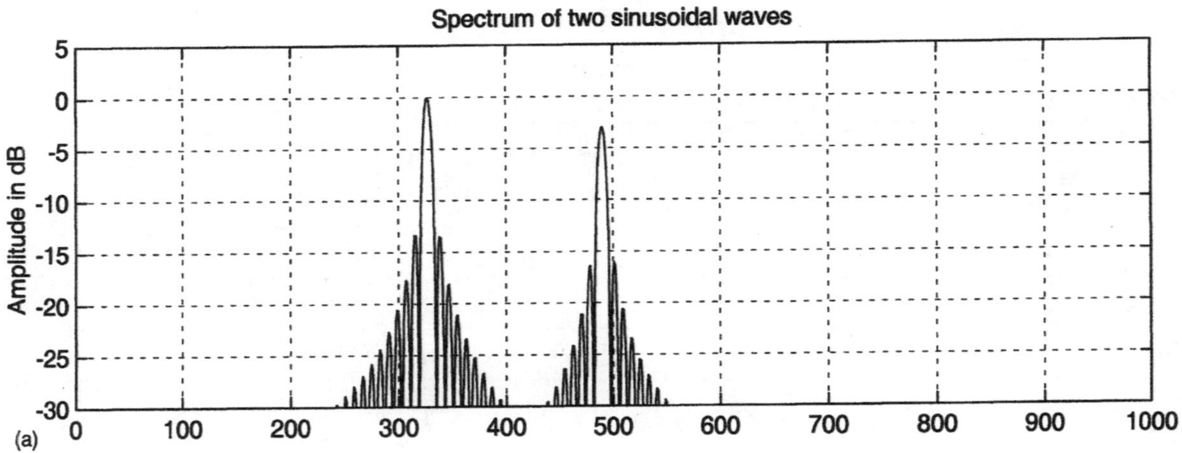

Figure 12.3 Effect of hard limiter (1-bit ADC): (a) input signal and (b) digitized signal.

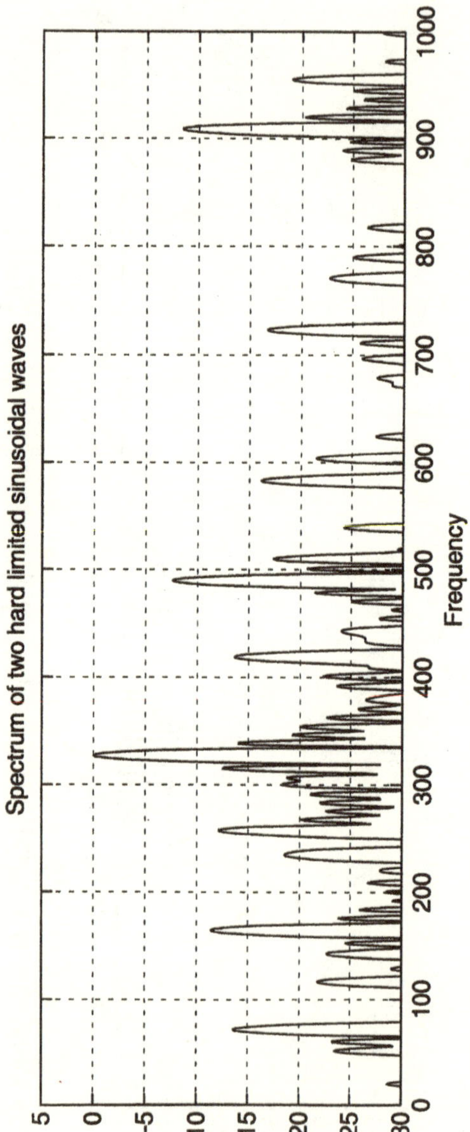

(b)

Figure 12.3 (continued).

shows the result after it digitized by the 1-bit ADC. The amplitudes of the two signals are separated by about 7 dB and many harmonics are generated.

From this figure one can see that the receiver is not expected to receive two signals separated very far in amplitude. In other words, the instantaneous dynamic range of the receiver is lower. This is the major deficiency of the monobit receiver.

12.5 RECEIVER COMPONENTS

The monobit receiver can be divided into five major portions: the RF chain, the ADC, the demultiplexer, the FFT, and the frequency encoder. Figure 12.4 shows the five components. The ADC operates at 2.5 GHz with only 2 bits. The demultiplexer is 1 to 16, which connects the input to 16 parallel outputs. Each data point contains 2 bits and each bit needs a demultiplexer. In order to simplify the discussion, only the data points rather than the bit number will be used in later discussion.

The FFT operation uses 256 data points and it takes 16 parallel input data every 6.4 ns (16×0.4). It takes total 16 cycles or 102.4 ns (16×6.4) to collect all the data. The FFT operation is performed every 102.4 ns. The encoder determines the number of input signals and their frequencies; thus, it is referred to as a frequency encoder. A general encoder usually provides frequency, pulse amplitude, pulse width, and time of arrival (TOA) information, as discussed in Section 2.6. To generate a time mark from the encoder is relatively easy; however, it is not included in the present chip design. If a time mark is included in the design, the resolution will be 102.4 ns, which will be used to generate pulse width and TOA. The amplitude information is lost through the 2-bit ADC. Thus, this receiver cannot provide pulse amplitude information. This information, if desired, must be obtained from some other circuit such as a log video amplifier placed in parallel with the monobit circuit.

In the present design only the FFT operation and the frequency encoder are integrated on one chip. In this chapter the discussion will be concentrated on this chip design. Because the ADC is very simple, it appears that the ADC can be fabricated on the same chip in later designs.

12.6 RF CHAIN, ADC, AND DEMULTIPLEXER

The RF chain design in a wideband receiver is very important. As discussed in Chapter 7, in order to obtain the desired sensitivity and dynamic range, the gain must be equal to a certain value and the third-order intercept point must be above a certain value. In the monobit receiver, the instantaneous dynamic range is rather low at about 5 dB, which means that when two input signals are separated by more than 5 dB, the receiver will only process the strong one.

With this low dynamic range the RF chain can be very simple. Instead of using linear amplifiers, a limiting amplifier can be used. The input versus output of a

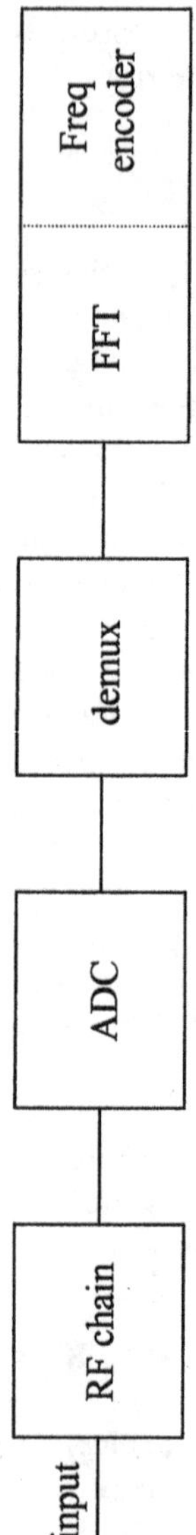

Figure 12.4 The five major portions of the receiver.

limiting amplifier is shown in Figure 12.5. The performance of a limiting amplifier is similar to an amplifier operating at saturation. The difference between a conventional amplifier operating at saturation and a limiting amplifier is that the output from a limiting amplifier is constant and the output from a conventional amplifier may vary. In many designs the limiting amplifier and conventional amplifier can be interchangeable. It is a common practice to use limiting amplifiers in front of an instantaneous frequency measurement (IFM) receiver.

The RF chain in front of the ADC is shown in Figure 12.6. The center frequency of the two filters is at the center of the second alias zone as shown in Figure 12.7. The first filter is used to reject out-of-band interference. The second filter is used to limit the noise generated by the amplifier. Without the second filter, the noise from 0 to 2.5 GHz will appear at the input of the ADC. Figure 12.8 shows a picture of the RF chain. In this figure two limiting amplifiers are used. Each amplifier has about 30-dB gain. The overall gain is about 60 dB.

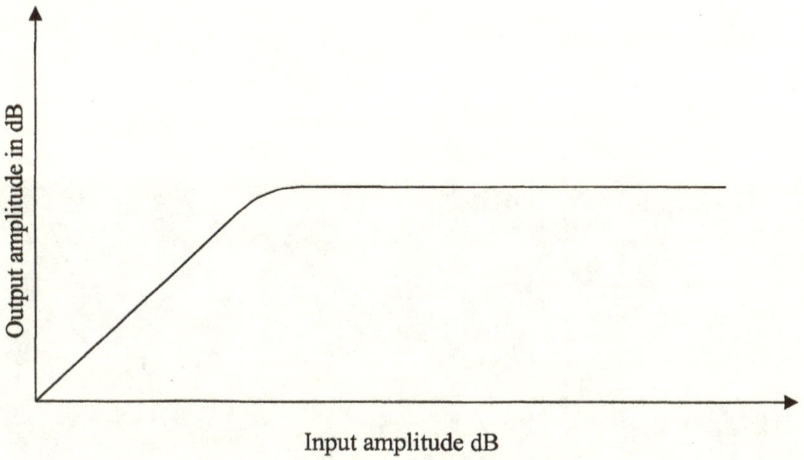

Figure 12.5 Input versus output of a limiting amplifier.

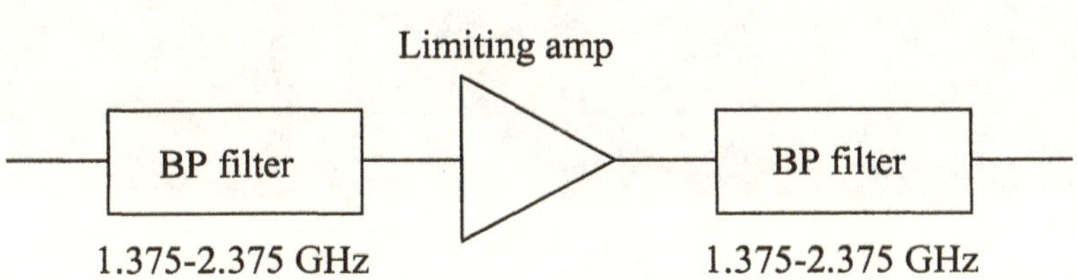

Figure 12.6 RF chain arrangement.

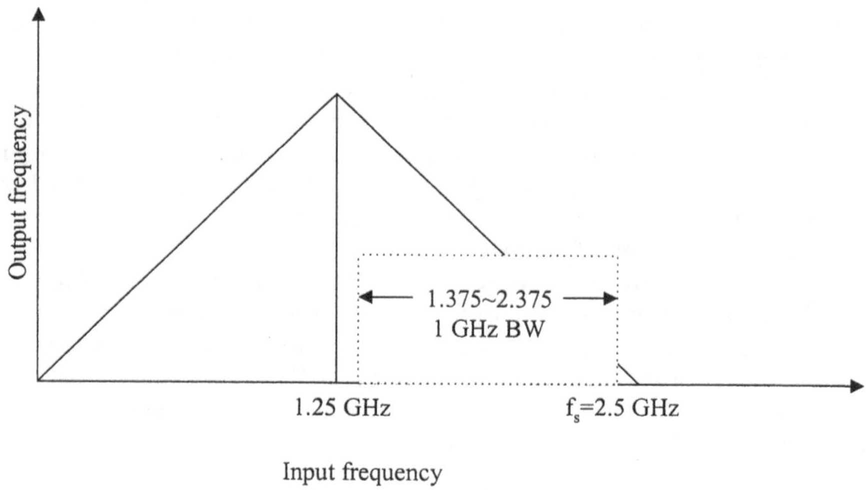

Figure 12.7 Frequency plan of the RF chain.

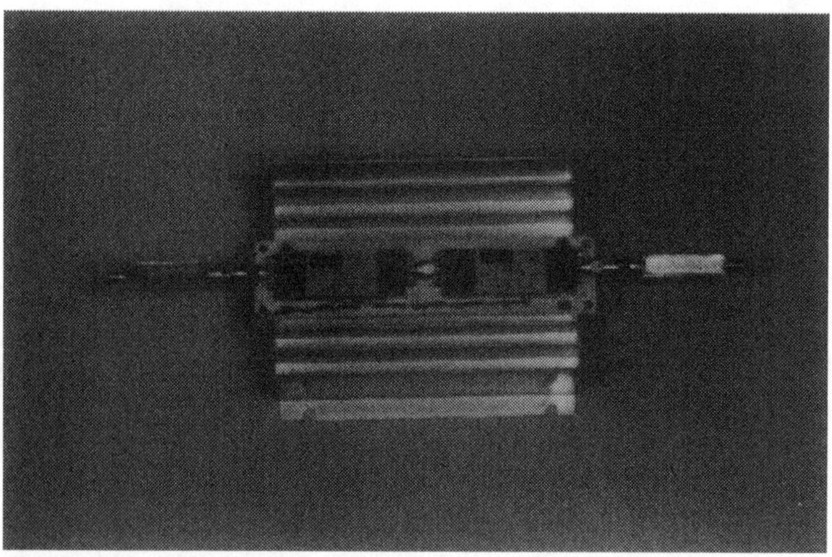

Figure 12.8 Picture of the RF chain.

The FFT chip cannot accept the input signal at 2.5 GHz. If the data are decimated by 16 and fill a buffer of 16 bits wide, the equivalent sampling rate is 156.25 MHz (2500/16) and the chip can accept this transfer rate. The ADC has two bits (outputs) and each output is connected to a demultiplexer as shown in Figure 12.9. The chip will accept 16 bits in parallel at 156.25 MHz. Figure 12.10 shows a picture of the actual receiver on a board including the ADC, the demultiplexers, and the FFT/frequency encoder chip. This board is built using existing components that are specifically designed for the monobit receiver. On the left of the board is the ADC and it has 3 bits. Following the ADC are three demultiplexers. Only two of the three demultiplexers are used. The large chip on the right lower corner of the board is the FFT/frequency encoder chip. The components in the middle of the board are the level translators, which change the voltage level of the emitter coupled logic (ECL) to the voltage level of the complementary metal oxide semiconductor (CMOS) logic. The ADC and the demultiplexers are ECL and the FFT/frequency encoder chip is CMOS. Because the voltage level used in these two technologies is different, level translators are used to match the voltage levels. In the future the same logic types will be used to build the ADC, the demultiplexers, and FFT chip; thus, the level translators will no longer be needed. Many RF inputs are shown on the edge of the board and they are used to supply clocks to the

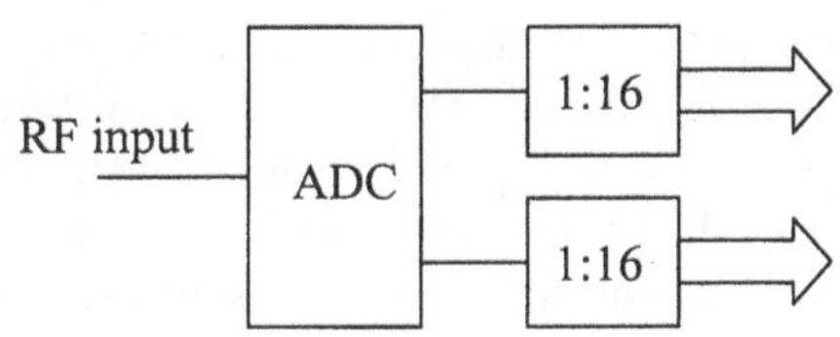

Figure 12.9 ADC and demultiplexers.

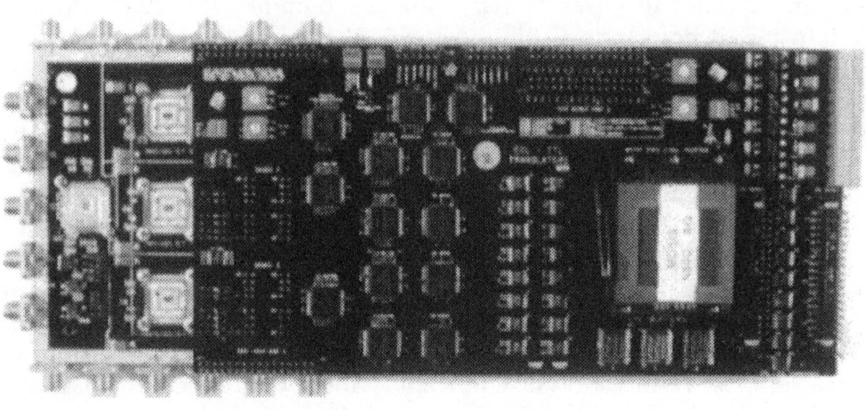

Figure 12.10 An experimental monobit receiver.

different components such as the demultiplexers. Because all the clocks used on the chip are phase locked, only one clock input is necessary. This experimental model is rather complicated, and future versions of the board could be simplified tremendously.

The required gain of the RF chain can be determined experimentally. In the experimental setup, an 8-bit ADC is used to collect data and the data are converted into 2 bits through a software program. There are several possible ways to convert the outputs of the 8-bit ADC to 2 bits. It appears that when the outputs of the four levels of the 2-bit ADC generate approximately the same number of outputs, the receiver produces a better result. This experiment can be used to adjust the RF gain to match the 2-bit ADC.

12.7 BASIC FFT CHIP DESIGN [2, 3]

The basic chip design follows the FFT operation. As discussed in Section 4.7, the FFT uses the butterfly technique to pass the input data from one layer to another. The operation can be shown symbolically as in Figure 12.11. Because the input has 256 data points, there are 8 layers of processors because $2^8 = 256$. The operations between the layers are the values obtained from the Kernel function. Because the Kernel function has only four values: ± 1, $\pm j$, only additions are needed between layers. A computer program is developed to trace the signal flow. From the signal flow, a table is generated to determine all the adders and inverters used. The final results obtained from inverters and adders are compared with the results calculated from (12.1) and (12.2) to ensure that the design is correct.

A 256-point FFT generates 256 outputs. Because 128 outputs carry redundant information, only 128 outputs are kept as outputs. These 128 outputs cover 1.25-GHz bandwidth. In order to cover 1 GHz, only 104 outputs ($128/1.25 = 102.4$) are needed. In the demonstration chip, however, all 128 outputs are available.

The input data to the FFT chip have only 2 bits. Because the operation uses addition and subtraction, the input values can be either positive or negative. In order to accommodate these operations, the one sign bit is added to the 2 bits of the ADC at the input of the FFT operation. Thus, it can be considered as 3 bits to the input of the FFT. After the first butterfly operation, the outputs of the first layer become 4 bits including 3 amplitude bits and one sign bit. The number of bits increases by one from one layer to the next. The final outputs should have 11 bits including the sign bit. In order to simplify the chip design, variations are tested experimentally to reduce the number of bits in the higher layers. After the fifth layer, the output has 8 bits. It is found that if after layers 6, 7, and 8 only 8 bits are kept, the results show insignificant change. To truncate the output from 9 to 8 bits, the least significant bit is ignored. This operation is applied to the outputs of layers 6, 7, and 8.

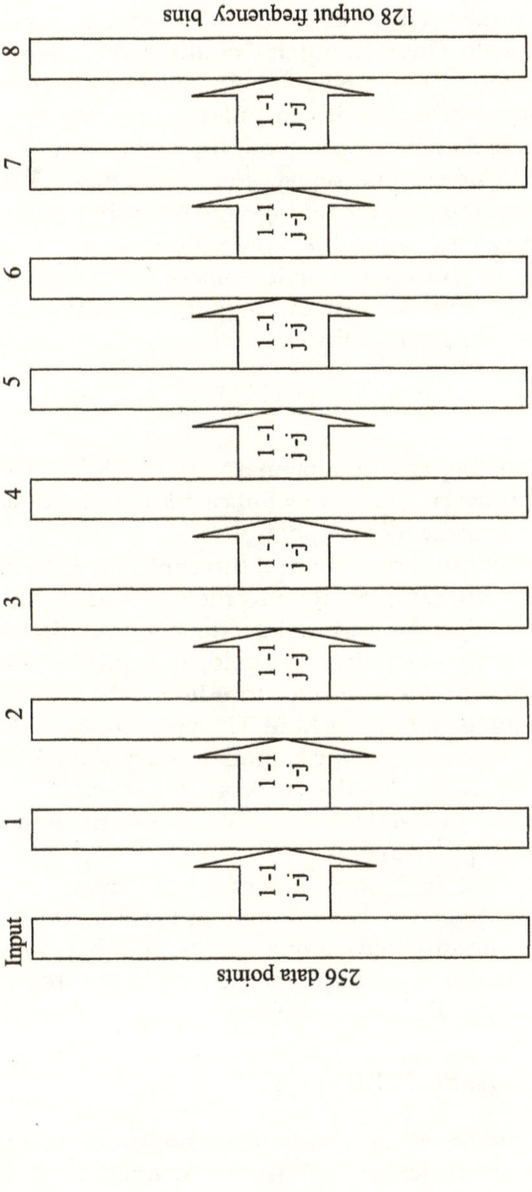

Figure 12.11 Basic FFT layout.

12.8 FREQUENCY ENCODER DESIGN [2, 3]

The main purpose of the frequency encoder design is to determine the number of input signals and their frequency. The 128 complex outputs of the FFT are input to the frequency encoder. In the following discussion the values used are either from simulated results or from processing actual digitized data.

From simulated results it is observed that the monobit receiver can process several simultaneous signals. This means that if multiple input signals of comparable amplitude are present, the FFT outputs will have peaks of the correct frequencies. However, in order to simplify the chip design, only two input signals will be processed by the receiver. With this design goal in mind, the receiver output should be one of three possibilities: no signal, one signal, and two signals. This limited output possibility makes the frequency encoder design relatively simple in comparison with a conventional frequency encoder with a large number of unknown signals.

One obvious approach of the frequency encoder is to find the amplitudes of all 128 frequency components. Because the outputs from the FFT are complex, it is necessary to find their amplitudes through the relation

$$|X(k)| = \sqrt{X_r^2(k) + X_i^2(k)} \tag{12.3}$$

where $X_r(k)$ and $X_i(k)$ are the real and imaginary parts of the kth frequency components. This operation, however, is rather complicated in a chip design. It is difficult to perform on every one of the 128 outputs.

To avoid this complicated operation, a threshold is set at the FFT outputs. Because the FFT outputs are complex, the thresholds are set for both the real and imaginary parts. The detail of the threshold setting will be discussed in the next section. If an FFT output crosses the threshold, it might be an eligible signal. Experimental results show that with proper thresholds, the maximum number of outputs that can cross the thresholds is four. The amplitudes of these outputs are calculated from (12.3). Because the total number is equal or less than four, the calculation is manageable in the chip design. Another threshold is used to compare with the amplitude of the outputs. If a frequency component crosses the threshold, it will be declared as an input signal.

Figure 12.12 shows the basic functions of the frequency encoder chip. The final output from the frequency encoder could be 0, which means that no input signal is detected. The output could be one specific number, which presents the frequency of the input signal. The output could be two numbers, which presents the frequencies of two input signals.

12.9 SELECTION OF THRESHOLDS [1–3]

The determination of the threshold can be considered one of the most difficult issues in the monobit receiver design. Because the system is nonlinear, it is difficult

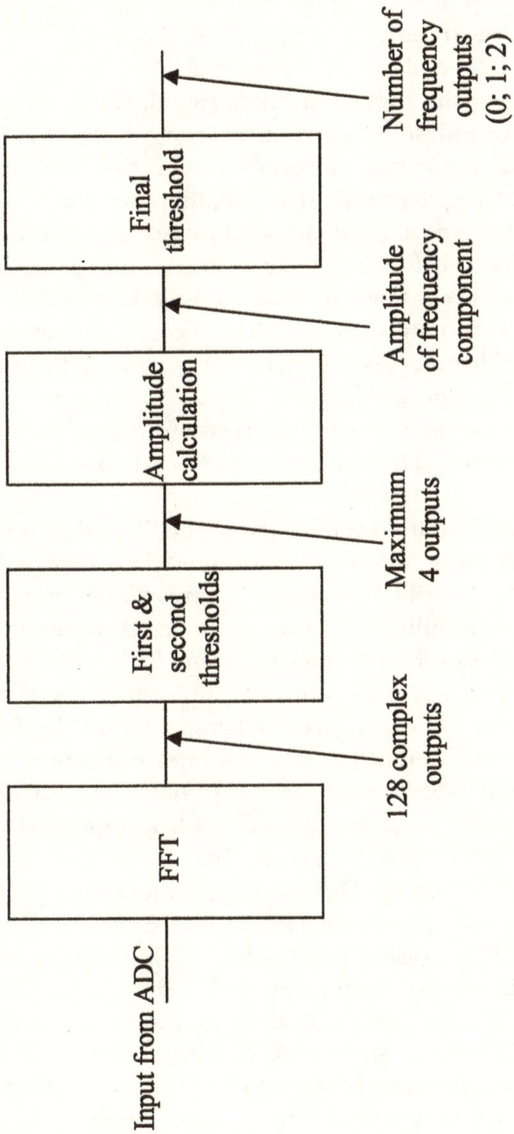

Figure 12.12 The functional block of the frequency encoder.

to analyze it. One way to determine the threshold is through a large number of trials of experimental data.

Before discussing details of the threshold selection, let us present the basic requirements. The basic philosophy is as follows: 1) The receiver should only rarely produce a false frequency report, when only noise is present. This is often referred to as the probability of false alarm of the receiver. 2) The receiver should report only one frequency when the input is a single signal. If the receiver reports more than one frequency, the additional signals are often referred to as spurious signals. 3) The receiver should report two frequencies when the input has two signals.

Consider the first requirement. It is usually unacceptable for a receiver to continue reporting false signals. If the threshold is set high enough, the false alarm rate is low; however, the receiver sensitivity is also reduced, which is undesirable. Usually, one false report every few tens of seconds is acceptable. Consider requirements 2 and 3. There is usually a compromise. If one wants to reduce the chance of a spurious response, this increases the chance of missing the second signal. The general consensus is that one would rather miss a signal than report a false one. A false signal may cause the signal sorting processor following the receiver to identify the false signal as a real one. This operation wastes valuable resources and time in the processor.

In determining the threshold of false alarm, the FFT outputs must be measured without input signal. A large amount of data must be processed to estimate the noise spectrum outputs. Each FFT operation utilizes 256 digitized data sampled at 2.5 GHz. The maximum amplitude of the frequency components on each trial is stored. A total of 350,000 sets are processed in about 72 hours. Each 256 data points represent about 100 ns; thus, 350,000 records represent 35 ms of data. This process stores 350,000 maximum values. A threshold higher than the highest maximum value must be used to ensure that the receiver does not report a false alarm. Let us refer to this threshold as the preliminary threshold. This threshold is not used as an actual threshold in the encoder design but it will be used to determine the actual threshold. In this test, it only ensures that the receiver will not generate a false alarm during this 35-ms trial. This is a big problem with design by simulation because so much time is spent to obtain a few results. If the actual hardware receiver is used to monitor the false alarms, it takes only 35 ms to process 35 ms of data.

Once the preliminary threshold is selected, a single signal is used as input to test the receiver response. The desired result is to receive one output for one input signal. The test result, however, shows that the receiver may generate more than one output signal report for one input signal. This means that the preliminary threshold is too low. In order to reduce the spurious response, the actual threshold must be higher than the preliminary threshold. This level is referred to as the first threshold level. In determining this threshold, experimental data are used. The experimental data are generated with various input power levels. In general, strong input signals are used for this testing. This threshold is used in the encoder design.

This first threshold is chosen so that most of the time one input signal generates only one output signal report.

Once the first threshold is determined in the encoder design, two input signals are applied and tested. Because the front end is highly nonlinear, caused by the limiting amplifier and the 2-bit ADC, the two signals can interfere with each other as discussed in Section 12.6. As a result, sometimes the first threshold is not crossed, even if two strong input signals are applied to the input of the receiver. This condition is obviously not acceptable. To remedy this problem a second threshold is selected. This threshold is lower than the first one but higher than the preliminary. Figure 12.13 shows the two-threshold arrangement. Figure 12.13(a) shows the single signal condition. In this figure, one signal crosses the first threshold and a spur crosses the second threshold. If the second threshold is used alone, the receiver will report one false signal. Figure 12.13(b) shows the two-signal condition. In this figure, both signals are under the first threshold but cross the second one. If the first threshold is used alone, the receiver will miss both signals. Because the FFT outputs are complex, both the first and the second thresholds have real and imaginary parts. Figure 12.13 shows only the real portion of the threshold. The spectrum is the real part also. The imaginary part displays similar results.

The operation of the two-threshold arrangement is as follows. Initially, the first threshold is tested. If the FFT outputs cross this level, these outputs are kept and the second threshold will not be tested. If the first threshold is not crossed, the second threshold is tested. The FFT outputs crossing this threshold are kept. Limited experimental results show that most of the time four or less than four FFT outputs can cross the thresholds. When two input signals are exactly the same amplitude, a very small percent of time more than four outputs will cross the threshold. If this situation occurs, the first four outputs from the output order (not the largest four) are selected. Under this condition a real input signal might be missed.

The amplitudes of the FFT outputs crossing the thresholds are calculated from (12.3). These amplitudes will be compared with a threshold, which is referred to as the final threshold. If no signal is present at the input of the receiver, the first and second thresholds will not be crossed. If one or two FFT outputs cross the final threshold, the receiver will report the number and the frequency of the signals. If more than two FFT outputs cross the final threshold, only the two largest values are kept as the measured signals. This is the way to limit the receiver to process only two signals. Once the number of signals is determined, their corresponding frequencies can be obtained from the numerical FFT frequency bins.

12.10 PRELIMINARY PERFORMANCE OF A MONOBIT RECEIVER

In order to test a receiver, a computer must be connected to the outputs of the encoder. The results from the encoder are compared with the input signal to

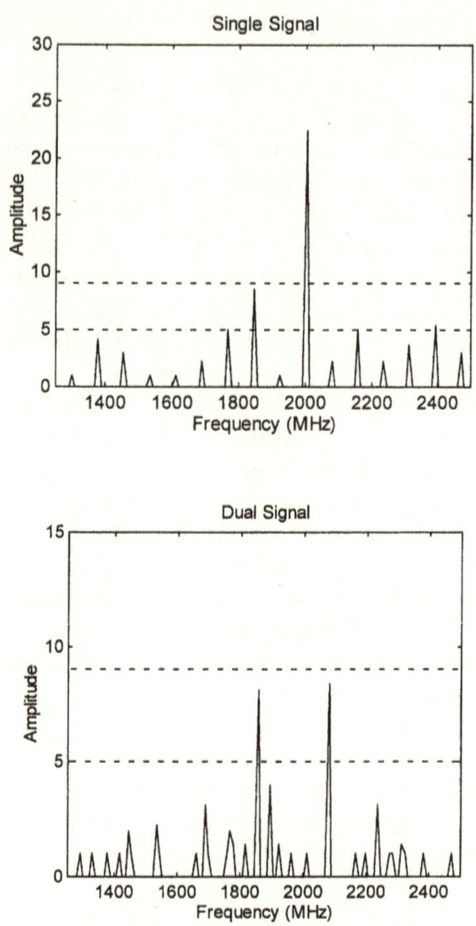

Figure 12.13 Two-threshold arrangement: (a) single input signal and (b) two input signals.

determine the errors measured. The outputs are usually measured in a statistical manner by repeating the same input many times, say 100 to 1,000 times. A percentage of erroneous reports can be obtained using this testing method. The monobit receiver outputs, however, cannot be read by a computer yet. Only a limited number of input conditions can be evaluated; therefore, the performance of this is referred to as preliminary performance data.

First the RF chain is not included in the test. The purpose of this test is to test the ADC, the FFT, and frequency encoder chip. The ADC can digitize very high frequency input signals such as 10 GHz. Because the ADC will down convert the input to the baseband as part of the ADC process, the FFT chip can process the signal. It is demonstrated that ADC and FFT chip can process input signals up to 10 GHz. This might be a very important factor in designing the wideband RF chain. The term "wideband RF chain" is used to distinguish from the RF chain shown in Figure 12.6. As mentioned in Section 2.4 for EW applications, the frequency of interest is from 2 to 18 GHz and the baseband receiver has only 1-GHz bandwidth. The conventional approach is to divide the 2–18 GHz frequency range into 1-GHz bands. The frequency of each band is shifted to the input of the baseband receiver through frequency conversion. In accomplishing this, mixers and local oscillators are needed. Because the ADC can take inputs from 2 to 10 GHz directly, filters and wideband amplifiers might be enough to build the wideband RF chain. This approach can eliminate the mixers and local oscillators.

The monobit receiver complete with the RF chain and the FFT chip has been tested. In this case the RF chain limits the input bandwidth to 1 GHz. The performance can be listed in Table 12.1.

The input frequency range is equal to the design goal. When two signals are separated by about 10 MHz, the receiver is able to measure them. The single signal frequency resolution is 9.77 MHz, which is obtained from (1,250/128). The sensitivity of the receiver is usually measured across the input frequency range of 1 GHz. For this receiver, however, the sensitivity is measured over a few frequency values and the approximate value is about –70 dBm. There is no probability of false alarm nor probability of detection associated with this sensitivity. The single signal dynamic range is 80 dB because the receiver can process a strong input signal at 10 dBm. The two-signal spur-free (or third-order intermodulation) dynamic range is rather high compared with a conventional receiver. The reason is that the receiver can process only two simultaneous signals; thus, it cannot detect the third signal, which determines the lower limit of the spur-free dynamic range.

The instantaneous dynamic range is about 5 dB, which usually means that when two signals are within 5 dB in amplitude, the receiver can measure both of

Table 12.1
Preliminary Performance of Monobit Receiver

Input frequency (GHz)	1.375 ~ 2.375
Two-frequency resolution (MHz)	10
Single-frequency resolution (MHz)	10
Sensitivity (dBm)	–70
Dynamic range (dB):	
Single signal	75
Two-signal spur-free	70
Instantaneous	5
Minimum pulse (ns)	200
Time of arrival (ns)	102.4
Number of signal capability	2

them. This definition does not apply to the monobit receiver. When two signals are of the same amplitude, the receiver does not report them all the time. Only about 24% of the time the receiver reports both signals. About 76% of the time the receiver only reports one signal. When two signals are present at the input of the receiver, it does not report erroneous frequency as does an instantaneous frequency measurement (IFM) receiver. The receiver either reports one frequency correctly or reports both frequencies correctly. When two input signals are separated by more than 5 dB, the receiver will report only the frequency of the strong one. The above definition is referred to as the instantaneous dynamic range, which is different from conventional definition. In the conventional definition, the receiver must measure both signals, which is more stringent than the definition used here. When three signals of equal amplitude are presented at the input of the receiver, the receiver often reports one or two frequencies correctly. If the third signal is weak, it usually will not affect the frequency measurement.

Very limited data are collected from the above test. Some of the results are obtained from the design stage, and the input data are collected from an 8-bit ADC and converted to 2 bits through a software program. Two input frequencies are randomly selected and their amplitude difference is kept constant. For each power difference, 1,000 sets of data are collected on the signals with random frequencies. These data are listed in Table 12.2. The purpose is to provide a rough idea of the receiver performance.

The receiver occasionally misses pulses and sometimes generates erroneous signals. An erroneous signal is defined as a signal whose measured frequency is off by 6 MHz, which is slightly larger than half the frequency resolution bin of 9.77 MHz. A rigorous test on a hardware monobit is needed to generate quantitative results.

The minimum pulse width is about twice the FFT frame time (204.8 ns) because this pulse width will guarantee to fill one FFT frame (102.4 ns). Test results indicate that a 100-ns pulse can be detected by the receiver but not 100% of the

Table 12.2
Performance on Two Simultaneous Signals

Amplitude of 2nd Signal vs. 1st (dB)	*Found 1st Signal (%)*	*Found 2nd Signal (%)*	*Found Both Signals (%)*	*Found Neither Signal (%)*	*Found Erroneous Signals (%)*
0	65.1	59.3	24.4	0	0
−1	78.9	45.0	23.9	0	0.45
−2	89.2	29.9	20.9	0	0.38
−3	93.9	18.0	12.0	0.13	0.38
−4	97.9	9.5	7.6	0.13	0.25
−5	99.8	3.3	3.0	0	0.13

time. A 150-ns pulse can be detected most of the time. Because only limited trials can be accomplished without an automatic testing setup, the minimum pulse capability is given as 200 ns. The time resolution is 102.4 ns.

The two major deficiencies of this receiver are the limited instantaneous dynamic range and the two-signal capability. It is desirable to have higher instantaneous dynamic range and the ability to process more than two simultaneous signals.

12.11 POSSIBLE IMPROVEMENTS

It is desirable to eliminate the two deficiencies of the receiver. Possible approaches are to increase the bit number of the ADC or change the Kernel function of the receiver. Tests have indicated that increasing the input number of bits with 1 bit Kernel function of ± 1, $\pm j$, results in little change to the FFT outputs. Changing the Kernel function should improve the performance. Increasing the number of bits in the Kernel function, however, requires multiplication, which defeats the concept of minimal hardware and processing used in the monobit receiver.

One possible approach is to increase the number of bits in the Kernel function and still keep the FFT operation limited to additions. Increasing the Kernel function from four points to eight points will increase four points at $(1 + j)/\sqrt{2}$, $(1 - j)/\sqrt{2}$, $(-1 - j)/\sqrt{2}$, and $(-1 + j)/\sqrt{2}$. These four points are on the unit circle. Because the amplitude has the factor $1/\sqrt{2}$, multiplication is needed to produce the FFT outputs. If these four points are moved toward the corner of a square with unity sides, the additional four points become $1 + j$, $1 - j$, $-1 - j$, and $-1 + j$. This operation is illustrated in Figure 12.14. Because the factor $1/\sqrt{2}$ is eliminated, multiplication is no longer needed in producing the FFT outputs.

Simulated data are used to test this concept. These eight points are obtained from Figure 12.15. The values of the 256 points of the Kernel function can be divided into eight regions. There are 32 points in one region. All the values in one region can be represented by one point. After this modification the input data are increased to 3 bits. Limited trials are tested and the results indicate that this arrangement can improve the dynamic range slightly.

From observing the FFT outputs, it appears that the peaks of the spurious signals are slightly lower (about 3 dB) than that of the monobit receiver. Because of the sidelobe limit of the rectangular window and the low number of bits of the ADC, the dynamic range cannot be expected to improve drastically. As mentioned in Section 6.6, every bit can provide about 6 dB in dynamic range. It also appears that three signals of the same amplitude can produce three peaks at the correct frequency bins. This phenomenon indicates that the modified kernel function still can process three signals.

If this idea is implemented in a receiver design, the FFT will be more complicated because of the additional terms in the Kernel function. The frequency encoder might also be more complicated if the receiver is designed to process three signals because the final results could be 0, 1, 2, and 3. However, one can also design a

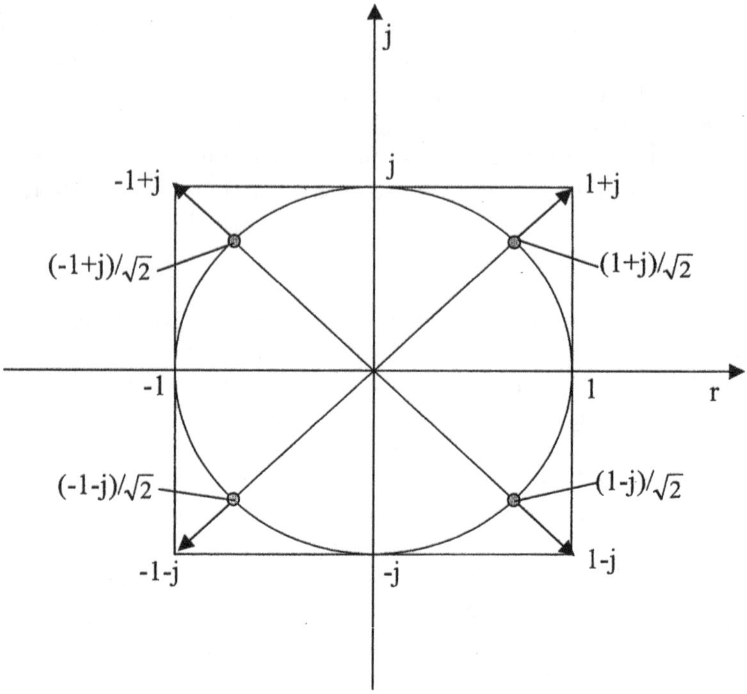

Figure 12.14 Moving the Kernel function values.

receiver to process two signals with this idea to improve instantaneous dynamic range. Because of the potentially limited performance improvement and the complication of the chip design this idea has not been investigated thoroughly.

12.12 CHIP LAYOUT [2, 3]

This last section provides some information on the chip layout. Although it is possible to build the chip on a field programmable gate array (FPGA), the first chip is built with the application-specific integrated circuit (ASIC) technology. The ASIC uses double metal 0.5 μm scalable CMOS technology and builds in an 84-pin package. It operates at 156.25 MHz (2.5 GHz/16) because the demultiplexer is 1 to 16. The chip contains 812,931 transistors and has a die size of approximately 15 × 15 mm. The chip is broken down into five subsystems as discussed in previous sections. They are the input stage, FFT block, initial sorting, squaring and addition, and final sorting. The process in each subsystem must be completed within 102.4 ns (0.4 × 256), which is the time of accumulating 256 data points. The timing of the subsystems is simulated and the results are listed in Table 12.3.

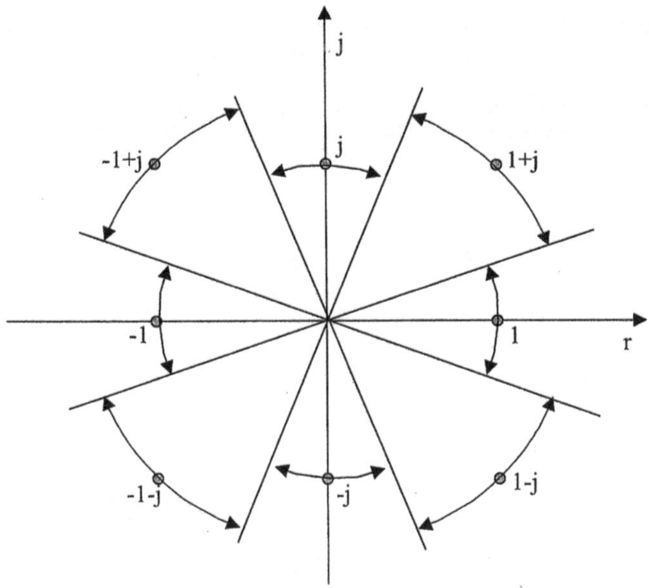

Figure 12.15 Digitizing the Kernel function values.

Table 12.3
Timing Analysis of Each Subsystem

Subsystem	*Critical Path (ns)*
Input stage	99.5
FFT block	48.02
Initial sorting	90.11
Squaring and addition	28.95
Final sorting	34.42

The timing analysis includes the delay of each pipelined flip-flop.

Although the input stage shows the longest time, it is not the time of concern because this is the time for the chip to take all the input data. It takes about 102.4 ns to collect the 256 points of data. The longest time required is in the initial sorting. After the chip is fabricated, it performs satisfactorily. From this brief discussion it can be shown that the chip is rather simple.

REFERENCES

[1] Tsui, J., Schamus, J., Kaneshiro, D., "Monobit Receiver," presented at IEEE MTT International Conference, Denver, Colorado, June 9–13, 1997.

[2] Pok, D., Chen, H., Schamus, J., Motgomery, C., Tsui, J., "ASIC Design for Monobit Receiver," presented at 10th Annual IEEE International ASIC Conference and Exhibit. Portland, Oregon, September 7–10, 1977.

[3] Pok, D., Chen, H., Schamus, J., Tsui, J., Motgomery, C., "Chip Design for Monobit Receiver," *IEEE Trans. on Microwave Theory and Techniques*, December 1997, pp. 2283–2295.

CHAPTER 13

Processing Methods After Frequency Channelization

13.1 INTRODUCTION

Chapters 11 and 12 described how the input signals are separated in the frequency domain through channelization. This chapter describes how, after the frequency channelization, further processing is performed to determine the number of frequencies and the frequencies themselves. The frequency data resolution (or frequency bin width) calculated through an FFT operation determines the frequency precision measured on the input signal. It is often desirable to obtain better frequency precision than the FFT operation can provide. This is especially true for the 32-point FFT discussed in Chapter 11, which produces a frequency data resolution of 93.75 MHz with a 3-GHz sampling frequency. Another problem with such a wide frequency bin is that when two signals fall into one frequency output channel, the receiver cannot effectively separate them. In other words the two-signal frequency resolution is only about 93.75 MHz.

The traditional way to separate signals by frequency is through an analog filter bank. Although the main emphasis of this book is digital, one still can consider the analog filtering approach. In an analog channelized receiver, after the filter bank and amplifier, crystal video detectors are used to convert the radio frequency (RF) into video signals. The video signals are digitized by ADCs and are further processed to determine the number of signals and their frequencies. In converting RF into video signal, some information is lost. If two signals fall into one channel, it is difficult to separate them.

In a digital receiver ADCs can be used after an analog filter bank to obtain digitized RF information. This information can be further processed; for example, the monobit receiver idea can be used to separate two signals in one channel.

A discussion of finer frequency estimation and the capability of processing two signals close in frequency are the two main goals in this chapter. To solve these two problems; the concepts of the instantaneous frequency measurement (IFM) receiver and the monobit receiver will be used after channelization. The IFM receiver concept is discussed in Chapter 10 as the phase measurement method. Both the analog filter bank and digital channelization will be considered. For digital channelization the output sampling rate discussed in Sections 11.5, 11.12, and 11.13 becomes a very important issue, and this subject will be further discussed in this chapter.

13.2 BASIC CONSIDERATIONS OF CHANNELIZED APPROACH

A receiver is often designed to match the minimum pulse width the receiver is expected to process. If the minimum desirable pulse width is 100 ns, the general rule is to select a filter bandwidth of 10 MHz (1/100 ns), which can be referred to as the minimum required filter bandwidth. A narrower filter will degrade the signal-to-noise ratio and disturb the pulse width measurement because the transient effect of the filter may last longer than the minimum pulse width. Under this condition, a pulse having the minimum pulse width is actually extended by the transient effect of the filter. Pulse width, however, is not a very reliable parameter in electronic warfare (EW) receivers because of the multipath problem. Multipath means that a signal reaches the receiver from many different paths. The direct path is the signal reaching the receiver directly. The signal can also reach the receiver by reflecting from some objects. The direct signal and the reflected signals may interfere and change the pulse width.

Another more serious problem with the transient effect is that during the transient period the output frequency will shift toward the center of the filter. The transient effect is the filter response to a step function, which causes a damped oscillation at the center frequency of the filter. If the transient period is longer than the pulse width, the output from the filter will not contain the input frequency, especially when the signal is near the edge of the filter. Once this phenomenon occurs, the frequency encoding circuit following the filter may generate erroneous frequency information.

A short pulse passing through a filter bank can generate outputs in many adjacent filters because the leading and trailing edges appear as a step function to every filter. It is not only difficult to determine the center frequency of the signal, but it also difficult to determine the number of input signals. From previous receiver design experience, the filter bandwidth is usually chosen to be much wider than the value obtained from the reciprocal of the minimum pulse width. Sometimes the selected bandwidth may be five or more times wider than this value.

There is a significant difference between designing a single-channel narrowband receiver and a wideband receiver with narrowband channels. In the narrowband receiver the signal can be tuned to the center of the filter by changing the frequency of the local oscillator. Once the signal is shifted to the center of the filter, the transient effect will be minimized. In a wideband channelized receiver the filters are fixed in frequency, as are the frequencies of the local oscillators. A signal can fall in the center of a filter as well as in between two channels. When there is more than one signal in one channel, the problem becomes more complicated. Figure 13.1 shows such a problem. In this figure three adjacent filters A, B, and C are shown with two input signals. Both signals are at the edge of filter B. Because of the finite slope of the filters, these two signals will be processed by all three channels. Theoretically, one would like to determine the frequency of signal 1 by channels A and B and signal 2 by channels B and C. Determining the number of signals under this condition, however, can be difficult. That is why in many receiver designs, the minimum separation in frequency is specified wider than the bandwidth of the filters in order to avoid the need to separate signals in the same channel. It is highly desirable to design a receiver that can measure two signals falling into one channel especially for wideband channels.

13.3 FILTER SHAPE SELECTION

Selection of the filter shape is one of the major tasks in designing a channelized receiver. In Section 11.13, filters are selected by the ripple factors in the pass band as well as the stop band. There are, however, more criteria involved in selecting a filter shape. First let us consider the bandwidth of the filters. Figure 13.2 shows the shapes of three different filters. In these figures two bandwidths are shown. One is the 3-dB bandwidth, which is equal to the total bandwidth of the receiver divided by the number of channels, and the other one is the 60-dB bandwidth. The name of the 60-dB bandwidth is arbitrarily chosen because using a numerical value is easier to refer to in later discussion. Using this definition, the maximum instantaneous dynamic of the receiver is limited to 60 dB, which is a very large value for a wideband receiver. In actual receiver design this number depends on the filter selected.

In Figure 13.2(a) the 60-dB bandwidth is double the 3-dB bandwidth. With this arrangement a signal will fall into two filters most of the time except right at the center of a channel where it falls into only one filter. This can be referred to as the maximum 60-dB bandwidth allowed. If the skirt of the filter is wider than this value, as shown in Figure 13.2(b), one signal can fall into three channels most of the time. This is highly undesirable because one needs to compare the outputs of three channels to determine a signal.

It is desirable to keep the probability of one signal falling into two adjacent channels small, as shown in Figure 13.2(c). If a signal falls into one channel, the logic circuit of that channel will process the signal. If a signal falls into two adjacent channels, both circuits of the adjacent channels will process the signal. It is easier to measure one signal with one encoder circuit. If two channels measure one signal, the results must be compared to determine whether there are one or two signals.

This last requirement in Figure 13.2(c) needs filters with a very sharp skirt. A sharp filter will have a relatively long transient time when the leading and trailing

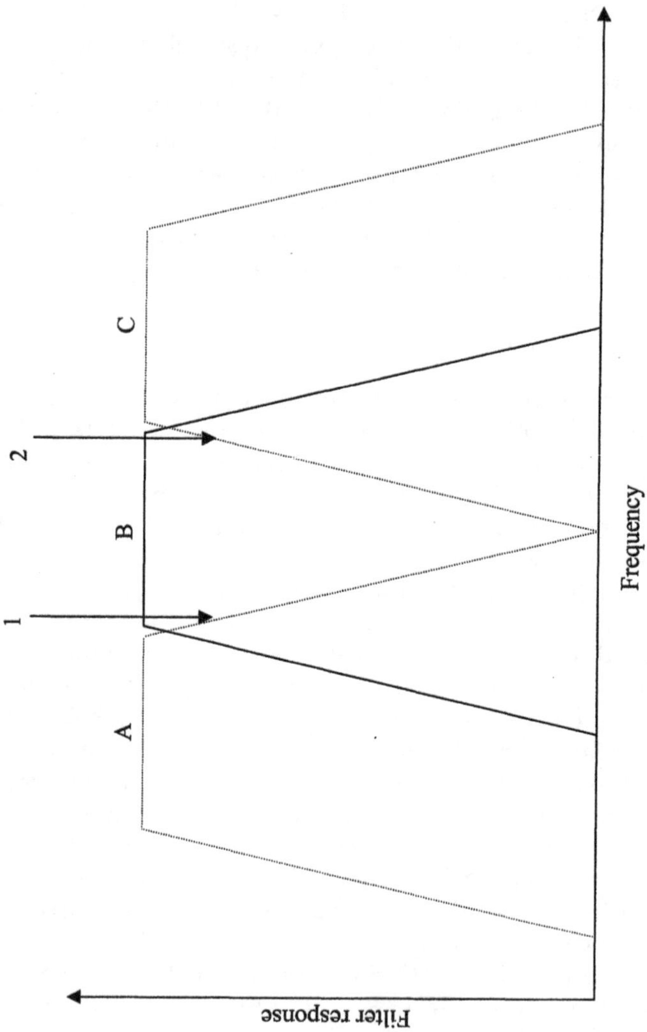

Figure 13.1 Two signal conditions of a filter bank.

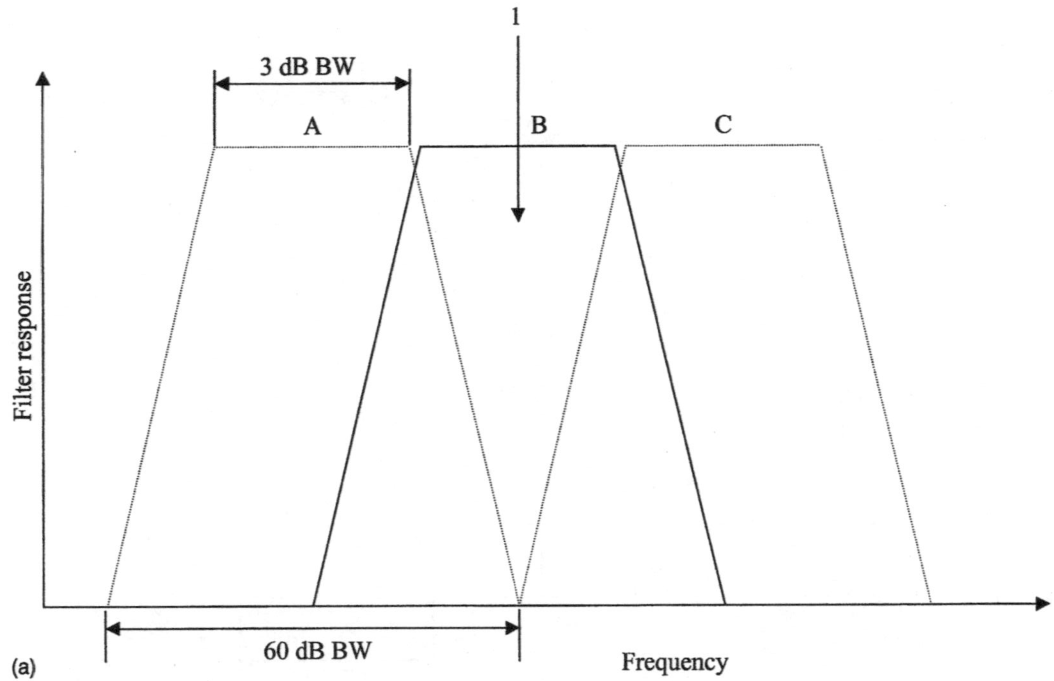

Figure 13.2 Filter shapes: (a) filters with limiting skirt, (b) filter with wider skirt, and (c) filters with narrower skirt.

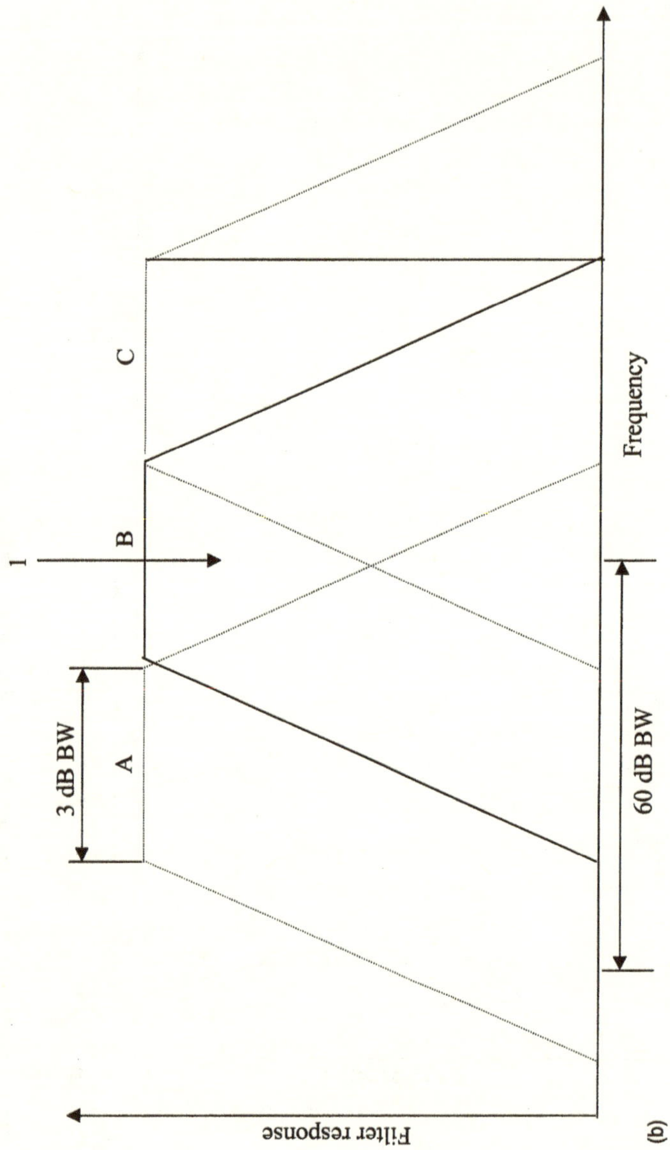

Figure 13.2 (continued).

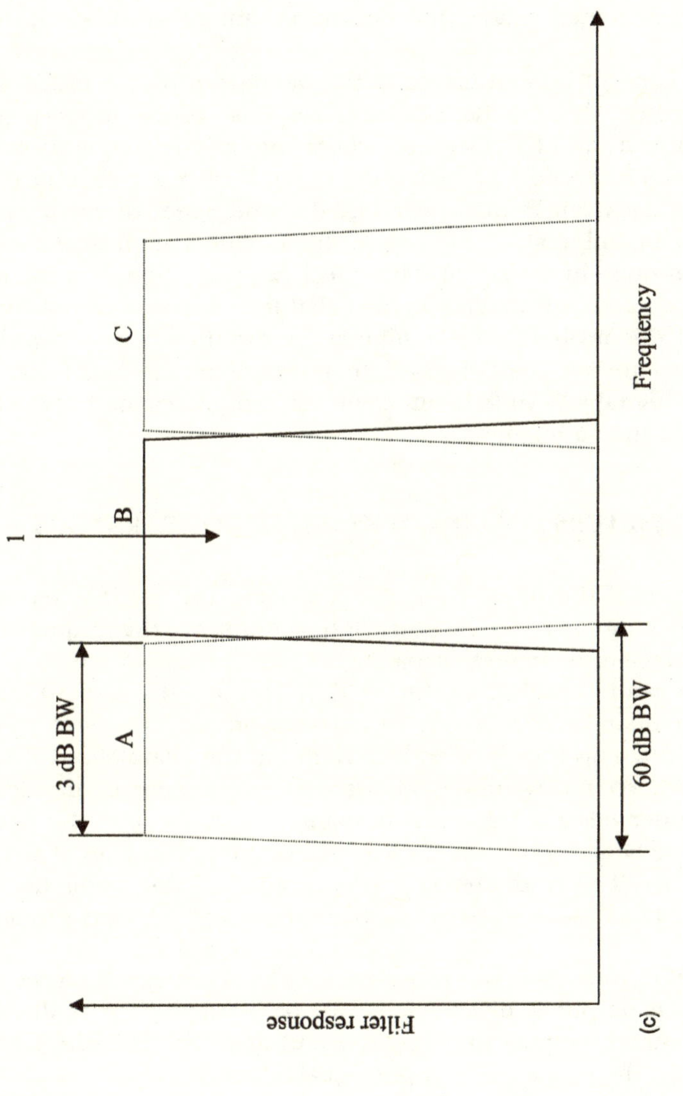

Figure 13.2 (continued).

edges of the signal pass through it. During the transient period both the amplitude and the frequency of the output signal change as mentioned before. If the duration of the transient time is comparable to the minimum pulse width, the steady state of the pulse might be too short to provide an estimation of the signal frequency. The long transient time eliminates the selection of very sharp filters. The transient effect must be evaluated against the desired minimum pulse width in selecting filters.

Another issue to be considered is the bandwidth of the processing circuit following the filter. Usually, the frequency response of the processing circuit is repetitive such as in an FFT operation where the unique frequency is from 0 to $f_s/2$. As discussed in Section 4.5, when the signal is near the edge of the band, it can be assigned to a wrong frequency. If the 3-dB band is used for further processing, there is usually an ambiguity at the edge of the circuit. The problem is because of the presence of noise in the signal. This effect can put a signal on the wrong side of the filter and cause a frequency error equal to the bandwidth of the filter. In order to avoid this problem, the minimum bandwidth of the processing circuit should be equal to the 60-dB bandwidth rather than the 3-dB one. With this bandwidth, if a signal falls in between channels, the processing bandwidth is wide enough to avoid the ambiguity.

13.4 ANALOG FILTERS FOLLOWED BY PHASE COMPARATORS

The performance of the phase comparison scheme (or the IFM receiver in the analog scheme) is discussed in Chapter 10. It can improve the frequency measurement accuracy. It cannot, however, separate two signals easily, although theoretically the two signals can be resolved as discussed in Sections 10.6 and 10.7. It appears that an analog filter bank followed by narrowband IFM receivers can produce satisfactory results. This idea is often referred to as the channelized IFM receiver, which has not been successfully demonstrated yet because of problems in the frequency encoder design. The main problem is to determine the frequency of one signal falling between two channels as well as the frequencies of two signals in adjacent channels. This is an analog approach and will not be discussed further. An analog filter bank followed by narrowband phase comparators should produce similar results.

Figure 13.3 shows such an arrangement. The input signal passes through a filter bank. Each output is divided into two paths through a 90-degree hybrid because the complex outputs with in-phase and quadrant (I and Q) outputs are needed to obtain the phase of the input signal.

Let us use an example to illustrate the phase comparison configuration. Assume that the input bandwidth is 1 GHz and divided into eight parallel channels. Each channel has a 3-dB bandwidth of 125 MHz (1,000/8). Let us also assume that the 60-dB bandwidth is 250 MHz, which is the maximum allowed bandwidth as

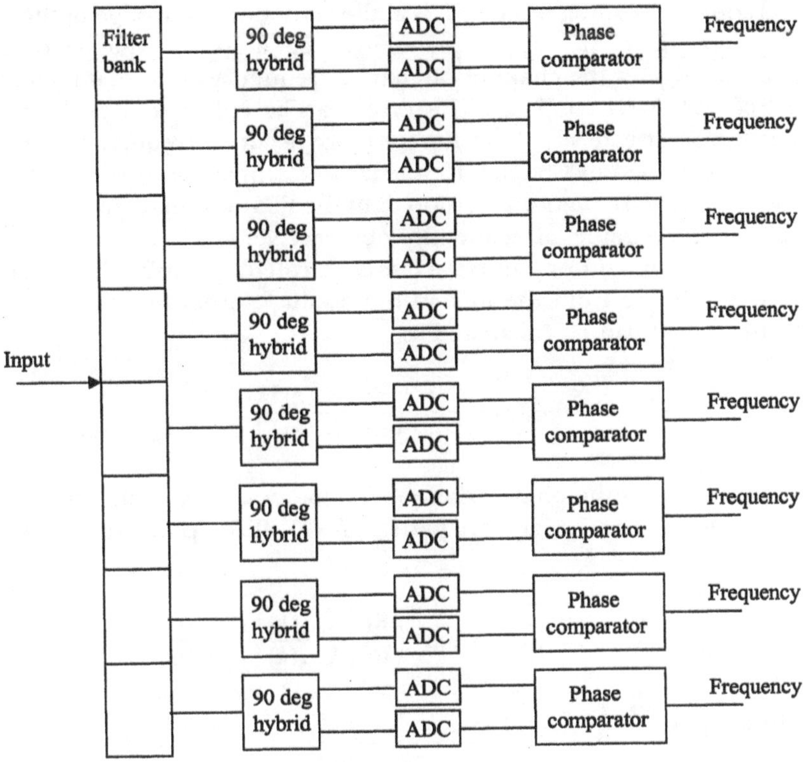

Figure 13.3 A filter bank followed by phase comparators.

discussed in the previous section. Because the inputs to the phase comparator are complex (I and Q channels), the sampling frequency can be 250 MHz to cover the 250-MHz bandwidth. This approach theoretically can process two input signals per channel, as mentioned in Section 10.7. If the I and Q channels are not perfectly balanced the measured response of a single signal can appear as two simultaneous signals. This is one of the fundamental limitations of the phase measurement method.

The concept of using phase comparators after analog filters has not been attempted experimentally, perhaps because of the hardware complexity of the implementation.

13.5 MONOBIT RECEIVER FOLLOWED BY PHASE COMPARATORS

The wide monobit receiver discussed in Chapter 12 has a relatively narrow output band, which is about 9.77 MHz (2,500/256). A signal in a certain channel will have the same frequency reading whether it is at the center or at the edge of the channel. Thus, a signal at the center of the channel can get a good frequency reading but a signal on the edge of the channel can have a frequency error that is close to half a channel of 4.89 MHz (9.77/2). This error can be considered as the frequency digitization error. It is desirable to have a more accurate frequency reading than this value. As discussed in Chapter 10, the phase comparison method can provide a much better single frequency precision than the FFT outputs. Thus, it is desirable to have phase comparators after the monobit receiver.

The phase comparator can be achieved through the following approach. If the highest amplitude from the FFT with a 4-value Kernel function is $|X(k)|$, the phase of the output can be calculated as

$$\theta_n(k) = \tan^{-1}\left(\frac{\mathrm{Re}[X_n(k)]}{\mathrm{Im}[X_n(k)]}\right) \tag{13.1}$$

where $\theta_n(k)$ represents the phase of frequency component k at time n, and Re and Im represent the real and imaginary parts of $X_n(k)$. If the phase measured at time $n + 1$ is

$$\theta_{n+1}(k) = \tan^{-1}\left(\frac{\mathrm{Re}[X_{n+1}(k)]}{\mathrm{Im}[X_{n+1}(k)]}\right) \tag{13.2}$$

the fine frequency can be found as

$$f = \frac{\theta_{n+1}(k) - \theta_n(k)}{2\pi t_o} \tag{13.3}$$

where t_o is the output sampling time. The single frequency precision improvement is discussed in Section 10.4.

The filter bandwidth of 9.77 MHz can be considered as the 3-dB bandwidth, which is also the frequency data resolution of the monobit receiver. Two signals separated by less than 9.77 MHz in frequency are difficult to separate by the phase comparators. Because this bandwidth is relatively narrow, the probability of this situation occurring is low and will not be considered here.

The outputs from the monobit receiver are complex, which is equivalent to the I and Q channels; therefore, the phase can be obtained directly from these outputs. The output sampling time is 102.4 ns, as discussed in Section 12.4, and the corresponding sampling rate is 9.77 MHz (1/102.4 ns). Because the outputs are complex, the equivalent bandwidth is also 9.77 MHz, which is the same as for the 3-dB bandwidth. Using (13.1), (13.2), and (13.3) to calculate fine frequency, the phase calculated will have 2π ambiguity. The 2π-phase range covers the frequency bandwidth of 9.77 MHz. If a signal falls between two adjacent channels, it also falls on the edge of the phase comparator. The noise in the receiver may determine that the signal is 9.77 MHz away because of the 2π ambiguity. This effect may cause catastrophic error in the frequency reading.

In order to eliminate the catastrophic error, two approaches can be taken. The first approach is to compare the amplitudes of two neighboring channels of the output channel. Figure 13.4 shows such an arrangement. In this figure, one signal is between two channels B and C. Let us assume that the output from channel B is higher than that from channel C. Therefore, the output from channel B will be used to find the fine frequency. Because of the noise in the input data, it is possible to put the fine frequency at position 2, which is the wrong frequency. In this approach, the amplitudes of channels A and C (the neighboring channels of B) are compared. Because the signal is between B and C, the output from channel C should be higher than from channel A. This condition can be used to determine that the frequency of the input signal should be between B and C rather than A and B.

The phase comparators are used after the monobit receiver to improve the frequency accuracy of the input signal. Although the monobit receiver only has 2 bits as input, the FFT outputs have 8 bits (including one sign bit). Thus, the phase value can be calculated rather accurately. A computer is used to calculate the results of (13.1) through (13.3). If the phase calculated is near the edge of a phase comparator, the amplitudes of the two neighboring channels are compared to determine the direction of the frequency shift. Frequency error of 0.5 MHz can be achieved without catastrophic error. Thus, this method improves the single frequency precision by about 10 times. In order to use this method, two consecutive outputs from the monobit receiver must be used; therefore, the minimum pulse width must be more than 204.8 ns.

The second approach to eliminate the 2π ambiguity is to increase the output sampling rate. One can double the output sampling rate of the monobit receiver to 19.53 MHz, which corresponds to a sampling time of 51.2 ns. With this arrangement, a signal falling between two adjacent channels will not fall on the edge of the phase comparator. Thus, the 2π ambiguity problem at the edge of a phase comparator will not occur. This implies that the monobit receiver output rate will be increased to 19.53 MHz, which will require doubling the monobit receiver clock rate. This approach will increase the complexity of the design significantly. One advantage of this approach is that the minimum pulse width is still at 102.4 ns because this method uses two cycles of 51.2-ns outputs.

13.6 DIGITAL FILTERS FOLLOWED BY PHASE COMPARATORS [1]

The digital filter obtained from the FFT operation as discussed in Section 11.15 is similar to the wideband monobit receiver. For example, the outputs are also complex. The major difference is the bandwidth of the filter. As discussed in Chapter 11, the sampling rate is 3 GHz and a 32-point FFT is performed with a window

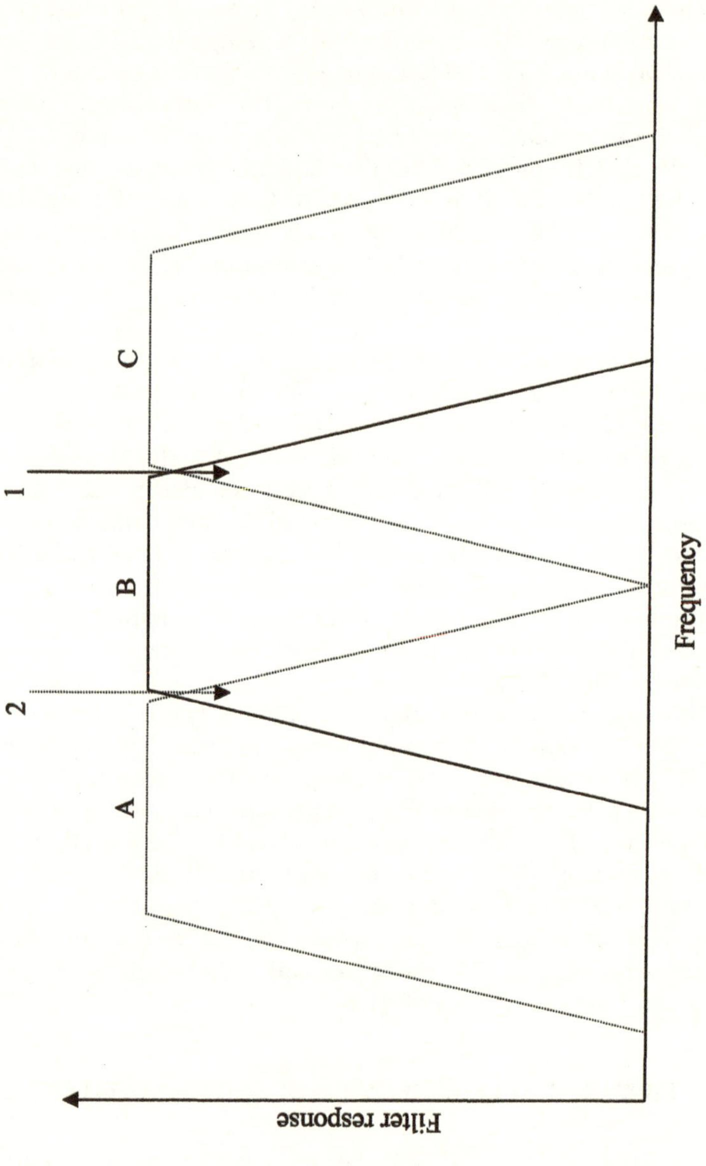

Figure 13.4 One signal falls between two channels.

function; the resulting filter bandwidth is about 93.75 MHz (3,000/32). With this wideband filter the frequency data resolution is 93.75 MHz, which is too coarse for signal sorting applications. The minimum frequency precision allowed is about 10 MHz. A better frequency precision is always desirable as long as the minimum pulse width stays the same. Therefore, it is essential to improve this frequency precision. The approach is identical to the method discussed in the previous section, which can improve the frequency precision many times. In these wideband filters, simultaneous signals should also be considered because the probability of two signals falling into one channel is high.

Let us use the digital filter bank discussed in Section 11.13 as an example. The filter bank shape is shown in Figure 11.15. If there are two signals and both are in channel 5, signal 1 is close to channel 4, and signal 2 is close to channel 6. The signals are separated by 1/4 of the channel width. Signal 1 is 20 dB stronger than signal 2. In this case the outputs from all three channels 4, 5, and 6 are used to calculate the phase. The corresponding frequencies from the three channels are shown in Figure 13.5. Channel 4 shows the frequency of signal 1, and channel 6 shows the frequency of signal 2. Channel 5 has both signals and the output has ripple on it. Because signal 1 is stronger than signal 2, the average frequency

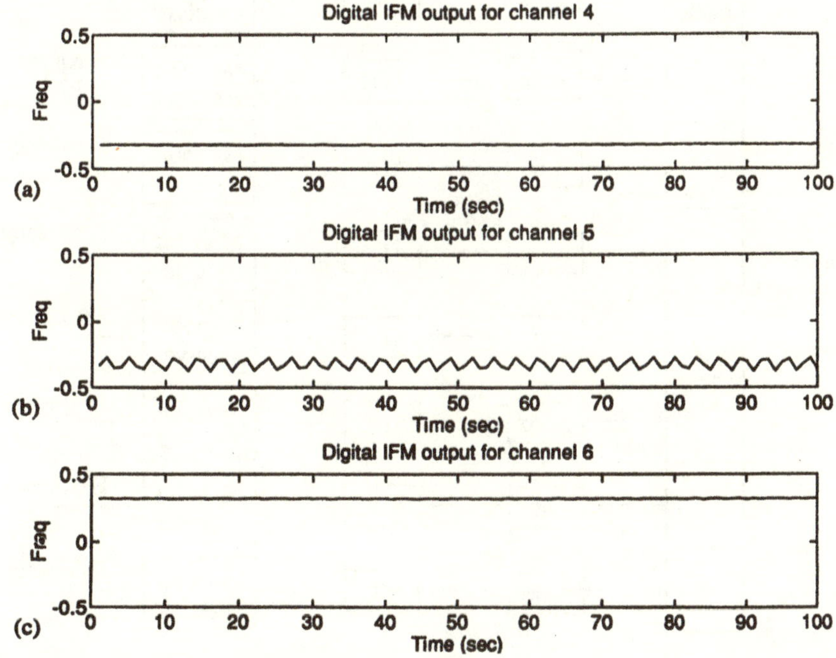

Figure 13.5 Frequency outputs from channels (a) 4, (b) 5, and (c) 6.

measured in this channel is the frequency of signal 1 as discussed in Section 10.7. Comparing frequencies in channels 4 and 5, one can see this effect. The ripple indicates that two simultaneous signals are present in the channel. The frequency of the ripple represents the difference frequency between the two signals. In this design, channels 4 and 6 can measure the two frequencies, making it unnecessary to obtain the frequency from channel 5.

13.7 ANALOG FILTERS FOLLOWED BY MONOBIT RECEIVERS [2]

There are two major deficiencies in the wideband monobit receiver discussed in Chapter 12. The first one is the low instantaneous dynamic range, which indicates that the receiver cannot process two signals separated more than 5 dB in amplitude. The receiver usually misses the weaker signal when two simultaneous signals are present. The second deficiency is that the receiver can process only two simultaneous signals in order to keep the design simple.

In order to remedy these deficiencies, a channelized approach can be used, as shown in Figure 13.6. The channelization can be accomplished using analog

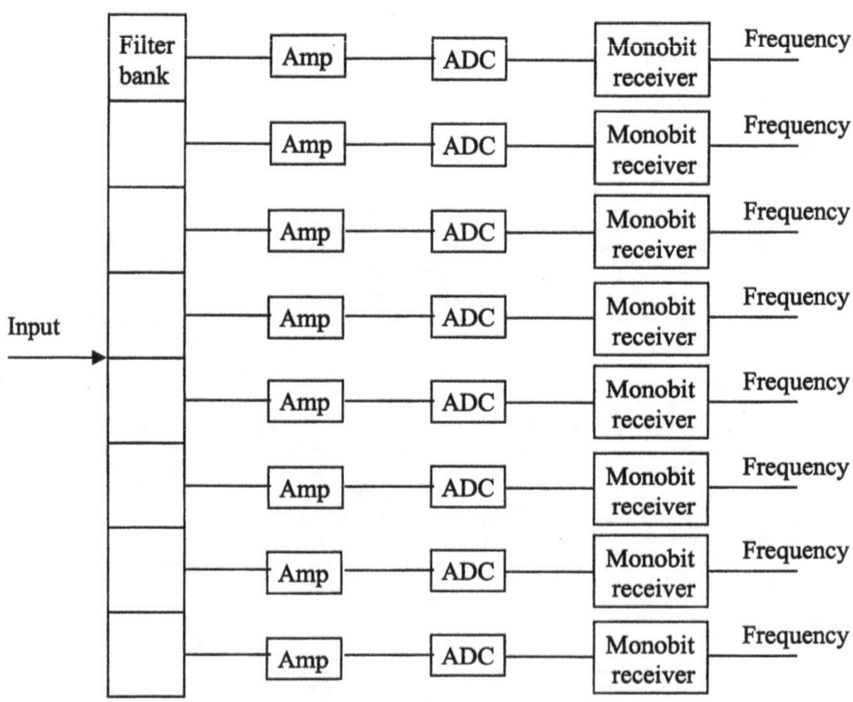

Figure 13.6 Analog filter bank followed by monobit receivers.

filters, which will separate the signals into different channels according to the input signal frequency. In each channel an amplifier can be used to increase the input signal level. These amplifiers can be limiting amplifiers as discussed in Section 12.6. Narrowband monobit receivers can be used at the output of the amplifiers. Each narrowband monobit receiver can process two simultaneous signals. Theoretically, this approach can process 16 simultaneous signals. In each channel the instantaneous dynamic range is limited to about 5 dB. Because the filters limit out-of-band signals, the instantaneous dynamic range of the receiver depends on the analog filter shape.

Because the narrowband filters limit the input band to the ADC, each ADC only covers a portion of the receiver input band. Using the example in Section 13.4 and Figure 13.6, the 3-dB bandwidth is 125 MHz (1,000/8). The 60-dB bandwidth is about 250 MHz. In order to eliminate ambiguity between channels, each channel should cover 250 MHz. The Nyquist sampling criterion requires the ADC operating at two times the bandwidth for real signals; thus, the ADC should operate at 500 MHz (2 × 250), which is much slower than the 2.5-GHz sampling rate used in the wideband monobit receiver. Operation at this sampling rate can cause a band overlapping problem, as shown in Figure 13.7. In this figure let us assume that the 60-dB bandwidth matches the second alias region. The 3-dB bandwidth of channels A, B, and C are also shown. Channels A and C are at the centers of alias regions $f_s/2$ to f_s and f_s to $3f_s/2$, respectively. Under this condition the output bandwidth equals the input bandwidth and there is no ambiguity problem. The center of channels B and D are at f_s and $3f_s/2$, respectively.

Figure 13.8 shows the bandwidth folding of channels B and D. It shows that the output bandwidth of channels B and D is only half of the input bandwidth for both the 3-dB and 60-dB cases. Therefore, a signal on either side of f_s in channel B can be aliased to the same output frequency and cause ambiguity in the frequency reading. The same situation happens when an input signal is on either side of $3f_s/2$ in channel D.

One can see from Figure 13.7 that channels A, C, E, and G do not have this problem, although channel G is not shown. The other four channels B, D, F, and H have the band overlapping problem, although channels F and H are not shown. One way to eliminate this problem is to redesign the front end of the receiver. The channelized monobit can only accept input in the A, C, E, and G bands. The input frequency range of an EW receiver system is basically from 2 to 18 GHz; the EW receiver discussed in this book is a baseband receiver, which covers a bandwidth of about 1 GHz as discussed in Section 2.5. The input signal must be converted to the input of the baseband receiver. One way to convert the input is shown in Figure 13.9. The input signal is channelized into eight consecutive frequency bands. The outputs from the odd channels 1, 3, 5, and 7 are converted to the frequency bands A, C, E, and G. The outputs from the even channels 2, 4, 6, and 8 are also converted to the frequency bands A, C, E, and G. With this arrangement there is no band overlapping.

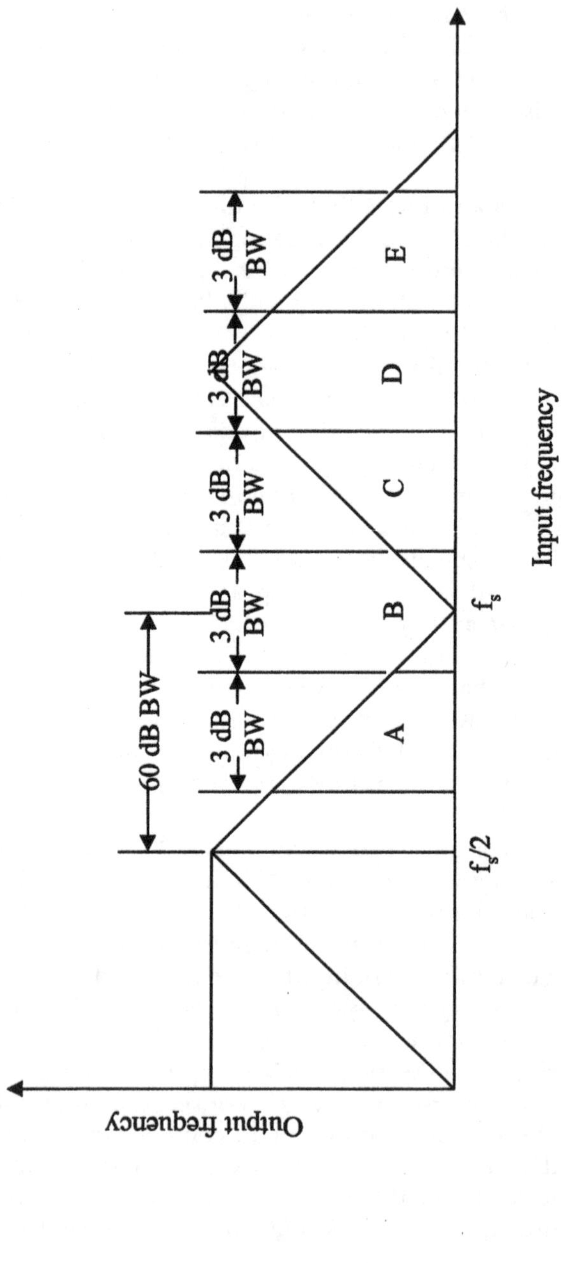

Figure 13.7 Adjacent channels aliased into baseband.

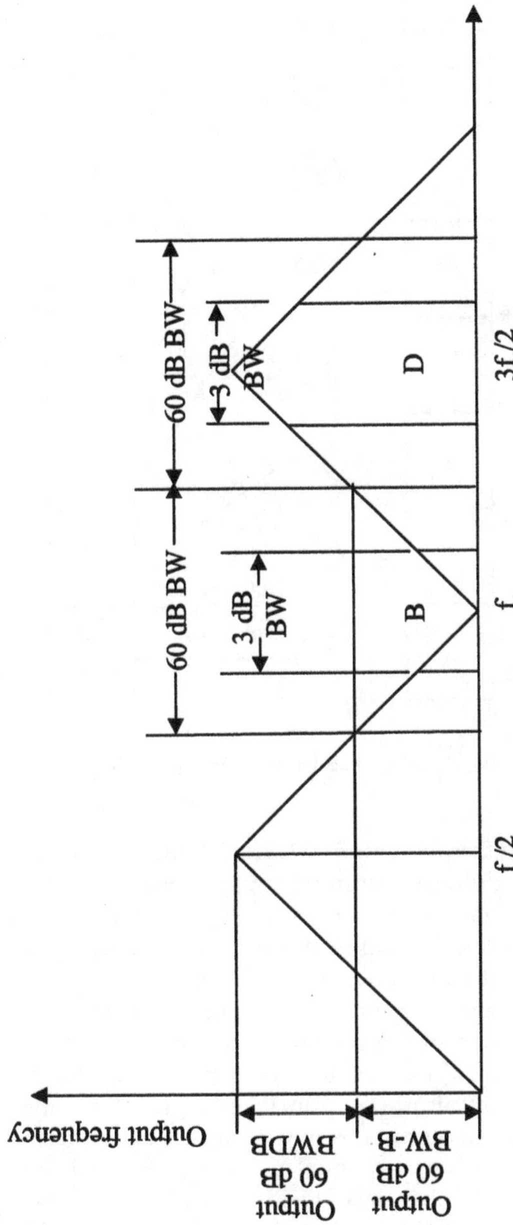

Figure 13.8 Band overlapping caused by aliasing.

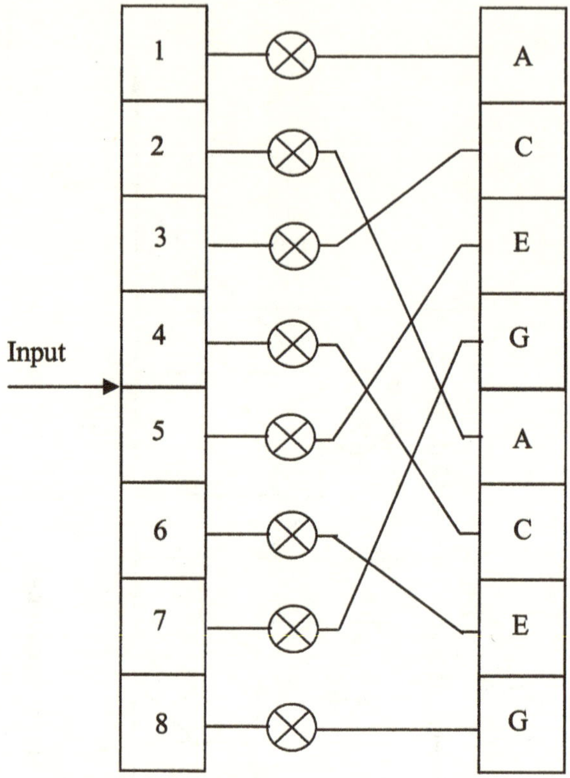

Figure 13.9 Front-end arrangement of analog channelized monobit receiver.

Because the sampling frequency is 500 MHz, 64 samples will last 128 ns, which is slightly longer than the desired minimum pulse width of 100 ns. In this design the FFT is 64 points long and the input data rate is 500 MHz. If eight narrowband monobit receiver chips are used to build the receiver, each chip is rather simple. It is possible to put all eight monobit receivers on one chip. Simulated results using the front end in Figure 13.9 indicate that the receiver can process several simultaneous signals with high instantaneous dynamic range.

One possible approach [2] to enhance the outputs from the monobit receiver outputs is to add together amplitude outputs from adjacent channels, as shown in Figure 13.10. In this figure the outputs from two adjacent channels are shown. The bandwidth of the monobit receiver is equal to the 60-dB bandwidth of the filter. If one signal falls between two channels, it will be read by both monobit receivers A and B. If the amplitudes of the outputs of these two monobit receivers are summed together, the strength of the signal should be enhanced. This approach will improve the detection sensitivity.

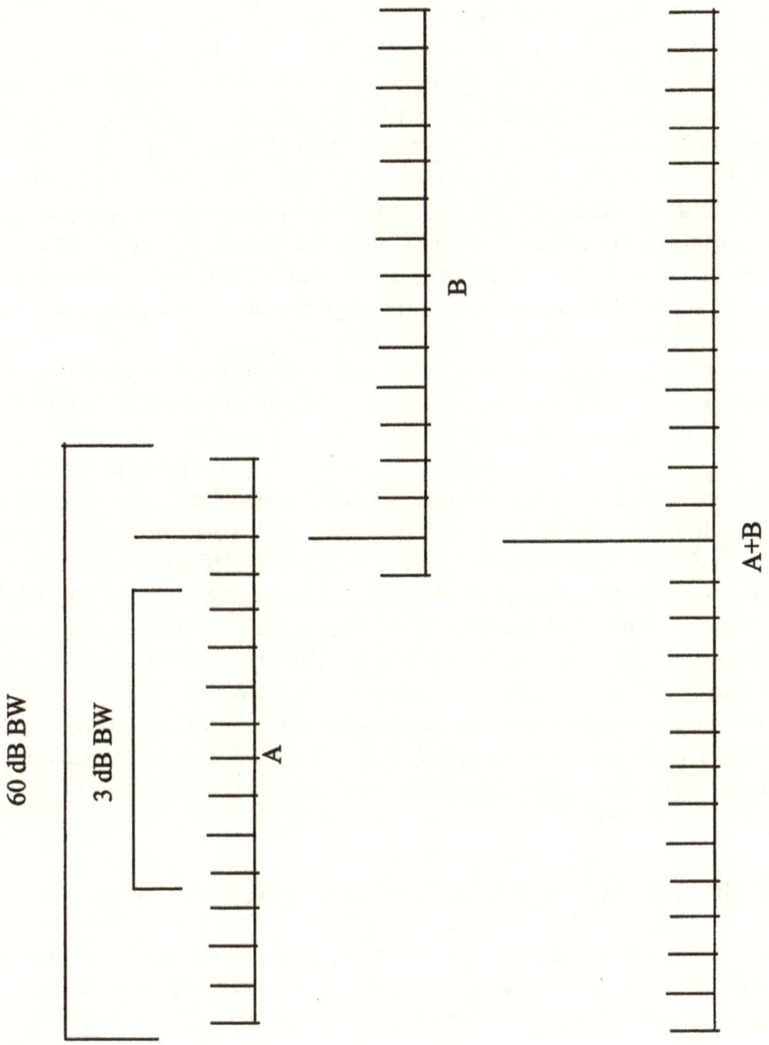

Figure 13.10 Summing outputs from two adjacent monobit receivers.

13.8 CONSIDERATIONS OF DIGITAL FILTERS FOLLOWED BY MONOBIT RECEIVERS

The digital filters usually have wide bandwidth, which is limited by the operating speed of the FFT chip. A 32-point windowed FFT with a sampling frequency of 3 GHz will have a channel bandwidth of 93.75 MHz (3,000/32), which is also the output sampling rate and the FFT operation rate. With this bandwidth there is a higher chance of two simultaneous signals falling in the same channel. It is desirable to separate two signals in one channel, and a narrow monobit receiver can fulfill this requirement. As discussed in Section 13.5, two signals in one channel can be measured by the two neighboring channels. The basic operation is still one phase comparator measuring one signal. In this arrangement two adjacent channels always process one signal. If the filter shape is chosen as shown in Figure 13.2(c), it is most likely that one channel processes only one signal. With this type of arrangement it is desirable to separate two signals in one channel because a phase comparator in one channel cannot separate two signals effectively.

The basic idea of using a narrowband monobit receiver after a digital filter bank is to perform an FFT operation on the output of the FFT. The monobit receivers used at the outputs of the FFT can be considered as part of the encoding circuit. One important factor in building this receiver is the bandwidth of the monobit receiver. In order to process signals falling between channels, the monobit receiver should have a bandwidth wider than the 3-dB bandwidth. It is desirable to have the bandwidth equal to the 60-dB bandwidth. If the channelization is accomplished in software, changing the output sampling rate is rather simple (Section 11.12). The input data can be shifted by any desired value through software operation. Although the software approach is flexible, it is limited to low frequency operation. It is impractical to operate on input data digitized at 3 GHz. The channelization can only be achieved in hardware for a wideband digital receiver. The output sampling rate must be increased to match the desired bandwidth required for the narrowband receivers. The following section will discuss the increase of the output sampling rate in hardware.

13.9 INCREASE THE OUTPUT SAMPLING RATE BY TWO [3–6]

A general discussion on increasing the output sampling rate can be found in [3]. Chapter 11 discusses the critical sampling rate. Under this critical sampling condition the relation between the output frequency bin number and the output sampling rate is [3]

$$K = M \tag{13.4}$$

where K is number of output frequency bins and M is the number of data points shifting per FFT operation, which is related to the output sampling rate. Using the sample in Chapter 11, $K = M = 32$. The input sampling rate is 3,000 MHz; the output sampling rate is 93.75 MHz (3,000/32). Because the FFT outputs are complex, with a 93.75-MHz sampling rate the bandwidth is also 93.75 MHz, which is equal to the 3-dB bandwidth of the filter. As discussed in Section 13.3, the bandwidth of the monobit receiver should be equal to the 60-dB bandwidth of the filter. If the output sampling rate is increased to 187.5 MHz, which is double the 3-dB bandwidth, the monobit receiver can process signals from the filter bank as shown in Figure 13.2(a). Therefore, it is only required to increase the output sampling rate by a factor of two.

The relation in (13.4) can be modified as [3]

$$K = MI \tag{13.5}$$

where I is an integer, which is referred to as the oversampling ratio. In this section only the oversampling ratio of 2 (or $I = 2$) will be discussed. If $K = 32$ and $I = 2$, then $M = 16$. This means the data will shift 16 points per FFT operation, which is the desired result.

When the filter reaches steady state, the input to the FFT operator is the same as in (11.23), which is rewritten here as

$$\begin{aligned} y(0) &= x(0)h(0) + x(32)h(32) + \ldots + x(224)h(224) \\ y(1) &= x(1)h(1) + x(33)h(33) + \ldots + x(225)h(225) \\ &\ldots \\ y(31) &= x(31)h(31) + x(63)h(63) + \ldots + x(255)h(255) \end{aligned} \tag{13.6}$$

In the second cycle that the FFT operates, however, the input is different from the result of (11.27); the desired results are

$$\begin{aligned} y(0) &= x(16)h(0) + x(48)h(32) + \ldots + x(240)h(224) \\ y(1) &= x(17)h(1) + x(49)h(33) + \ldots + x(241)h(225) \\ &\ldots \\ y(31) &= x(47)h(31) + x(79)h(63) + \ldots + x(287)h(255) \end{aligned} \tag{13.7}$$

The output starts from $x(16)$ instead of starting from $x(32)$. In order to perform this operation, the arrangement in Figure 11.11 should be modified. This discussion is similar to the discussion in [4]. Figure 13.11 shows that the input is decimated into 16 outputs rather than 32. These 16 inputs are fed into filters 0 to 15. The inputs to filters 16 to 31 are obtained from delaying the first 16 outputs by one clock cycle. Some of the inputs are listed in this figure. For example, the input to filter 16 is obtained by delaying input to filter 0 by one clock cycle, thus,

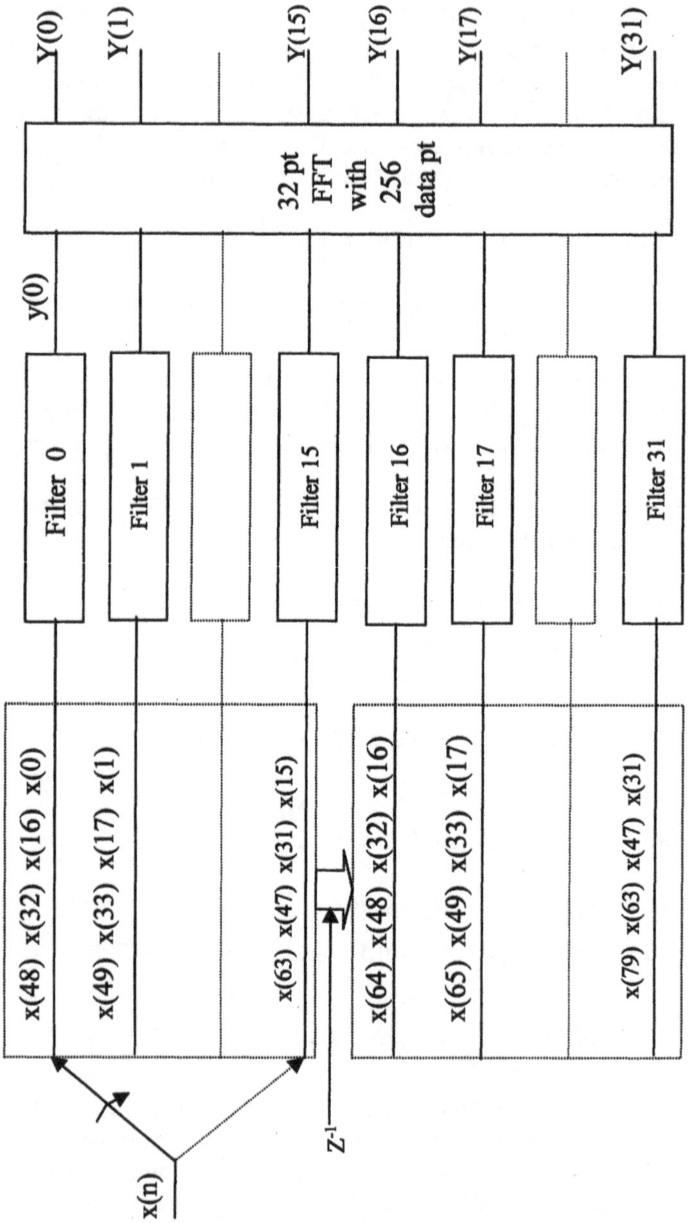

Figure 13.11 Arrangement of oversampling by two.

this input is $x(16)$, $x(32)$, ... When the input is decimated by 16 the output speed is doubled, as compared with being decimated by 32.

The filters in this figure are also modified as shown in Figure 13.12. Between each output there are two delay cycles rather than one. The first two consecutive outputs are shown, and they match the result of the first equation in (13.6) and (13.7). From this arrangement one can see that the output rate to generate $Y(k)$ from the FFT is doubled. The output sampling rate is 187.5 MHz. This is the input rate to the monobit receiver. Because the outputs from the FFT are complex, the monobit receiver has a bandwidth of 187.5 MHz, which is double the 3-dB bandwidth of the output filter.

13.10 DIGITAL FILTERS FOLLOWED BY MONOBIT RECEIVERS

A simulation is used to evaluate the performance of the digital filter followed by the monobit receivers. Digital filters are generated through FFT operation. The filter shape is the same as shown in Figure 11.15. The only difference is that the output rate is doubled because each FFT is performed by shifting the input 16 points rather than 32 points. This sampling rate change, however, cannot be shown in Figure 11.15. A threshold should be placed at the outputs of each filter. If the output is lower than the threshold, one should consider that there is no signal output from this channel and its output will not be processed. A possible arrangement is shown in Figure 13.13. In this figure the threshold at the outputs of the filters is not shown. Although the 32-point FFT generates 32 outputs, only 16 of them carry independent information; thus, only 16 narrowband monobit receivers are needed.

The channels with signals are processed with the monobit receivers. Only the highest 2 bits from the real and imaginary parts are used as the input of the monobit receiver. This operation should be equivalent to putting a limiting amplifier at the filter output, although a real limiting amplifier cannot operate on complex signals. This limiting action destroys the amplitude information on the input signals; thus, the threshold is needed at the channel output. In the time domain, every 16 outputs from a certain frequency bin are used as one input frame of the monobit receiver. Because the output sampling rate is 187.5 MHz, the time to collect 16 samples is about 85 ns ($16/187.5 \times 10^6$). This is the same time needed to collect 256 input data points because $256/3 \times 10^9$ is also equal to 85 ns. Thus, both the digital filter and the monobit receiver process 256 input signals. Because there are only 16 inputs to a monobit receiver, the design can be very simple. There are 16 outputs because the inputs are complex. These 16 outputs cover a bandwidth of 187.5 MHz; thus, each individual output of the monobit receiver is about 11.72 MHz (187.5/16). This implies that the receiver can separate two signals separated by about 12 MHz. However, if the amplitudes of the two signals from a wideband digital filter are separated by more the 5 dB, the receiver will miss the weak signal.

Because the FFT operation only takes 16 input points, both the Kernel functions in Figures 12.2 and 12.15 are used in the simulation. Although two signals in

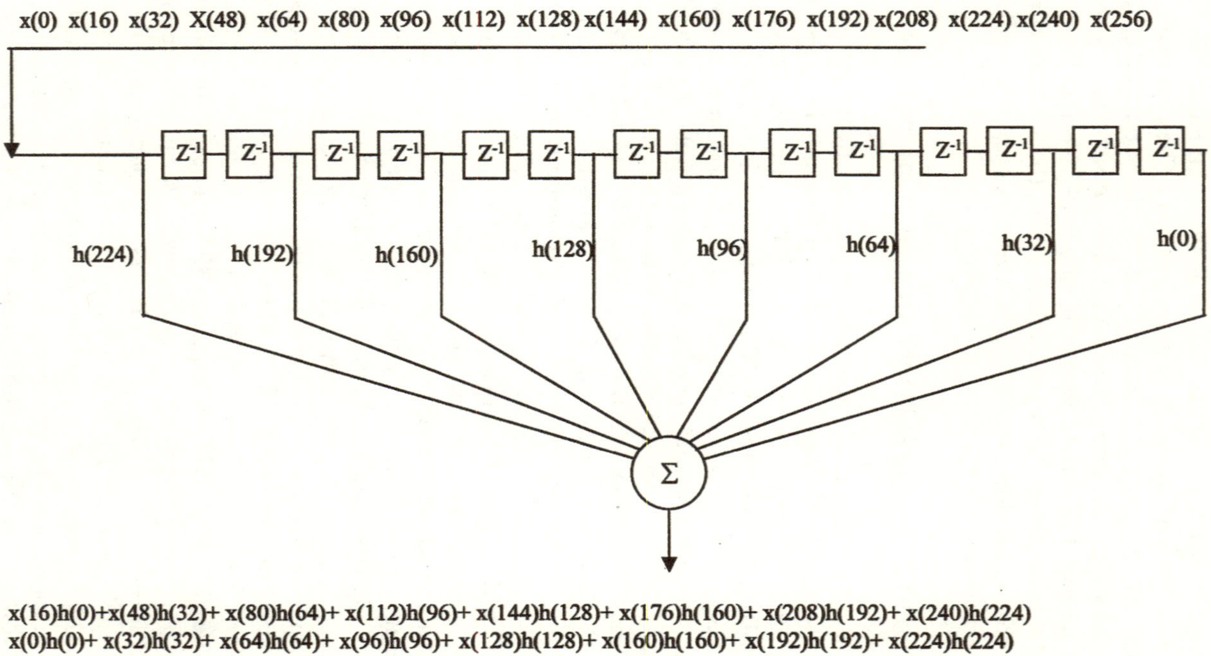

Figure 13.12 Modified filter structure.

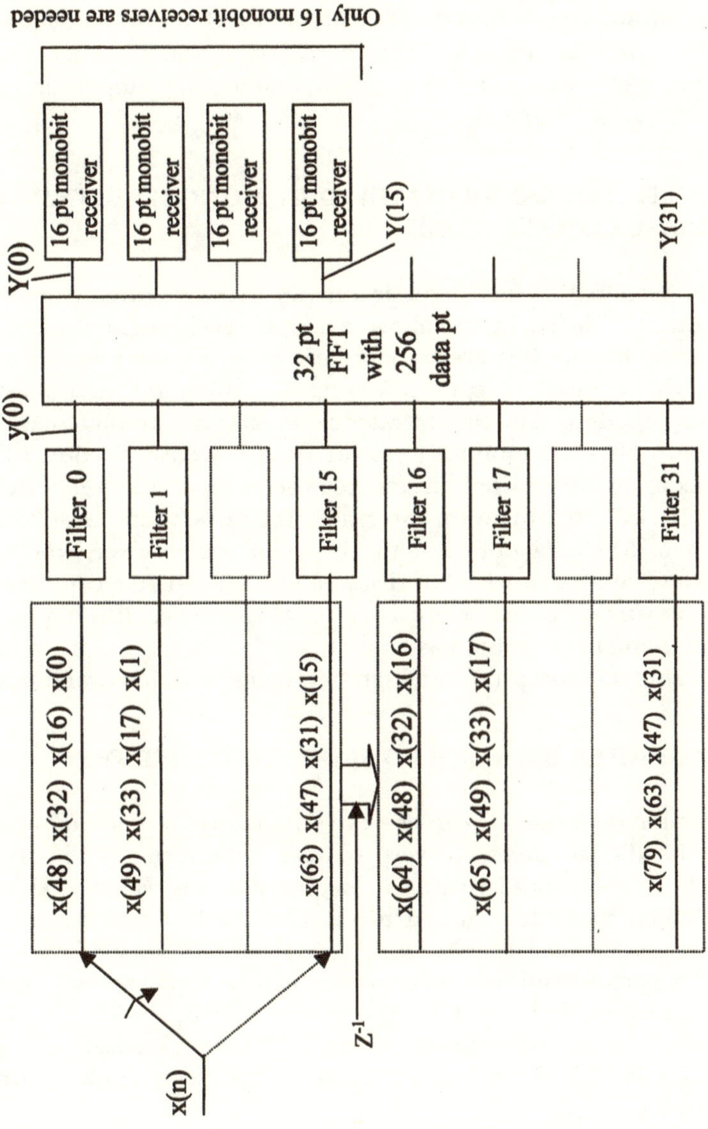

Figure 13.13 Oversampling filters followed by monobit receivers.

one channel are shown in both approaches, the eight-point Kernel function provides a better result, which means higher instantaneous dynamic range. The monobit receiver limits the instantaneous dynamic range; thus, two signals in one channel must have comparable amplitudes to be detected by the receiver. The instantaneous dynamic range of two signals in different channels is determined by the shape of the digital filter. In this simulation the threshold is not included. In order to integrate the monobit receiver into this system, a practical configuration and more thorough simulation are needed.

13.11 DIGITAL FILTER BANK FOLLOWED BY MONOBIT RECEIVERS AND PHASE COMPARATORS

As discussed in Section 13.5, the single frequency data resolution from a monobit receiver is about 9.77 MHz. In the above example, each channel has a width of 11.72 MHz. These bandwidths are narrow enough to separate two simultaneous signals but not fine enough to report a frequency reading. It is desirable to report a finer frequency reading. The fine frequency reading can be obtained by using a phase comparator after the monobit receiver, as discussed in Section 13.5.

Using this approach, a single frequency precision of about 0.5 MHz should be achievable. In order to obtain this precision, the pulse width must be increased to a minimum of 512 data points at the input of the receiver, which is about 171 ns. The phase comparator idea as discussed in Section 10.5 can be applied to obtain even finer single frequency precision on longer pulse. It is highly desirable to measure the frequency precision as a function of pulse width (i.e., to obtain low frequency precision on short pulse and high frequency precision over longer pulse).

13.12 DIGITAL FILTER BANK FOLLOWED BY ANOTHER FFT

From the discussion in Section 13.10, one can see that narrowband monobit receivers following the digital filters are very simple. It performs a 16-point FFT at 187.5 MHz with only 2 bits of real and imaginary inputs. Because the monobit receiver has limited dynamic range, if two signals are in one filter the monobit receiver may miss the weaker one. With the advance in digital signal processing, it appears that the narrowband monobit receiver can be replaced by a regular FFT operation, which will be referred to as the second FFT operation (or chip). This FFT chip will take the channel outputs as input without bit truncation. It performs a 16-point FFT at 187.5 MHz and generates 16 independent outputs, because the inputs are complex.

With this kind of arrangement, the receiver should have higher instantaneous dynamic range for signals in the same digital channel. The number of signals per channel is not limited to two as in the case of the monobit receivers. Theoretically, it should be able to process 16 simultaneous signals per channel. Another potential advantage is that because the inputs to the second FFT are not truncated, the amplitude information on the signals is retained. Thresholds can be set at the outputs of the second FFT, and the detection circuit suggested in Section 13.10 may no longer be needed at the first channel outputs. The frequency data resolution generated from the second FFT operation is about 11.7 MHz (187.5/16). Detecting the signals at these outputs should result in higher sensitivity. It is anticipated that this approach may be considered in future wideband digital receiver designs.

REFERENCES

[1] Fields, T. W., Sharpin, D. L., and Tsui, J. B. Y. "Digital Channelized IFM Receivers," *IEEE MTT-S Digest*, Vol. 3, 1994, pp. 1667–1670.

[2] Private communication with McCormick, W., Professor of Electrical Engineering, Wright State University, Dayton, OH.

[3] Crochiere, R. E., and Rabiner, L. R., "Multirate Digital Signal Processing," Englewood Cliffs, NJ: Prentice Hall, 1983, p. 311.

[4] Zahirniak, D. R., Sharpin, D. L., and Fields, T. W., "A Hardware-Efficient Multirate, Digital Channelized Receiver Architecture," *IEEE Trans. Aerospace and Electronic Systems*, Vol. 34, No. 1, January 1998.

[5] Vary, P., and Heute, U., "A Short-Time Spectrum Analyzer with Polyphase-Network and DFT," *Signal Processing*, Vol. 2, January 1980, pp. 55–65.

[6] Vaidyanathan, P. P., "Multirate Systems and Filter Banks," Englewood Cliffs, NJ: Prentice Hall, 1993.

CHAPTER 14

High-Resolution Spectrum Estimation

14.1 INTRODUCTION

In previous chapters, most of the discussion on spectrum estimation has concentrated on fast Fourier transform (FFT) because it has been used in digital receiver designs. In this chapter, some other spectrum estimation approaches will be introduced that will be referred to as high-resolution spectrum estimations. Their major advantage is that they can provide higher frequency resolution than FFT, especially on simultaneous signals.

If there are two signals with frequencies very close, an FFT operation may generate one peak containing both signals. High-resolution spectrum estimation may separate the two signals by generating two sharp peaks. The major drawback of applying high-resolution spectrum estimation to digital microwave receivers is the complexity of the operation. Because of the large number of operations required to estimate the frequencies, they might not be implemented for real-time application in the near future. However, they might be implemented for special applications in the near future. For example, if the peak of an FFT output appears to contain more than one signal, high-resolution spectrum estimation may be used to find the frequencies. Thus, the operation may not be required on all the input data.

Many different high-resolution approaches can be used to estimate frequencies from digitized input data. In this chapter, seven high-resolution methods will be discussed, and they are as follows:

1. Linear predication (or autoregressive (AR)) method;
2. Prony's method;
3. The least squares Prony's method;
4. The multiple signal classification (MUSIC) method;
5. The estimation of signal parameters via rotational invariance techniques (ESPRIT) method;

6. The minimum norm method;
7. The minimum norm with discrete Fourier transform (DFT) method.

The input data can be manipulated before the linear prediction is applied. Depending on the manipulation of the data, the AR method can be briefly subdivided into the forward, backward, and Burg methods.

Finally, an adaptive method will be introduced. This method will take the input data and strip off one signal at a time.

Some of these methods can produce satisfactory visual displays so that one can determine the frequencies by observing the output result. It has been emphasized numerous times in the previous chapters that an EW receiver must generate digital words as output. The outputs from these spectrum estimation methods must be converted into digital words with real-time processing. In some methods, determining the order of operation must also be implemented in real time.

Many of the references can be found in [1, 2]. These are two books that contain a collection of papers. Computer programs used to generate some of the figures are listed in the appendix of this chapter.

14.2 AUTOREGRESSIVE (AR) METHOD [1–18]

In time series, a powerful model is called the prediction method. It is assumed that the present value can be predicted from past values. For example, prediction can be used in many areas (i.e., in environmental trend, weather forecasting, stock market movement), although the reliability is questionable.

If it is used in spectrum estimation, the present value can be written as a linear combination of input and output

$$x(n) = -\sum_{i=1}^{p} a_i x(n-1) + Gu(n) + G\sum_{l=1}^{q} b_l u(n-l) \tag{14.1}$$

where $x(n)$ is the digitized data, a_i and b_l are constants, G is the gain of the system, and $u(n)$ represents white noise. In statistics, this equation is called the autoregressive moving average (ARMA) model.

If one takes the z transform of this equation, the result is

$$X(z) = -\sum_{i=1}^{p} a_i X(z) z^{-i} + GU(z) + G\sum_{l=1}^{q} b_l U(z) z^{-l} \tag{14.2}$$

In this equation, the white noise is usually considered as input and the data are the output. Thus, the transfer function $H(z)$ of this equation is defined as the output divided by the input. The result is

$$H(z) = \frac{X(z)}{U(z)} = G\frac{1 + \sum_{l=1}^{q} b_l z^{-l}}{1 + \sum_{i=1}^{p} a_i z^{-i}} \tag{14.3}$$

This equation is called the general pole-zero form because the transfer function has both zeros and poles. The zeros are the z values that cause the numerator to be zero and the poles are the z values that cause the denominator to be zero.

If all the a_i values are zero in (14.1), the equation becomes

$$x(n) = Gu(n) + G\sum_{l=1}^{q} b_l u(n-1) \tag{14.4}$$

This equation is called the moving average (MA) model. Its corresponding transfer function is

$$H(z) = \frac{X(z)}{U(z)} = G\left(1 + \sum_{l=1}^{q} b_l z^{-l}\right) \tag{14.5}$$

This is an all-zero model. In filter design, this is referred to as the finite impulse response (FIR) filtering. To solve the constants b_l from $x(n)$, nonlinear equations will be formed; therefore, it is difficult to solve.

If all the b_l values are zero in (14.1), the equation becomes

$$x(n) = -\sum_{i=1}^{p} a_i x(n-1) + Gu(n) \tag{14.6}$$

This is called the AR model. It is also called the linear prediction model because the present value can be predicted by a linear combination of past output values. Its corresponding transfer function is

$$H(z) = \frac{X(z)}{U(z)} = \frac{G}{1 + \sum_{i=1}^{p} a_i z^{-i}} \tag{14.7}$$

This is an all-pole model. In filter design, this model is also known as the infinite impulse response (IIR) filter.

The AR model will be studied in this chapter for two reasons. First, the spectrum generated by (14.7) will be narrowband since $H(z)$ has a very sharp peak when the denominator approaches zero (i.e., a pole is close to unit circle). Second, to obtain the constants a_i in (14.6), the processing is linear.

14.3 YULE-WALKER EQUATION [1–21]

The linear prediction defined in (14.6) and (14.7) can be considered as a filter with noise $u(n)$ as input and $x(n)$ as output. This filter is shown in Figure 14.1. Figure 14.1(a) shows the overall function of the filter and Figure 14.1(b) shows the feedback circuit. The first step is to find the coefficients of the filter. If the input $u(n)$ is assumed to be an unknown response, the signal $x(n)$ can be predicted only approximately from a linearly weighted summation of the past terms. As a result, the linear prediction expression can be written as

$$\hat{x}(n) = -\sum_{i=1}^{p} a_i x(n-i) \tag{14.8}$$

where $\hat{x}(n)$ represents the estimate of $x(n)$. For simplicity, $x(n)$ is used for $\hat{x}(n)$ in later discussion. Replacing $x(n)$ by $x(n-1)$, $x(n-2)$, . . . , $x(n-p+1)$, one can obtain a set of p linear equations. For example, if there are four data points $x(1)$, $x(2)$, $x(3)$ and $x(4)$, and $p = 2$, there will be two equations with constants a_1 and a_2. These equations can be written as

$$\begin{aligned} -x(4) &= a_1 x(3) + a_2 x(2) \\ -x(3) &= a_1 x(2) + a_2 x(1) \end{aligned} \tag{14.9}$$

Since all the $x(n)$ are known, theoretically, it is possible to determine the a values. This result will be equivalent to the Prony's method, which originates from a slightly different point of view.

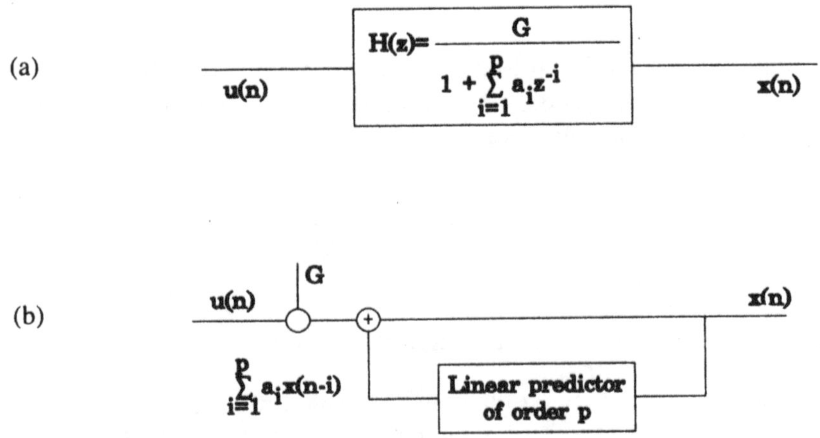

Figure 14.1 AR model: (a) plane result, (b) equivalent circuit.

The above equation has one possible shortcoming in that the data points are usually contaminated with noise. The usual way to solve this problem is by least squares. This approach can be demonstrated through the following procedure. Equation (14.6) can be rewritten here as

$$x(n) = -\sum_{i=1}^{p} a_i x(n-i) + u(n) \tag{14.10}$$

where the gain $G = 1$ is assumed. Multiplying both sides by $x^*(n-k)$ and taking the expectation of them, the result is

$$E[x(n)x^*(n-k)] = -E\left[\sum_{i=1}^{p} a_i x(n-i)x^*(n-k)\right] + E[u(n)x^*(n-k)] \tag{14.11}$$

The following "sample" autocorrelation is used in place of the expectation value as

$$E[x(n)x(n-k)] = \frac{1}{N}\sum_{n=0}^{N-k-1} x(n)x(n-k) \tag{14.12}$$

From the definition of autocorrelation $R(k-i)$, the expectation value can be written as

$$R(k-i) = E[x(n-i)x^*(n-k)] \tag{14.13}$$

The data point $x(n)$ is composed of two parts, the signal and the noise, which can be written as

$$x(n) = x_s(n) + u(n) \tag{14.14}$$

where $x_s(n)$ is the signal without noise. Thus

$$E[u(n)x^*(n-k)] = E[u(n)x_s^*(n-k)] + E[u(n)u^*(n-k)] \tag{14.15}$$

The first part of this equation is zero because the signal and the noise are uncorrelated. The second part can be written as

$$\begin{aligned} E[u(n)u^*(n-k)] &= 0 \qquad \text{when } k \neq 0 \\ &= \sigma^2 \qquad \text{when } k = 0 \end{aligned} \tag{14.16}$$

because the noise is uncorrelated. When $k = 0$, the expectation value equals the noise power. From all these equations (14.12–14.16), (14.11) can be written as

$$R(k) = -\sum_{i=1}^{p} a_i R(k-i) \qquad \text{for } k \neq 0 \tag{14.17}$$
$$R(k) = -\sum_{i=1}^{p} a_i R(k-i) + \sigma^2 \qquad \text{for } k = 0$$

These two equations are the Yule-Walker equations. If p is given, these two equations can be written explicitly as

$$\begin{aligned} -R(0) &= a_1R(-1) + a_2R(-2) + \ldots + a_pR(-p) - \sigma^2 && \text{for } k = 0 \\ -R(1) &= a_1R(0) + a_2R(-1) + \ldots + a_pR(-p+1) && \text{for } k = 1 \\ -R(p) &= a_1R(p-1) + a_2R(p-2) + \ldots + a_{pR}(0) && \text{for } k = p \end{aligned} \tag{14.18}$$

These equations are linear equations of a_i and they can be rearranged and written in matrix form as

$$\begin{bmatrix} R(0) & R(-1) & \ldots & R(-p) \\ R(1) & R(0) & \ldots & R(-p+1) \\ . & . & . & . \\ R(p) & R(p-1) & \ldots & R(0) \end{bmatrix} \begin{bmatrix} 1 \\ a_1 \\ . \\ . \\ . \\ a_p \end{bmatrix} = \begin{bmatrix} \sigma^2 \\ 0 \\ . \\ . \\ . \\ 0 \end{bmatrix} \tag{14.19}$$

If the first equation in (14.18) is removed, the remaining equations can be written in matrix form as

$$\begin{bmatrix} R(0) & R(-1) & \ldots & R(-p+1) \\ R(1) & R(0) & \ldots & R(-p+2) \\ . & . & . & . \\ R(p-1) & R(p-2) & \ldots & R(0) \end{bmatrix} \begin{bmatrix} a_1 \\ a_2 \\ . \\ . \\ . \\ a_p \end{bmatrix} = -\begin{bmatrix} R(1) \\ R(2) \\ . \\ . \\ . \\ R(p) \end{bmatrix} \tag{14.20}$$

The Yule-Walker equations can also be obtained from the least mean square (LMS) approach. This approach is illustrated as follows. From (14.8), the error can be written as

$$e(n) = x(n) - \hat{x}(n) = x(n) + \sum_{i=1}^{p} a_i x(n-1) \tag{14.21}$$

The coefficients a_i can be obtained by minimizing the summation of the error square, which can be written as

$$\sum_{n=0}^{N-i-1} |e(n)|^2 = \sum_{n=0}^{N-i-1} |x(n) + \sum_{i=1}^{p} a_i x(n-i)|^2 \tag{14.22}$$

To minimize the above equation, one can take the derivative with respect to a_i and set the result to zero. Since a_i is a complex number, taking the derivative with respect to it will break it into real and imaginary parts, respectively. This is known as the result of the LMS solution [4, 19]

$$\begin{aligned}
&\sum_{n=0}^{N-2}\left(x(n) + \sum_{i=1}^{p} a_i x(n-i)\right) x^*(n-1) = 0 \\
&\sum_{n=0}^{N-3}\left(x(n) + \sum_{i=1}^{p} a_i x(n-i)\right) x^*(n-2) = 0 \\
&\qquad \cdots \\
&\sum_{n=1}^{N-p-1}\left(x(n) + \sum_{i=1}^{p} a_i x(n-i)\right) x^*(n-p) = 0
\end{aligned} \tag{14.23}$$

Using the definition of autocorrelation, these results are the same as those shown in (14.18) if all the expected values are replaced by estimates except the first one.

The first equation in (14.18) can be obtained as follows. By substituting these results into (14.22), the minimum of the summation of error square is equal to the noise, which can be written as

$$\begin{aligned}
\frac{1}{N}\sum | e_{\min}(n)|^2 &= \frac{1}{N}\sum_{n=1}^{N} |x(n) + \sum_{i=1}^{p} a_i x(n-i)|^2 \\
&= \frac{1}{N}\sum_{n=1}^{N}\left(x(n) + \sum_{i=1}^{p} a_i x(n-i)\right)\left(x(n) + \sum_{i=1}^{p} a_i x(n-i)\right)^* \\
&= \frac{1}{N}\sum_{n=1}^{N} x(n)x^*(n) + \frac{1}{N}\sum_{n=1}^{N}\sum_{i=1}^{p} a_i x(n-i)x^*(n)
\end{aligned} \tag{14.24}$$

After multiplying the two terms in this equation, the end result will contain four terms. The last two terms are zero if the relation of (14.23) is used. By using the definition of autocorrelation, this equation is the same as the first one in (14.18). Therefore, the Yule-Walker equation can be obtained from a different approach.

14.4 LEVINSON-DURBIN RECURSIVE ALGORITHM [1–21]

In (14.19), all the diagonal elements including the off diagonal of the R matrix are equal. This kind of matrix is called Toeplitz. A Toeplitz matrix equation can be solved through the Levinson-Durbin algorithm, which is a recursive approach and more calculation-efficient than directly solving from the Yule-Walker equation. The results of the recursive equation can be written as

$$\sigma_0^2 = R(0)$$
$$\sigma_j^2 = (1 - a_{j,j}^2)\sigma_{j-1}^2 \tag{14.25}$$
$$a_{j,j} = \frac{-\left[R(j) + \sum_{i=1}^{j-1} a_{i,j-1} R(j-i)\right]}{\sigma_{j-1}^2}$$
$$a_{i,j} = a_{i,j-1} + a_{j,j} a_{j-i,j-1}^*$$

where σ_j is called the prediction error power, which may be used to determine the order of the AR method. The first subscript in $a_{j,i}$ is the numerical order to the constants and the second subscript is the number of recursions. The value of σ_j should decrease when j increases. When the correct order is reached theoretically, its value will stay constant.

The illustration of using (14.25) can be shown as follows. For a final order of $p = j = 2$, the first order is $j = 1$, there are $R(0)$ and $R(1)$, with $\sigma_0 = R(0)$ and $a_{1,1} = -R(1)/R(0)$. For the second order, there are $R(0)$, $R(1)$, and $R(2)$ $a_{2,2} = -[R(2) + a_{1,1}R(1)]/\sigma_1^2$, where $\sigma_1 = (1 - a_{1,1}^2)\sigma_0^2$, and $a_{1,2} = a_{1,1} + a_{2,2}a_{1,1}^*$. The final results are $a_{1,2} = a_1$ and $a_{2,2} = a_2$, where a_1 and a_2 are the constants in (14.20).

In the above discussion, the autocorrelations are used to calculate the constants a_i. These quantities are obtained from averaging the input data. Therefore, the coefficients a_i calculated should be better than the results obtained from (14.8).

After the values of a_i are obtained, the results can be substituted into (14.7) to find the spectrum response. The gain of the system equals to the variance of the noise power and the variable z is replaced by

$$z = e^{j2\pi f t_s} \tag{14.26}$$

The power spectrum obtained from the AR model can be obtained from (14.7) as

$$P_{AR}(f) = |H(e^{j2\pi f t_s})|^2 = \frac{\sigma^2}{\left|1 + \sum_{i=1}^{p} a_i e^{-j2\pi f}\right|^2} \tag{14.27}$$

where t_s is assumed. The values of σ^2 can be obtained as discussed in the Levinson-Durbin recursive equation.

Let us use an example to demonstrate the AR model. Assume that the input signal consists of three sinusoidal waves without noise. The data are generated by

$$x(n) = \cos(2\pi \cdot 0.21n + 0.1) + 2\cos(2\pi \cdot 0.36n) + 1.9\cos(2\pi \cdot 0.38n) \quad (14.28)$$

where $n = 0, 1, \ldots, 31$ for a total of 32 data points. This input contains three signals, two of them with frequencies close together, but there is no noise.

The data are padded with 4,064 zeros, to a total length of 4,096 points, and FFT is performed. The result is shown in Figure 14.2. In this figure, the two frequencies 0.36 and 0.38 cannot be separated and they form a single peak.

Figure 14.3 shows the result obtained using the AR modeling approach and (14.27). In Figures 14.3(a–c), the orders of the process are $p = 14$, 20, and 30, respectively. From these figures one can see that when the order is low, signals with frequencies close together may not show. When the order is too high, spurious signals may appear. This problem may also occur in other high-resolution spectrum estimation methods. Hence, correct order determination is an important issue in AR modeling or any high-resolution method.

14.5 INPUT DATA MANIPULATIONS [3, 4, 13, 19, 22]

In (14.13), the R matrix is obtained from the autocorrelations with different lags. However, this is not the only way to form the R matrix. The data can be manipulated

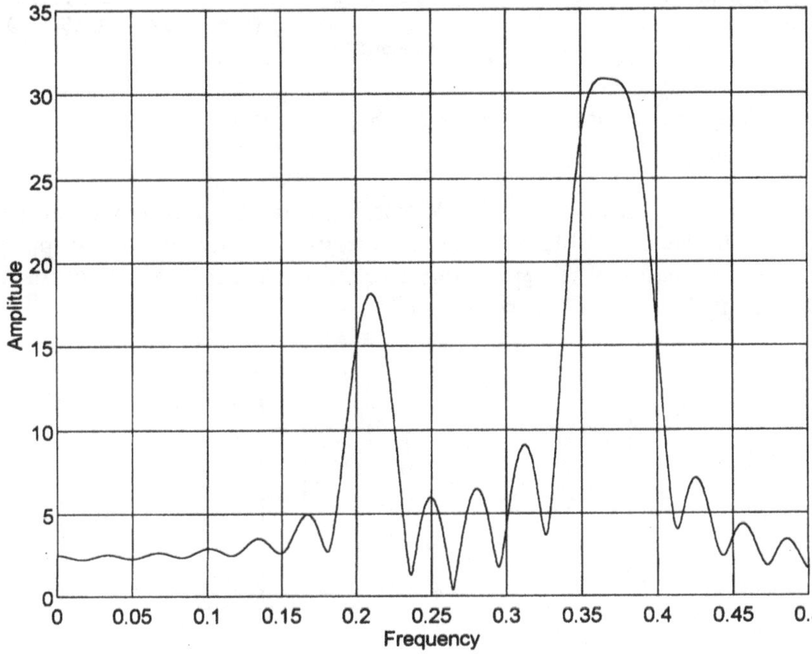

Figure 14.2 FFT output of the input signal.

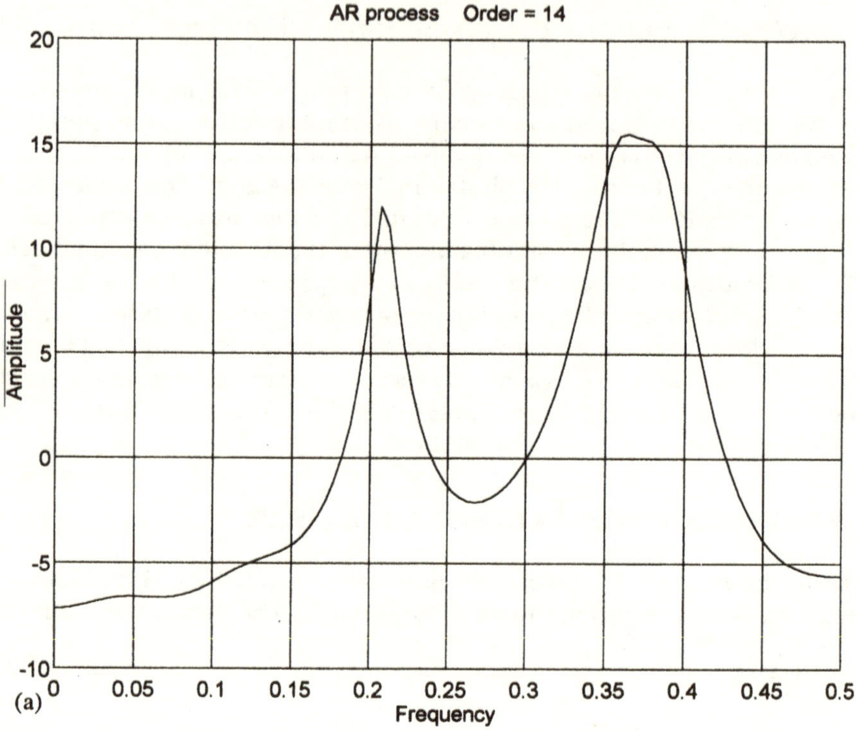

Figure 14.3 Results from the AR process: (a) $p = 14$, (b) $p = 20$, (c) $p = 30$.

differently to obtain different results. In other words, the autocorrelation matrix is not the only approach to obtain the a_i constants in (14.18). Some of the approaches can improve the quality of the spectrum estimation. In order to use other ways to adopt the input data, (14.19) is rewritten as

$$\begin{bmatrix} T_{00} & T_{01} & \cdots & T_{0p} \\ T_{10} & T_{11} & \cdots & T_{1p} \\ \cdot & \cdot & \cdot & \cdot \\ T_{p0} & T_{p1} & \cdots & T_{pp} \end{bmatrix} \begin{bmatrix} 1 \\ a_1 \\ \cdot \\ \cdot \\ \cdot \\ a_p \end{bmatrix} = \begin{bmatrix} \sigma^2 \\ 0 \\ \cdot \\ \cdot \\ \cdot \\ 0 \end{bmatrix} \tag{14.29}$$

This equation is identical to (14.19), except that the matrix T is used to replace R.

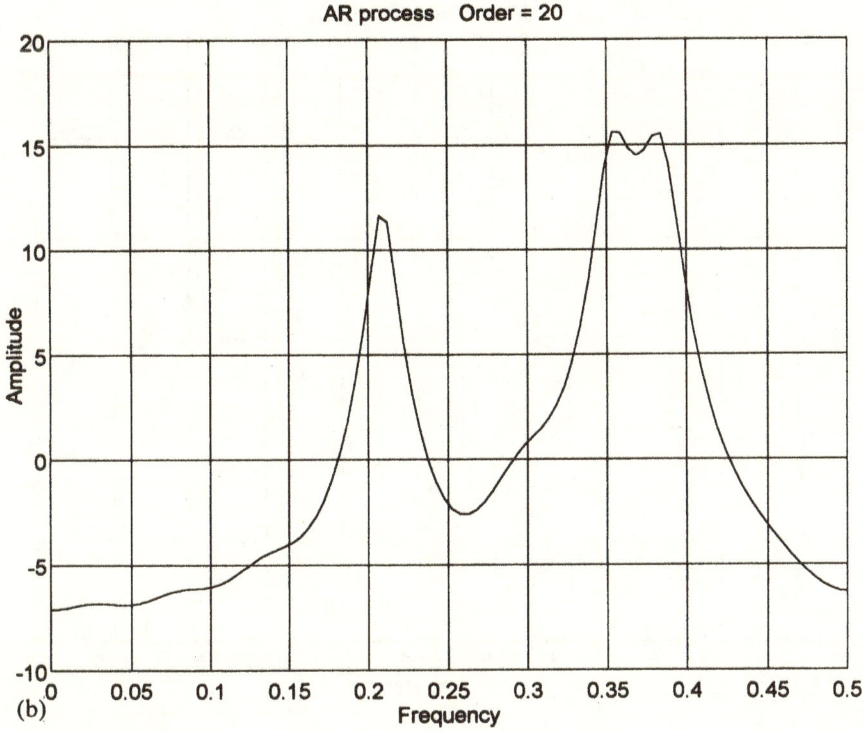

(b)

Figure 14.3 (continued).

Different ways of generating the T matrix from the same input data can produce quite different results. Two different ways to obtain T will be discussed. Assume there are N input data points from $x(0)$ to $x(N-1)$.

14.5.1 Covariance Method

In this method the input data are related to matrix T as

$$\begin{bmatrix} T_{00} & T_{01} & \cdots & T_{0p} \\ T_{10} & T_{11} & \cdots & T_{1p} \\ \cdot & \cdot & \cdot & \cdot \\ T_{p0} & T_{p1} & \cdots & T_{pp} \end{bmatrix} = \frac{1}{N-p} \begin{bmatrix} x^*(p) & x^*(p+1) & \cdots & x^*(N-1) \\ x^*(p-1) & x^*(p) & \cdots & x^*(N-2) \\ \cdot & \cdot & \cdot & \cdot \\ x^*(0) & x^*(1) & \cdots & x^*(N-p-1) \end{bmatrix} \begin{bmatrix} x(p) & x(p-1) & \cdots & x(0) \\ x(p+1) & x(p) & \cdots & x(1) \\ \cdot & \cdot & \cdot & \cdot \\ x(N-1) & x(N-2) & \cdots & x(N-1-p) \end{bmatrix} \quad (14.30)$$

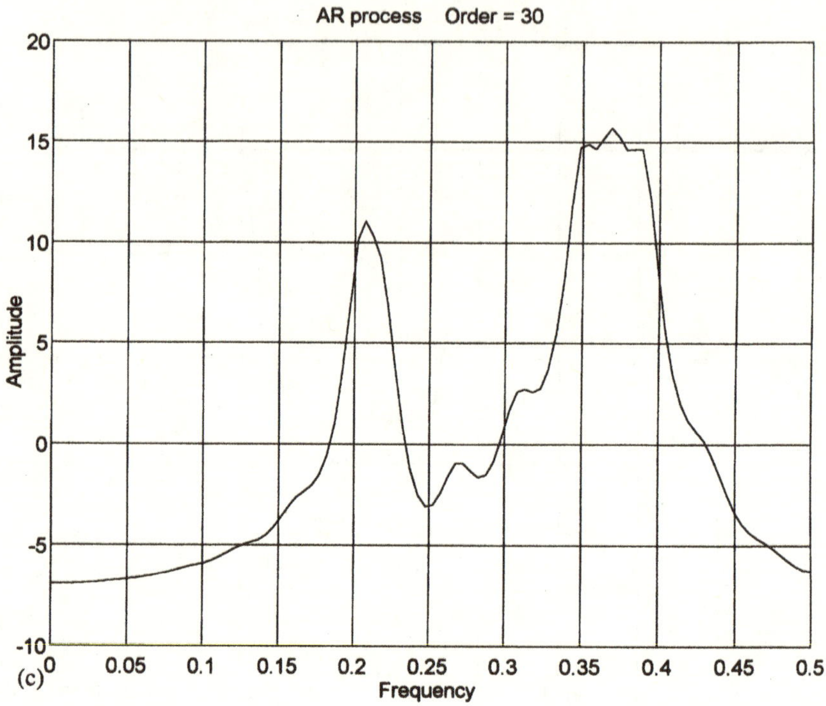

Figure 14.3 (continued).

This matrix is not Toeplitz. Although a special recursive method can be developed for this approach, the Levinson-Durbin method does not apply directly. In this approach, all the elements in the T matrix contain the same number of terms from the data points.

If $p = 1$ and $N = 3$, the data are $x(0)$, $x(1)$, and $x(2)$, and the T matrix is

$$\begin{bmatrix} T_{00} & T_{01} \\ T_{10} & T_{11} \end{bmatrix} = \frac{1}{2}\begin{bmatrix} x^*(1) & x^*(2) \\ x^*(0) & x^*(1) \end{bmatrix}\begin{bmatrix} x(1) & x(0) \\ x(2) & x(1) \end{bmatrix} \tag{14.31}$$

If $p = 1$ and $N = 4$, the data are $x(0)$, $x(1)$, $x(2)$, and $x(4)$, and the T matrix is

$$\begin{bmatrix} T_{00} & T_{01} \\ T_{10} & T_{11} \end{bmatrix} = \frac{1}{3}\begin{bmatrix} x^*(1) & x^*(2) & x^*(3) \\ x^*(0) & x^*(1) & x^*(2) \end{bmatrix}\begin{bmatrix} x(1) & x(0) \\ x(2) & x(1) \\ x(3) & x(2) \end{bmatrix} \tag{14.32}$$

This is a common way to obtain the R matrix. In this chapter, some of the examples use this method to obtain the R matrix. This approach provides accurate frequency estimates for no noise and correct order.

14.5.2 Autocorrelation Method

In this approach, the result of the T matrix is identical to that obtained from the autocorrelation function and that is why the name is used. In this approach, the data points outside of the range $x(0)$ to $x(N-1)$ are assumed to be zeros, and this implies a window function is applied. The T matrix is related to the input data as

$$\begin{bmatrix} T_{10} & T_{11} & \cdots & T_{1p} \\ \cdot & \cdot & \cdot & \cdot \\ T_{p0} & T_{p1} & \cdots & T_{pp} \end{bmatrix} = \frac{1}{N} \begin{bmatrix} 0 & x^*(0) & \cdots & & x^*(N-1) & \cdots & 0 \\ & & \cdots & & & & \\ 0 & \cdots & 0 & x^*(0) & & \cdots & x^*(N-1) \end{bmatrix} \begin{bmatrix} x(0) & 0 & \cdots & 0 \\ x(1) & x(0) & \cdots & 0 \\ \cdot & \cdot & \cdot & \cdot \\ x(N-1) & x(N-2) & & x(0) \\ 0 & x(N-1) & \cdots & x(1) \\ 0 & \cdots & & \\ 0 & 0 & \cdots & x(N-1) \end{bmatrix} \tag{14.33}$$

The number of zeros added in front of $x(0)$ and after $x(N-1)$ equals to p. The T matrix obtained this way is Toeplitz; thus, the Levinson-Durbin algorithm can be used to solve for the coefficients in (14.29).

If $p = 1$ and $N = 3$, the data are $x(0)$, $x(1)$, and $x(2)$, and the T matrix is

$$\begin{bmatrix} T_{00} & T_{01} \\ T_{10} & T_{11} \end{bmatrix} = \frac{1}{3} \begin{bmatrix} x^*(0) & x^*(1) & x^*(2) & 0 \\ 0 & x^*(0) & x^*(1) & x^*(2) \end{bmatrix} \begin{bmatrix} x(0) & 0 \\ x(1) & x(0) \\ x(2) & x(1) \\ 0 & x(2) \end{bmatrix} \tag{14.34}$$

If $p = 1$ and $N = 4$, the data are $x(0)$, $x(1)$, $x(2)$, and $x(4)$, and the T matrix is

$$\begin{bmatrix} T_{00} & T_{01} \\ T_{10} & T_{11} \end{bmatrix} = \frac{1}{4}\begin{bmatrix} x^*(0) & x^*(1) & x^*(2) & x^*(3) & 0 \\ 0 & x^*(0) & x^*(1) & x^*(2) & x^*(3) \end{bmatrix}\begin{bmatrix} x(0) & 0 \\ x(1) & x(0) \\ x(2) & x(1) \\ x(3) & x(2) \\ 0 & x(3) \end{bmatrix} \tag{14.35}$$

The same number of zeros is added as in the above case. This approach may not produce the correct frequency, even though there is no noise and the correct order is selected.

14.6 BACKWARD PREDICTION AND MODIFIED COVARIANCE METHOD [3, 4, 13, 19, 22]

The two methods discussed in the next section can be considered as subcategories of the AR method. The purpose of introducing the backward prediction and modified covariance method is to prepare for the discussion of the Burg method, because this concept is used in the Burg method. The linear prediction discussed in previous sections is referred to as the forward prediction, since past data are used to predict the present data. In the backward prediction, the past data are predicted from the previous data. This may sound a little absurd, because the previous data are already known. However, if one considers the time series as a data set rather than a sequence occurring in time, one can linearly predict in either direction. The backward prediction can be written as

$$\hat{x}(n-p) = -\sum_{i=1}^{p} c_i x(n-p+i) \tag{14.36}$$

where c_i is the backward prediction coefficient.

The relation between the backward and forward coefficients can be found as follows. The Yule-Walker equation can be obtained from the above equation by multiplying both sides by $x^*(n-p-k)$ and taking the expectation value. The result is

$$E[x(n-p)x^*(n-p-k)] = -\sum_{i=1}^{p} c_i E[x(n-p+i)x^*(n-p-k)] \tag{14.37}$$

From the definition of autocorrelation, this equation can be written as

$$R(k) = -\sum_{i=1}^{p} c_i R(k+i) \tag{14.38}$$

Written in separate equations, the result is

$$
\begin{aligned}
-R(-1) &= c_1R(0) + c_2R(1) + \ldots + c_pR(p-1) && k=-1\\
-R(-2) &= c_1R(-1) + c_2R(0) + \ldots + c_pR(p-2) && k=-2 \qquad (14.39)\\
& && \ldots\\
-R(-p) &= c_1R(-p+1) + c_2R(-p+2) + \ldots + c_pR(o) && k=-p
\end{aligned}
$$

In matrix form, it is

$$
\begin{bmatrix} R(0) & R(1) & \ldots & R(p-1) \\ R(-1) & R(0) & \ldots & R(p-2) \\ \ldots & & & \\ R(-p+1) & R(-p+2) & \ldots & R(0) \end{bmatrix}
\begin{bmatrix} c_1 \\ c_2 \\ \cdot \\ c_p \end{bmatrix} = -\begin{bmatrix} R(-1) \\ R(-2) \\ \cdot \\ R(-p) \end{bmatrix} \tag{14.40}
$$

Now, let us find the relation between the forward prediction coefficients a_i and the backward prediction coefficients c_i. Taking the complex conjugate on both sides of the above equation, the result is

$$
\begin{bmatrix} R^*(0) & R^*(1) & \ldots & R^*(p-1) \\ R^*(-1) & R^*(0) & \ldots & R^*(p-2) \\ \ldots & & & \\ R^*(-p+1) & R^*(-p+2) & \ldots & R^*(0) \end{bmatrix}
\begin{bmatrix} c_1^* \\ c_2^* \\ \cdot \\ c_p^* \end{bmatrix} = -\begin{bmatrix} R^*(-1) \\ R^*(-2) \\ \cdot \\ R^*(-p) \end{bmatrix} \tag{14.41}
$$

Applying the relation $R^*(i) = R(-i)$, the above equation is

$$
\begin{bmatrix} R(0) & R(-1) & \ldots & R(p+1) \\ R(1) & R(0) & \ldots & R(-p+2) \\ \ldots & & & \\ R(p-1) & R(p-2) & \ldots & R(0) \end{bmatrix}
\begin{bmatrix} c_1^* \\ c_2^* \\ \cdot \\ c_p^* \end{bmatrix} = -\begin{bmatrix} R(1) \\ R(2) \\ \cdot \\ R(p) \end{bmatrix} \tag{14.42}
$$

Comparing with (14.20), it is obvious that

$$
c_i^* = a_i \tag{14.43}
$$

This is the relation between the coefficient of the forward and backward prediction.

A backward prediction error $b(n)$ similar to the forward prediction error $e(n)$ in (14.21) can be defined from (14.36) as

$$b(n) = x(n-p) - \hat{x}(n-p) = x(n-p) + \sum_{i=1}^{p} c_i x(n-p+i) \tag{14.44}$$
$$= x(n-p) + \sum_{i=1}^{p} a_i^* x(n-p+i)$$

which can be written in a slightly different form as

$$b(n) = x(n) - \hat{x}(n) = x(n) + \sum_{i=1}^{p} c_i x(n+i) \tag{14.45}$$
$$= x(n) + \sum_{i=1}^{p} a_i^* x(n+i)$$

The square of the backward error can be minimized to generate an equation similar to the Yule-Walker equations.

The modified covariance method minimizes the sum of the square of the average linear prediction error. The square of the average linear prediction error $\epsilon(n)$ is defined in terms of the forward and backward linear prediction errors as

$$\epsilon(n) = \frac{1}{2(N-p)}\left[\sum_{n=p}^{N-1} e^2(n) + \sum_{n=0}^{N-p-1} b^2(n)\right] \tag{14.46}$$

where $e(n)$ and $b(n)$ are the forward and backward prediction errors, respectively, defined in (14.21) and (14.44). By minimizing $\epsilon(n)$ by taking the derivative with respect to a_i and setting the result to zero, an equation identical to (14.29) can be obtained. The corresponding T matrix can be written as

$$\begin{bmatrix} T_{00} & T_{01} & \cdots & T_{0p} \\ T_{10} & T_{11} & \cdots & T_{1p} \\ \cdots & & & \\ T_{p0} & T_{p1} & \cdots & T_{pp} \end{bmatrix} = \left[\frac{1}{2(N-p)}\right]$$
$$\left\{\begin{bmatrix} x^*(p) & x^*(p+1) & \cdots & x^*(N-1) \\ x^*(p-1) & x^*(p) & \cdots & x^*(N-2) \\ \cdots & & & \\ x^*(0) & x^*(1) & \cdots & x^*(N-p-1) \end{bmatrix}\right. \tag{14.47}$$
$$\begin{bmatrix} x(p) & x(p-1) & \cdots & x(0) \\ x(p+1) & x(p) & \cdots & x(0) \\ \cdots & & & \\ x(N-1) & x(N-2) & \cdots & x(N-p-1) \end{bmatrix}$$

$$+\begin{bmatrix} x(0) & x(1) & \dots & x(N-p-1) \\ x(1) & x(2) & \dots & x(N-p-2) \\ \dots & & & \\ x(p) & x(+1) & \dots & x(N--1) \end{bmatrix} \left.\begin{bmatrix} x^*(0) & x^*(1) & \dots & x^*(p) \\ x^*(1) & x^*(2) & \dots & x^*(p+1) \\ \dots & & & \\ x^*(N-p-1) & x^*(N-p) & \dots & x^*(N-1) \end{bmatrix}\right\}$$

The modified covariance method uses the data more times than either the forward or backward prediction method. The modified covariance method appears to be less sensitive to noise because of the enlarged data set. It is also less sensitive to the initial phase of the input signal than the autocorrelation method. Maple [22] developed a relatively efficient method to solve for the coefficients.

14.7 BURG METHOD [3, 4, 13, 23–32]

One of the most popular approaches in linear prediction spectrum estimation is the Burg method, and it has also been called the maximum entropy method (MEM). If there are n points of data from $x(0)$ to $x(N-1)$ and $p+1$ lags of autocorrelations from $R(0)$ to $R(p)$, Burg suggested that the unknown autocorrelation lags from $R(p+1)$, $R(p+2)$, ... can be extrapolated from the input data points. There are an infinite number of ways to extrapolate the autocorrelations. Burg further suggested that the extrapolation of the autocorrelations should not add any new information arbitrarily to the sequence. The information is measured in terms of entropy from Shannon's theorem. Maximizing the entropy implies that the time series is in the most random state and no new information is arbitrarily added to the series. Thus, the name MEM is used.

Later investigators showed that in order to use the MEM, the autocorrelation of the time series must be known. However, the data obtained from most experiments are, in general, a series of real or complex values as a function of time. In other words, the only known data are the time series but not the autocorrelations of the input signal. The autocorrelations calculated using the time series are not the true values, but only estimations. Thus, the promise of the MEM is never delivered in practice. As a result, the term maximum entropy method is no longer popular; instead, Burg method is used to honor the inventor.

Instead of estimating the autocorrelation lags from $R(p+1)$, $R(p+2)$, ... directly, Burg devised a new method. This method is quite similar to the modified covariance method. The sum of the squares of the average linear prediction error is minimized. The only difference between these methods is that the Burg method uses a constraint in the process of the minimization to assure the filter is stable. The constraint is the Levinson recursive condition in (14.25).

$$a_{i,j} = a_{i,j-1} + a_{j,j} a^*_{j-i,j-1} \tag{14.48}$$

In this section, only the results of the Burg method will be presented here. Let us define

$$\begin{aligned} e_{0,n} &= x(n) \qquad \text{for } n = 1, 2, \ldots, N-1 \\ b_{0,n} &= x(n) \qquad \text{for } n = 0, 1, \ldots, N-2 \\ e_{i,n} &= e_{i-1,n} + \Gamma_i b_{i-1,n-1} \\ b_{i,n} &= b_{i-1,n-1} + \Gamma^*_i e_{i-1,n-1} \end{aligned} \tag{14.49}$$

where $e_{i,n}$ and $b_{i,n}$ represent the forward and backward errors, respectively. The coefficient a values can be obtained as

$$\begin{aligned} a_{j,j} &= \Gamma_j = \frac{-2\sum_{n=j}^{N-1} e_{n,j-1} b^*_{n-1,j-1}}{\sum_{n=j}^{N-1} [|e_{n,j-1}|^2 + |b_{n-1,j-1}|^2]} \\ a_{i,j} &= a_{i,j-1} + a_{j,j} a^*_{j-i,j-1} \qquad \text{for } j = 1, 2, \ldots, i-1 \\ \epsilon_j &= (1 - |\Gamma_j|^2)\,\epsilon_{j-1} \end{aligned} \tag{14.50}$$

where Γ_j is referred to as the jth reflection coefficient and ϵ_j is the jth error. The double subscripts are used here because the approach is recursive. The first subscript represents the ith order of the coefficient $a_{i,j}$ and the second subscript represents the recursive order.

The Burg method can produce very high spectrum peaks with short data lengths and can resolve signals close in frequency. However, the frequency bias depends on the initial phase of the input signal and the length of the data. Another problem with this method is spectral line splitting. This means that when there is only one signal, the spectrum may show two very close spectra. Spectral line splitting depends on the initial phase of the input signal.

Windows can be added to the Burg method to reduce bias error and spectral line splitting. The window function can be inserted into the reflection coefficient as

$$\Gamma_j = \frac{-2\sum_{n=j}^{N-1} w_{n,j} e_{n,j-1} b^*_{n-1,j-1}}{\sum_{n=j}^{N-1} w_{n,j}[|e_{n,j-1}|^2 + |b_{n-1,j-1}|^2]} \tag{14.51}$$

Here, $w_{n,j}$ is the window function.

The result of the Burg method can be found in MATLAB with instructions "lpc" and "freqz." The result of the Burg method is shown in Figure 14.4.

In these figures, five different p values (14, 15, 26, 29, and 32) are chosen. When $p < 14$, the two frequencies at 0.36 and 0.38 are difficult to identify. With $p = 15$ and 26, all three frequencies can be identified. When $p = 29$, an additional frequency is generated. When $p = 32$, five peaks can be found. This phenomenon is referred to as the spectral splitting.

14.8 ORDER SELECTION [3, 4, 13, 27, 33, 34]

From the previous section, it is clear that when the wrong order in the Burg method is selected, the spectrum does not reflect the real input signal. If the order is too low, closely spaced frequencies cannot be detected. If the order is too high, spurious frequencies will be generated. Both situations cannot be tolerated in receiver designs. Thus, it is important to select the correct order of the linear model, but this is a difficult task.

One intuitive approach is to use the recursive approach to find the coefficients and monitor the prediction error. If the data can be truly described by a finite-

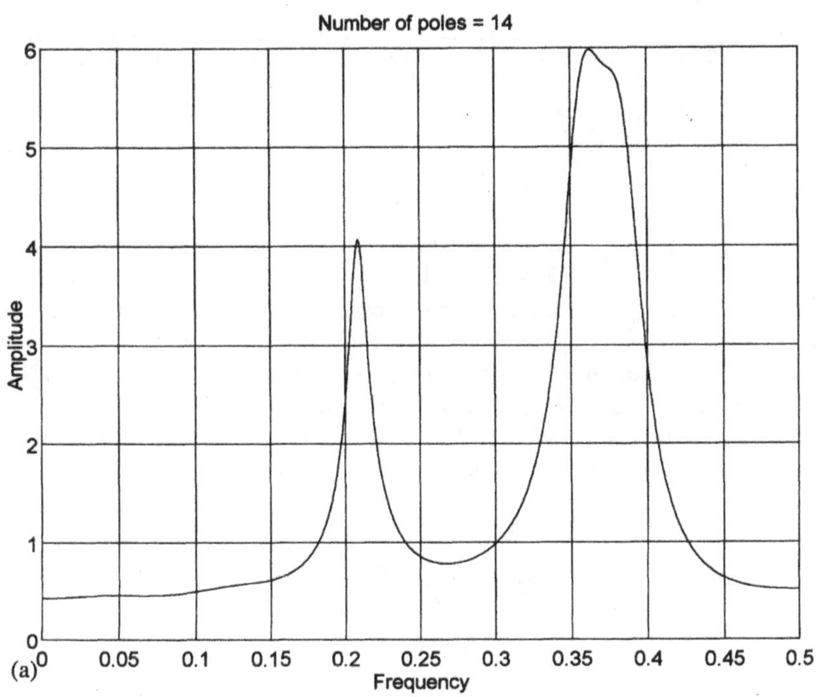

(a)

Figure 14.4 Spectrum generated by Burg method: (a) $p = 14$, (b) $p = 15$, (c) $p = 26$, (d) $p = 29$, (e) $p = 32$.

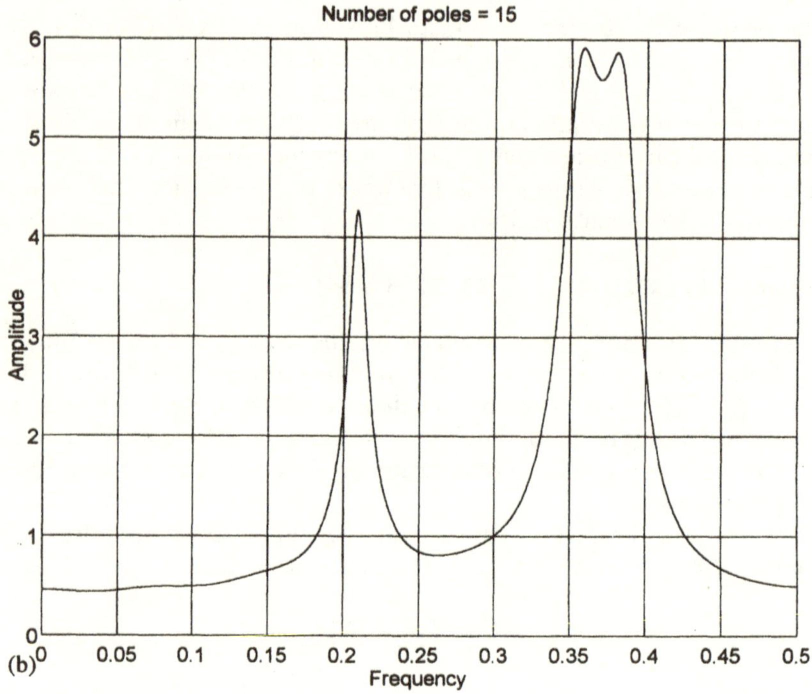

Figure 14.4 (continued).

order linear model, when the correct order is reached the error will either reach a minimum or stay constant. However, this approach may not work. The prediction error may not converge or change monotonically. As a result, there is no easily detected minimum.

The four common methods to choose the order of the linear model are 1) the final prediction error (*FPE*), 2) the Akaike information criterion (*AIC*), 3) the criterion autoregression transfer (*CAT*), and 4) the minimum description length (*MDL*). The results of the four approaches are listed below.

$$
\begin{aligned}
FPE_p &= \frac{\sigma_p^2(N+p+1)}{N-p-1} \\
AIC_p &= N\ln(\sigma_p^2) + 2p \\
CAT_p &= \frac{1}{N}\left(\sum_{i=1}^{p}\frac{1}{\sigma_i^2}\right) - \frac{1}{\sigma_p^2} \\
MDI_p &= N\ln(\sigma_p^2) + p\ln(N)
\end{aligned}
\tag{14.52}
$$

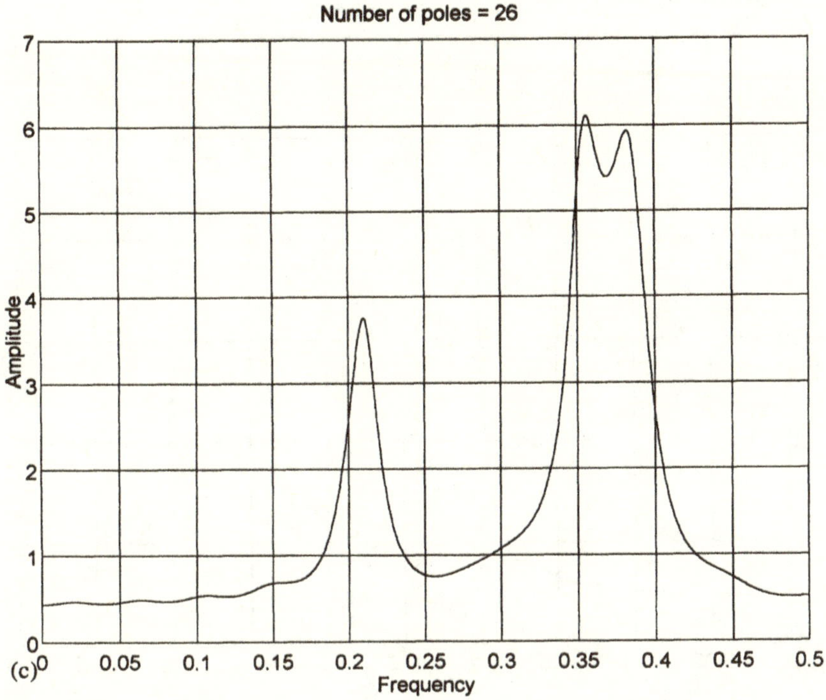

Figure 14.4 (continued).

In order to obtain the p value, one of above equations will be minimized. For example, the *AIC* method determines the p value by minimizing AIC_p. If the data do not fit an AR model, the above approaches are not very useful.

Ulrych [27] suggests an empirical approach, meaning that when

$$\frac{N}{3} < p < \frac{N}{2} \tag{14.53}$$

satisfactory results can be obtained. Using this relation to check the result from the example in Figure 14.4 where $N = 32$, the result is $11 < p < 16$. It appears the lower bound is too low, but the upper bound is satisfactory.

For digital receivers, the important issue is to determine the number of signals without generating spurious signals. The maximum number of anticipated signals can be considered as four sinusoidal waves. This is a common requirement for EW applications and it is determined from the possibility of pulse overlap. This requirement may create a different criterion to select the order of the linear prediction methods.

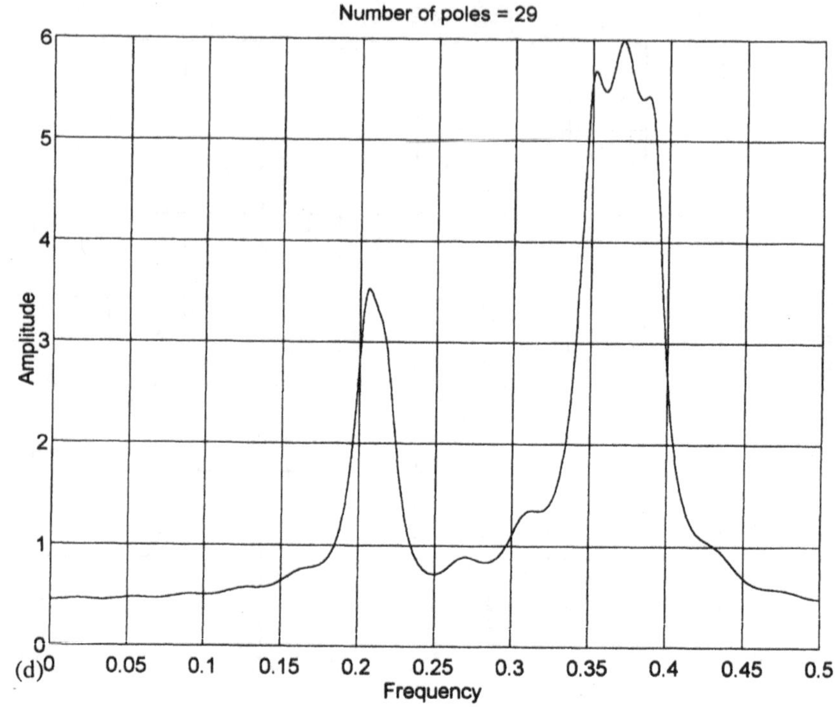

Figure 14.4 (continued).

14.9 PRONY'S METHOD [3–5, 13, 35–40]

There are different ways to derive Prony's method. Only one approach will be presented in this section, which is based on Hildebrand [35]. Prony's method solves a particular set of simultaneous nonlinear equations, which must fit in a certain form. An input signal consisting of only sinusoidal waves can be written in the desired form. In order to simplify the discussion, a simple example will be used to illustrate the basic idea and then a general case will be presented.

Let us assume that the input signal contains two complex sinusoidal waves without noise. The signal can be written as

$$x(t) = A_1 e^{j(2\pi f_1 t+\theta_1)} + A_2 e^{j(2\pi f_2 t+\theta_2)} \tag{14.54}$$

where A_1 and A_2, f_1 and f_2, and θ_1, and θ_2 are the amplitude, the frequency, and the initial phase of the two signals, respectively. The amplitude and the frequency are of primary interests. The amplitudes and initial phases A_1 and θ_1 and A_2 and θ_2 can

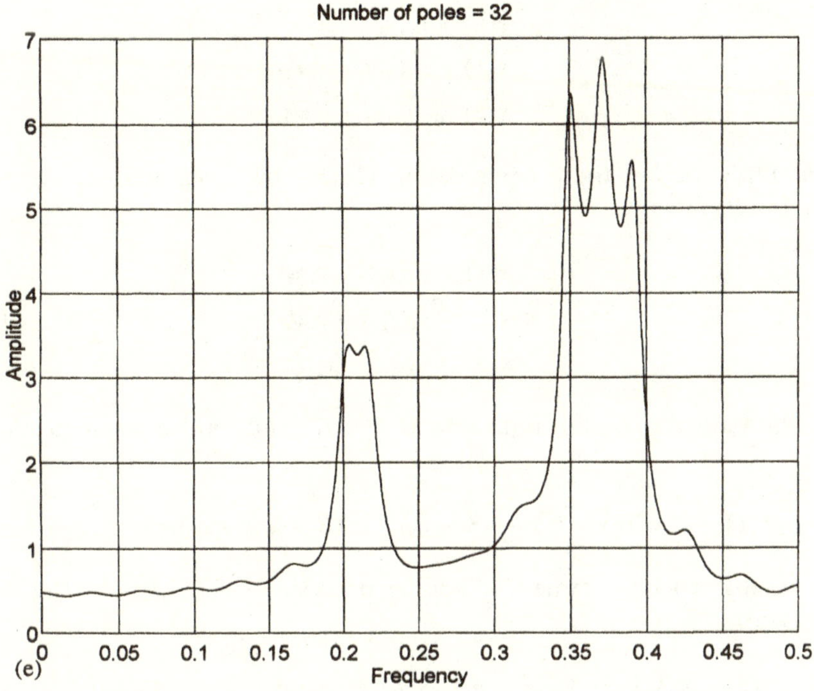

Figure 14.4 (continued).

be combined into two unknowns. In order to solve for these unknowns, four equations are required. If these signals are digitized at $t = 0$, 1, 2, and 3, the results are

$$\begin{aligned} x(0) &= A_1 e^{j\theta_1} + A_2 e^{j\theta_2} \equiv c_1 + c_2 \\ x(1) &= c_1 e^{j2\pi f_1} + c_2 e^{j2\pi f_2} \equiv c_1 z_1 + c_2 z_2 \\ x(2) &= c_1 e^{j2\pi f_1 2} + c_2 e^{j2\pi f_2 2} \equiv c_1 z_1^2 + c_2 z_2^2 \\ x(3) &= c_1 e^{j2\pi f_1 3} + c_2 e^{j2\pi f_2 3} \equiv c_1 z_1^3 + c_2 z_2^3 \end{aligned} \tag{14.55}$$

where

$$\begin{aligned} c_1 &= A_1 e^{j\theta_1} \qquad & c_2 &= A_2 e^{j\theta_2} \\ z_1 &= e^{j2\pi f_1} \qquad & z_2 &= e^{j2\pi f_2} \end{aligned} \tag{14.56}$$

Prony developed a clever method to solve the above equation by converting this essentially nonlinear problem to a linear one. First, multiply the first three equations by a_2, a_1, and -1, respectively, where a_2 and a_1 are unknown. The results are

$$\begin{aligned} a_2 x(0) &= a_2(c_1 + c_2) \\ a_1 x(1) &= a_1(c_1 z_1 + c_2 z_2) \\ -x(2) &= -(c_1 z_1^2 + c_2 z_2^2) \end{aligned} \tag{14.57}$$

Next, multiply the last three equations in (14.55) by a_2, a_1, and −1, respectively, and the results are

$$\begin{aligned} a_2 x(1) &= a_2(c_1 z_1 + c_2 z_2) \\ a_1 x(2) &= a_1(c_1 z_1^2 + c_2 z_2^2) \\ -x(3) &= -(c_1 z_1^3 + c_2 z_2^3) \end{aligned} \tag{14.58}$$

Add all the left side and the right side of (14.57) and set the result to zero. The result is

$$-x(2) + a_1 x(1) + a_2 x(0) = c_1(z_1^2 + a_1 z_1 + a_2) + c_2(z_2^2 + a_1 z_1 + a_2) = 0 \tag{14.59}$$

Similar results can be obtained by adding both sides of (14.58) and setting the result to zero,

$$-x(3) + a_1 x(2) + a_2 x(1) = c_1 z_1(z_1^2 + a_1 z_1 + a_2) + c_2 z_2(z_2^2 + a_1 z_1 + a_2) = 0 \tag{14.60}$$

From these operations, two linear simultaneous equations can be formed and they are

$$\begin{aligned} -x(2) + a_1 x(1) + a_2 x(0) &= 0 \\ -x(3) + a_1 x(2) + a_2 x(1) &= 0 \end{aligned} \tag{14.61}$$

In these equations, a_1 and a_2 are the unknowns, and the $x(i)$ with $i = 0, 1, 2$, and 3 are the known digitized data. Thus, a_1 and a_2 can be solved from this equation. This is a similar equation to that shown in (14.9). Thus, this operation leads to linear prediction.

In order to make (14.59) and (14.60) equal zero, the following relation must hold:

$$z_i^2 - a_1 z_i - a_2 = 0 \qquad \text{for } i = 1, 2 \tag{14.62}$$

Since a_1 and a_2 are known, by solving (14.61) one can solve for z_i. Once z_i is obtained, the frequency of the signal f_i can be obtained from (14.56) and the constant c_i can be found from (14.55).

The same approach can be extended to more than two signals. If there are M signals, the result can be written as

$$x(t) = \sum_{i=1}^{M} A_i e^{j(2\pi f_i t + \theta_i)} \tag{14.63}$$

This equation can be written as

$$x(t) = \sum_{i=1}^{M} c_i e^{j2\pi f_i t} \tag{14.64}$$

where $c_i = A_i e^{j\theta i}$ is the complex amplitude. There are $2M$ unknowns in this equation; therefore, $2M$ points of sampling data are needed and they can be written as

$$\begin{aligned} x(0) &= c_1 + c_2 + \ldots + c_M \\ x(1) &= c_1 z_1 + c_2 z_2 + \ldots + c_M z_M \\ &\ldots \\ x(2M-1) &= c_1 z_1^{2M-1} + c_2 z_2^{2M-2} + \ldots + c_M z_M^{2M-1} \end{aligned} \tag{14.65}$$

To solve these equations, M unknowns (a_1 to a_M) are introduced and a set of linear equations of a_i can be formed.

The first equation can be obtained from the following procedure. Take the first $M + 1$ equations, from $x(0)$ to $x(M)$, in the above equation and multiply the first one by a_M, the second one by a_{M-1}, and so forth. The last equation that starts with $x(M)$ is multiplied by -1. These $M + 1$ equations, after the above operation, will be

$$\begin{aligned} a_M x(0) &= a_M c_1 + a_M c_2 + \ldots + a_M c_M \\ a_{M-1} x(1) &= a_{M-1} c_1 z_1 + a_{M-1} c_2 z_2 + \ldots + a_{M-1} c_M z_M \\ &\ldots\ldots \\ -x(M) &= -c_1 z_1^M - c_2 z_2^M - \ldots - c_M z_M^M \end{aligned} \tag{14.66}$$

Add both sides of the equations and set the results equal to zero. From the left side, the result is

$$a_M x(0) + a_{M-1} x(1) + \ldots + a_1 x(M) - x(M+1) = 0 \tag{14.67}$$

This is one of the equations needed to solve for the constant value of a_i. It is interesting to notice that this is a linear prediction equation (i.e., the value of $x(M + 1)$ can be written as a linear combination of $x(0)$ to $x(M)$ with an unknown coefficient of a_i).

The rest of the $M - 1$ equations can be obtained in a similar way. For example, to obtain the second equation, another set of $M + 1$ equations from $x(1)$ to $x(M + 1)$ in (14.65) will be selected. Multiply the first equation by a_M, the second one by a_{M-1}, and so on, and the last one by -1. When the left side of all the equations is added and set to zero, the result is

$$a_M x(1) + a_{M-1} x(2) + \ldots + a_1 x(M+1) - x(M+2) = 0 \tag{14.68}$$

The right side of the equations will be discussed separately. If one considers all the equations, the result is written in matrix form as

$$\begin{bmatrix} x(0) & x(1) & \ldots & x(M) \\ x(1) & x(2) & \ldots & x(M+1) \\ \cdot & \cdot & \cdot & \cdot \\ x(M) & x(M+1) & \ldots & x(2M-1) \end{bmatrix} \begin{bmatrix} a_M \\ a_{M-1} \\ \cdot \\ \cdot \\ a_1 \end{bmatrix} = \begin{bmatrix} x(M+1) \\ x(M+2) \\ \cdot \\ \cdot \\ x(2M) \end{bmatrix} \tag{14.69}$$

From the equation, the coefficients a_i can be solved linearly.

Now let us look at the right side of (14.66). The result is

$$\begin{aligned} & c_1(a_M + a_{M-1} z_1 + \ldots + a_1 z_1^{M-1} - z_1^M) \\ & + c_2(a_M + a_{M-1} z_2 + \ldots + a_1 z_2^{M-1} - z_2^M) \\ & + \ldots \\ & + c_M(a_M + a_{M-1} z_M + \ldots + a_1 z_1^{M-1} - z_M^M) = 0 \end{aligned} \tag{14.70}$$

This equation can be written in matrix form as

$$\begin{bmatrix} c_1 & c_2 & \ldots & c_M \\ c_1 z_1 & c_2 z_2 & \ldots & c_M z_M \\ \ldots & & & \\ c_1 z_1^{M-1} & c_2 z^{M-1} & \ldots & c_M z_1^{M-1} \end{bmatrix} \begin{bmatrix} a_M + a_{M-1} z_1 + \ldots - z_1^M \\ a_M + a_{M-1} z_2 + \ldots - z_2^M \\ \cdot \\ \cdot \\ \cdot \\ a_M + a_{M-1} z_M + \ldots - z_M^M \end{bmatrix} = 0 \tag{14.71}$$

To fulfill this equation, the following relation must be true:

$$a_M + a_{M-1} z_i + \ldots + a_1 z_i^{M-1} - z_i^M = 0 \qquad i = 1, 2, \ldots, M-1 \tag{14.72}$$

Or, in a slightly different form,

$$z^M - a_1 z^{M-1} - \ldots - a_{M-1} z - a_M = 0 \tag{14.73}$$

where z_i is the root of the equation. These are all the required equations for the Prony's method. In these equations, $x(0), x(1), \ldots, x(2M)$ are the measured values.

For the above discussion, the Prony's method can be summarized into four steps as follows.

1. From (14.69), the coefficients a_i, $i = 1, 2, \ldots, M$ can be obtained.
2. From (14.73), z_i, $i = 1, 2, \ldots, M$ can be obtained.
3. From (14.56), $z_i = e^{j2\pi f_i}$, the frequency of the input signal, can be obtained.
4. The amplitude and the initial phase of the input signals, which are expressed as $c_i = A_i e^{j\theta_i}$, can be found from (14.65) once z_i is obtained.

When the signal-to-noise ratio (S/N) is high, Prony's method can produce very accurate results. However, when the input S/N is low, the error generated through the Prony's method can be rather high.

A different approach to derive the Prony's method through the z transform can be found in [40].

14.10 PRONY'S METHOD USING THE LEAST SQUARES APPROACH [3–5, 13, 19, 35–40]

To improve the performance of the Prony's method, one can use more data points. The data points will be used in the least squares way to generate the desired a_i coefficients. Let us use a simple example to demonstrate this idea first, then a general case will be presented.

Assume that there are only two signals which require a minimum of four complex data points. However, in order to obtain better accuracy, six data points are acquired. The result is

$$\begin{aligned} x(0) &= c_1 + c_2 \\ x(1) &= c_1 z_1 + c_2 z_2 \\ x(2) &= c_1 z_1^2 + c_2 z_2^2 \\ x(3) &= c_1 z_1^3 + c_2 z_2^3 \\ x(4) &= c_1 z_1^4 + c_2 z_2^4 \\ x(5) &= c_1 z_1^5 + c_2 z_2^5 \end{aligned} \tag{14.74}$$

Since there are only two signals (or the Prony's method is limited to a second-order equation), only two coefficients a_1 and a_2 will be introduced to generate the linear functions as mentioned in the previous section. The results are

$$\begin{aligned} x(2) &= a_1 x(1) + a_2 x(0) \\ x(3) &= a_1 x(2) + a_2 x(1) \\ x(4) &= a_1 x(3) + a_2 x(2) \\ x(5) &= a_1 x(4) + a_2 x(3) \end{aligned} \tag{14.75}$$

There are four linear equations, but with only two unknowns. To solve these equations, the least squares method is used and the result is [4, 19]

$$\begin{bmatrix} x^*(1) & x^*(2) & x^*(3) & x^*(4) \\ x^*(0) & x^*(1) & x^*(2) & x^*(3) \end{bmatrix} \begin{bmatrix} x(1) & x(0) \\ x(2) & x(1) \\ x(3) & x(2) \\ x(4) & x(3) \end{bmatrix} \begin{bmatrix} a_1 \\ a_2 \end{bmatrix} \tag{14.76}$$

$$= \begin{bmatrix} x^*(1) & x^*(2) & x^*(3) & x^*(4) \\ x^*(0) & x^*(1) & x^*(2) & x^*(3) \end{bmatrix} \begin{bmatrix} x(2) \\ x(3) \\ x(4) \\ x(5) \end{bmatrix}$$

In this equation, all six known values are used to solve for the two unknowns a_1 and a_2. Once the a_i are obtained, the corresponding z values can be obtained from (14.72) and the input frequency can be obtained from the definition of z. The only difference is in the calculation of a_i. All the other steps are the same as mentioned at the end of the previous section.

In general, assume that there are N known $x(0)$ to $x(N-1)$, and M signals in order to solve for the input signal frequencies $N \geq 2M$. If $N > 2M$, the least squares method can be used to solve for a_i and the result is

$$\begin{bmatrix} x^*(M-1) & x^*(M) & .. & x^*(N-2) \\ x^*(M-2) & x^*(M-1) & .. & x^*(N-3) \\ \cdots & & & \\ x^*(0) & x^*(1) & .. & x^*(N-M-1) \end{bmatrix} \tag{14.77}$$

$$\begin{bmatrix} x(M-1) & x(M-2) & .. & x(0) \\ x(M) & x(M-1) & .. & x(1) \\ .. & & & \\ x(N-2) & x(N-3) & .. & x(N-M-1) \end{bmatrix} \begin{bmatrix} a_1 \\ a_2 \\ \cdot \\ \cdot \\ \cdot \\ a_M \end{bmatrix}$$

$$= \begin{bmatrix} x^*(M-1) & x^*(M) & .. & x^*(N-2) \\ x^*(M-2) & x^*(M-1) & .. & x^*(N-3) \\ \cdots & & & \\ x^*(0) & x^*(1) & .. & x^*(N-M-1) \end{bmatrix} \begin{bmatrix} x(M) \\ x(M+1) \\ \cdot \\ \cdot \\ \cdot \\ x(N-1) \end{bmatrix}$$

After a_i ($i = 1, 2, \ldots, M$) are obtained, one nonlinear equation of z (14.72) can be solved to obtain z and the individual frequency of the input signal can be found. The performance of the least squares method should produce a better result because more data points are used in the calculation.

14.11 EIGENVECTORS AND EIGENVALUES [3–5]

In this section the concept of eigendecomposition, eigenvectors, and eigenvalues will be introduced. This concept will be used in later sections to estimate frequencies.

If A is a given square matrix, a constant λ and a vector $\mathbf{X}$ can be found such that

$$A \cdot \mathbf{X} = \lambda \mathbf{X} \tag{14.78}$$

where λ is called an eigenvalue and $\mathbf{X}$ is called its corresponding eigenvector. This process is called the eigendecomposition of A. To find λ and $\mathbf{X}$, the above equation can be written as

$$(A - \lambda I) \cdot \mathbf{X} = 0 \tag{14.79}$$

where I is the identity matrix. In order to have a nontrivial solution (i.e., $\mathbf{X} \neq 0$), the determinant of $A - \lambda I$ should equal to zero. For example, if

$$A = \begin{bmatrix} 1 & 2 \\ 3 & 4 \end{bmatrix} \quad \text{then} \quad A - \lambda I = \begin{vmatrix} 1-\lambda & 2 \\ 3 & 4-\lambda \end{vmatrix} = (1-\lambda)(4-\lambda) - 6 = 0 \tag{14.80}$$

The eigenvalues solved are $\lambda_1 = -0.3723$ and $\lambda_2 = 5.3723$, which can be found using the MATLAB "eig" command. With each eigenvalue there is an eigenvector. The corresponding eigenvector $\mathbf{X}_i = [\mathrm{x}_{i1}\ \mathrm{x}_{i2}]^T$ where superscript T representing the transpose of a matrix can be solved from (14.78). The eigenvectors are found from MATLAB as $\mathbf{X}_1 = [-.8246\ .5658]^T$ and $\mathbf{X}_2 = [-.4160\ -.9094]^T$ with the restriction $\mathrm{x}_{i1}^2 + \mathrm{x}_{i2}^2 = 1$, $i = 1, 2$.

Let us use a simple example to demonstrate the application of eigenvectors and eigenvalues. If the input signal is

$$\mathrm{x}(i) = Ae^{j(2\pi fi+\phi)} + u(i) \tag{14.81}$$

where A, f, and ϕ are the amplitude, frequency, and initial phase of a sine wave, respectively, and $u(i)$ is white Gaussian noise, the autocorrelation with lag k is

$$\begin{aligned} R(k) &= E[x(i+k)x(i)^*] \\ &= E[\{Ae^{j[2\pi f(i+k)+\phi]} + u(i+k)\}\{Ae^{-j[2\pi f(i+k)+\phi]} + u(i)^*\}] \end{aligned} \tag{14.82}$$

where $E[\ ]$ is the expectation value. Since the cross product of signal and noise is zero and the noise is uncorrelated, the results will be

$$R(k) = A^2 e^{j2\pi fk} + \sigma^2 \delta_{0k} \tag{14.83}$$

where σ^2 is the variance of the noise and δ_{ok} is the Kronecker delta, which has the following property:

$$\begin{aligned} \delta_{ij} &= 1 \qquad i = j \\ &= 0 \qquad i \neq j \end{aligned} \tag{14.84}$$

The correlation matrix R can be found as

$$\begin{bmatrix} R(0) & R(-1) \\ R(1) & R(0) \end{bmatrix} = \begin{bmatrix} A^2 + \sigma^2 & A^2 e^{-j2\pi f} \\ A^2 e^{j2\pi f} & A^2 + \sigma^2 \end{bmatrix} \tag{14.85}$$

The eigenvalue and eigenvector of R can be written as

$$\begin{bmatrix} R(0) & R(-1) \\ R(1) & R(0) \end{bmatrix} \begin{bmatrix} 1 \\ a \end{bmatrix} = \lambda \begin{bmatrix} 1 \\ a \end{bmatrix} \tag{14.86}$$

The minimum eigenvalue corresponds to the noise power and this can be proven as follows. Multiplying both sides of the above equation by $[1\ a^*]$, the result is

$$[1\ a^*] \begin{bmatrix} R(0) & R(-1) \\ R(1) & R(0) \end{bmatrix} \begin{bmatrix} 1 \\ a \end{bmatrix} = \lambda [1\ a^*] \begin{bmatrix} 1 \\ a \end{bmatrix} = \lambda (1 + |a|^2) \tag{14.87}$$

Substitute the $R(k)$ values into the left side of above equation, and the result is

$$\begin{aligned} [1\ a^*] \begin{bmatrix} R(0) & R(-1) \\ R(1) & R(0) \end{bmatrix} \begin{bmatrix} 1 \\ a \end{bmatrix} &= [R(0) + a^* R(1) \quad R(-1) + a^* R(0)] \begin{bmatrix} 1 \\ a \end{bmatrix} \\ &= R(0) + a^* R(1) + aR(-1) + aa^* R(0) \\ &= A^2 + \sigma^2 + A^2 a^* e^{j2\pi f} + A^2 a e^{-j2\pi f} + |a|^2 A^2 + \sigma^2 |a|^2 \\ &= \sigma^2 (1 + |a|^2) + A^2 (1 + a^* e^{j2\pi f} + a e^{-j2\pi f} + |a|^2) \\ &= \sigma^2 (1 + |a|^2) + A^2 |1 + a e^{-j2\pi f}|^2 \end{aligned} \tag{14.88}$$

Comparing (14.87) and (14.88), one can obtain the following results:

$$\lambda (1 + |a|^2) = \sigma^2 (1 + |a|^2) + A^2 |1 + a e^{-j2\pi f}|^2 \tag{14.89}$$

The right side of this equation contains two non-negative terms. The minimum value occurs when $A^2|1 + ae^{-j2\pi f}|^2 = 0$; thus, one can conclude that the minimum eigenvalue is

$$\lambda_{\min}^2 = \sigma^2 \tag{14.90}$$

This proves that the minimum eigenvalue is equal to the noise power. This simple example can be generalized to multiple signal cases.

14.12 MUSIC METHOD [5, 41–43]

The MUSIC method was developed by Schmidt in 1981. The word MUSIC stands for MUltiple SIgnal Classification. The basic idea of the MUSIC method is to separate signal from noise through eigenvalue decomposition. The approach is to find all the eigenvalues and eigenvectors of the R matrix and the result can be written as

$$\begin{bmatrix} R(0) & R(1)^* & .. & R(p)^* \\ R(1) & R(0) & .. & R(p-1)^* \\ ... & & & \\ R(p) & R(p-1) & .. & R(0) \end{bmatrix} \begin{bmatrix} v_{00} & v_{01} & .. & v_{0p} \\ v_{10} & v_{11} & .. & v_{1p} \\ ... & & & \\ v_{p0} & v_{p1} & .. & v_{pp} \end{bmatrix} = \begin{bmatrix} \lambda_0 v_{00} & \lambda_1 v_{01} & .. & \lambda_p v_{0p} \\ \lambda_0 v_{10} & \lambda_1 v_{11} & .. & \lambda_p v_{1p} \\ ... & & & \\ \lambda_0 v_{p0} & \lambda_1 v_{p1} & .. & \lambda_p v_{pp} \end{bmatrix} \tag{14.91}$$

The first subscript represents the individual element of the matrix. The second subscript represents numerical order to the eigenvector. For example, the first eigenvector is the first column of the **V** matrix and the second eigenvector is the second column of the **V** matrix, and so on.

The eigenvalues are represented by λ_i. If there are M signals, there are M eigenvalues $\lambda_0 \ldots \lambda_{M-1}$ corresponding to the signals and the rest of the eigenvalues $\lambda_M \ldots \lambda_P$ corresponding to noise. These eigenvalues can be listed in decreasing order as $\lambda_0 > \lambda_1 > \ldots > \lambda_{M-1} > \lambda_M = \ldots = \lambda_p = \sigma^2$.

The eigenvectors **V** with eigenvalues $\lambda_0 \ldots \lambda_{M-1}$ are referenced as the signal subspace, which is represented by $\mathbf{V}_s$. Eigenvectors **V** with eigenvalues $\lambda_M \ldots \lambda_P$ are referenced as the noise subspace, which is represented by $\mathbf{V}_n$. These two subspaces $\mathbf{V}_s$ and $\mathbf{V}_n$ matrices can be written as

$$\mathbf{V}_s = \begin{bmatrix} v_{00} & v_{01} & \cdot\cdot & v_{0M-1} \\ v_{10} & v_{11} & \cdot\cdot & v_{1M-1} \\ \cdots & & & \\ v_{p0} & v_{p1} & \cdot\cdot & v_{pM-1} \end{bmatrix} \quad \mathbf{V}_n = \begin{bmatrix} v_{0M} & v_{0M+1} & \cdot\cdot & v_{0p} \\ v_{1M} & v_{1M+1} & \cdot\cdot & v_{1p} \\ \cdots & & & \\ v_{pM} & v_{pM+1} & \cdot\cdot & v_{pp} \end{bmatrix} \tag{14.92}$$

The signal subspace is orthogonal to the noise subspace.

The basic idea of the MUSIC method is to use the orthogonal property of the signal and noise subspaces. Assume that the input vector **s** is

$$\mathbf{s} = [1 \quad e^{-j2\pi f} \ldots e^{-j2\pi(N-1)f}] \tag{14.93}$$

This vector is orthogonal to the noise subspace. One can write a function P_{MUS} of variable frequency f as

$$P_{MUS}(f) = \frac{1}{\mathbf{s}\mathbf{V}_n\mathbf{V}_n^H\mathbf{s}^H} \tag{14.94}$$

because **s** is a function of f. If the value of f is equal to the frequency of the input signal, the denominator of $P_{MUS}(f)$ is zero (actually it has a minimum) because the signal **s** is orthogonal to the noise subspace $\mathbf{V}_n$. Thus, by plotting the $P_{MUS}(f)$ as a function f, the peaks represent the input frequencies. One can use a root method to find the peak of the $P_{MUS}(f)$ and this method is discussed in [5].

As a summary, the MUSIC method can be obtained from the following steps.

1. From the input data $x(0), x(1), \ldots, x(N-1)$, create the R matrix. The common approach to form the R matrix is through the modified covariance method as shown in (14.47). In this step, the value of p should be selected. Since p should be larger than the maximum number of signals M_{max}, a p value greater than M_{max} is required. If there are N points of data, $p = 2N/3$ seems to be a good selection.
2. Use eigendecomposition on the R matrix to find the eigenvalues λ_i and the **V** matrix. If the number M of the input signals is known, the first large M eigenvalues will be chosen. The rest are the noise eigenvalues. However, if the number of input signals is unknown, one must examine the eigenvalues to determine the number of signals. Large eigenvalues correspond to signals and small eigenvalues correspond to noise. However, the determination is somewhat subjective and S/N-dependent. If two input signals are close in frequency, sometimes it is difficult to separate the signal eigenvalues from the noise eigenvalues. Once the eigenvalues are selected, the corresponding noise eigenvectors $\mathbf{V}_n$ can be obtained from (14.92).
3. Use (14.94) to find the plot of $P_{MUS}(f)$ as a function of f. The peaks in the $P_{MUS}(f)$ plot represent the input frequencies.

A computer program is used to perform the MUSIC calculation. This program does not compare the amplitudes of the eigenvalues to determine the number of signals. This program requires two inputs: the number of signals M and the order of the filter p (or the covariance matrix). The order of the filter must be equal to or greater than $2M + 1$.

The result of the MUSIC is shown in Figure 14.5. The data used are the same as in previous examples. In Figures 14.5(a, b), $M = 4$ and $p = 9$ and 27, respectively. The results show the desired three peaks. These peaks are sharper than those obtained from the Burg method as shown in Figure 14.4. In Figure 14.5(c), $M = 4$ and $p = 28$, and the result does not represent the true input signal. In Figure 14.5(d), $M = 6$ and $p = 13$, and the result shows four peaks. The lowest peak is a spurious response. In Figure 14.5(e), $M = 3$ and $p = 7$, and the plot does not represent the input at all.

From these results, it appears that order of p is not very critical, because in Figures 14.5(a, b), the p value varies from 9 to 27 and the results are satisfactory. Thus, $p = 2N/3 \approx 21$, $(N = 32)$ seems to be a reasonable choice. However, the

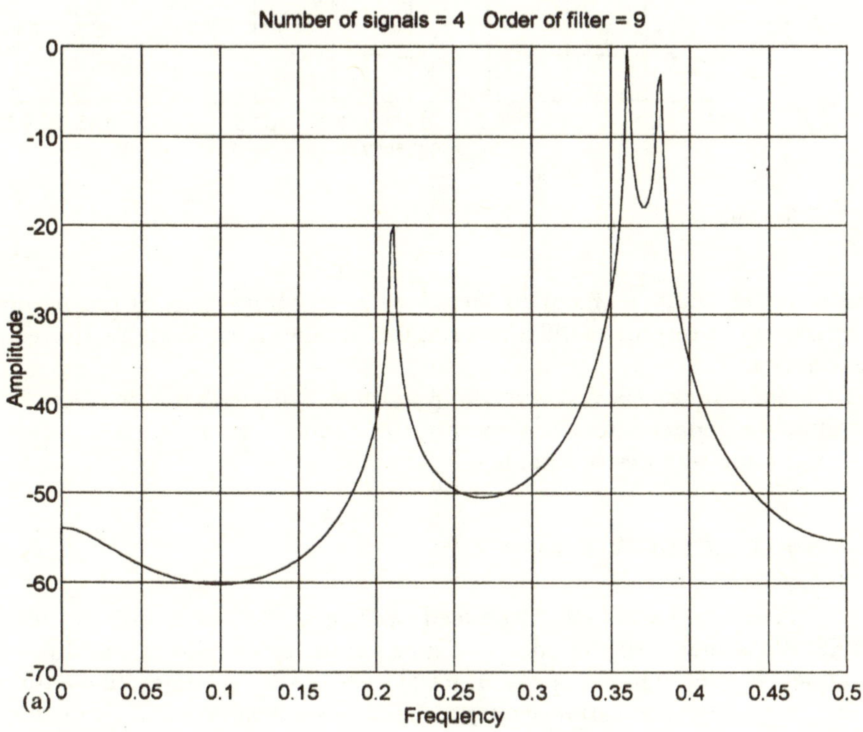

Figure 14.5 Result from MUSIC method: (a) $M = 4$, $p = 9$, (b) $M = 4$, $p = 27$, (c) $M = 4$, $p = 28$, (d) $M = 6$, $p = 13$, (e) $M = 3$, $p = 7$.

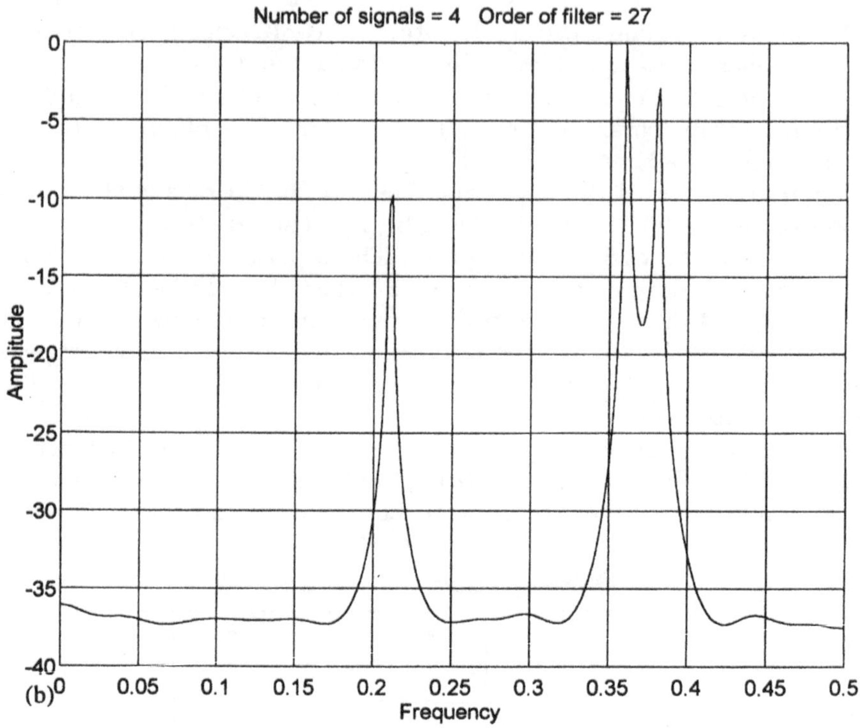

Figure 14.5 (continued).

selection of the number of signals seems to be very important. If the number is too high, spurious responses will appear. If the number is too low, the desired peak will not appear.

In a wideband receiver, it is very important to determine the number of input signals. In the eigendecomposition methods, one must develop an effective method to determine the number of signals.

14.13 ESPRIT METHOD [5, 44–48]

The ESPRIT method was first introduced by Paulraj, Roy, and Kailath [44]. The name ESPRIT stands for Estimation of Signal Parameters via Rotational Invariance Techniques. In the original approach, sometimes the R matrix was ill-conditioned and the eigenvalues were not accurate. Later several approaches were used to improve the performance such as the total least squares ESPRIT [46, 48] and the procrustes rotation [47]. In this section, the original ESPRIT will be presented and the discussion will emphasize the procedure.

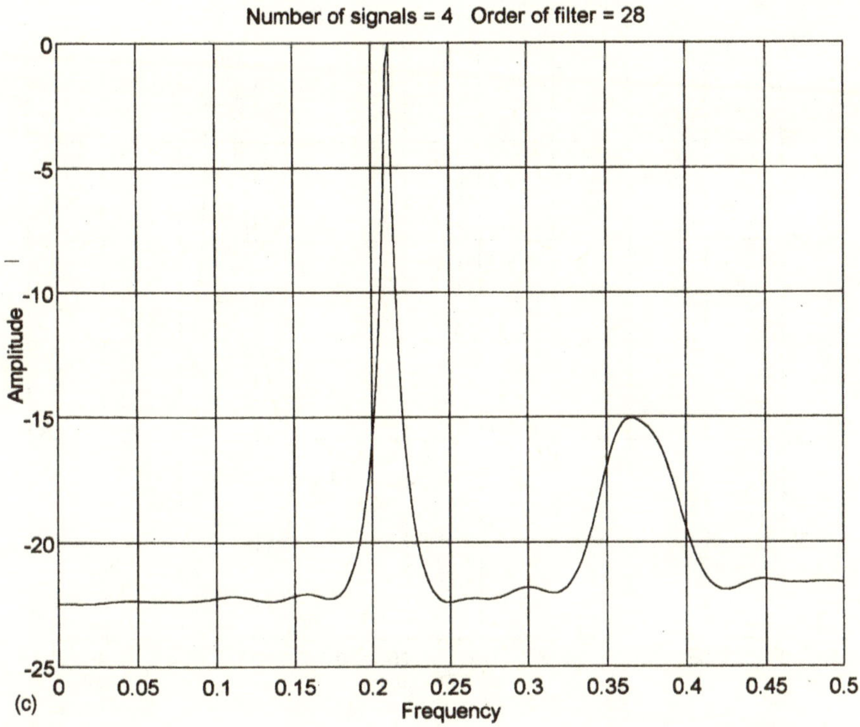

Figure 14.5 (continued).

1. Assume that there are N points of data represented by $x(n)$ where $n = 0$ to $N-1$. These data points can be divided into two groups, each consisting of $N-1$ points. The first group G_1 contains data from $x(0)$ to $x(N-2)$ and the second group G_2 contains data from $x(2)$ to $x(N-1)$, which can be written as

$$
\begin{aligned}
G_1 &= x(0),\ x(1),\ x(2),\ \ldots,\ x(N-2) \\
G_2 &= x(1),\ x(2),\ x(3),\ \ldots,\ x(N-1)
\end{aligned}
\tag{14.95}
$$

These data will be used to make two R matrices R_{yy} and R_{yz} through the covariance approach. The matrix R_{yy} is obtained from data in G_1 and R_{yz} is obtained from data in G_1 and G_2. In obtaining R_{yz}, the first matrix in (14.30) uses the data from G_1 and the second matrix uses data from G_2.

As an example, if N is even and the order $p = N/2 - 1$, the data can be arranged into the following two matrices y and z as

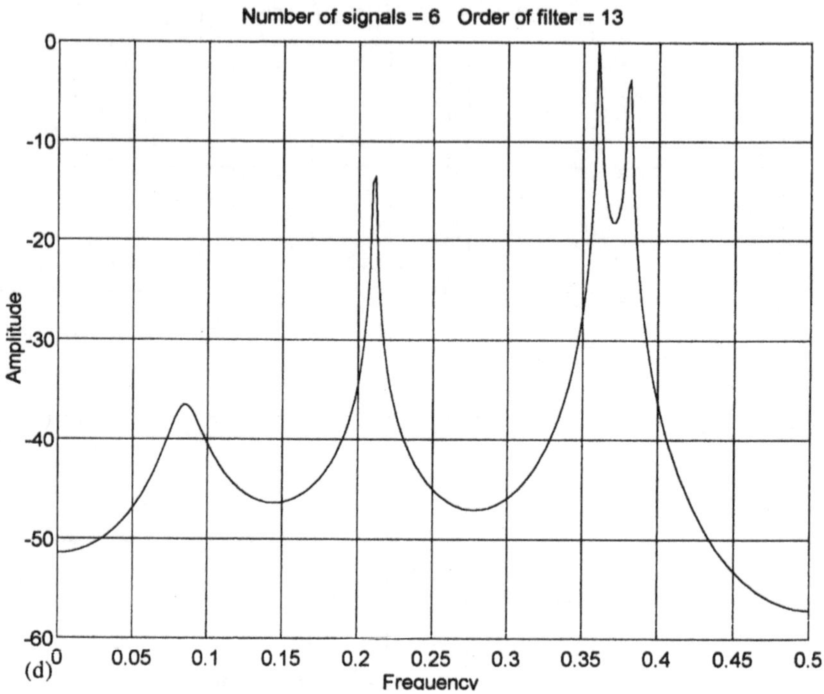

(d)

Figure 14.5 (continued).

$$y = \begin{bmatrix} x(0) & x(1) & \cdots & x\left(\frac{N}{2} - 1\right) \\ x(1) & x(2) & \cdots & x\left(\frac{N}{2}\right) \\ & \cdots & & \\ x\left(\frac{N}{2} - 1\right) & x\left(\frac{N}{2}\right) & \cdots & x(N-2) \end{bmatrix} \tag{14.96}$$

$$z = \begin{bmatrix} x(1) & x(2) & \cdots & x\left(\frac{N}{2}\right) \\ x(2) & x(3) & \cdots & x\left(\frac{N}{2} + 1\right) \\ & \cdots & & \\ x\left(\frac{N}{2}\right) & x\left(\frac{N}{2} + 1\right) & \cdots & x(N-1) \end{bmatrix}$$

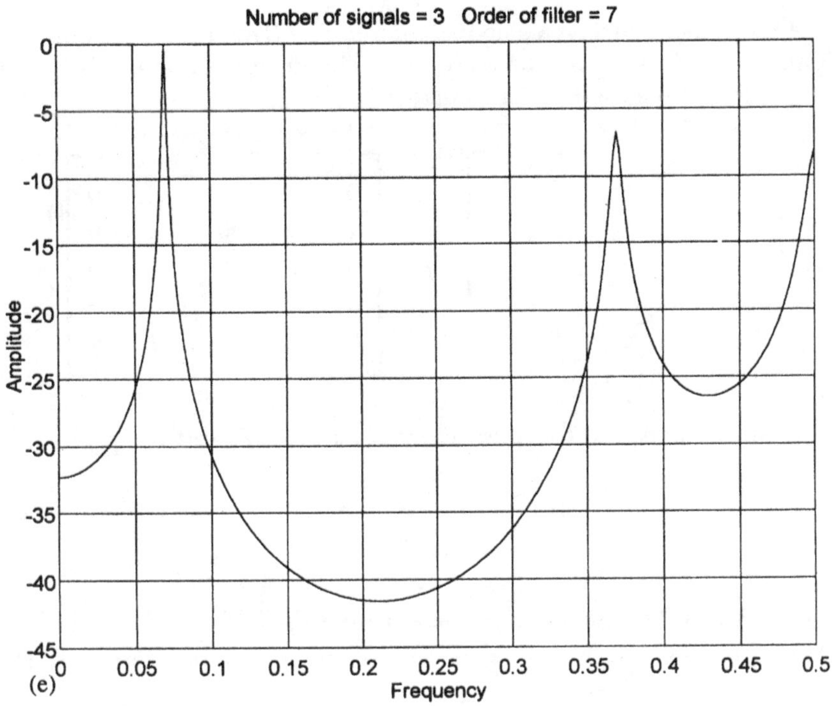

Figure 14.5 (continued).

From these two matrices, one can form two R matrices as

$$R_{yy} = yy^H \qquad R_{yz} = yz^H \tag{14.97}$$

where the superscript H represents the hermitian of a matrix, which is the transpose and conjugate of the matrix. The value R_{yy} can be considered as the autocorrelation matrix and R_{yz} as the cross-correlation matrix.

2. The second step is to perform eigendecomposition on R_{yy} to find the eigenvectors and eigenvalues as

$$R_{yy}\mathbf{e}' = \lambda'\mathbf{e}' \tag{14.98}$$

where $\mathbf{e}'$ and λ' are the eigenvectors and eigenvalues, respectively. From the distribution of eigenvalues, the number of signals can be determined. Large eigenvalues correspond to signal and small eigenvalues correspond to noise. This is the same argument used in the MUSIC method.

3. The next step is to define two matrices—the I and D matrices. Both matrices have a dimension of $N/2$ by $N/2$ to match the dimensions of R_{yy} and R_{yz} in

(14.96). The I matrix is a diagonal matrix. The D matrix has only one diagonal line of 1's; all the other data are zero. The diagonal line of 1's is under the main diagonal axis. These two matrices are

$$I = \begin{bmatrix} 1 & 0 & 0 & \cdots & 0 \\ 0 & 1 & 0 & \cdots & 0 \\ 0 & 0 & 1 & \cdots & 0 \\ & & \cdots & & \\ 0 & \cdots & 0 & 0 & 1 \end{bmatrix} \quad D = \begin{bmatrix} 0 & 0 & 0 & \cdots & 0 \\ 1 & 0 & 0 & \cdots & 0 \\ 0 & 1 & 0 & \cdots & 0 \\ & & \cdots & & \\ 0 & \cdots & 0 & 1 & 0 \end{bmatrix} \tag{14.99}$$

4. The fourth step is to form two more R matrices R_s and R_t

$$\begin{aligned} R_s &= R_{yy} - \lambda_{\min} I \\ R_t &= R_{yz} - \lambda_{\min} D \end{aligned} \tag{14.100}$$

   where $\lambda_{\min}$ is the smallest eigenvalue from step 2.
5. The fifth step is to find the generalized eigendecomposition of R_s and R_t as

$$R_s\mathbf{e} = \lambda R_t\mathbf{e} \tag{14.101}$$

   where $\mathbf{e}$ is the eigenvector and λ is the eigenvalue.
6. The last step is to find the input frequencies. The procedure is to find the λ values close to unit circle. Once these λ_i values are found, the input frequency can be found as

$$f_i = \frac{1}{2\pi}\tan^{-1}\left(\frac{Im\ \lambda_i}{Re\ \lambda_i}\right) \tag{14.102}$$

   where Im and Re represent the imaginary and real parts of λ_i.

In this approach, the eigendecomposition is performed twice, which is computationally intensive. One major advantage is that once the eigenvalues of (14.101) are found, only the eigenvalues close to the unit circle will be selected and the corresponding signal frequencies can be found. It is not necessary to search the entire frequency range, as in the AR and MUSIC methods. This method has the same problem as mentioned before, which is to determine the number of signals from the values of the eigenvalues obtained from (14.102). An ESPRIT program with $p = N/2$ is listed at the end of this chapter.

14.14 MINIMUM NORM METHOD [49, 50]

The minimum norm method was derived by Kumaresan and Tufts [50]. This method and the MUSIC method are somewhat alike. The basic idea is to find a vector **d** that is a linear combination of eigenvectors in the noise subspace. This vector **d** can be written as

$$\mathbf{d} = [\mathbf{d}_0 \quad \mathbf{d}_1 \ldots \mathbf{d}_P]^T \equiv [1 \quad \mathbf{d}_1 \ldots \mathbf{d}_P]^T \tag{14.103}$$

where the superscript T represents a transpose of a matrix. In this equation $\mathbf{d}_0$ is assigned to be unity ($\mathbf{d}_0 \equiv 1$). The square of the norm of this equation is

$$|\mathbf{d}|^2 = \sum_{i=0}^{P} \mathbf{d}_i^2 \tag{14.104}$$

This method minimizes this norm and that is why it is named as such.

The procedure of the minimum norm method is presented here. The first two steps are identical to the MUSIC method.

1. The first step is to find the eigenvectors **V** of the R matrix. The common approach to form the R matrix is through the modified covariance method as shown in (14.47).
2. Perform eigendecomposition on the R matrix to find the eigenvalues λ_i and the **V** matrix. If the number M of the input signals is known, the first large M eigenvalues will be chosen. The rest are the noise eigenvalues. However, if the number of input signals is unknown, one must examine the eigenvalues to determine the number of signals. Large eigenvalues correspond to signals and small eigenvalues correspond to noise. The vector **d** can be found from either the noise subspace $\mathbf{V}_n$ or from the signal noise subspace $\mathbf{V}_s$.
3. Find the vector **d** from the noise subspace $\mathbf{V}_n$. The noise subspace $\mathbf{V}_n$ can be written as

$$\mathbf{V}_n = \begin{bmatrix} v_{0M} & v_{0M+1} & \cdot\cdot & v_{0p} \\ v_{1M} & v_{1M+1} & \cdot\cdot & v_{1p} \\ \cdots & & & \\ v_{pM} & v_{pM+1} & \cdot\cdot & v_{pp} \end{bmatrix} \equiv \begin{bmatrix} c^H \\ \mathbf{V}_n' \end{bmatrix} \tag{14.105}$$

where the superscript H is the hermitian, C^H represents the first row of $\mathbf{V}_n$, and $\mathbf{V}_n'$ represents the rest of the $\mathbf{V}_n$ matrix, which can be written explicitly as

$$c^H = [v_{0M} \quad v_{0M+1} \quad .. \quad v_{0p}] \qquad \mathbf{V}'_n = \begin{bmatrix} v_{1M} & v_{1M+1} & .. & v_{0p} \\ ... & & & \\ v_{pM} & v_{pM+1} & .. & v_{pp} \end{bmatrix} \tag{14.106}$$

4. From c_H and $\mathbf{V}'_n$, the **d** vector can be found as

$$\mathbf{d} = \begin{bmatrix} 1 \\ ------- \\ \mathbf{V}'_n c/(c_H c) \end{bmatrix} \tag{14.107}$$

5. The **d** vector can also obtained from the signal subspace. The signal subspace $\mathbf{V}_s$ can be written as

$$\mathbf{V}_s = \begin{bmatrix} v_{00} & v_{01} & .. & v_{0M-1} \\ v_{10} & v_{11} & .. & v_{1M-1} \\ ... & & & \\ v_{p0} & v_{p1} & .. & v_{pM-1} \end{bmatrix} \equiv \begin{bmatrix} g^H \\ \mathbf{V}'_s \end{bmatrix} \tag{14.108}$$

where g^H is the first row of the $\mathbf{V}_s$ matrix and $\mathbf{V}'_s$ is the rest of the matrix. These two matrices can be written as

$$g^H = [v_{00} \quad v_{01} \quad .. \quad v_{0M-1}] \qquad \mathbf{V}'_s = \begin{bmatrix} v_{10} & v_{11} & .. & v_{1M-1} \\ ... & & & \\ v_{p0} & v_{p1} & .. & v_{pM-1} \end{bmatrix} \tag{14.109}$$

6. The **d** vector can be obtained as

$$\mathbf{d} = \begin{bmatrix} 1 \\ --------- \\ -\mathbf{V}'_s g/(1 - g^H g) \end{bmatrix} \tag{14.110}$$

It should be noted that if steps 3 and 4 are carried out, then steps 5 and 6 are not necessary. The converse is also true (i.e., steps 5 and 6 can replace steps 3 and 4).

7. Once the **d** vector is obtained, a function $P_{MN}(f)$ can be defined as

$$P_{MN}(f) = \frac{1}{\mathbf{sdd}^H\mathbf{s}^H} \tag{14.111}$$

since **s** is a function of the input frequency f, as shown in (14.93), and $P_{MN}(f)$ is a function of f. Plot $P_{MN}(f)$ the peaks represent the input frequencies. This step is similar to the MUSIC method. One can also use a root method to find the frequencies of $P_{MN}(f)$.

The result of the minimum norm is shown in Figure 14.6. The data used are the same as in previous examples. In this figure, the number of signals is selected as 4 and the p value is chosen as 20. The program uses the noise subspace matrix to find the vector **d**. There are some low-level peaks in this approach.

14.15 MINIMUM NORM METHOD WITH DISCRETE FOURIER TRANSFORM [51]

This method was derived by Shaw and Xia and this discussion is based on [51]. It can be considered as a modified approach to the minimum norm discussed in the previous section. The main difference is to replace the eigendecomposition with discrete Fourier transform (DFT). The DFT might be easier to carry out than the eigendecomposition from a hardware point of view. The difference is to replace

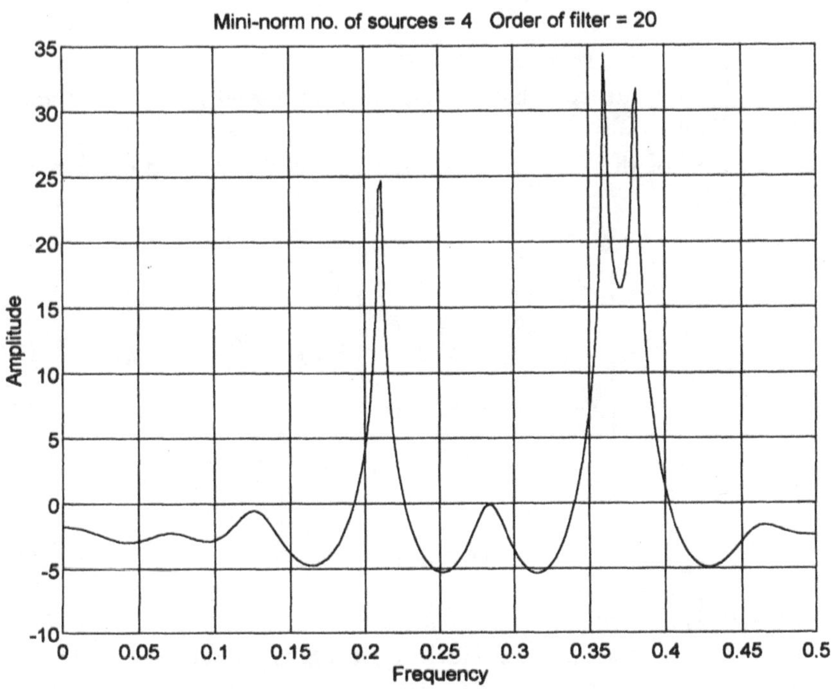

Figure 14.6 Frequency response from minimum norm method.

steps 3 and 4 or 5 and 6 in the previous section. Instead of repeating all the steps in the previous sections, only the changes will be discussed here.

The correlation matrix R has a dimension of $(1 + p) \times (1 + p)$. The following steps are used to obtain the minimum norm.

1. Let us define a Fourier matrix E of dimension of $(1 + p) \times (1 + p)$ as

$$E = \begin{bmatrix} 1 & 1 & 1 & \cdots & 1 \\ 1 & e^{\frac{j2\pi}{p+1}} & e^{\frac{j4\pi}{p+1}} & \cdots & e^{\frac{j2\pi p}{p+1}} \\ 1 & e^{\frac{j4\pi}{p+1}} & e^{\frac{j8\pi}{p+1}} & \cdots & e^{\frac{j4\pi p}{p+1}} \\ & & \cdots & & \\ 1 & e^{\frac{j2\pi p}{p+1}} & e^{\frac{j4\pi p}{p+1}} & \cdots & e^{\frac{j2\pi p}{p+1}} \end{bmatrix} \tag{14.112}$$

From this result, a matrix $\mathbf{F}$ is constructed as

$$\mathbf{F} = RE \tag{14.113}$$

where R is the autocorrelation matrix. This $\mathbf{F}$ matrix has a dimension of $(1 + p) \times (1 + p)$ also.

2. Each column of the $\mathbf{F}$ matrix is considered as a vector, so the $\mathbf{F}$ matrix can be written as

$$\mathbf{F} = \begin{bmatrix} F_{00} & F_{01} & \cdots & F_{0p} \\ F_{10} & F_{11} & \cdots & F_{1p} \\ & & \cdots & \\ F_{p0} & F_{p1} & \cdots & F_{pp} \end{bmatrix} \equiv [f_0 \quad f_1 \quad \cdots \quad f_p] \tag{14.114}$$

where

$$f_i = [F_{0i} \quad F_{1i} \quad \cdots \quad F_{pi}]^T \tag{14.115}$$

and $(i = 0, 1, \ldots, p)$.

3. Find the norm of all the $\mathbf{f}_i$ vectors and represent them by $|\mathbf{f}_i|$. The $|\mathbf{f}_i|$ with large values represent the signals. One has to determine the number of signals from the amplitude of $|\mathbf{f}_i|$. If one selects M signals, the largest M column in (14.114) will be used as the signal matrix. Thus, as

$$\mathbf{V}_s = \begin{bmatrix} F'_{00} & F'_{01} & \cdots & F'_{0,M-1} \\ F'_{10} & F'_{11} & \cdots & F'_{1,M-1} \\ & & \cdots & \\ F'_{p0} & F'_{p1} & \cdots & F'_{p,M-1} \end{bmatrix} \tag{14.116}$$

where $M < p$, it is important to note that the columns of the $\mathbf{V}_s$ matrix are the columns of matrix $\mathbf{F}$ with largest $M\,|f_i|$ values.

For example, if there are two large norms and they are elements 2 and 4 from (14.114) and (14.115), the signal subspace matrix is

$$\mathbf{V}_s = \begin{bmatrix} F_{01} & F_{03} \\ F_{11} & F_{13} \\ \cdots & \\ F_{p1} & F_{p3} \end{bmatrix} \tag{14.117}$$

Once $\mathbf{V}_s$ is obtained, (14.108) to (14.110) can be used to find vector $\mathbf{d}$. The frequency response can be found from (14.111) once $\mathbf{d}$ is known. An example is shown in Figure 14.7. The result is similar to the result in Figure 14.6, though no eigendecomposition is required.

14.16 ADAPTIVE SPECTRUM ESTIMATION [52–54]

The general idea of the adaptive method will not be discussed in this section. There are many books on this subject [52, 53]. In this section, only sinusoidal inputs are considered since this input is suited for our application. This discussion is based on [54].

Let us assume that the input signal contains M sinusoidal waves. The digitized data can be written as

$$x(n) = \sum_{i=1}^{M} A_i e^{j2\pi f_i n} + u(n) \tag{14.118}$$

where A_i and f_i are the amplitude and frequency of the input sinusoidal wave and $u(n)$ is noise. The problem is, as usual, to find the amplitude and the frequency of the input signal given $x(n)$.

The basic relationship used in this approach is

$$e^{j2\pi fn} = 2\cos(2\pi f)\, e^{j2\pi f(n-1)} - e^{j2\pi f(n-2)} \tag{14.119}$$

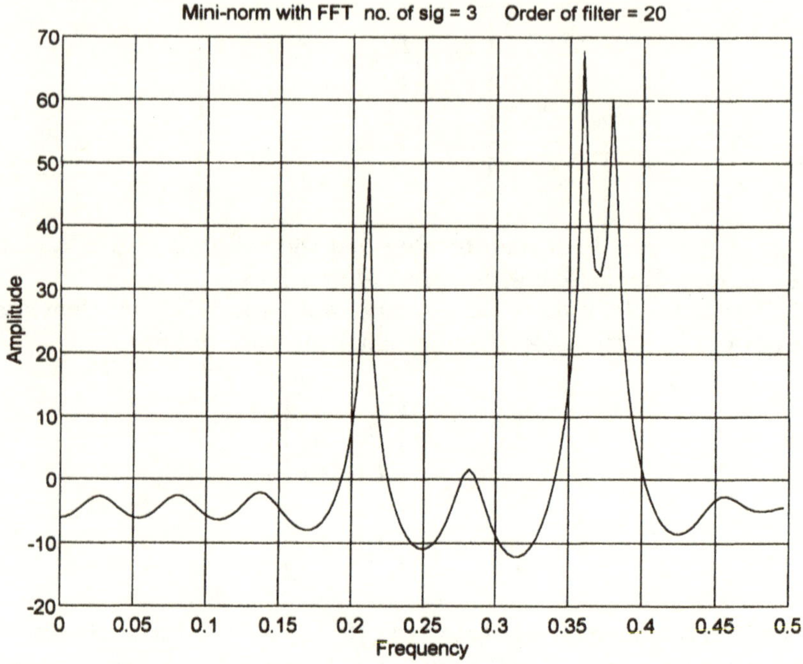

Figure 14.7 Frequency response of minimum norm method with DFT.

This relation can be proved easily if one substitutes

$$\cos(2\pi f) = \frac{e^{j2\pi f} + e^{-j2\pi f}}{2} \tag{14.120}$$

into (14.119). A similar equation is used in Section 10.8 to find zero crossing. Equation (14.119) indicates that the recent sample output $x(n)$ can be predicted by the two previous samples if the frequency f is known.

Separating the first term in (14.118), the following equations can be obtained:

$$\begin{aligned} x(n) &= A_1 e^{j2\pi f_1(n)} + \sum_{i=2}^{M} A_i e^{j2\pi f_i(n)} + u(n) \\ x(n-1) &= A_1 e^{j2\pi f_1(n-1)} + \sum_{i=2}^{M} A_i e^{j2\pi f_i(n-1)} + u(n-1) \\ x(n-2) &= A_1 e^{j2\pi f_1(n-2)} + \sum_{i=2}^{M} A_i e^{j2\pi f_i(n-2)} + u(n-2) \end{aligned} \tag{14.121}$$

Let us define a new quantity $x^{(1)}(n)$ as

$$x^{(1)}(n) \equiv x(n) - 2\cos(2\pi f_1)\, x(n-1) + x(n-2) \qquad (14.122)$$
$$\equiv x(n) - 2a_{1,n-1}\, x(n-1) + x(n-2)$$

where

$$a_{1,n-1} = \cos(2\pi f_1) \qquad (14.123)$$

The first subscript of a represents the numerical order of the signal and the second subscript represents the number of iterations. Substituting the results of (14.121) into (14.122), the result can be written as

$$x^{(1)}(n) = \sum_{i=2}^{M} \alpha_{i,1} A_i e^{j2\pi f_i n} + u^{(1)}(n) \qquad (14.124)$$

where

$$\cos(2\pi f_1)\, \alpha_{k,m} = 2[\cos(2\pi f_k) - \cos(2\pi f_m)] \qquad (14.125)$$

and $u^{(1)}(n)$ represents the residue noise. This operation removes the first sinusoidal signal from the $x(n)$ data and generates a second signal sequence $x^{(1)}(n)$.

This procedure can be applied to the new sequence $x^{(1)}(n)$ generated from (14.124). The result is

$$x^{(2)}(n) \equiv x^{(1)}(n) - 2\cos(2\pi f_2)\, x^{(1)}(n-1) + x^{(1)}(n-2)$$
$$\equiv x^{(1)}(n) - 2a_{2,n-1}\, x^{(1)}(n-1) + x^{(1)}(n-2) \qquad (14.126)$$
$$= \sum_{i=3}^{M} \alpha_{i,1}\, \alpha_{i,2} A_i e^{j2\pi f_i n} + u^{(2)}(n)$$

where

$$a_{2,n-1} = \cos(2\pi f_2) \qquad (14.127)$$

This operation removes the second signal.

This process can be continued in a repetitive form until there are no further sinusoidal components, but only noise in the series. The general extraction process can be written as

$$
\begin{aligned}
x^{(k)}(n) &= x^{(k-1)}(n) - 2\cos(2\pi f_k)\, x^{(k-1)}(n-1) + x^{(k-1)}(n-2) \\
&= x^{(k-1)}(n) - 2a_{k,n-1}\, x^{(k-1)}(n-1) + x^{(k-1)}(n-2) \\
&= \sum_{i=k+1}^{M} \left(\prod_{l=1}^{k} \alpha_{i,l} \right) A_i e^{j2\pi f_i n} + u^{(k)}(n)
\end{aligned} \tag{14.128}
$$

$$
\begin{aligned}
&\text{for} \quad k = 1, 2, \ldots, M \\
&x^{(0)}(n) = x(n) \\
&x^{(k)}(0) = x(o)
\end{aligned}
$$

where

$$a_{k,n-1} = \cos(2\pi f_k) \tag{14.129}$$

The final data $x^{(M)}(n)$ contain only noise. All the other terms $x^{(1)}(n)$, $x^{(2)}(n)$, $\ldots$, $x^{(M-1)}(n)$ obtained through the intermediate steps contain signals.

It should be noted that $a_{k,n}$ is only related to $x^{(k)}$ and $x^{(k-1)}$. Once the final values of $a_{k,n}$ are obtained, the frequency can be obtained through (14.129) as

$$f_k = \frac{\cos^{-1}(a_{k,n})}{2\pi} \tag{14.130}$$

The value of $a_{k,n}$ can be calculated from the gradient equation using the method of the steepest slope. Let us define the error as

$$\epsilon(n) \equiv x(n) - \hat{x}(n) \tag{14.131}$$

To find the gradient of the squared error is to take the derivative of $\epsilon^{(k)}(n)\epsilon^{(k)}(n)^*$ or

$$
\begin{aligned}
\nabla[\epsilon(n)\epsilon(n)^*] &= -2[\epsilon x^{(k-1)}(n-1)^* + \epsilon^* x^{(k-1)}(n-1)] \\
&= -2[x^M(n)\, x^{(k-1)}(n-1)^* + x^M(n)^* x^{(k-1)}(n-1)]
\end{aligned} \tag{14.132}
$$

where the superscript * represents complex conjugate and $x^M(n)$ represents the noise, because there are only M signals.

To find $a_{k,n}$ from $a_{k,n-1}$ is done through the gradient of the squared error. The result is

$$
\begin{aligned}
a_{k,n} &= a_{k,n-1} + \mu\left(\frac{1}{-2}\right)\nabla[\epsilon^{(k)}(n)\epsilon^{(k)}(n)^*] \\
&= a_{k,n-1} + \mu[x^M(n)^* x^{(k)}(n-1) + x^M(n)\, x^{(k)}(n-1)^*]
\end{aligned} \tag{14.133}
$$

This equation is obtained from the result of (14.132). The constant $-1/2$ is introduced to absorb the -2 into the constant μ where μ is the step size. If μ is chosen too large, the result may not converge. If μ is chosen too small, the result will converge very slowly. The actual process is to calculate a_k values one at a time.

Let us put all the equations used in this calculation together and then use an example to demonstrate this operation. Two sets of equations are needed with initial conditions. If there are M signals, the initial conditions are

$$x^{(0)}(n) = x(n) \qquad x^{(k)}(0) = x(0) \tag{14.134}$$

The first set of equations are

$$\begin{aligned}
x^{(1)}(n) &= x^{(0)}(n) - 2a_{1,n-1}x^{(0)}(n-1) + x^{(0)}(n-2) \\
x^{(2)}(n) &= x^{(1)}(n) - 2a_{2,n-1}x^{(1)}(n-1) + x^{(1)}(n-2) \\
&\cdots \\
x^{(M)}(n) &= x^{(M-1)}(n) - 2a_{M,n-1}x^{(M-1)}(n-1) + x^{(M-1)}(n-2)
\end{aligned} \tag{14.135}$$

The second set of equations are

$$\begin{aligned}
a_{1,n} &= a_{1,n-1} + \mu[x^{(M)}(n)^*x^{(0)}(n-1) + x^{(M)}(n)\,x^{(0)}(n-1)^*] \\
a_{2,n} &= a_{2,n-1} + \mu[x^{(M)}(n)^*x^{(1)}(n-1) + x^{(M)}(n)\,x^{(1)}(n-1)^*] \\
&\cdots \\
a_{M,n} &= a_{M,n-1} + \mu[x^{(M)}(n)^*x^{(M-1)}(n-1) + x^{(M)}(n)\,x^{(M-1)}(n-1)^*]
\end{aligned} \tag{14.136}$$

Now let us use an example to illustrate the applications of these equations. The input data are $x(0)$, $x(1)$, . . . , $x(N-1)$ and there are two signals ($M = 2$).

1. Arbitrarily choose $a_{1,0} = a_{2,0} = 0.25$ as the initial values and the initial conditions from (14.134) are

$$x^{(0)}(n) = x(n) \quad x^{(k)}(0) = x(0)$$

2. Use (14.135) to obtain $x^{(k)}(1)$ as

$$\begin{aligned}
x^{(1)}(1) &= x^{(0)}(1) - 2a_{1,0}x^{(0)}(0) = x(1) - 2a_{1,0}x(0) \\
x^{(2)}(1) &= x^{(1)}(1) - 2a_{2,0}x^{(0)}(0)
\end{aligned}$$

In these equations, $x(-1)$ does not exist; thus, only two terms are shown.

3. Use (14.136) to obtain $a_{k,1}$ as

$$a_{1,1} = a_{1,0} + \mu[x^{(2)}(1)^* x^{(0)}(0) + x^{(2)}(1)x^{(0)}(0)^*]$$
$$a_{2,1} = a_{2,0} + \mu[x^{(2)}(1)^* x^{(1)}(0) + x^{(2)}(1)x^{(1)}(0)^*]$$

4. Use (14.135) again to obtain $x^{(k)}(2)$

$$x^{(1)}(2) = x^{(0)}(2) - 2a_{1,1}x^{(0)}(1) + x^{(0)}(0)$$
$$x^{(2)}(2) = x^{(1)}(2) - 2a_{1,1}x^{(1)}(1) + x^{(1)}(0)$$

5. Use (14.136) again to obtain $a_{k,2}$ as

$$a_{1,2} = a_{1,1} + \mu[x^{(2)}(2)^* x^{(0)}(1) + x^{(2)}(2)\,x^{(0)}(1)^*]$$
$$a_{2,1} = a_{2,0} + \mu[x^{(2)}(2)^* x^{(1)}(1) + x^{(2)}(2)\,x^{(1)}(1)^*]$$

These two equations, (14.135) and (14.136), can be used alternatively, to find the value of $a_{1,n}$ as a function of n. When $a_{1,n}$ converges, the value can be used to find the input frequency f_1 from (14.130).

If the k value in $a_{k,n}$ is smaller than the number of input signals M, the error $x^{(k)}(n)$ will oscillate, which means the remaining sinusoidal waves are still in the error signal. When the k value in $a_{k,n}$ is equal to the number of input signals M, the error $x^{(k)}(n)$ will approach a very small value, which is the noise level.

This method is sensitive to noise. If the input S/N level is low, the $a_{k,n}$ value will not converge fast or may not converge. It is also sensitive to step size and number of iterations. A program to predict up to three signals is listed in the end of the chapter. Figure 14.8 shows the result of three signals with frequencies at 0.1, 0.2, and 0.4 with the same amplitude and signal-to-noise ratio of 40 dB. Five hundred iterations are performed with $\mu = 0.001$.

This specific case starts to converge to the desired frequencies at about 200 iterations.

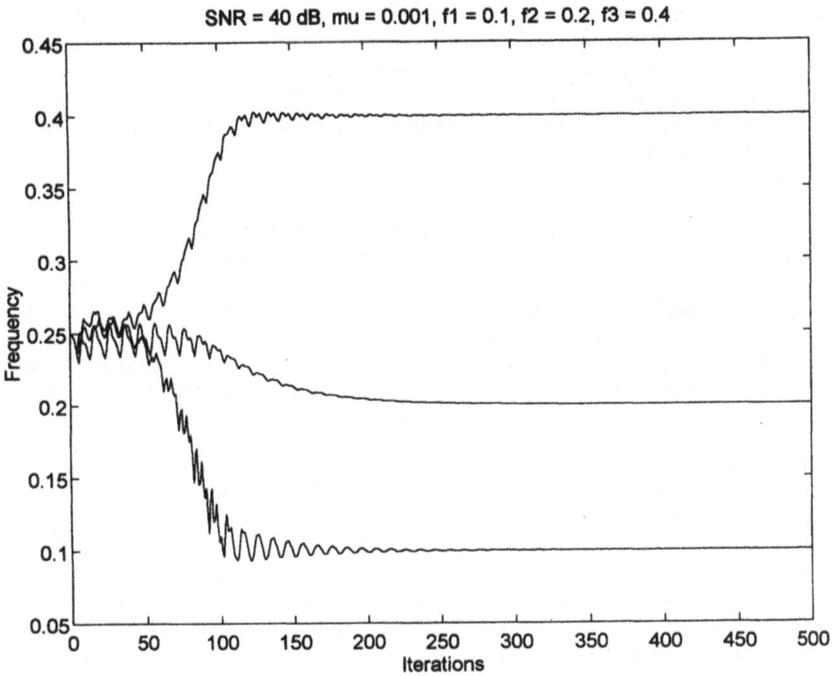

Figure 14.8 Adaptive frequency estimation.

REFERENCES

[1] Childers, D. G., Editor. *Modern Spectrum Analysis I,* Piscataway, NJ: IEEE Press, 1978.

[2] Kesler, S. B., Editor. *Modern Spectrum Analysis II,* Piscataway, NJ: IEEE Press, 1986.

[3] Marple, S. L. *Digital Spectral Analysis With Applications,* Englewood Cliffs, NJ: Prentice Hall, 1987.

[4] Kay, S. M. *Modern Spectral Estimation, Theory and Application,* Englewood Cliffs, NJ: Prentice Hall, 1987.

[5] Therrien, C. W. *Discrete Random Signals and Statistical Signal Processing,* Englewood Cliffs, NJ: Prentice Hall, 1992.

[6] Mitra, S. K., and Kaiser, J. F., Editors. *Handbook for Digital Signal Processing,* New York, NY: John Wiley & Sons, 1993.

[7] Parzen, E. "Some Recent Advances in Time Series Modeling," *IEEE Trans. Automatic Control,* Vol. AC-19, Dec. 1974, pp. 723–730.

[8] Makhoul, J. "Linear Predication: a Tutorial Review," *IEEE Proc.,* Vol. 63, April 1975, pp. 561–580.

[9] Jackson, L. B., and Wood, S. L. "Linear Prediction in Cascade Form," *IEEE Trans. Acoustics, Speech, and Signal Processing,* Vol. ASSP-26, Dec. 1978, pp. 518–528.

[10] Kay, S. M. "The Effects of Noise on the Autoregressive Spectral Estimator," *IEEE Trans. Acoustics, Speech, and Signal Processing,* Vol. ASSP-27, Oct. 1979, pp. 478–485.

[11] Kay, S. M. "Noise Compensation for Autoregressive Spectral Estimate," *IEEE Trans. Acoustics, Speech, and Signal Processing,* Vol. ASSP-28, June 1980, pp. 292–303.

[12] Toomey, J. P. "High Resolution Frequency Measurement by Linear Prediction," *IEEE Trans. Aerospace and Electronic Systems,* Vol. AES-16, July 1980, pp. 517–525.

[13] Kay, S. M., and Marple, S. L. "Spectrum Analysis - a Modern Perspective," *IEEE Proc.*, Vol. 69, Nov. 1981, pp. 1380–1419.

[14] Kay, S. M. "Robust Detection by Autoregressive Spectrum Analysis," *IEEE Trans. Acoustics, Speech, and Signal Processing*, Vol. ASSP-30, April 1982, pp. 256–268.

[15] Kumaresan, R. "Accurate Frequency Estimation Using an All-Pole Filter With Mostly Zero Coefficients," *IEEE Proc.*, Vol. 70, Aug. 1982, pp. 873–875.

[16] Cakzow, J. A. "Spectral Estimation: an Overdetermined Rational Model Equation Approach," *IEEE Proc.*, Vol. 70, Sept. 1982, pp. 907–939.

[17] Kay, S., and Makhoul, J. "On the Statistics of the Estimated Reflection Coefficients As an Autoregressive Process," *IEEE Trans. Acoustics, Speech, and Signal Processing*, Vol. ASSP-31, Dec. 1983, pp. 1447–1455.

[18] Tretter, S. A. "Estimating the Frequency of a Noisy Sinusoid by Linear Regression," *IEEE Trans. Information Theory*, Vol. IT-31, Nov. 1985, pp. 831–835.

[19] Tsui, J. B. Y. *Digital Microwave Receivers, Theory and Concepts*, Norwood MA: Artech House, 1989.

[20] Levinson, N. "The Wiener rms (root mean square) Error Criterion in Filter Design and Prediction," *Journal of Mathematical Physics*, Vol. 25, 1947, pp. 261–278.

[21] Durbin, J. "The Fitting of Time Series Models," *Review of the International Statistical Institute*, Vol. 28, 1960 NTIS Accession Number: AD-230 528/2, pp. 233–243.

[22] Haykin, S. *Adaptive Filter Theory*, Englewood Cliffs, NJ: Prentice Hall, 1986.

[23] Burg, J. P. "Maximum Entropy Spectral Analysis," *Proc. of 37th Meeting of the Society of Exploration Geophysicists*, 1967.

[24] Burg, J. P. "New Analysis Technique for Time Series Data," NATO Advanced Study Institute on Signal Processing with Emphasis on Underwater Acoustics, Aug. 1968, pp. 12–23.

[25] Chen, W. Y., and Stegen, G. R. "Experiments With Maximum Entropy Power Spectra of Sinusoids," *J. Geophysical Research*, Vol. 79, July 1974, pp. 3019–3022.

[26] Burg, J. P. "Maximum Entropy Spectral Analysis," Ph.D. dissertation, Stanford University, 1975.

[27] Ulrych, T. J., and Clayton, R. W. "Time Series Modeling and Maximum Entropy," *Phys. Earth Planetary Interiors*, Vol. 12, Aug. 1976, pp. 188–200.

[28] Fougere, P. F., Zawalick, E. J., and Radoski, H. R. "Spontaneous Line Splitting in Maximum Entropy Power Spectrum Analysis," *Phys. Earth Planetary Interiors*, Vol. 12, Aug. 1976, pp. 201–207.

[29] Swingler, D. N. "A Comparison Between Burg's Maximum Entropy Method and a Nonrecursive Technique for the Spectral Analysis of Deterministic Signals," *J. Geophysical Research*, Vol. 84, Feb. 1979, pp. 679–685.

[30] Swingler, D. N. "A Modified Burg Algorithm for Maximum Entropy Spectral Analysis," *IEEE Proc.*, Vol. 67, Dec. 1979, pp. 1368–1369.

[31] Kaveh, M., and Lippert, G. A. "On Optimum Tapered Burg Algorithm for Linear Prediction and Spectral Analysis," *IEEE Trans. Acoustics, Speech, and Signal Proc.*, Vol. ASSP-31, April 1983, pp. 438–444.

[32] Helme, B. I., and Nikias, C. L. "A High Resolution Modified Burg Algorithm for Spectral Estimation," *Proc. Int. Electr. Clectron. Conf.*, Toronto, Canada, Sept. 26–28, 1983, pp. 348–351.

[33] Akaike, H. "Statistical Predictor Identification," *Ann. Inst. Statistic Math.*, Vol. 22, 1970, pp. 203–217.

[34] Akaike, H. "A New Look at the Statistical Model Identification," *IEEE Trans. Automatic Control*, Vol. AC-19, Dec. 1976, pp. 716–723.

[35] Hildebrand, F. B. *Introduction to Numerical Analysis*, New York, NY: McGraw Hill Book Co., 1956.

[36] Chuang, C. W., and Moffatt, D. L. "Natural Resonances of Radar Targets via Prony's Method and Target Discrimination," *IEEE Trans. Aerospace and Electronic Systems*, Vol. AES-12, Sept. 1976, pp. 583–589.

[37] Blaricum, M. L., and Mittra, R. "Problems and Solutions Associated With Prony's Method for Processing Transient Data," *IEEE Trans. Antennas and Propagation*, Vol. AP-26, Jan. 1978, pp. 174–182.

[38] Kumaresan, R., and Tufts, D. W. "Improved Spectral Resolution III: Efficient Realization," *IEEE Proc.*, Vol. 68, Oct. 1980, pp. 1354–1355.

[39] Kumaresan, R., Tufts, D. W., and Scharf, L. L. "A Prony Method for Noisy Data: Choosing the Signal Components and Selecting the Order in Exponential Signal Models," *IEEE Proc.*, Vol. 72, Feb. 1984, pp. 230–233.

[40] Kumaresan, R. "Spectral Analysis," Ch. 16 of Mitra, S. K., and Kaiser, J. F., Editors. *Handbook for Digital Signal Processing,* New York, NY: John Wiley & Sons, 1993.

[41] Tufts, D. W., and Kumaresan, R. "Singular Value Decomposition and Improved Frequency Estimation Using Linear Prediction," *IEEE Trans. Acoustics, Speech, and Signal Processing,* Vol. ASSP-30, Aug. 1982, pp. 671–675.

[42] Schmidt, R. "Multiple Emitter Location and Signal Parameter Estimation," *Proc. of the RADC Spectrum Estimation Workshop,* Rome Air Development Center, 1979, pp. 243–258. Reprinted in *IEEE Trans. Antennas and Propagation,* Vol. AP-34, March 1986, pp. 276–290.

[43] Schmidt, R. "A Signal Subspace Approach to Multiple Emitter Location and Spectral Estimation," Ph.D. thesis, Stanford University, CA, Aug. 1981.

[44] Paulraj, A., Roy, R., and Kailath, T. "Estimation of Signal Parameters via Rotational Invariance Techniques-ESPRIT," *Proc. 19th Asilomar Conference on Circuits, Systems and Computers,* Asilomar, CA, Nov. 1985, pp. 83–89.

[45] Roy, R., Paulraj, A., and Kailath, T. "ESPRIT - A Subspace Rotation Approach to Estimation of Parameters of Cisoids in Noise," *IEEE Trans. Acoustics, Speech, and Signal Processing,* Vol. ASSP-34, Oct. 1986, pp. 1340–1342.

[46] Roy, R. H. "ESPRIT - Estimation of Signal Parameters via Rotational Invariance Techniques," Ph.D. thesis, Stanford University, Stanford, CA, Aug. 1987.

[47] Zoltowski, M. D., and Stavrinides, D. "Sensor Array Signal Processing via a Procrustes Rotations Based Eigenanalysis of the ESPRIT Data Pencil," *IEEE Trans. Acoustics, Speech, and Signal Processing,* Vol. ASSP-37, June 1989, pp. 832–860.

[48] Roy, R. H., and Kailath, T. "ESPRIT-Estimation of Signal Parameters via Rotational Invariance Techniques," *IEEE Trans. Acoustics, Speech, and Signal Processing,* Vol. ASSP-37, July 1989, pp. 984–995.

[49] Tufts, D. W., and Kumaresan, R. "Estimation of Frequencies of Multiple Sinusoids: Making Linear Prediction Behave Like Maximum Likelihood," *IEEE Proc.*, Vol. 70, Sept. 1982, pp. 975–989.

[50] Kumaresan, R., and Tufts, D. W. "Estimating the Angles of Arrival of Multiple Plane Waves," *IEEE Trans. Acoustics, Speech, and Signal Processing,* Vol. ASSP-19, Jan. 1983, pp. 134–139.

[51] Shaw, A. K., and Xia, W. "Minimum-Norm Method Without Eigendecomposition," *IEEE Signal Processing Letters,* Vol. 1, Jan. 1994, pp. 12–14.

[52] Grant, P. M. *Adaptive Filters,* Englewood Cliffs, NJ: Prentice Hall, 1985.

[53] Widrow, B., and Stearns, S. D. *Adaptive Signal Processing,* Englewood Cliffs, NJ: Prentice Hall, 1985.

[54] Cheung, J. Y. "A Direct Adaptive Frequency Estimation Technique," *Presented to the 30th Midwest Symposium on Circuits and Systems,* Syracuse, NY, Aug. 1987.

APPENDIX 14.A

```
% df12_2.m generate input data
clear
N = 4096;
t = [0:1:31];
x = cos(2*pi*.21*t+.1) + 2*cos(2*pi*.36*t) + 1.9*cos(2*pi*.38*t);
xp = [x zeros(1,N-32)];
xpf = fft(xp);
xpa = abs(xpf);
xaxis = linspace(0,.5,N/2);
```

```
plot(xaxis, xpa(1:N/2))
xlabel('Frequency')
ylabel('Amplitude')
grid
```

APPENDIX 14.B

```
% ******** df12_3 Autocorrelation method ********
% ******** Autocorrelation method (or the Yule-Walker approach)
clear
load xx.mat
N = length(x);
ip=input(' The order of the model ip = ?  ');
% ******** Generate the autocorrelation coefficient vector r_xx from x
r_xx=zeros(1,ip+1);
for k=1:ip+1,
     for n=1:N-k+1,
        r_xx(k)=r_xx(k)+conj(x(n))*x(n+k-1);
     end
end
r_xx=r_xx./N;
% ******** Generate the autocorrelation coefficient matrix R from r_xx
for ii=1:ip,
     for jj=ii:ip,
        R(jj,ii)=r_xx(jj-ii+1);
     end
end
for ii=1:ip-1,
     for jj=ii+1:ip,
        R(ii,jj)=R(jj,ii)';
     end
end

% ******** Find the autoregressive model coefficient vector a
a=inv(R)*(-r_xx(2:ip+1)).';
% ******** Find the power spectral density S_xx of the autogressive model
f = linspace(0,.5,100);
for ii=1:100,
     s=0;
     for jj=1:ip,
        s=s+a(jj).'*exp(-i*2*pi*f(ii)*jj);
     end
```

```
    S_xx(ii)=1/(abs(1+s)^2);
end
% ******* plot ************
S_xxlog = 10*log10(S_xx);
xaxpt = length(S_xx);
xax = linspace(0,.5,xaxpt);
plot(xax, S_xxlog)
ips = conv_vs(ip);
title(['AR Process    order =', ips])
xlabel('Frequency')
ylabel('Amplitude')
grid
```

APPENDIX 14.C

```
% df12_4.m Burg method
clear
load xx.mat
p = input('enter # of poles = ');
a = lpc(x,p);
H = freqz(1, a, 1000);
xaxis = linspace(0,.5,1000);
plot(xaxis, abs(H))
xlabel('Frequency')
ylabel('Amplitude')
plab = conv_vs(p);
title(['Number of poles = ', plab])
grid
```

APPENDIX 14.D

```
% df12_5 MUSIC algorithm
% ************************************************************
clear
eval(['load xx.mat ']);
sig = x;

% ************************************************************
numpts = 512; % input('Enter Number of Output Data Points: ');
Ns = input('Enter Number of Sources: ');
mf  = input('Enter Desired Filter Order min = 2*source#+1 : ');
[a,b] = size(sig);
```

```
K = b;   % Number of snapshots
% ************************************************************
% Compute the R matrix using the covariance method
% ************************************************************
C2 = [];
for k = 1:mf,
     C1 = sig(k:K-mf+k);
     C1 = C1(:);
     C2 = [C2 C1];
end
Ca = fliplr(C2);
C3 = C2';
Cb = flipud(C3);
rmat = Cb*Ca*1/(K);
% ***********************************************************
[v d] = eig(rmat);                                  % find eigenvalues
[lambda,k1]=sort(diag(d));
E=v(:,k1);
nspace = v(:,1:mf-2*Ns);
% nspace = v(:,1:mf-Ns);
for complex data for k = 1:numpts+1,
          w(k) = (k-1)*2*pi/(numpts);
          l = 1:mf;
          ss = exp(j*w(k)*(l-1));
          s = ss(:);
          pmu(k) = 1.0/(s'*nspace*nspace'*s);
end
pmu = abs(pmu);
pmu = 10.0*log10(pmu/max(pmu));
% ***********************************************************
xaxpt = length(pmu)/2;
xax = linspace(0,.5,xaxpt);
plot(xax, pmu(1:xaxpt))
grid;
Nss = conv_vs(Ns);
mfs = conv_vs(mf);
title(['Number of signals = ', Nss, '  Order of filter = ', mfs]);
xlabel('Frequency');
ylabel('Amplitude');_
```

APPENDIX 14.E

```
% ESPRIT method
```

```
clear
load xx.mat
N=length(x);
x=x.';
mm = input('enter the number of signals = ');
l=N/2;
I=diag(diag(ones(l,l)));
% ******** Generate D1 matrix ********
% D1=tril(ones(l,l));
d2 = diag(diag(ones(l-1,l-1)));
d3 = zeros(l-1,1);
d4 = zeros(l,1);
d5 = [d2 d3];
d6 = [d4 d5'];
D1 = d6';
% ******** Generate Ryy Ryz matrices ********
for k=1:l,
    x_k(:,k)=x(k:k-1+l,1);
end
Ryy_est=x_k*x_k'/l;
y=x(2:N);
for k=1:l,
    y_k(:,k)=y(k:k-1+l,1);
end
Ryz_est=x_k*y_k'/l;
% ******** Generate Rs Rss matrices ********
[u_est,q]=eig(Ryy_est);
for ii=1:l,
    m(ii)=q(ii,ii);
end
[y1,g1]=min(abs(m));
var0=q(g1,g1);
Rs=Ryy_est-I.*var0;
Rss=Ryz_est-D1.*var0;
% ******** Find eigenvalues of Rs Rss ********
[u_est,q]=eig(Rs,Rss);
for ii=1:l,
    m(ii)=q(ii,ii);
end
% ******** Find Frequency ********
for ii=1:mm,
    [y1,g(ii)]=min(abs(1-abs(m)));
    m(g(ii))=0;
```

```
    z(ii) = q(g(ii),g(ii));
end
r=angle(z);
f=r/pi/2;
a = find(f>0);
f(a)
```

APPENDIX 14.F

```
% df12_6.m Minimum_Norm
% ***** Input the input vector size and load input data *****
clear
eval(['load xx.mat']);
X = x;
end
points = 512;  % input('Enter Number of Output Data Points: ');
Ns = input('Enter the Number of Sources: ');
Fs = input('Enter Desired Filter Size: ');
[a,b] = size(X);
K = b;   % Number of snapshots
% ******* Compute the R matrix using covariance method ********
C2 = [];
for k = 1:Fs,
   C1 = X(k:K-Fs+k);
   C1 = C1(:);
   C2 = [C2 C1];
end
Ca = fliplr(C2);
C3 = C2';
Cb = flipud(C3);
Rx = Cb*Ca*1/(K);
% ***** Compute the eigenvalues and eigenvectors of Rx *****
[Ea,L] = eig(Rx);
EL = length(Ea);
[lambda,k1]=sort(diag(L));
E=Ea(:,k1);
En = E(:,1:Fs-2*Ns);
% En = E(:,1:Fs-Ns); % for complex data
Enp = En(2:EL,:);
c = En(1,:);
% ******** Compute the vector d ********
d = 1/(c*c')*(En*c');
```

```
% ******** Compute Pseudospectrum ********
for k=1:points + 1,
  w(k) = 2*pi*(k-1)/(points);    % 2pi coverage
  [v,t]=size(d);
  l=1:v;
  B1 = exp(j*w(k)*(l-1));
  B = B1(:);
  Pmn(k) = 1.0/(B'*d*d'*B);
end
Pmn = Pmn(:);
Pmn = 10*log10(abs(Pmn));
% ******** Plot ********
xaxpt = length(Pmn)/2;
xax = linspace(0,.5,xaxpt);
plot(xax, Pmn(1:xaxpt))
xlabel('Frequency')
ylabel('Amplitude')
nss = conv_vs(Ns);
fss = conv_vs(Fs);
title(['Nimi-Norm No. of Sources = ', nss, '  Order of filter = ', fss])
grid
```

APPENDIX 14.G

```
% df12_7.m  DFT - min norm method (three real signals)
clear
load xx.mat
l= input('enter order of filter = ');
ns = input('enter number of signals = ');
N=length(x);
p=256; % (number of output data points)
f=0:1/p:1-1/p;
for ii=1:p/2,
    for jj=1:l,
        e_f(jj,ii)=exp(j*2*pi*(jj-1)*f(ii));
    end
end

for ii=1:l,
    for jj=1:l,
        D(ii,jj)=exp(i*(2*pi/l)*(ii-1)*(jj-1));
    end
end
```

```
for k=1:N-l+1,
     x_k(:,k)=x(k:k-1+l).';
end

for k=1:l,
     x_kk(k,:)=x_k(l-(k-1),:);
end
x_kk=conj(x_kk);

Rxx_est=zeros(l,l);
for k=1:N-l+1,
     Rxx_est=x_k(:,k)*x_k(:,k)'+x_kk(:,k)*x_kk(:,k)'+Rxx_est;
end
Rxx_est=Rxx_est./2./l;
v=Rxx_est*D;
for ii=1:l,
     m(ii)=norm(v(:,ii));
end

for ii=1:ns*2,
     [y,g(ii)]=max(abs(m));
     m(g(ii))=0;
     Es(:,ii)=v(:,g(ii));
end

g=Es(1,:)';
Es_=Es(2:l,:);
d_s(1,1)=1;
d_s(2:l,1)=-Es_*inv(Es_'*Es_)*g;

for ii=1:p/2,
     P_est(ii)=1/(abs(e_f(:,ii)'*d_s)^2);
end
plot([0:1/p:1/2-1/p],20*log10(P_est))
ylabel('Amplitude')
xlabel('Frequency')
nss = conv_vs(ns);
ls = conv_vs(l);
title(['Mini Norm with FFT  # of sig = ', nss, '    filter order = ', ls])
grid
```

APPENDIX 14.H

```
% df12_8.m
```

```
% This program uses the direct adaptive frequency estimation
% technique (DAFE) by John Y. Cheung, using the least-mean-square
% algorithm (LMS).
% Use u = .001 and iter = 500.

%          Steven Nunes
%          5/29/92

clear

% Input Parameters. 'order' signifies the number of frequencies detected
% above the number present. 'avnum' is the number of iterations, over
% which, a given estimate is to be to be averaged. 'char' and 'blank'
% are used for printing input frequencies on plots. 'u' is a constant
% used in the adaptive process for convergence.

M=input('Enter the number of frequencies present (max 3): ');
order=M; avnum=49; char=4; blank='      ';
u=input('Enter value of u (max .01): ');
SNR=input('Enter the SNR: ');
iter=input('Enter # of iterations(min of 100): ');

% Set input signal with noise. Set random numbers to normal distribution.

nampl=1/(sqrt(2)*10^(SNR/20));                  % amplitude of the noise
noise=nampl*(randn(1,iter) + j*randn(1,iter));      % and produce sequence.

f=[.1 .2 .4] ;          % Vectors containing input frequencies and
A=[2 2 2];              % amplitudes. M are used.
w=2*pi*f;               % Convert Hz to radians.
x3=zeros(1,iter);           % Initialize temp variable x3 to zeros.
k=linspace(1,iter,iter);         % Initialize time vector.
x0=zeros(1,M);              % Initialize phase vector x0 to zeros.
x0(1)=j*2*pi*0.125;         % Initialize first phase to its value.

for i=1:M,                            % Loop through the number of
    x1=A(i)*exp(j*w(i)*k + x0(i));           % frequencies present adding
    x3=x3 + x1;                       % the next signal to the
end                              % previous.
x=x3 + noise;                         % Add the noise to signal.

% ******** Iterate through the LMS algorithm ********
```

```
M1=order;        % Set M1 equal to the order of the LMS algorithm
f_est=zeros(M1,iter);            % used. Initialize the frequency estimate, f_est,
e=zeros(M1+1,iter+1);            % the partial signal error, e, and the adaptation
a=zeros(M1,iter);       % variable, a, to zero. Set the first partial
e(1,1:iter)=x;          % signal, e(1,:), to the sampled sequence, x.

for l=1:M1,                              % Initialize the first values of
    e(l+1,1)=x(1);                       % e and f_est, needed in
    f_est(l,1)=acos(a(l,1))/2/pi;              % calculations below.
    e(l+1,2)=e(l,2) -2*cos(2*pi*f_est(l,1))*e(l,1);
end

for i=2:iter,           % For each sample and for each signal, find next
    for l=1:M1,         % values of a and e through iteration.
        a(l,i)=a(l,i-1) + u*(e(M1+1,i)'*e(l,i-1) + e(M1+1,i)*e(l,i-1)');
        f_est(l,i)=acos(a(l,i))/2/pi;                % Calculate the estimates.
        if i==2,                         % The e(l,i-1) term is zero.
        e(l+1,i+1)=e(l,i+1) - 2*cos(2*pi*f_est(l,i))*e(l,i);
        else
        e(l+1,i+1)=e(l,i+1) - 2*cos(2*pi*f_est(l,i))*e(l,i)+ e(l,i-1);
        end
    end
end

% ******** Convert SNR to string for plot printing ********
plot(f_est')                     % Plot the frequency estimates.
xlabel('Iterations')
ylabel('Frequency')
title(['SNR = ' num2str(SNR) ' dB, mu = ' num2str(u) ', f1 = ' num2str(f(1)) ', f2 = '
num2str(f(2)) ', f3 = ' num2str(f(3)) ])
```

APPENDIX 14.I

```
% CONV_VS converts from a vector to string
% JT April 29 1992

function str = convs(r)
str=[];
for i = 1:length(r)
    eval(['str=[str, ''',num2str(r(i)),' ''];']);
end
```

CHAPTER 15

Angle of Arrival Measurements

15.1 INTRODUCTION [1–6]

As mentioned in Chapter 2, the angle of arrival (AOA) is the most valuable information that can be obtained from an enemy radar because the radar cannot change its position drastically in a very short time frame (i.e., a few milliseconds). Unfortunately, the AOA information is also the most difficult information to obtain. It requires several antennas with receivers. The two common approaches to measure AOA are based on amplitude and phase comparisons. Another approach is to use the Doppler frequency shift generated by the aircraft movement. However, this approach is closely related to the phase measurement system. If the requirement is to measure AOA on simultaneous signals, the problem becomes even more complicated because receivers with multiple signal capability are needed.

In an amplitude comparison system, the amplitude of all the receivers must be matched from the antennas through the outputs of the receivers. This approach usually generates an AOA resolution of ±15 deg, which is less than the desired value of ±1 deg required in modern electronic warfare (EW) applications. In a phase measurement system, the phase of all the receivers must be matched from the antennas through the outputs of the receivers, which is a very difficult task in receiver design. The phase comparison system usually can generate AOA resolution of ±1 deg, which satisfies the modern EW requirement. The accuracy obtained from an amplitude comparison system is not theoretically limited. In most of the amplitude comparison systems, the AOA coverage is rather wide; thus, the antenna beams are wide, which results in poor AOA accuracy. If antennas have narrow beams (e.g., in monopulse radar), highly accurate AOAs can be obtained.

Theoretically in a system, if the antennas/receivers are not properly balanced, a calibration table can be used to minimize the difference. If a system is far from matched, many calibration points are needed, which is often difficult to implement. In the practical design of an AOA measurement system, primary effort usually is placed on balancing the antennas/receivers among different channels. A calibration table is used only to remedy the remaining hardware mismatch.

If a digital approach is used to measure the AOA, the data generated will be two-dimensional—in time and in space. Obviously, the processing will be more complicated than a one-dimensional case in time. To collect AOA data, multiple analog-to-digital converters (ADCs) must be used. In front of the ADCs, the antennas/receivers must be phase-matched as in a phase measurement system. The digitization of the ADCs should be operated in a synchronized manner. The number of antennas that can be used in an AOA system is very low in comparison with electronically scanning antennas in a radar system. In an airborne system, the maximum number of antennas might be 10. In a shipborne application, the number might be larger because more room is available.

Amplitude comparison AOA measurement for digital signal processing will not be discussed in this chapter because the approach will not be much different from an analog amplitude comparison system. The discussion will concentrate on the phase comparison or related approaches, such as Doppler frequency measurement. Most recent digital AOA studies are concentrated on high-resolution approaches, such as the MUSIC, ESPRIT, and minimum norm methods. These approaches are quite similar to the frequency measurements discussed in the previous chapter; thus, they will not be included here. However, due to the small data size collected in the space domain, these methods should be useful for special applications.

First, a queuing concept will be discussed. This concept is not limited to digital AOA systems, but it may provide some ideas for digital receiver designs. Secondly, data generated from linear and circular arrays will be presented. A simple approach based on zero crossing which can process only one signal will be discussed. A multiple signal approach using fast Fourier transform (FFT) will be discussed. The Chinese remainder theorem will be presented. The Chinese remainder theorem can be used to solve antenna location problems. Finally, a simple AOA data collection system that can be used in conjunction with a digital system will be presented.

15.2 QUEUING CONCEPT [7]

The queuing concept is probably one of the few AOA measurement approaches that can process simultaneous signals with a reasonable amount of hardware. This concept has been successfully investigated in analog receiver designs. The queuing concept is to measure some quantity in a gross manner over a wide instantaneous bandwidth. The information obtained is used to direct some narrowband measurement systems to obtain the fine information on the input signals. There are many different types of queuing arrangements, one of which will be discussed here.

Figure 15.1 shows a frequency queuing system. The input signal is received by an omnidirectional antenna with a wideband receiver of coarse-frequency resolu-

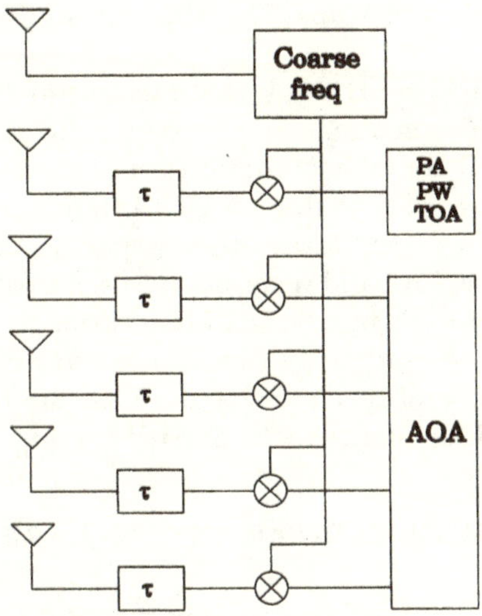

Figure 15.1 A simple queuing AOA system.

tion. The measured frequency of the input signal is used to tune a set of narrowband receivers. One of the narrowband receivers is usually dedicated to measuring the fine-grain information (i.e., frequency, pulse amplitude, pulse width (PW), and time of arrival (TOA)). Other narrowband receivers can be used to measure AOA. The AOA can be measured either through amplitude or phase comparison schemes. Wideband radio frequency (RF) delay lines are used in front of the narrowband receivers to delay the input signal for the frequency tuning to settle.

Although the wideband receiver can process simultaneous signals, this arrangement can provide fine-grain information, including AOA on only one input signal. If more than one signal needs to be processed, additional hardware is required. Each set of narrowband receivers can process only one signal. The number of narrowband systems matches the number of signals to be processed. If the number is high (i.e., four simultaneous signals), the logic circuit to assign the narrowband systems to the input signals can become very complicated.

The advantage of this arrangement is that these narrowband receivers are required to process only one signal, and therefore they are relatively easy to build. In addition, narrowband receivers are easier to match in amplitude and phase among different channels than a wideband system. It is important to note that this approach obtains the AOA and fine-grain information on a pulse-by-pulse basis.

The design may have some possible technical problems in that the wideband delay lines are difficult to build and the insertion loss is usually high. The amplitude and phase of the delay line may be temperature-sensitive. A signal passing a long delay line has many phase variations. Therefore, it is difficult to match the phase among long delay lines.

A disadvantage of this approach is that the sensitivity of the system might be low. The antenna is omnidirectional, and therefore the gain is low. A receiver with coarse channel resolution contains more noise, and therefore the sensitivity is low. The initial signal detection through this low-gain antenna and low-sensitivity receiver combination produces an overall low-sensitivity system. Although high-gain antennas and high-sensitivity narrowband receivers are used to obtain fine-grain information, they are not used for the initial signal detection.

Theoretically, digital technology can be used to build the wideband or the narrowband receiver if digital receivers are used for the narrowband applications where the wide RF delay lines might not be needed to temporarily store information.

15.3 DIGITAL DATA FROM A LINEAR ANTENNA ARRAY [7–9]

In this section, the data obtained from a linear antenna array will be presented. These data can be used for different AOA measurement schemes. In an EW application, the antenna array usually has few elements. A linear antenna array with three elements is not uncommon in practice. Usually a linear array is placed in the horizontal direction to measure the azimuth angle. One linear array can theoretically cover up to π radians or 180 deg of azimuth angle, although it is often limited to 120 deg to avoid operating in the end-fire mode. If the elevation angle is also of interest, a linear array in the vertical direction has to be added. As far as AOA measurement is concerned, the two linear arrays are usually treated separately.

Figure 15.2 shows a linear array of antennas with Q elements from $q = 0, 1, \ldots, Q - 1$ along the x direction. The elements in this array are uniformly spaced and the distance is d. Let us assume that there is only one plane wave and the equiphase plane is shown. If the input signal is a sinusoidal wave, the antenna of the qth element will have an output of

$$x(q, t) = A_1 \cos[2\pi f_1(t - \tau_q)] \tag{15.1}$$

where A_1 and f_1 are the amplitude and the frequency of the input signal and τ_q is a phase delay time on the qth element with respect to the first antenna element ($q = 0$). Since the distance between the antenna element is d, this delay time can be written as

$$\tau_q = \frac{-qd \sin \theta_1}{c} \tag{15.2}$$

where θ_1 is the incident angle of the input signal shown in Figure 15.2 and c is the speed of light. The minus sign in this equation is from the fact that the equiphase

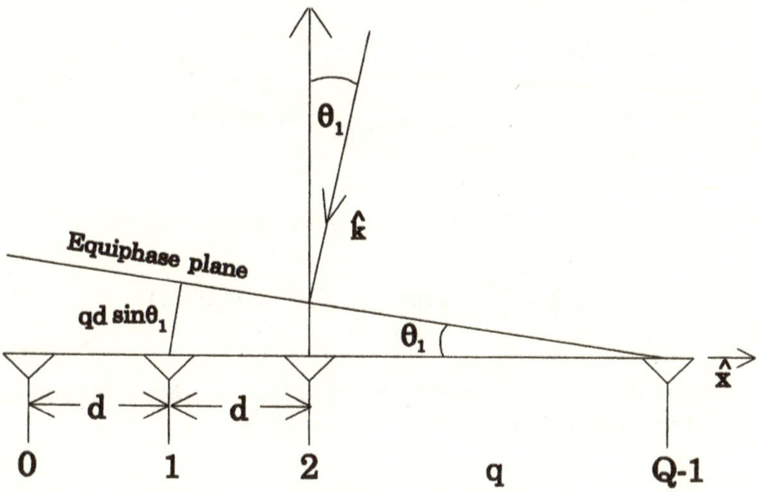

Figure 15.2 A linear array and a plane incident wave.

plane arrives at the qth element before it reaches the zero antenna element. From this figure, one can see that a linear array cannot distinguish a signal coming from above or below the array; thus, in order to avoid ambiguity, the angle θ_1 is usually limited from $-\pi/2$ to $\pi/2$.

Substituting the result in (15.2) into (15.1), one obtains

$$x(q, t) = A_1 \cos\left[2\pi f_1\left(t + \frac{qd \sin \theta_1}{c}\right)\right] \tag{15.3}$$

Often this equation is written in a different form. Let $\hat{k}_1$ be a unit vector pointing in the direction of the wave propagation and $\hat{x}$ a unit vector along the direction of array. Do not confuse $\hat{x}$ with the input $x(q, t)$. Equation (15.3) can be written as

$$x(q, t) = A_1 \cos\left(2\pi f_1 t - \frac{2\pi f_1 qd\hat{k}_1 \cdot \hat{x}}{c}\right) \tag{15.4}$$

$$\text{where} \quad \hat{k}_1 \cdot \hat{x} = \cos\left(\frac{\pi}{2} + \theta_1\right) = -\sin \theta_1$$

If there are two signals, the output from the qth element of the antenna array is the sum of the two signals.

$$x(q, t) = A_1 \cos\left(2\pi f_1 t - \frac{2\pi f_1 q d\hat{k}_1 \cdot \hat{x}}{c}\right) + A_2 \cos\left(2\pi f_2 t - \frac{2\pi f_2 q d\hat{k}_2 \cdot \hat{x}}{c}\right) \tag{15.5}$$

If there are M signals, the output from the qth antenna element is

$$x(q, t) = \sum_{m=1}^{M} A_m \cos\left(2\pi f_m t - \frac{2\pi f_m q d\hat{k}_m \cdot \hat{x}}{c}\right) \tag{15.6}$$

where $f_{\hat{m}}$ and $k_{\hat{m}}$ are the frequency and direction of the mth signal. Often exponential form is used to express this result as

$$x(q, t) = \sum_{m=1}^{M} A_m e^{j\left(2\pi f_m t - \frac{2\pi f_m q d\hat{k}_m \cdot \hat{x}}{c}\right)} \tag{15.7}$$

In order to obtain digitized data, the time t will be replaced by integers as $n = 0, 1, \ldots, N-1$.

All the frequency measurement methods discussed in previous chapters can be used to find the AOA. For example, when the fast Fourier transform (FFT) is applied to the time domain outputs, the outputs are in the frequency domain. If FFT is applied to data in the space domain that are collected from different antennas at one specific time instance, the output will represent AOA (given knowledge of the frequency f). The major difference is that the time domain data contain many data points and the space domain data points are equal to the number of antennas in the linear array. The Fourier transform of data in the space domain will be further discussed in Section 15.8.

15.4 OUTPUTS FROM A CIRCULAR ANTENNA ARRAY [10–13]

This section presents the output of a circular array. A circular array is a two-dimensional arrangement; therefore, it can be used to measure both the azimuth and elevation angles. These types of arrays occupy more space than the linear array and they might be suitable for some special aircraft and shipboard applications. A circular antenna array is shown in Figure 15.3. In this figure, the radius of the circle is R and there are Q elements from $q = 0$ to $Q - 1$. The array is in the xy plane and the first antenna element is the zero element and it is on the x axis; thus ϕ is zero. The locations of the antennas are at multiples of the angle ϕ_q.

Assume that a plane wave with amplitude A_1 and frequency f_1 is incident on this array with angle ϕ_1 and θ_1. The output induced at the qth element can be written as

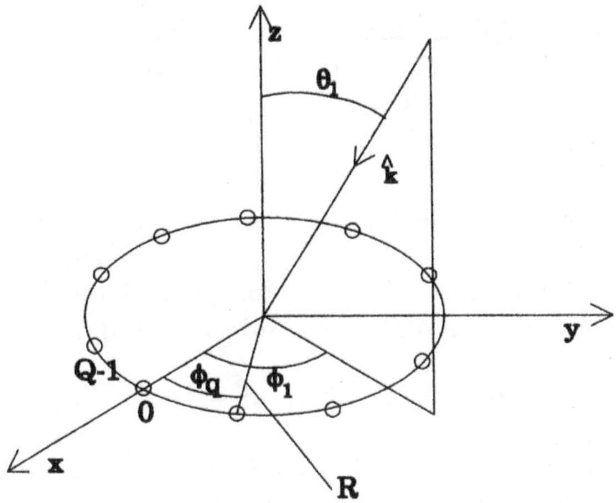

Figure 15.3 Circular antenna array.

$$e(q,\ t) = A_1 e^{j2\pi f_1\left(t - \frac{\tau_{q1}}{c}\right)} \tag{15.8}$$

where the delay time τ_{q1} referenced to the center of the array is

$$\tau_{q1} = -R \sin\ \theta_1 \cos\ \phi_{q1} \tag{15.9}$$

where R is the radius of the array and θ_1 is the angle between the incident ray and the normal of the array. In this case, because the antenna locations are discrete, the angle ϕ_{q1} can be written as

$$\phi_{q1} = q\phi_q - \phi_1 \tag{15.10}$$

where $q = 0, 1, \ldots, Q - 1$, ϕ_1 is the azimuth angle of the incident wave, and ϕ_q is the angle between two adjacent antennas, which is

$$\phi_q = \frac{2\pi}{Q} \tag{15.11}$$

Substituting (15.9) through (15.11) into (15.8), the result is

$$\begin{aligned} e(q,\ t) &= A_1 e^{j2\pi f_1\left(t + \frac{R \sin\theta_1 \cos\left(\frac{2\pi q}{Q} - \phi_1\right)}{c}\right)} \\ &= A_1 e^{j2\pi\left(f_1 t + \frac{R \sin\theta_1 \cos\left(\frac{2\pi q}{Q} - \phi_1\right)}{\lambda_1}\right)} \end{aligned} \tag{15.12}$$

For example, if the signal is incident from the x direction (i.e., $\theta_1 = 90^0$), $\phi_1 = 0$. The output from the first antenna element ($q = 0$) is

$$e(0, t) = A_1 e^{j2\pi f_1\left(t + \frac{R}{c}\right)} \tag{15.13}$$

This means the output is leading from the center of the circle by R/c, which is the anticipated result.

This result can be written in vector form. The incident direction of the input signal can be represented by a unit vector $\hat{k}_1$ in Cartesian coordinate as

$$\hat{k}_1 = -(\sin\,\theta_1\,\cos\,\phi_1\,\hat{x} + \sin\,\theta_1\,\sin\,\phi_1\hat{y} + \cos\,\theta_1\hat{z}) \tag{15.14}$$

where $\hat{x}$, $\hat{y}$, and $\hat{z}$ are unit vectors in the Cartesian system. The minus sign in front means the vector is pointing to the center of the origin. The position of the qth antenna element is in the $\hat{r}_q$ direction as

$$\hat{r}_q = \cos(q\phi_q)\hat{x} + \sin(q\phi_q)\hat{y} \tag{15.15}$$

The dot product of $\hat{k}_1 \cdot \hat{r}_q$ is

$$\begin{aligned}\hat{k}_1 \cdot \hat{r}_q &= -[\sin\,\theta_1\,\cos\,\phi_1\,\cos(q\phi_q) + \sin\,\theta_1 \sin\,\phi_1 \sin(q\phi_q)] \\ &= -\sin\,\theta_1 \cos(q\phi_q - \phi_1)\end{aligned} \tag{15.16}$$

Thus, (15.12) can be written as

$$\begin{aligned}e(q, t) &= A_1 e^{j2\pi f_1\left(t - \frac{R\,\hat{k}_1 \cdot \hat{r}_q}{c}\right)} \\ &= A_1 e^{j\left(2\pi f_1 t - \frac{2\pi f_1 R\,\hat{k}_1 \cdot \hat{f}_q}{c}\right)}\end{aligned} \tag{15.17}$$

Extending this equation to cover M signals,

$$\begin{aligned}e(q, t) &= \sum_{m=1}^{M} A_1 e^{j2\pi f_m\left(t + \frac{R\,\sin\theta_m \cos\left(\frac{2\pi q}{Q} - \phi_m\right)}{c}\right)} \\ &= \sum_{m=1}^{M} A_m e^{j2\pi\left(f_m t + \frac{R\,\sin\theta_m \cos\left(\frac{2\pi q}{Q} - \phi_m\right)}{\lambda_m}\right)} \\ &= \sum_{m=1}^{M} A_m e^{j2\pi f_m\left(t - \frac{R\,\hat{k}_m \cdot \hat{r}_q}{c}\right)} \\ &= \sum_{m=1}^{M} A_m e^{j2\pi\left(f_m t - \frac{R\,\hat{k}_m \cdot r_q}{\lambda_m}\right)}\end{aligned} \tag{15.18}$$

This result can be considered as identical to (15.7).

For digitized data, t is replaced by $n = 0, 1, \ldots, N-1$. The number of antennas $q = 0, 1, \ldots, Q-1$. Therefore, the data are two-dimensional in n and q. If signal processing is applied in n domain (time), the result will be frequency. If signal processing is applied to q, the result is AOA.

The results from the linear array and the circular array are very much alike. In later discussion, both data can be used and the results will be quite similar. The linear array can only provide azimuth angle information because the corresponding $\hat{k}$ is two-dimensional. The circular array can provide both azimuth and elevation information because the corresponding $\hat{k}$ is three-dimensional. For simplicity and airborne consideration, the linear array will be used in the following sections.

15.5 TWO-ELEMENT PHASE ARRAY ANTENNA

The phase measurement discussed in Chapter 10 can be used to measure AOA information. In order to keep this discussion simple, a linear antenna array with only two elements will be presented, as shown in Figure 15.4. In this figure, the θ_1 is in the positive direction. It is also assumed that this is a narrowband system. In other words, there is only one input signal. With two elements, the outputs are

$$\begin{aligned} e(0, t) &= A_1 e^{j2\pi f_1 t} \equiv A_1 e^{j\psi_0(t)} \\ e(1, t) &= A_1 e^{j\left(2\pi f_1 t + 2\pi f_1 \frac{d \sin\theta_1}{c}\right)} \equiv A_1 e^{j\psi_1(t)} \end{aligned} \tag{15.19}$$

where A_1 and θ_1 are the amplitude and incident angle of the input signal and d is the distance between the two antenna elements. The phase angles are represented by $\psi_1(t)$ and $\psi_2(t)$. In this equation, continuous time is considered.

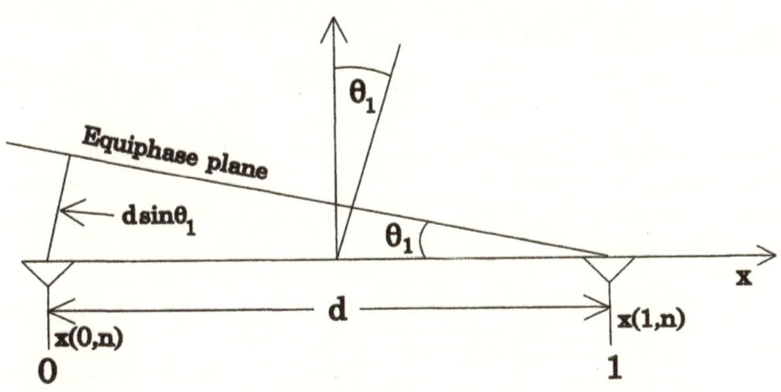

Figure 15.4 A two-antenna element AOA system.

In an analog phase measurement system, $\psi_1(t)$ and $\psi_2(t)$ are the measured quantities. The difference between the two phases can be used to measure the AOA information as

$$\psi \equiv \Delta\psi_{10}(t) = \psi_1(t) - \psi_0(t) = \frac{2\pi f_1 d \sin\theta_1}{c} = \frac{2\pi d \sin\theta_1}{\lambda_1} \tag{15.20}$$

Two interesting points are noted in this equation. First, ψ is independent of time t. Secondly, there are two unknowns, f_1 and θ_1.

Since ψ is independent of time t, it needs only one measurement at any time instant to obtain the AOA. This is true only theoretically; however, due to the noise contamination, the result obtained from one or few data samples may be inaccurate. In general, many samples are needed to generate the AOA information.

Since there are two unknowns, the frequency f_1 must be found first. This is always true in a phase comparison system where the frequency of the input signal must be measured first. This frequency can be measured by using the data from one antenna element (say, the 0 element). Since this is a narrowband system containing only one input signal, the frequency can be found from the phase comparison system discussed in Chapter 10 as

$$f_1 = \frac{\psi_0(n) - \psi_0(0)}{nt_s} \equiv \frac{\Delta\psi_0(n)}{nt_s} \tag{15.21}$$

In order to avoid frequency ambiguity, the phase angle $\Delta\psi_0(n)$ must be less than 2π. As discussed before, a small n value that represents a short delay time is used to cover a wide-frequency bandwidth and a large value of n that represents a long delay time is used to generate fine-frequency resolution.

Once the frequency of the input signal is obtained, the AOA information can be obtained from (14.20) as

$$\sin\theta_1 = \frac{c\psi}{2\pi f_1 d} = \frac{\lambda_1 \psi}{2\pi d} \tag{15.22}$$

As discussed before, the incident angle is limited to $-\pi/2$ to $\pi/2$ and the corresponding difference in $\sin\theta_1$ is $\sin(\pi/2) - \sin(-\pi/2) = 2$. Since the maximum of ψ is 2π, then from the above equation, the maximum separation between the two antenna elements is

$$2 = \frac{\lambda_1 \psi_{max}}{2\pi d_{max}} = \frac{2\pi\lambda_1}{2\pi d_{max}} \quad \text{or} \quad d_{max} = \frac{\lambda_1}{2} \tag{15.23}$$

This is the well-known relation in linear antenna array that the maximum separation between two elements must be less than or equal to $\lambda/2$.

Let us summarize the approach discussed above.

1. In a phase measurement AOA system, the maximum separation between the two antenna elements is $\lambda/2$, where λ is the wavelength of the input signal.
2. From the time domain output of any one of the antenna elements, the frequency can be obtained by measuring the phase difference as a function of time. This is discussed in detail in Chapter 10. The input frequency (or wavelength) is needed to calculate the AOA.
3. From the phase difference between the two antennas, the AOA can be measured through (15.22). Theoretically, only one pair of phase measurements can determine the AOA; however, the AOA accuracy is rather poor.

In order to improve the measurement accuracy, the phase measurement can be obtained as

$$\psi = \Delta\psi_{10} = \frac{1}{N}\sum_{n=0}^{N-1}[\phi_1(n) - \phi_0(n)] \tag{15.24}$$

This is taking the average of phase difference between the two antennas at many time intervals.

15.6 AOA MEASUREMENT THROUGH ZERO CROSSING

Zero crossing, discussed in Chapter 10, to measure the frequency of an input signal can also be used to measure the AOA information. Similar to the phase measurement system, this approach can process only one input signal. If more than one input signal arrives at the antenna/receivers, the measured AOA could be erroneous. As mentioned in Chapter 10, the zero crossing approach requires only one receiver with one ADC per antenna and an *I-Q* channel arrangement is not required. The detailed approach to improve the zero crossing measurement is presented in Chapter 10 and will not be repeated here. Only the concept will be discussed.

A two-element zero crossing AOA system is shown in Figure 15.5. The positions of the two antennas are represented by two circles and the incident angle is θ_1. The two sinusoidal waves represent the input signal. The first zero crossing occurs at the left antenna at t_0. At the time instant t_0, the zero crossing is $d \sin \theta_1$ from the second antenna. Therefore, the difference of zero crossing time between the two antennas can be written as

$$\Delta t = \frac{d \sin \theta_1}{c} \quad \text{or} \quad \sin \theta_1 = \frac{c\Delta t}{d} \tag{15.25}$$

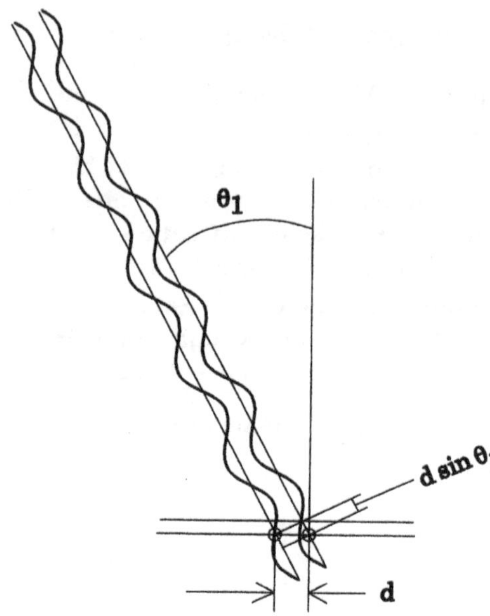

Figure 15.5 Zero crossing of a two-antenna system.

It is interesting to note that in this measurement, the AOA information obtained is independent of the signal frequency because the time difference rather than the phase is measured.

In this arrangement, the two antennas cannot be too far apart; otherwise, the system will have ambiguity. The maximum separation can be determined when the incident angle $\theta_1 = 90$ deg. Since there are two zero crossings per cycle, the maximum distance between two zero crossings is $\lambda/2$. The maximum separation between the two antennas is $\lambda/4$ to avoid comparing the zero crossings between two different zero crossing points.

However, if one can track the sign change before and after the zero crossing, the length between two ambiguous zero crossings is λ. Under this condition, the maximum separation between the two antenna elements is $\lambda/2$ to avoid comparing the zero crossings between two different zero crossing points.

Although this discussion is based on the zero crossing of one data point, in actual measurement many zero crossings should be averaged to reduce the noise effect.

15.7 PHASE MEASUREMENT IN AOA SYSTEMS WITH MULTIPLE ANTENNAS [14, 15]

In the previous section, the accuracy of the AOA measured by a two-antenna system is usually quite limited. In (15.22), a large value of d means a smaller value of θ_1.

Physically, this means a longer base line d can provide more accurate AOA information at the cost of narrower unambiguous incident angle coverage. For a two-antenna linear array, the maximum distance must be less than $\lambda_1/2$ to avoid AOA ambiguity. If d is longer than this value, the measured angle ψ will be greater than 2π and the angle θ_1 will be ambiguous.

In order to avoid ambiguity and provide accurate AOA information, a linear array with multiple antennas can fulfill this need. Figure 15.6 shows a four-element system. In this figure, the shortest distance between the two antennas is d, which is less than $\lambda_1/2$. The other two antenna elements are at $4d$ and $16d$ from the first element. The input angle coverage will be less than π $(-\pi/2 < \theta_1 < \pi/2)$.

For simplicity, let us assume that the angle coverage is equal to π. The antennas with the shortest distance will provide the coarse AOA information and the resolution must be finer than $\pi/4$ (such as $\pi/8$). The second antenna pair has an ambiguous AOA range of $\pi/4$, since the distance between them is $4d$. This pair should provide AOA resolution of at least $\pi/16$ because the longest antenna pair has an ambiguous AOA range of $\pi/16$. The longest antenna pair will provide the final AOA resolution.

This same idea can be applied to zero crossing measurement. Antennas with short separation can be used to resolve ambiguity and antennas with long separation provide better AOA resolution.

15.8 FOURIER TRANSFORM OVER SPACE DOMAIN [8, 9, 15]

In this section, the data are obtained from an antenna array at a certain time instant. Fourier transform will be performed on the data to obtain the AOA information. Let

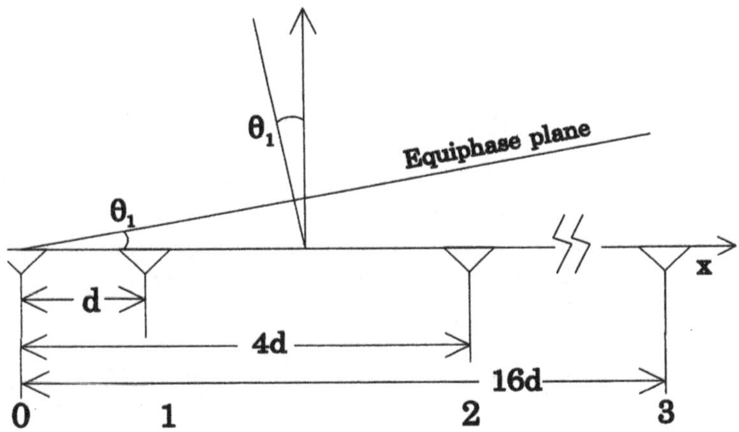

Figure 15.6 A four-element linear antenna array.

us assume that the linear antenna array has uniform spacing and there is only one input signal. Also assume that the data collected are at $t = 0$. This assumption simplifies the calculation, but does not lose generality. Under this condition, from (15.7), the input signal is

$$x(q) = A_1 e^{\frac{-j2\pi f_1 q d\hat{k}_1 \cdot \hat{x}}{c}} \tag{15.26}$$

where A_1, f_1, and $\hat{k}_1$ are the amplitude, frequency, and incident direction of the incoming signal, respectively; q is the number of antenna elements and d is the distance between two adjacent antennas.

If there are Q antenna elements, the Fourier transform in the space domain can be written as

$$\begin{aligned} X(\hat{k}) &= \sum_{q=0}^{Q-1} x(q) e^{\frac{-j2\pi f q d \sin\theta}{c}} = \sum_{q=0}^{Q-1} x(q) e^{\frac{j2\pi f q d \hat{k} \cdot \hat{x}}{c}} \\ &= \sum_{q=0}^{Q-1} A_1 e^{\frac{j2\pi f q d(\hat{k}-\hat{k}_1) \cdot \hat{x}}{c}} = A_1 \frac{1 - e^{\frac{j2\pi f Q d(\hat{k}-\hat{k}_1) \cdot \hat{x}}{c}}}{1 - e^{\frac{j2\pi f d(\hat{k}-\hat{k}_1) \cdot \hat{x}}{c}}} \end{aligned} \tag{15.27}$$

The amplitude of this function can be written as

$$|X(\hat{k})| = A_1 \frac{\left|1 - e^{\frac{j2\pi f Q d(\hat{k}-\hat{k}_1) \cdot \hat{x}}{c}}\right|}{\left|1 - e^{\frac{j2\pi f d(\hat{k}-\hat{k}_1) \cdot \hat{x}}{c}}\right|} = A_1 \frac{\left|\sin\left(\frac{2\pi f Q d(\hat{k} - \hat{k}_1) \cdot \hat{x}}{2c}\right)\right|}{\left|\sin\left(\frac{2\pi f d(\hat{k} - \hat{k}_1) \cdot \hat{x}}{2c}\right)\right|} \tag{15.28}$$

This is the well-known response of a linear array [1, 2]. The peak of $|X(\hat{k})|$ occurs at

$$\hat{k} = \hat{k}_1 \qquad \text{or} \qquad \Delta\theta = 0 \tag{15.29}$$

because $(\hat{k} - \hat{k}_1) \cdot \hat{x} = -\sin(\Delta\theta)$. The peak is in the direction of the incident wave, and this is the expected result.

One can eliminate the frequency dependence in $X(\hat{k})$ by assuming that d and λ are linearly related (i.e., $d = \lambda/2$). Under this specialized condition, the above equation can be simplified as

$$|X(\hat{k})| = A_1 \frac{\left|\sin\left(\frac{\pi Q(\hat{k} - \hat{k}_1) \cdot \hat{x}}{2}\right)\right|}{\left|\sin\left(\frac{\pi(\hat{k} - \hat{k}_1) \cdot \hat{x}}{2}\right)\right|} = A_1 \frac{\left|\sin\left(\frac{\pi Q \sin(\Delta\theta)}{2}\right)\right|}{\left|\sin\left(\frac{\pi \sin(\Delta\theta)}{2}\right)\right|} \tag{15.30}$$

It is easy to see that $X(\hat{k})$ has a peak at $\hat{k} = \hat{k}_1$, the same result as in (15.29). The first zero occurs at

$$\frac{\pi Q \sin(\Delta\theta)}{2} = \pi \quad \text{or} \quad \Delta\theta = \sin^{-1}\left(\frac{2}{Q}\right) \tag{15.31}$$

The beamwidth θ_{bw} can be considered as

$$\theta_{bw} = 2\Delta\theta = 2\sin^{-1}\left(\frac{2}{Q}\right) \tag{15.32}$$

This equation indicates that a large value of Q, the number of elements in the antenna array, produces a narrow beam and a high AOA resolution. A small number of elements produce a wide beam and poor AOA resolution. Since $d = \lambda/2$, the larger the Q value, the longer the antenna. It is well known that a long linear antenna array will provide fine beamwidth.

Figure 15.7(a) shows the response of a 10-antenna element case while Figure 15.7(b) shows the response of a 3-element case. The amplitude is plotted in log scale and the angle coverage is π (from $-\pi/2 < \Delta\theta < \pi/2$). It is clearly shown that a larger number of antenna elements provides finer AOA resolution. Due to the few terms (also the number of antenna elements) in the spatial Fourier transform, the sidelobes are very high.

If there are simultaneous signals, the high sidelobes make the instantaneous dynamic range low. The high sidelobes also make the parameter encoder designs more complicated. The levels of the sidelobes can be reduced if a proper weighting function is applied to the antenna array. One approach to add a weighting function to an antenna array is to reduce the outputs of some of the antennas. The outputs from the ends of the array have higher attenuation. Figure 15.8 shows a very simple example. The curve in the bottom represents the attenuation added and they are symmetrical with respect to the center antenna element. The element attenuations should be $A_1 > A_2 > A_3 > A_4 = 0$ where the center element does not have any attenuation. As discussed in Chapter 3, the weighting function will increase the width of the mainlobe. In addition to this problem, if there are only a few antenna elements, the weighting approach may not be very effective anyway.

It appears that if a few antenna elements are used in a system, such as in the airborne case, it is difficult to obtain fine AOA resolution through the Fourier transform in the spatial domain. If there are multiple inputs from different angles, theoretically the Fourier transform in the space domain should generate peaks at all the incident angles. However, if the beamwidth is wide due to a limited number of antenna elements and the sidelobes are high, it is difficult to separate simultaneous signals.

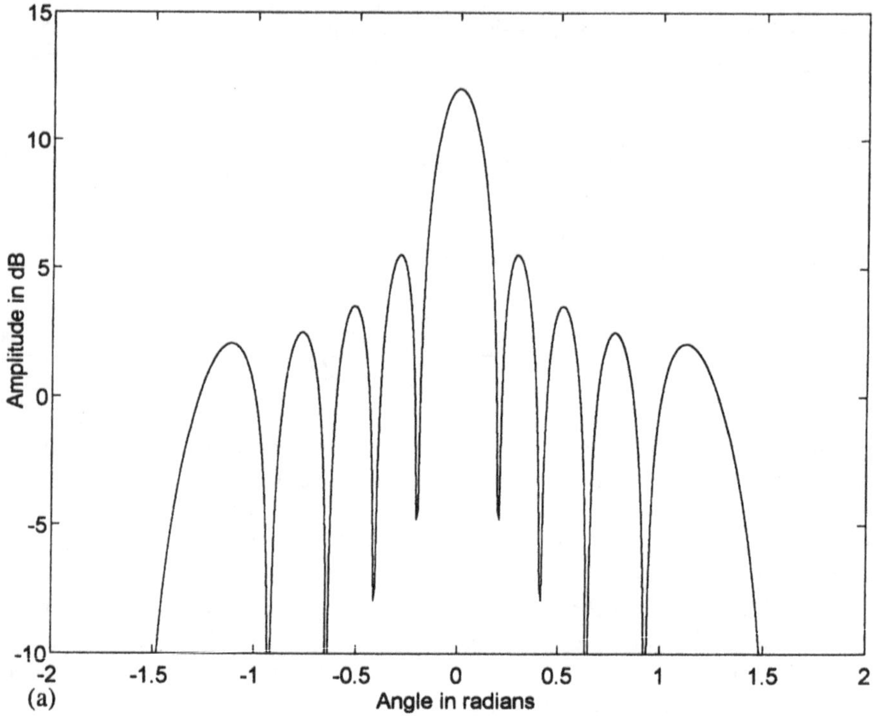

(a)

Figure 15.7 Outputs from spatial Fourier transform: (a) 10 elements, (b) 3 elements.

15.9 TWO-DIMENSIONAL FOURIER TRANSFORM [16, 17]

In this section, the two-dimensional Fourier transform will be presented. The input data from an antenna array are two-dimensional: one is in the frequency domain and the other one is in the space domain, as shown in Figure 15.9. It should be noted that the time intervals are uniform and all the antennas are sampled at the same time. The distances between antennas are also uniform.

The Fourier transform performed on the time domain data will provide frequency information and the Fourier transform performed on the space data will provide AOA information. Like the Fourier transform in the frequency domain, the Fourier transform in the space domain can process simultaneous signals with different input frequencies and incident angles.

A two-dimensional Fourier transform of input data $x(q,t)$ can be written as

$$X(\hat{k}, f) = \int_{-\infty}^{\infty}\int_{-\infty}^{\infty} x(q, t)\, e^{-j2\pi f t\left(t - \frac{qd\hat{k}\cdot\hat{x}}{c}\right)} dq dt \tag{15.33}$$

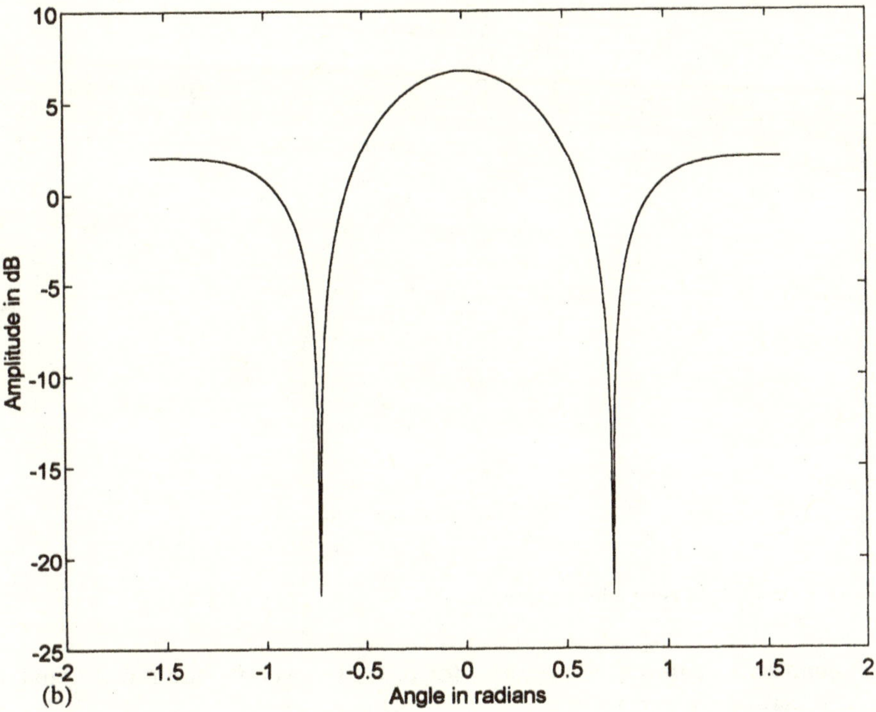

(b)

Figure 15.7 (continued).

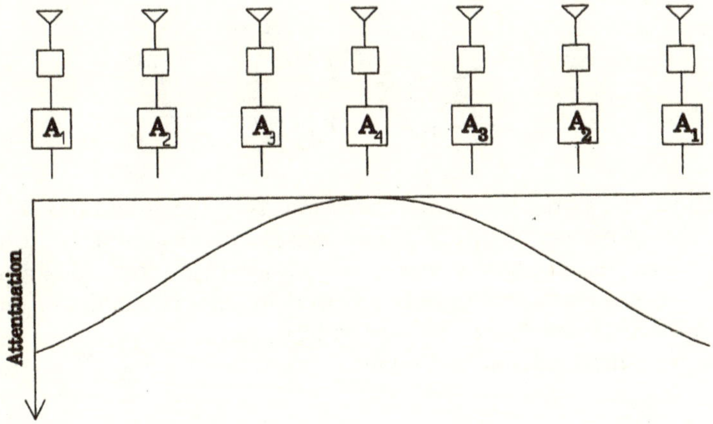

Figure 15.8 Antenna array with a weighting function.

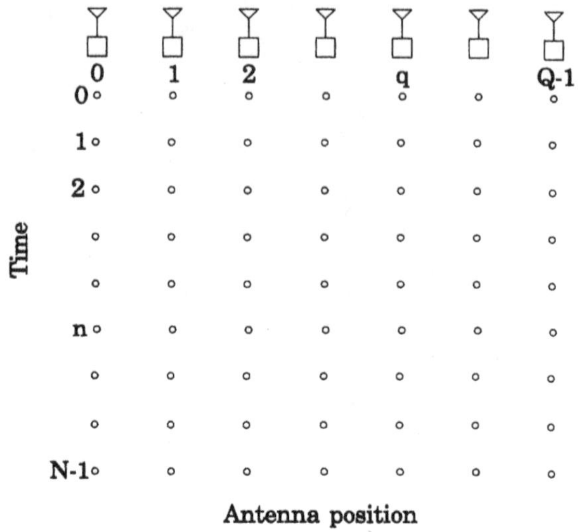

Figure 15.9 Data output from a linear antenna array.

This equation is written in continuous form. If there is only one input signal, from (15.7), it can be written as

$$x(q, t) = A_1 e^{j\left(2\pi f_1 t - \frac{2\pi f_1 q \hat{k}_1 \cdot \hat{x}}{c}\right)} \tag{15.34}$$

Substituting this result in (15.33), the result is

$$\begin{aligned} X(\hat{k}, f) &= \int_{-\infty}^{\infty}\int_{-\infty}^{\infty} A_1 e^{-j2\pi(f-f_1)t} e^{j\frac{2\pi f q d(\hat{k}-\hat{k}_1)\cdot\hat{x}}{c}} dt dq \\ &= \int_{-\infty}^{\infty} A_1 e^{-j2\pi(f-f_1)t} dt \int_{-\infty}^{\infty} e^{j\frac{2\pi f q d(\hat{k}-\hat{k}_1)\cdot\hat{x}}{c}} dq \end{aligned} \tag{15.35}$$

In this equation, the signal is assumed continuous in the time domain (i.e., in an analog receiver). However, it is difficult to consider the signal to be continuous in the space domain because the antennas are always in discrete locations. Even in an analog receiver system, the signals received in the space domain are discrete because the antennas are discretely located in space.

Considering that the space domain is discrete, the above equation can be written as

$$X(\hat{k}, f) = \sum_{q=0}^{Q-1}\int_{-\infty}^{\infty} x(q, t)\, e^{-j2\pi f t} e^{-j\frac{2\pi f q d \hat{k}\cdot\hat{x}}{c}} dt \tag{15.36}$$

This is a two-dimensional Fourier transform with continuous data in the time domain but discrete data in the space domain.

If the collected data are discrete in both the time and space domains as shown in Figure 15.9, the two-dimensional Fourier transform can be written as

$$X(\hat{k}, k) = \sum_{q=0}^{Q-1}\sum_{n=0}^{N-1} x(q, n) e^{\frac{-j2\pi nk}{N}} e^{-j\frac{2\pi kqd\hat{k}\cdot\hat{x}}{c}} \tag{15.37}$$

In this equation, $\hat{k}$ is a unit vector representing the AOA calculated from the spatial Fourier transform and k is the discrete frequency component in the frequency domain calculated from the Fourier transform in time domain.

The inverse Fourier transform in discrete form is

$$x(q, n) = \frac{1}{QN}\sum_{k=0}^{Q-1}\sum_{\hat{k}=0}^{N-1} x(k, \hat{k}) e^{\frac{j2\pi nk}{N}} e^{j\frac{2\pi kqd\hat{k}\cdot\hat{x}}{c}} \tag{15.38}$$

The two-dimensional discrete Fourier transform requires $(MN)^2$ complex multiplications. Thus, the calculation is rather complicated.

One can consider the two-dimensional Fourier as a combination of two one-dimensional Fourier transforms. For example, one can perform the time domain Fourier transform (one-dimensional) on all the antenna outputs to obtain the frequency domain information. Then, at each frequency component one can use all the outputs from the different antennas to perform the space Fourier transform (one-dimensional) to obtain the AOA information of that frequency component. Therefore, the final results will be the frequency of each incoming signal and the corresponding AOA information.

In the two-dimensional Fourier transform, in order to suppress the sidelobes, weighting functions can be added to both the space and time domains. However, due to the complication of the two-dimensional Fourier transform and the poor AOA resolution, it appears that this approach might not be very useful in digital receiver applications.

15.10 FREQUENCY SORTING FOLLOWED BY AOA MEASUREMENTS

In the previously discussed phase measurement system, it is assumed that the receiver processes only one input signal. Usually, the phase measurement system can provide fine AOA information. The actual receiver used in a single-signal phase comparison system must have narrow bandwidth to limit the probability of intercepting more than one signal. However a narrow bandwidth receiver has a low probability of intercept, which is not desirable for EW applications. A wideband phase measurement AOA system may be contaminated by simultaneous signals to produce erroneous information. Although theoretically the spatial Fourier transform can process simultaneous signals with different incident angles, the AOA resolution obtained is very coarse due to the limited antenna elements available in an EW system. It is highly questionable that a wideband AOA system using spatial Fourier transform can be used to separate signals with different incident angles.

In this section, a more promising approach will be presented. This is a two-step approach. The first step is to separate the incoming signals by frequency through Fourier transform in the time domain. If the data are collected over a long period of time, the frequency resolution can be reasonably narrow. For example, a 100-ns-long data string can provide approximately 10 MHz of frequency resolution. In EW applications, it is often suggested that simultaneous signals are not a severe problem when the frequency channel is 20 MHz wide. Sometimes even a 50- to 100-MHz channel width is considered in order to accommodate short pulses.

The first step is to use the data obtained from one antenna (say $q = 0$) to perform fast Fourier transform (FFT) in the time domain. From the power spectrum outputs, the frequencies of the input signals can be obtained. Suppose there are M input signals and their peaks occur at frequencies $k_1, k_2, \ldots, k_M$. The outputs can be written as $X(0, k_1), X(0, k_2), \ldots, X(0, k_M)$ where the first index is referenced to the antenna. Then the same peaks should be obtained from other antennas as

$$\begin{aligned} X(q, k_1) &= \sum_{n=0}^{N-1} x(q, n)\, e^{\frac{-j2pik_1 n}{N}} \\ X(q, k_2) &= \sum_{n=0}^{N-1} x(q, n)\, e^{\frac{-j2pik_2 n}{N}} \\ &\cdots \\ X(q, k_M) &= \sum_{n=0}^{N-1} x(q, n)\, e^{\frac{-j2pik_M n}{N}} \\ &\text{for} \quad q = 1, 2, \ldots, Q-1 \end{aligned} \tag{15.39}$$

In these equations, only the frequency components with peaks found from the $q = 0$ antenna are calculated at k_1, k_2, and so forth. Thus, there is less calculation than to find all the Fourier components from all the antenna outputs.

The second step is to find the AOA from $X(q, k_i)$ $i = 0, 1, 2, \ldots, M$ obtained from (14.39). Since the frequency information is known, many different methods can be used to find the AOA information at that frequency. A couple of simple approaches are presented here.

The first method is through phase measurement. This can be considered a straightforward approach because the phase information is already available in $X(q, k)$. The FFT output is complex, and can be written in terms of amplitude and phase as

$$X(q, k_i) = |X(q, k_i)| e^{j\phi(q, k_i)} \tag{15.40}$$

where $\phi(q, k_i)$ is the phase of the input signal at frequency k_i at antenna element q. The phase difference of $\phi(q, k_i)$ between two antenna elements is related to the incident angle of signal by

$$\phi(q_r, k_i) - \phi(q_s, k_i) = \frac{d_{rs} \sin(\theta_i)}{\lambda_i} \tag{15.41}$$

where q_r and q_s are the two antenna elements and d_{rs} is the distance between them; θ_i and λ_i are the incident angles and the wavelength of the signal at frequency k_i, respectively. As mentioned in the multiple antenna elements case, the antenna pair with the shortest distance between them can be used to resolve AOA ambiguity. The antenna pairs with longer distances between them can be used to provide fine AOA resolution.

A second approach to find the AOA is to use the FFT in the spatial domain. In this approach, the antenna elements should be uniformly located. The inputs to the FFT are the $X(q, k_i)$ values with the same frequency component k_i. This equation can be written as

$$X(\hat{k}, k_i) = \sum_{q=0}^{Q-1} x(q, k_i) e^{-j\frac{2\pi k q d \hat{k} \cdot \hat{x}}{c}} \tag{15.42}$$

The main beamwidth depends on the number of antenna elements and the weighting function used. The possible advantage of this approach over the phase measurement method is the ability to process simultaneous signals. When two incoming signals are of the same frequency but with different incident angles, the phase measurement method can produce erroneous results because the phase measurement cannot process simultaneous signals. While the spatial FFT approach may separate them, the two signals must be widely separated in AOA.

Other possible methods can be used to find the AOA information, such as Prony's method or the MUSIC method. Interestingly, many of the high-frequency resolution methods were originally developed for AOA estimation. These approaches need more calculations, but for small data length they might be applicable for some special cases.

15.11 MINIMUM ANTENNA SPACING [18–21]

In all the AOA measurement systems discussed in previous sections, the antennas cannot be spaced too far apart in order to avoid ambiguity in AOA information. The shortest distance between two antennas d_{min} must be less than half the wavelength of the highest frequency. It can be written as $d_{min} < \lambda_{min}/2$ where λ_{min} is the wavelength corresponding to the highest frequency.

In order to emphasize this requirement, it will be presented here again from the sampling point of view. Figure 15.10 shows a sine wave incident onto an antenna

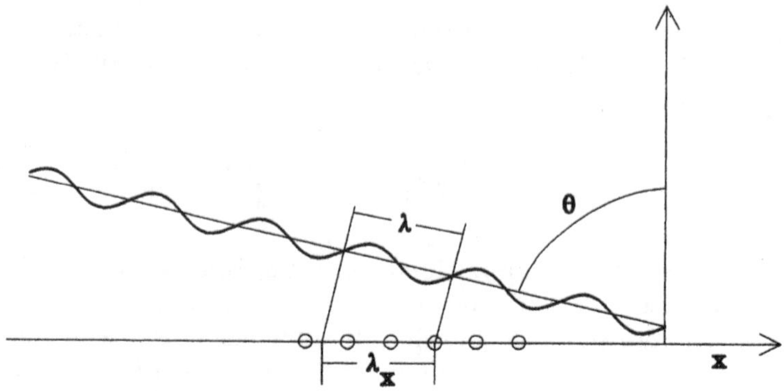

Figure 15.10 Wavelength along x direction.

array along the x direction. The wavelength of the input signal is λ and the corresponding wavelength along the x direction is

$$\lambda_x = \frac{\lambda}{\sin\theta} \tag{15.43}$$

where θ is the incident angle. When $\theta = 90$ deg (i.e., the incoming signal is along the x axis), then $\lambda_x = \lambda_{xmin} = 1$. In order to fulfill the Nyquist theorem, one must sample twice per cycle; that is, the shortest distance between two adjacent antennas d_{min} must be

$$d_{min} < \frac{\lambda_{min}}{2} \tag{15.44}$$

where λ_{min} is the signal with the shortest wavelength (the highest frequency) the antenna array is anticipated to intercept.

Now let us consider some practical problems in antennas. A spiral antenna is commonly used in EW applications, especially in airborne systems. This kind of antenna covers a wide frequency range and a wide angle. It can receive signals with either vertical or horizontal polarization. Only when the input signal is circularly polarized in the opposite direction of the spiral does the antenna have difficulty in intercepting it. A spiral antenna can cover 2 to 18 GHz. The diameter of the antenna is approximately equal to $\lambda_{max}/2$ where λ_{max} is the wavelength corresponding to the lowest frequency. For an antenna with 2- to 18-GHz frequency range, $\lambda_{max} = 3 \times 10^{10}/2 \times 10^9 = 15$ cm and $\lambda_{min} = 3 \times 10^{10}/18 \times 10^9 = 1.67$ cm. Thus, the diameter of the antenna is about 7 cm. The closest distance one can put two antennas together is 7 cm from center to center.

In order to meet the relation in (15.44), the minimum distance between two spiral antennas must be less than 0.8 cm (1.67 cm/2). Since the antenna diameter is 7 cm, it is impossible to install the antenna this close. As a result, if the spiral antennas are used in the array to cover a wide frequency, it is inevitable to have ambiguity problems at a high-input frequency. The ambiguity problem might be solved through the Chinese remainder theorem, which will be discussed in the next section.

15.12 CHINESE REMAINDER THEOREM [22, 23]

The story goes like this. In ancient times in China, there was a general who wanted to know how many bodyguard soldiers he had. Instead of counting the soldiers directly, he arranged the soldiers in rows several times. He did not count the number of rows, but only counted the remainder of each arrangement to find the total number of soldiers. Two examples are used to illustrate this concept.

Example 1

The general arranged his soldiers in rows of 7 and the remainder was 3, and he rearranged his soldiers in rows of 10 and the remainder was 2. The two numbers 7 and 10 must be relative prime of each other. If he had less than 70 (7×10) soldiers, he found the total number of soldiers from the two remainders 3 and 2.

If the unknown number $x < 70$ is divided by 7 with a remainder of 3, the possible values of x are

$$x = 3 + 7n = 3, 10, 17, 24, 31, 38, 45, 52, 59, \text{ and } 66$$

where n is an integer from 0 to 9 $(10 - 1)$. Similarly, the same unknown number divided by 10 with remainder of 2 can be

$$x = 2 + 10n = 2, 12, 22, 32, 42, 52, \text{ and } 62$$

where n is from 0 to 6 $(7 - 1)$. Comparing these two series, one can find that 52 is the correct answer. The number of arrangements can be extended to more than 2. For example, the soldiers can be rearranged by rows of 11. Under this condition, the total number of soldiers can be increased to $x < 770$ $(7 \times 10 \times 11)$.

The Chinese remainder theorem can be stated in a slightly formal way as follows. If an unknown x is divided by m_i and the remainder is a_i, the result can be written with the following notation

$$x = a_i \bmod(m_i) \qquad \text{where} \qquad i = 0, 1, \ldots, I - 1 \tag{15.45}$$

In the Chinese remainder theorem, the m_i are relative prime with different i values. The solution of this problem (without proof) can be found as

$$M \equiv \prod_{i=0}^{I-1} m_i \qquad t_i \equiv \frac{M}{m_i}$$
$$x = \left(\sum_{i=0}^{I-1} a_i t_i u_i\right) \mathrm{mod}(M) \tag{15.46}$$
$$t_i u_i \equiv 1 \ \mathrm{mod}(m_i)$$

where I is total number of arrangements and u_i is an integer. Let us use the above example to demonstrate the utilization of this equation.

Example 2

Given $a_0 = 3$, $m_0 = 7$; $a_1 = 2$, $m_1 = 10$; $a_2 = 5$, $m_2 = 11$, where 7, 10, and 11 are relative prime, find x. From the above equation, $M = 7 \times 10 \times 11 = 770$, $t_0 = 110$, $t_1 = 77$, and $t_2 = 70$.

$t_0\ u_0 = 110 \times 3 = 1\ \mathrm{mod}(7) \Rightarrow u_0 = 3$
$t_1\ u_1 = 77 \times 3 = 1\ \mathrm{mod}(10) \Rightarrow u_1 = 3$
$t_2\ u_2 = 70 \times 3 = 1\ \mathrm{mod}(11) \Rightarrow u_2 = 3$

The u_i are obtained through trial and error and it happens all $u_i = 3$. Then,

$$x = (3 \times 110 \times 3 + 2 \times 77 \times 3 + 5 \times 70 \times 3)\ \mathrm{mod}(770)$$
$$= (990 + 462 + 1{,}050)\mathrm{mod}(770) = 2{,}502\ \mathrm{mod}(770) = 192$$

Of course, this result can be obtained by using the same approach as in Example 1, by arranging all the possible numbers in three rows. The correct answer will appear in all three rows. With the advance in digital signal processing, this approach might be made reasonably simple.

15.13 APPLICATION OF CHINESE REMAINDER THEOREM TO AOA MEASUREMENTS

As mentioned in Section 15.11, one cannot put two antennas close together to fulfill (15.44) because of the large size of the antennas. For the spiral antennas (from 2 to 18 GHz) with a diameter of 7 cm, one can put the two antennas $5d_{min}$ = 8.35 cm apart. This arrangement produces an ambiguity problem. In order to resolve this ambiguity, another antenna pair is needed. Figure 15.11 shows such an arrangement. One antenna pair is $5\lambda_{min}$, the other pair is $8.5\lambda_{min}$, and the distance between the two end antennas is $15.5\lambda_{min}$ (5 + 8.5). All three numbers, 10, 17, and 27, in terms of $\lambda_{min}/2$ are prime ones. However, these three antennas can be considered as two pairs, and the third pair produces no additional information. In

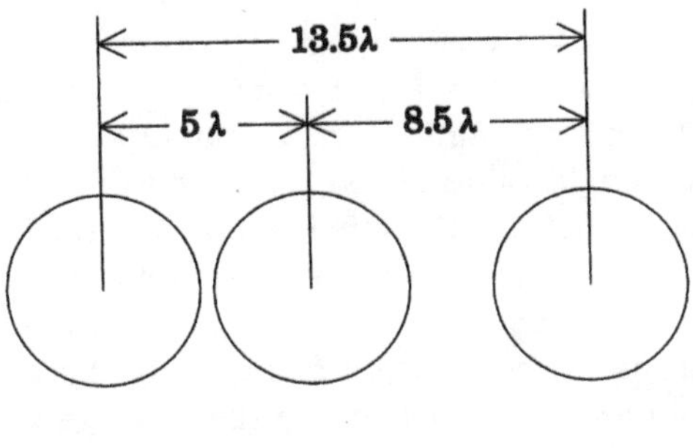

Figure 15.11 Three-element phase interferometric system.

this calculation, only two pairs of antennas will be considered and they are separated by 10 and 17 $\lambda_{min}/2$, respectively.

Although the remainders are used to find the desired AOA, the calculation is slightly different from the Chinese remainder mentioned in the previous section. First, let us study the case without ambiguity. Assume that the distance between the antenna elements is $\lambda/2$. Under this condition, there is no ambiguity and the relation in (15.20) can be written as

$$\psi = \pi \sin \theta \tag{15.47}$$

In this equation, ψ varies from $-\pi$ to π. As a result, $\sin \theta$ ranges from -1 to 1. This result is shown in Figure 15.12 and ψ versus $\sin \theta$ is a straight line. In this figure, there is no ambiguity. One value of ψ corresponds to one value of $\sin \theta$.

When $d = 5\lambda$, the result obtained from (15.20) is

$$\psi = 10\,\pi \sin \theta \tag{15.48}$$

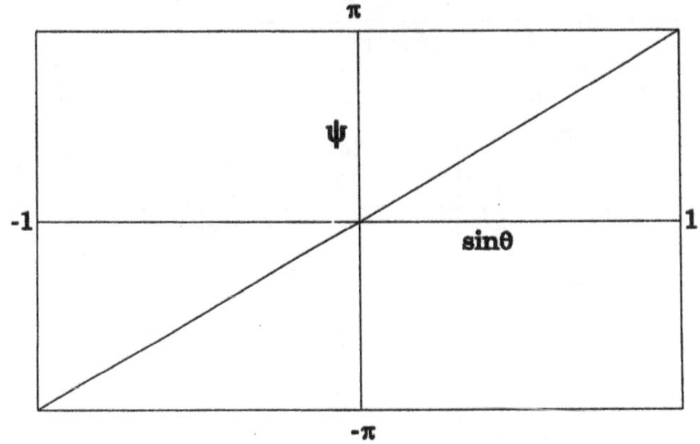

Figure 15.12 ψ versus $\sin \theta$ for $d = \lambda/2$.

In this equation, there should be 10 ambiguities. Since sin θ ranges from -1 to 1, the angle ψ can change from -10π to 10π. However, the value of angle measurement ψ is limited from $-\pi$ to π. Any angle over this range will be folded back into this range. That means any angle value ϕ less than $-\pi$ will be written as $\phi + n\pi$ and any angle value ϕ greater than π will be written as $\phi - n\pi$. The relation of (15.48) is shown in Figure 15.13. In this figure, for every value of ψ there are 10 possible values of sin θ, and this is the expected result. If a certain phase angle ψ is measured, it is impossible to determine the incident angle of the signal. In comparison with Figure 15.12, the resolution is better because the slope in this figure is steeper.

In order to resolve this ambiguity problem, another pair of antennas 8.5λ is required. The result is shown in Figure 15.14. In this figure at any incident angle

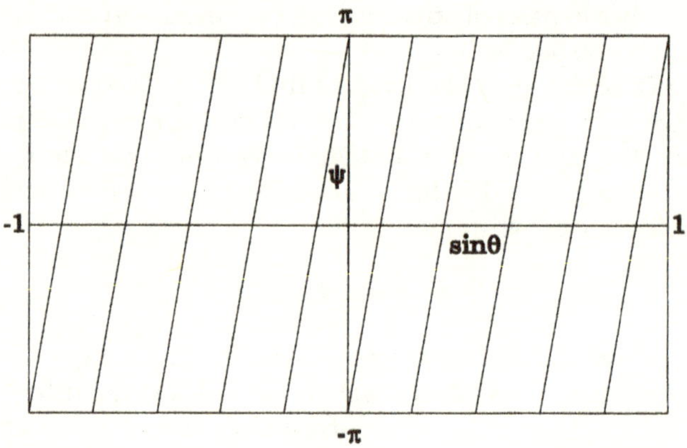

Figure 15.13 ψ versus sin θ for $d = 5\lambda$.

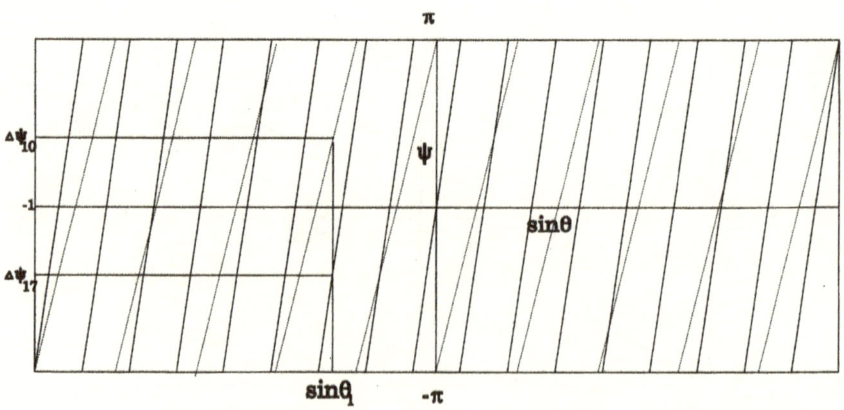

Figure 15.14 ψ versus sin θ for $d = 5\lambda$ and $d = 8.5\lambda$.

θ_1 there are two unique values ψ_{10} and ψ_{17}. Thus, if ψ_{10} and ψ_{17} are measured, the incident angle θ_1 can be obtained. In actual measurement, a conversion table can be generated; once ψ_{10} and ψ_{17} are obtained, the incident angle θ_1 can be found from the table.

15.14 PRACTICAL CONSIDERATIONS IN REMAINDER THEOREM [24]

It appears that the Chinese remainder theorem can be used to resolve the antenna array problem in a very neat way. In practice, when the remainder theorem is used to solve any type of problem, there are quite a few problems one must consider very carefully. Some of these problems will be presented as follows.

1. In the theoretical discussion in Section 15.12, the remainders are measured very accurately as integers. In practice, there are errors in the measured remainders. In conventional AOA measurement, an error in the final value only affects the accuracy of the measurement. However, if the remainders are used to determine the actual value, a measurement in the final value (the remainder) will cause catastrophic error, which means the error can be very large. This large error can be easily realized from the examples in Section 15.12. If one of the remainders is changed by one numerical number, the result will be totally different.
2. If noise in the remainder measurement is taken into consideration, the overall ambiguous range must be decreased. For example, if the dividers are 5 and 7, the maximum number should be 34 ($5 \times 7 - 1$). All numbers from 0 to 34 are possible solutions. If the errors are allowed in the remainders, the maximum number should be less than 34. Figure 15.15 provides a visual understanding of this problem.

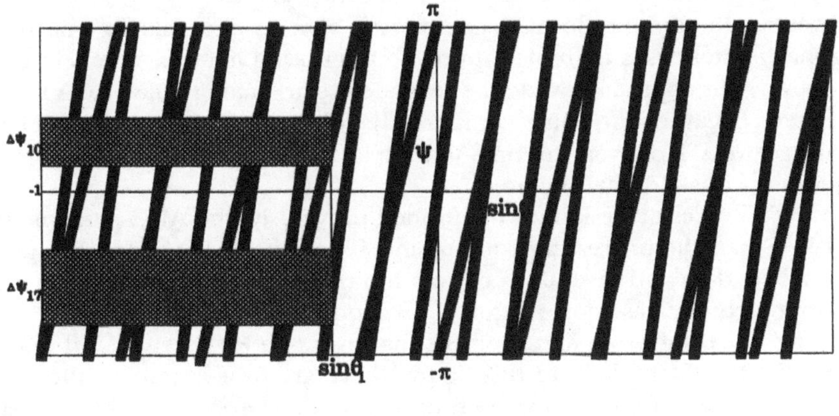

Figure 15.15 ψ versus $\sin\theta$ for $d = 5\lambda$ and $d = 8.5\lambda$ with noise.

This figure shows the same result as in Figure 15.14. The only difference is that there is noise in these remainders, and the noise increases the width of the lines. In this figure, at incident angle $q1$, the corresponding electrical angles measured as ψ_{10} and ψ_{17} cover a range of values rather than a single value as indicated in Figure 15.14. If one decreases the range of θ_1, this method may still be used to measure the AOA, but the total angle coverage will be decreased also.

3. The relationship between y and sin q is linear, as indicated by (15.20) and the straight lines in Figures 15.14 and 15.15. If actual data are collected from antennas, it is possible that these lines are no longer linear, but slightly curved. The deviation from a linear line will complicate the measurement procedure. It means it is difficult to generate the conversion table to obtain the incident θ_1 from ψ_{10} and ψ_{17}. The curvature in the curves must be included in the conversion table.
4. Additional hardware can be used to improve the measured results. For example, instead of using two antenna pairs and their remainders to provide the incident angle, one can add another antenna and use three antenna pairs to provide more redundancy in the measurement. This approach not only needs one more antenna element and receiver, but the conversion table will also be more complicated. Increasing the order of redundancy can reduce the probability of erroneous results at the cost of more hardware.

15.15 HARDWARE CONSIDERATIONS FOR DIGITAL AOA MEASUREMENTS [7, 25]

In analog AOA measurement systems, the most difficult problem is to build many parallel antennas/receivers and match the amplitude and phase among them. Some experimental systems have difficulty achieving amplitude and phase matching among different receivers, especially for wideband systems.

In an analog amplitude AOA comparison system, time-sharing one receiver among many antennas is a simpler approach than using many receivers in parallel [7]. However, this approach cannot be extended to an analog phase AOA measurement system because one must compare the phase simultaneously from many different receivers. However, this time-sharing idea can be used in a phase measurement system if it is a digital system.

In digital receiver design, as mentioned previously, the ADC performance is much faster than the processing components following it. Thus, one can use one ADC to collect data and several processors to process it. To measure AOA using a digital approach, data as shown Figure 15.9 should be collected.

Figure 15.16 shows an arrangement that uses one RF receiver and one ADC to collect the necessary data. In this figure, there are four antennas; the receiver and ADC are time-shared among the four antennas. There is a four-to-one switch that sequentially rotates through the four antennas. The switching time must match

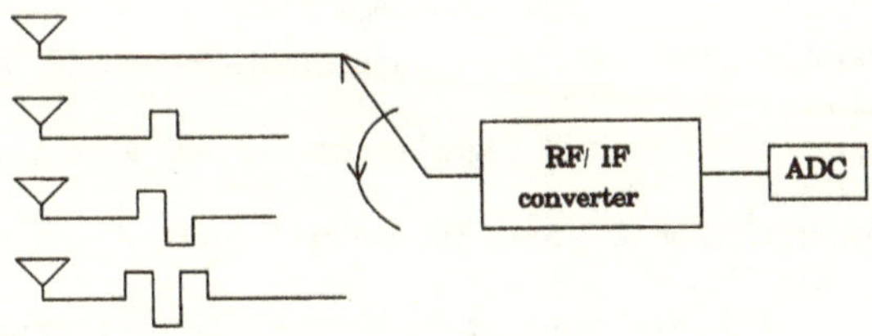

Figure 15.16 One receiver shared among four antennas.

the sampling time of the ADC. If the sampling time is t_s, the switch must operate at the same speed, switching to the next antenna at time t_s.

If it is desirable to have the data collected from all four antennas at the same time as shown in Figure 15.9, delay lines should be inserted to compensate for the switching delay. The delay time should be in the unit of t_s. In Figure 15.16, antenna 0 has no delay, antenna 1 has t_s delay, antenna 2 has $2t_s$ delay, and antenna 3 has $3t_s$ delay. Since an ADC can operate at 1 GHz, which corresponds to $t_s = 1$ ns, the delay time is not a big problem.

It is anticipated that the delay lines can be omitted in Figure 15.16. Under this condition, the data collected are shown in Figure 15.17. The data collected are not at the same time instant, but skewed in time. If this type of data are available, digital signal processing can be used to take the time skew into consideration and find the correct AOA.

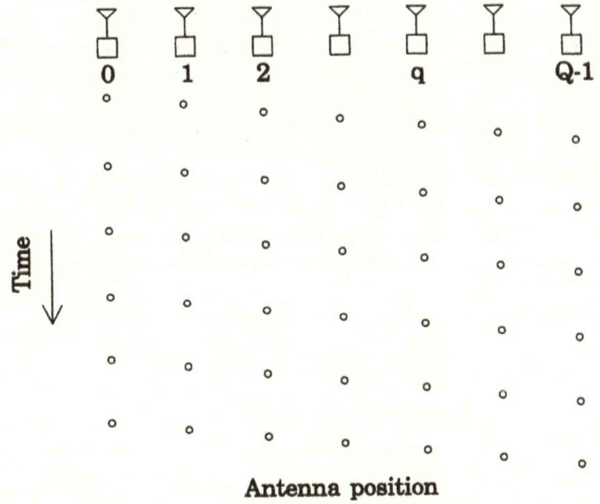

Figure 15.17 Digital data collected from a time-sharing system without delay correction.

REFERENCES

[1] Capon, J. "High Resolution Frequency Wavenumber Spectrum Analysis," *IEEE Proc.*, Vol. 57, Aug. 1969, pp. 1408–1418.

[2] Reddi, S. S. "Multiple Source Location-A Digital Approach," *IEEE Trans. Aerospace Electronic Systems*, Vol. AES-15, Jan. 1979, pp. 95–105.

[3] Johnson, D. H. "The Application of Spectral Estimation Methods to Bearing Estimation Problems," *IEEE Proc.*, Vol. 70, Sept. 1982, pp. 1018–1028.

[4] Kumaresan, R., and Tufts, D. W. "Estimating the Angles of Arrival of Multiple Plane Waves," *IEEE Trans. Aerospace Electronic Systems*, Vol. AES-19, Jan. 1983, pp. 134–138.

[5] Shan, T. J., Wax, M., and Kailath, T. "Spatial Smoothing Approach for Location Estimation of Coherent Sources," *IEEE Asilomar Conf. Circuits and Systems*, 1984, pp. 367–371.

[6] Li, F., and Vaccaro, R. V. "On Frequency Wavenumber Estimation by State-Space Realization," *IEEE Trans. Circuits and Systems*, Vol. 38, July 1991, pp. 800–804.

[7] Tsui, J. B. Y. *Microwave Receivers With Electronic Warfare Applications*, New York, NY: John Wiley & Sons, 1986.

[8] Jordan, E. C. *Electromagnetic Wave and Radiating Systems*, Englewood Cliffs, NJ: Prentice Hall, 1950.

[9] Kraus, J. D. *Electromagnetics*, New York, NY: McGraw Hill Book Co., 1953.

[10] Longstaff, I. D., Chow, P. E. K., and Davies, D. E. N. "Directional Properties of Circular Arrays," *IEEE Proc.*, Vol. 114, June 1967, pp. 713–718.

[11] Sheleg, B. "A Matrix-Fed Array for Continuous Scanning," *IEEE Proc.*, Vol. 56, Nov. 1968, pp. 2016–2027.

[12] King, W. P., and Harrison, C. W. *Antennas and Waves: a Modern Approach*, Cambridge, MA: The MIT Press, 1969.

[13] Ma, M. T. *Theory and Application of Antenna Arrays*, New York, NY: John Wiley & Sons, 1974.

[14] Shelton, J. P., and Kelleher, K. S. "Multiple Beams From Linear Arrays," *IRE Trans. Antennas and Propagation*, Vol. AP-9, March 1961, pp. 154–161.

[15] Jacobs, E., and Ralston, E. W. "Ambiguity Resolution in Interferometry," *IEEE Trans. Aerospace Electronic Systems*, Vol. AES-17, Nov. 1981, pp. 766–780.

[16] McClellan, J. H. "Multidimensional Spectral Estimation," *IEEE Proc.*, Vol. 70, Sept. 1982, pp. 57–67.

[17] Haykin, S., Editor. *Array Signal Processing*, Englewood Cliffs, NJ: Prentice Hall, 1985.

[18] Kraus, J. D. *Antennas*, New York, NY: McGraw Hill Book Co., 1950.

[19] Rumsey, V. *Frequency Independent Antennas*, New York NY: Academic Press, 1966.

[20] Stutzman, W. L., and Thiele, G. A. *Antenna Theory and Design*, New York, NY: John Wiley & Sons, 1981.

[22] Taylor, F. J. "Residue Arithmetic: a Tutorial With Examples," *IEEE Computer*, Vol. 17, No. 5, May 1984, pp. 50–62.

[23] Wolf, J. K. "The Chinese Remainder Theorem and Applications," Ch. 16 in Blake, J. F., and Poor, H. V. *Communications and Networks*, New York, NY: Springer Verlag, 1986.

[24] McCormick, W., Tsui, J.B.Y., and Bakke, V. "A Noise Insensitive Solution to a Simultaneous Congruence Problem in Spectrum Estimation," *IEEE Trans. Aerospace Electronic Systems*, Vol. 25, Sept. 1989, pp. 729–732.

[25] Tsui, D. C. WL/AAAI Wright Laboratory, Private communication.

CHAPTER 16

Receiver Tests

16.1 INTRODUCTION

This chapter discusses the performance of a digital receiver and the procedures of testing it. The receiver performance is one of the most important aspects in receiver research. One of the main impasses in receiver research is that there are no universally acceptable standards in the performance of electronic warfare (EW) receivers. Because of this shortcoming, researchers do not know where to improve the EW receiver performance. If one claims to have made some improvements in receiver performance, but cannot report the result quantitatively, it will be rather difficult to be accepted. Worse yet, people can make claims on certain performances of an EW receiver that cannot even be used in a system at all. For example, if a receiver misses many signals or produces a large number of spurious responses, in general it cannot be used in any system. Under this condition, no matter how good other performances are, the result should not be reported. There should be some minimum requirements a receiver must be measured against in order to qualify as a functioning receiver.

One interesting experience is in channelized receivers. The main research needed is in the parameter encoder to encode the channel outputs into frequency words. However, since there is no performance standard for the parameter encoder, very few people have worked in this area. Even if one obtains some results, they cannot be published because one may not be able to claim any improvement. On the other hand, the filter bank itself has sufficient performance standards (i.e., insertion loss, bandwidth, frequency and time domain spurious responses, etc.). As a result, a lot of research has concentrated in this area and many technical papers have been published, but very little effort has been devoted to the major problem of channelized receivers.

In order to stimulate research in EW receivers, not only are some performance standards needed, but the standards should also be publicized. In addition, researchers should be encouraged to publish their results so that engineers and scientists working in this area will know the problems and can seek solutions to them. In this chapter, all the performance issues discussed in Chapter 2 will be discussed.

This chapter will discuss different types of receiver tests. It will concentrate on the laboratory and anechoic chamber tests rather than field tests because these tests can be performed in a controlled manner and will generate results on EW receiver performance.

16.2 TYPES OF RECEIVER TESTS

An EW receiver should be tested under different situations. No matter how thoroughly a receiver is tested, it appears one can never cover all the possible input signal conditions. A receiver should be tested in the following order.

1. Laboratory tests.
   a. Preliminary test.
   b. Regular tests.
      (1) Single-signal test:
         (a) Frequency test;
         (b) Frequency standard deviation test;
         (c) False alarm test;
         (d) Sensitivity test;
         (e) Dynamic range test;
         (f) Pulse amplitude test;
         (g) Pulse width test;
         (h) Angle of arrival (AOA) test;
         (i) Time of arrival (TOA) test;
         (j) Shadow time, latency, and throughput rate tests;
         (k) Random frequency/amplitude test.
      (2) Two-signal tests:
         (a) Frequency resolution test;
         (b) Spurious free dynamic range test;
         (c) Instantaneous dynamic range test;
         (d) AOA resolution test;
         (e) Random frequency/amplitude test.
2. Anechoic chamber tests.
3. Simulator tests.
4. Field tests.

The laboratory tests are considered as the most important tests because the tests are performed under a controlled environment. Most of the important characteristics of a receiver can be obtained from these tests. Therefore, these tests results will determine the performance of the receiver. The preliminary tests should be carried out first to determine whether the receiver is good enough to undergo regular tests. In this chapter, the discussion will concentrate on all the laboratory tests.

The anechoic chamber tests are the same as the laboratory tests. If the RF input of the receiver can be applied through an RF cable as shown in Figure 16.1(a), the laboratory tests should be used because the input signals can be better controlled. If the input of the receiver is an antenna array as shown in Figure 16.1(b), the input cannot be applied through an RF cable. In this situation, anechoic chamber tests will be applied. The input signal is applied to the receiver through radiation in an anechoic chamber. The anechoic chamber tests generate all the performance results of a receiver as listed in the laboratory tests and should provide more reliable AOA measurements.

If a receiver has a receiving antenna array and an anechoic chamber is not available, one must simulate a planewave front as the input to the RF connectors. The planewave front is difficult to achieve, especially over a wide frequency and incident angle ranges. However, if a receiver has an amplitude AOA measurement system, it can be either tested in the laboratory or in the anechoic chamber because it is easy to simulate the signal input for an amplitude comparison AOA measurement system.

Simulator tests usually use special equipment to simulate the signal environment an EW receiver passes through. The environment can be static or dynamic, which changes with time. Under these tests, the receiver can be evaluated under very high signal density. Sometimes, the outputs of the receiver are processed by a real-time signal processor to determine the performance of the overall system, which includes the signal processor. If a real-time processor is not available, the

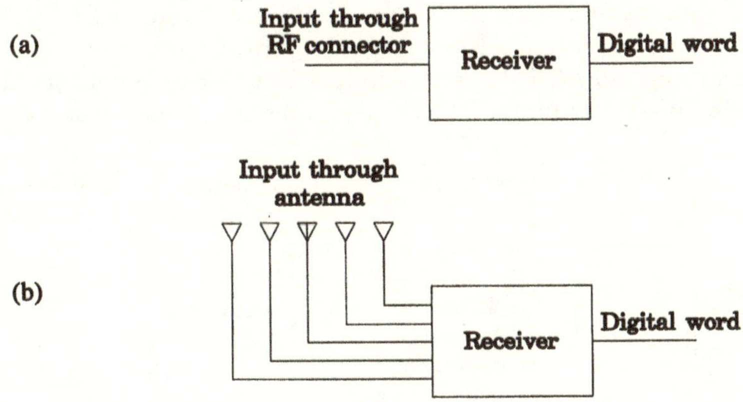

Figure 16.1 A receiver represented by a black box: (a) input through RF connector, (b) input through antenna.

output data from the receiver under test can be recorded and compared with the input signals.

Theoretically speaking, the input signal condition in the field tests is unknown. However, with an unknown input signal condition, the performance of the receiver cannot be determined. In order to remedy this shortcoming, two receivers can be tested at the same time to compare their results. If one can control some radars in the field, certain radars can be turned on to transmit. Under this situation, at least the signals of these radars are known and can be used to check the performance of the receiver.

16.3 PRELIMINARY CONSIDERATIONS IN LABORATORY RECEIVER TESTS

Before the discussion of the laboratory tests of an EW receiver, it is appropriate to consider the philosophy of receiver testing. The tests should be limited to receivers that can process simultaneous signals. Specific test procedures can be designed to evaluate receivers that can process only one signal. For example, the performance of an instantaneous frequency measurement (IFM) receiver and a superheterodyne receiver is often available.

1. The receiver under test will be considered as a black box with radio frequency (RF) as input and pulse descriptor words (PDWs) as outputs. Therefore, these tests can be applied to digital as well as analog receivers.
2. The receiver under test must have PDWs as output. Each PDW contains all the desired information (i.e., frequency, pulse amplitude (PA), pulse width (PW), TOA, and AOA). To emphasize this point, a digital EW receiver is represented by a block diagram as shown in Figure 16.2. In this figure the time-frequency and space-AOA conversions are through digital signal processing; thus, the output will be in digital form and represents the information in frequency and AOA domains. However, these digital data should not be considered as the outputs of the receiver. This information must be converted

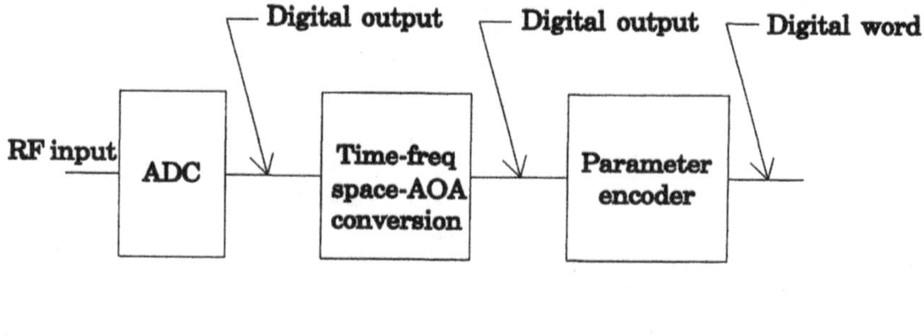

Figure 16.2 Three blocks of a digital receiver.

into a PDW containing the desired information. One should not use other information, even if it is in digital form (i.e., fast Fourier transform (FFT) outputs to evaluate the receiver performance) because it is not a trivial task to generate the PDW.

3. *Input conditions:* It is impractical to cover all the input conditions during the test. For example, if a receiver can receive two simultaneous signals and the input bandwidth is 1,000 MHz with a single-signal dynamic range of 70 dB (from −60 to 10 dBM) and an instantaneous dynamic range of 40 dB (from −60 to −20), the number of input conditions can be found as follows.

   a. *Single-signal test:* If the frequency data resolution of the receiver is 1 MHz and the amplitude data resolution is 1 dB, for the single-signal test, the required number of inputs is at least 70,000 (1,000 × 70) to test each output condition. However, the frequency and amplitude resolution steps may not be limited to 1-MHz and 1-dB steps. If the signal generator can change the frequency in steps of 1 kHz and the amplitude by 0.1 dB, the possible input conditions can be extended to 700 million (1,000,000 × 700) possibilities, which is impractical to carry out.
   b. *Two-signal test:* For the two-signal case, the number of input signals is even larger. For each signal, there are 40,000 (1,000 in frequency times 40 in amplitude) output cells. The total possible outputs are 799.98 million (40,000 × 39,999/2), which is unreasonably large. This is the minimum number of inputs required to test all the possible outputs. If the signal generators have finer frequency and amplitude steps than the receiver, the number of input conditions can be made unimaginably large.

   If the receiver can process more than two simultaneous signals, the number of input conditions would be even larger. In reality, it is impossible to test a receiver with several thousands of input conditions, even if the test setup is automated. From this discussion, it is apparent only a very small percentage of the input conditions can be tested. The limited tests may not reveal all the deficiencies in a receiver because even well-behaved receivers used in a real environment may generate unexpected data.

4. *Type of inputs:* The input signals can be applied to the receiver under test through two possible ways. First, the signal can be applied sequentially (i.e., the frequency changed from low to high values in uniform steps). Second, the signal can be applied in a random fashion (i.e., random frequency and amplitude).

   The advantages of using systematic inputs are that the results might be presented graphically and it is easy to repeat the same test. Sometimes, it is desirable to repeat a condition at which the receiver generates erroneous information so one can correct the receiver design. The major disadvantage of this type of test is that the input could be biased and limited to a very small region.

The advantage of random input signals is that the distribution can be made more uniform over a desired range. A true random input signal is difficult to repeat. A pseudorandom signal generated through a computer can be repeated. It might be difficult to present this type of test result graphically. Probably the simplest way to represent the data is through tabulation.

In a receiver test, both the systematic and the random input signal conditions should be used.

5. *The test results presentation:* The test results obtained from the receiver must be easily comprehensive from some simple display (i.e., graphic and tabulations). It is impractical to generate pages of printed data to locate some erroneous information. If the measured data cannot be presented in some simple forms, it does not make much sense to generate them in the first place.
6. *One and two-signal tests:* Based on the discussion in number 5, it is reasonable to limit the receiver tests to one signal and two signals. Most of the receiver performance definitions are based on one and two signals. If a receiver can process four simultaneous signals, the number of possible input signal conditions is very high. A small portion of input signal conditions used in a test may not be significant at all. Thus, it is reasonable only to test whether the receiver can really process four simultaneous signals.

16.4 RECEIVER TESTS THROUGH SOFTWARE SIMULATION

It is highly desirable to evaluate the performance of a receiver before it is fully fabricated. On average, development of a breadboard analog EW receiver takes more than three years from design to completion. Unfortunately, sometimes there are obvious design flaws, but they cannot be easily detected. However, once the receiver is built, these flaws are difficult to correct. If all the receiver tests can be performed through software simulation, some of these flaws can be discovered at a much earlier stage and can be corrected easily.

Simulations of analog receiver design have been carried out for some time. The effort can be divided into two parts: 1) the analog portion from RF to video output and 2) the parameter encoder from video to PDW as shown in Figure 16.3. The work on the first part, from RF to video signal, could be considered very successful. The video output matches the simulated output very closely. Due to the complicated nature of the parameter encoder design, only very limited simulations are achieved, if any at all. With the advances in computer technology, there is a chance the parameter encoder can be simulated and evaluated before it is actually built.

For a digital receiver, the time-frequency and space-AOA transforms are all performed in the digital domain. However, as mentioned before, these digital

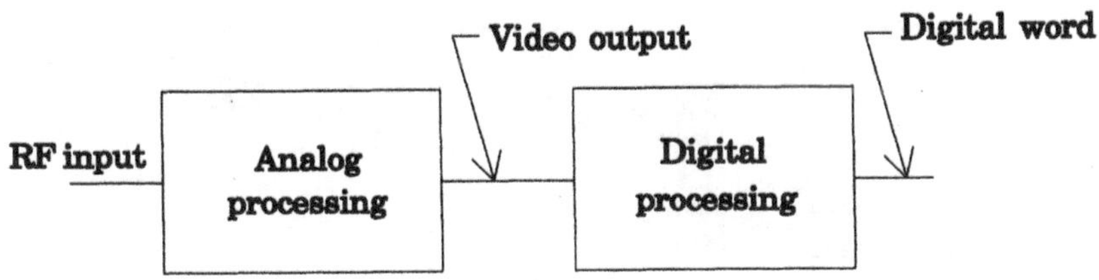

Figure 16.3 Two blocks of an analog receiver.

outputs can only be considered as intermediate outputs and should not be used as the receiver final outputs. Converting the frequency and AOA information into the PDWs is done through digital processing. Therefore, all these operations can be simulated by a computer. A receiver should be adequately simulated before actual fabrication. This simulation should be made equivalent to an actual receiver test procedure, which means that with a certain input the receiver should generate a certain output. If the simulation can be processed at very high speed, this simulation can be very useful in digital receiver design.

16.5 LABORATORY TEST SETUP

The laboratory test setup is shown in Figure 16.4. In this figure, there are two signal sources. Each signal source contains one signal generator followed by a pin modulator, which is controlled by a pulse generator to produce an RF signal with the desired PW. An attenuator is used to adjust the signal strength. It should be noted that in the arrangement, the input to the pin modulator is at a constant power level. Under this condition, when the input power level changes, the shape of the pulse generated does not change. If the attenuator is placed in front of the pin modulator, the input to the pin modulator changes. Under this condition, when the signal power changes, the shape of the signal may change.

The two signal generators are combined through a power combiner as the input to the receiver under test. A power meter, a spectrum analyzer, and an oscilloscope are used to check the output of the signals. The power level at the input of the receiver needs to be calibrated because the power loss from the signal generator to the input of the receiver is frequency-dependent. Figure 16.5 shows the uncalibrated and calibrated cases of the power output from 2 to 18 GHz. As expected, the uncalibrated power output is higher at the low-frequency range than at the high-frequency end, as shown in Figure 16.5(a). The calibrated output is obtained by increasing the attenuation at low frequency to equalize the output power across the entire frequency range, as shown in Figure 16.5(b). The outputs of the receiver are collected through a data collection instrument. The Tektronix's DAS 9200 can be used to collect data. However, the data collecting instrument can be a computer used in the direct memory addressing (DMA) mode.

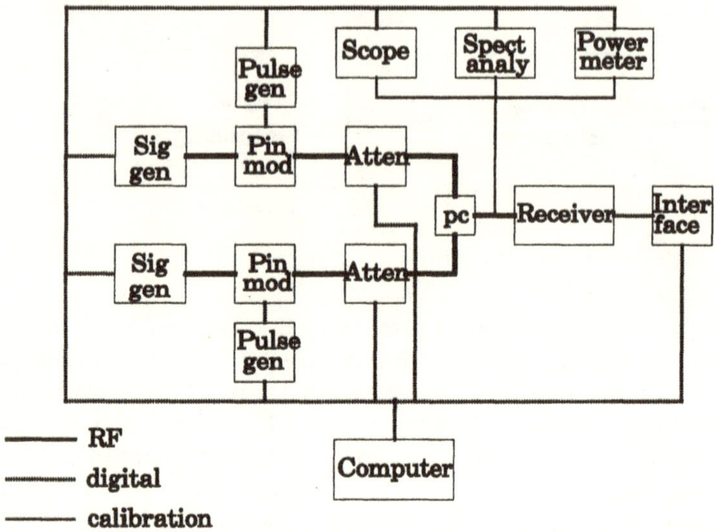

Figure 16.4 Laboratory receiver test setup.

All the equipment except the receiver under test is controlled by a computer. Some receivers are computer-controlled, but under tests the receiver should stay in a certain operation mode. The signal sources are computer-controlled to generate the desired signals. The output data will be compared with the input data. There are two ways to compare these data. One way is to compare each input immediately with the output. The other way is to collect all the data and compare with all the input data at the end of the test. It appears that the latter approach can save time.

16.6 ANECHOIC CHAMBER TEST SETUP

The anechoic chamber test setup is very much like the laboratory test setup. The only difference is that the signals are radiated to the receiving antenna of the receiver under test. Figure 16.6 shows such an arrangement. Two signal sources used in the test are identical to the ones discussed in the previous section. The only difference is that the two signals are not combined together; instead, each feeds a transmitting antenna. The antennas are directed at the receiver under test. The arrows in this figure indicate the direction of the antenna movement. The transmitting antennas can be manually placed in positions. The receiver with the receiving antenna is placed on a turning table. By rotating the receiving antenna, the AOA of the input signals can be changed. To change the incident angle between the two transmitting antennas, the positions of the transmitting antennas must be changed.

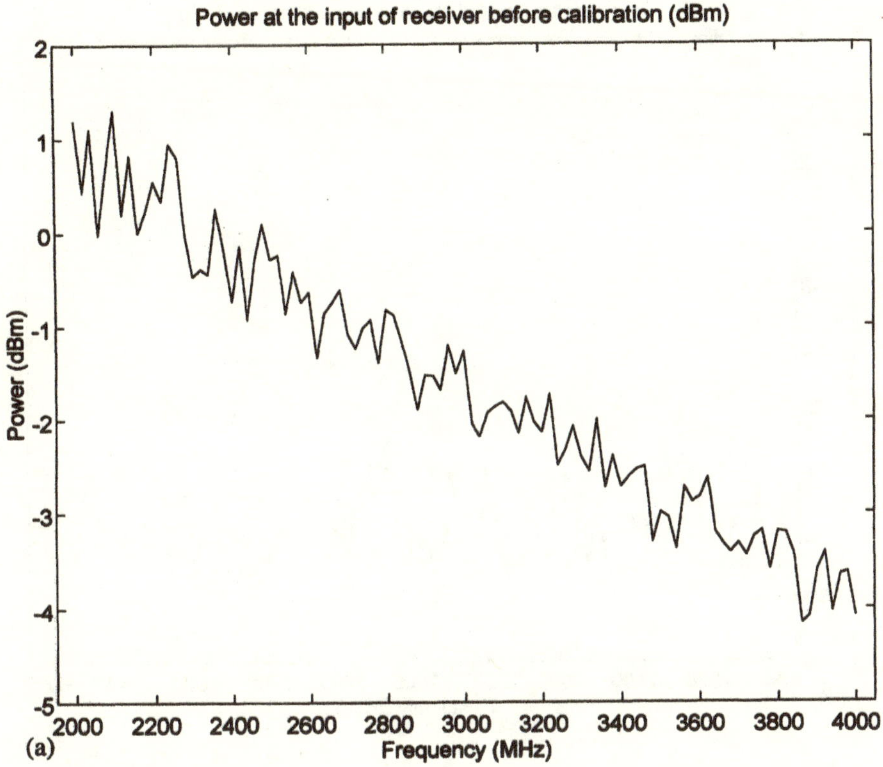

Figure 16.5 Signal power at input of receiver under test: (a) uncalibrated case, (b) calibrated case.

The power to the receiving antenna must be calibrated. It usually requires a standard horn antenna with known gain to measure the signal strength. The power measured is no longer in milliwatts or in dBM, but in milliwatts or dBM per unit area. In this calibration procedure, the receiving antenna must line up on bore sight with the transmitting antenna.

16.7 PRELIMINARY TESTS

A receiver must pass at least two simple tests to qualify for further tests. If the receiver is designed to cover 1,000 MHz (1,000 to 2,000 MHz) in bandwidth and can process two simultaneous signals in the range of 40 dB (from –60 to –20 dBM), simple one-signal and two-signal tests should be performed. The range of the input signals must be well within the desired performance range. For example, the input frequencies are limited within 800 MHz (from 1,100 to 1,900 MHz) and the ampli-

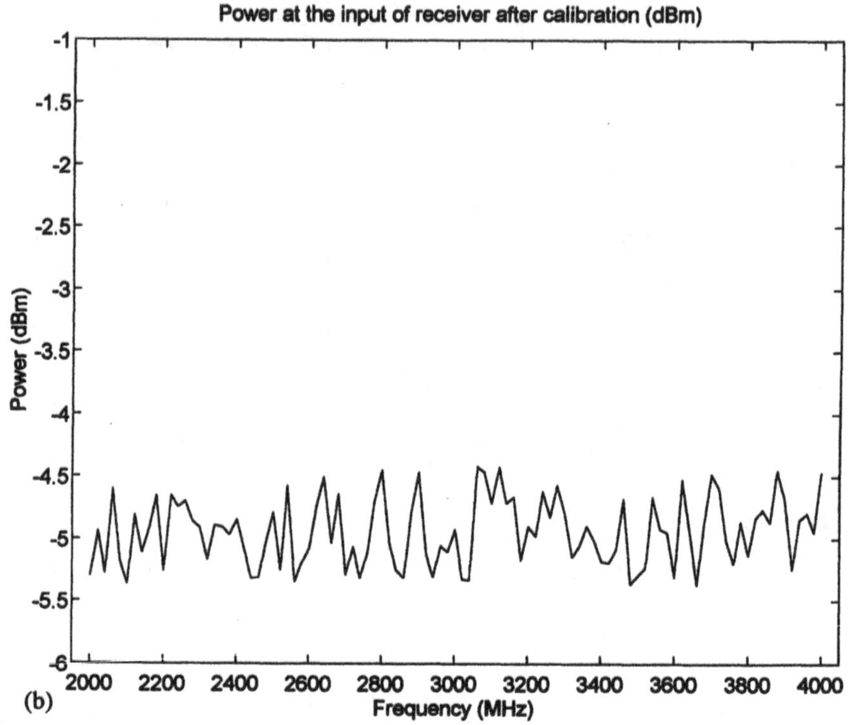

Figure 16.5 (continued).

tudes are limited to 20 dB (between –50 and –30 dBM). Within these limits, the well-designed receiver should perform satisfactorily.

1. *Single-signal test:* Randomly select one signal within the limited frequency and dynamic range to check for missing signals and spurious signals, which means more than one signal is detected. The random input signal can be selected at 1,000 different values to see what percentage of missing signals and spurs is detected. If the percentage of missing signals and spurious signals is high (i.e., over 1% missing signal and over 2% spurious response), further tests should not be performed.
2. *Two-signal test:* Randomly select two signals within the limited frequency and dynamic range to check for missing signals and spurious responses. The same criteria as mentioned in the one-signal test can be used to stop further tests.

Past experience indicates that under the single-signal test some poorly designed receivers often generate a high percentage of spurious signals. These receivers should not be tested for the two-signal condition. Even if a receiver passes

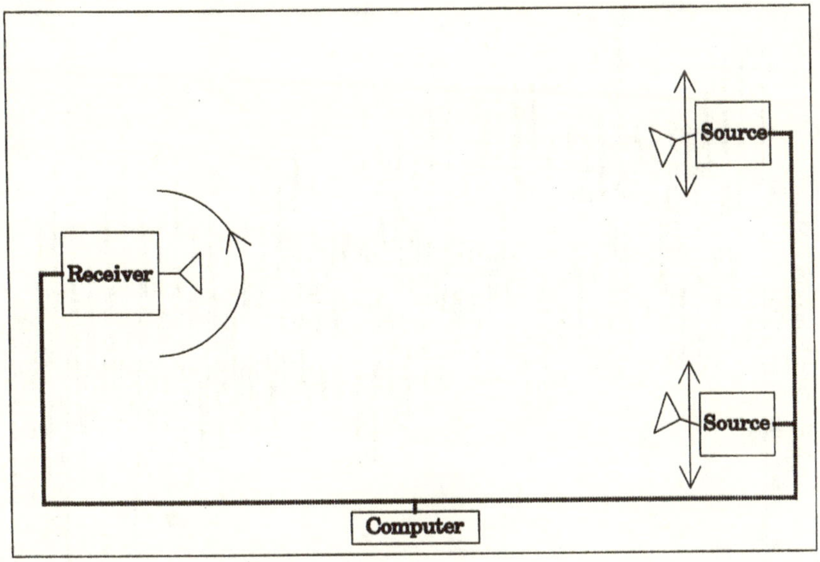

Figure 16.6 Anechoic chamber test setup.

the single-signal test, it may not pass the two-signal test. A poorly designed receiver under two-signal tests may miss one of the input signals, or worse yet, miss both signals.

16.8 SINGLE-SIGNAL FREQUENCY TEST

In this section two frequency tests will be discussed. First is the frequency accuracy test and the second is the standard deviation test on frequency.

16.8.1 Frequency Accuracy Test

In this test, the input frequency is considered as a variable, whereas the pulse amplitude and pulse width are constants. During the test, a certain pulse amplitude and PW are selected. The input frequency is varied from the lower frequency bound of the receiver to the upper frequency bound in predetermined steps. At each frequency interval, only one signal is sent to the receiver. If the input frequency is f_i and the measured frequency is f_m, the error frequency f_e is defined as

$$f_e = f_m - f_i \tag{16.1}$$

The frequency accuracy result is the plot of f_e versus f_i, as shown in Figure 16.7. In this figure, one can tell the frequency resolution and the bias error. The

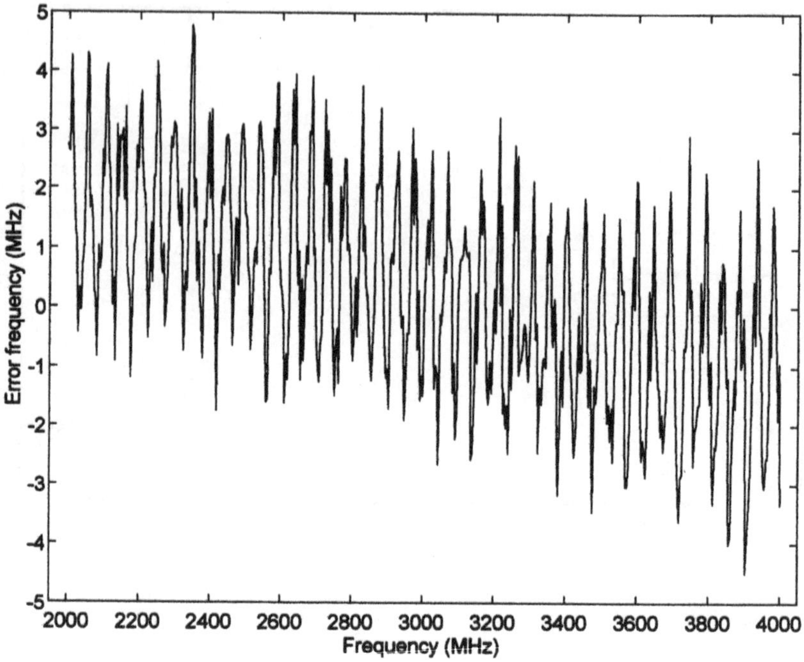

Figure 16.7 Frequency measurement.

frequency data resolution can be obtained from the width of the error frequency. From this figure, one can tell that the frequency data resolution is about 4 MHz. If the receiver is properly designed, the width of the error frequency is uniform. If N input frequencies are applied to the receiver, the bias frequency f_b is defined as

$$f_b = \sum_{i=0}^{N-1} f_{ei} = \sum_{i=0}^{N-1} (f_{mi} - f_{ii}) \tag{16.2}$$

where the second subscript i represents the ith data point. The first subscript m and i represent the measured and the input frequencies, respectively. If there is no bias, $f_b = 0$.

In Figure 16.7, at low frequency the bias is positive and at the high-frequency end the bias is negative. If the above equation is used to calculate the bias frequency, the result might not be accurate because the bias at the low frequency may offset the bias at high frequency. The width of the frequency variation is approximately equal to the frequency resolution cell. In this case, the frequency resolution cell is approximately 4 MHz wide.

16.8.2 Frequency Precision Test

The purpose of this test is to find the consistency of the receiver output data. In some receivers, for the same input condition the receiver may report different output. In this test, the input data are applied at the same value repetitively for N times and the output is measured N times. The root mean square (rms) value is defined as

$$f_s = \sqrt{\frac{1}{N}\sum_{i=0}^{N-1}(f_{mi} - f_i)^2} \tag{16.3}$$

where the second subscript i in m_i represents the ith data point and f_i is the input frequency. Figure 16.8 shows the rms values of frequency measurement for a receiver with $N = 100$. In this test, it appears that the frequency reading is consistent near the center of the channel where the rms values are equal or close to zero.

16.9 FALSE ALARM TEST [1–6]

The false alarm test is the most difficult test, especially if a numerical value is desired. In general, the false alarm rate is very low in an EW receiver and only a

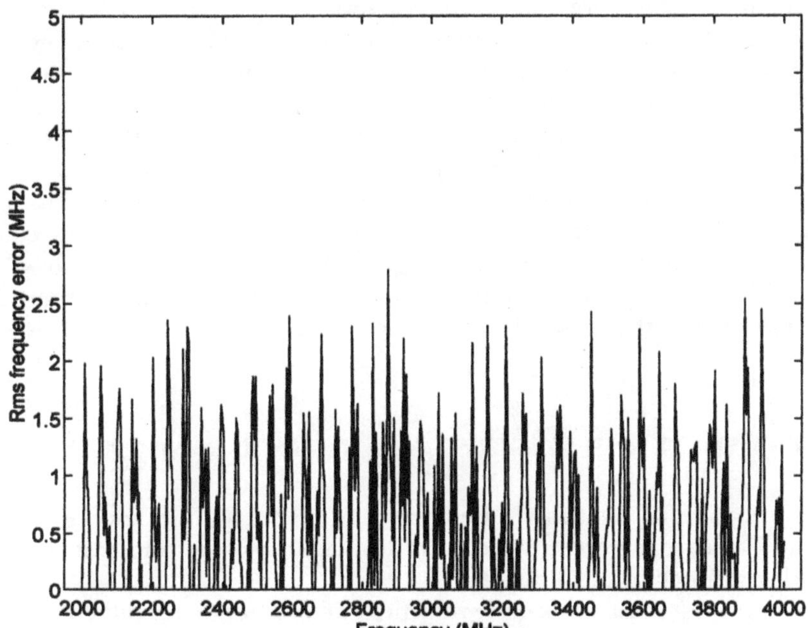

Figure 16.8 rms values of frequency measurement.

limited time can be used to measure it. For example, if a receiver is measured for one hour and zero false alarms are recorded, the only claim one can properly make is that the receiver has been tested for one hour and zero false alarms were recorded. Actually the receiver false alarm rate can possibly be a few in an hour to one in several days or even months. Thus, it is difficult to obtain a numerical value. However, it is desirable is to obtain a numerical value in false alarm measurement. If the receiver is tested for a long time, it is difficult to keep the environment (i.e., noise level) unchanged.

There are two approaches to obtain measurable false alarms. One is to decrease the threshold of the receiver. This approach is referred to as importance sampling [1–3]. The difficulty with this method in an EW receiver is that there might be many thresholds. For example, in a channelized receiver, at the output of each channel there is a threshold. To adjust all these thresholds is difficult. The second approach to increase the number of false alarms is to add noise in the front end of the receiver [4]. This method will be discussed here.

Noise can be added at the input of the receiver to obtain a measurable false alarm value without changing the threshold. In order to obtain a numerical value, the RF bandwidth B_R to video bandwidth B_V ratio $\beta = B_R/B_V$ and the noise figure of the receiver must be known. In addition, it must be known whether the receiver is dc or ac coupled. The derivation will not be discussed here, only the results will be presented. The measurement setup is shown in Figure 16.9. In this figure, the noise source is known and the attenuator can be adjusted to obtain a measurable false alarm (i.e., several false alarms in a second). Under this condition, if one measures over a relatively long time with respect to a second (say, for example, 10 min), a dependable quantity can be obtained. Once this amount of false alarm is obtained, the false alarm can be extrapolated from the measured value.

A computer program listed at the end of this chapter can be used to find the probability of the false alarm. The inputs to this program are 1) ac or dc coupling, 2) attenuator setting, 3) receiver noise figure, 4) noise source power in dBM, 5) RF bandwidth, 6) video bandwidth, and 7) measured false alarm time. If this information is furnished, the probability of false alarm of the receiver can be calculated.

16.10 SENSITIVITY AND SINGLE-SIGNAL DYNAMIC RANGE

The sensitivity of a receiver might be a function of PW, so it should be tested at a fixed PW. The sensitivity also varies with frequency, so the sensitivity should be measured as a function of frequency.

The procedure to measure the sensitivity at one frequency value is as follows. At a given PW and a fixed frequency, start the pulse amplitude at a very low value such that the receiver cannot detect the signal. Increase the pulse amplitude in steps, until the receiver can receive the signal (i.e., the signal triggers the data

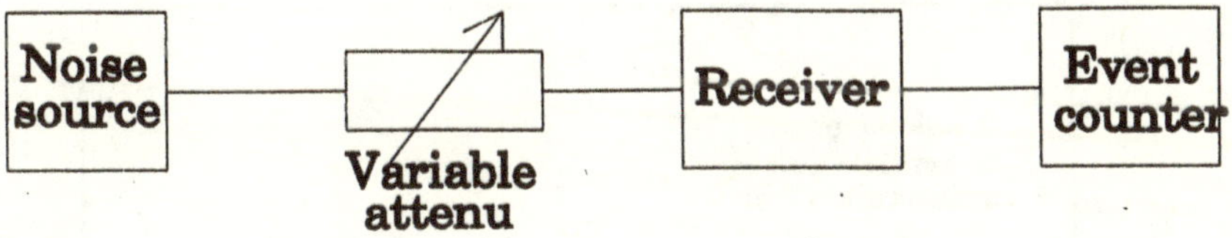

Figure 16.9 False alarm measurement setup.

reporting bit). In general, if a receiver is properly designed, at the sensitivity level the frequency reporting should be correct. Otherwise, the data reporting bit should not be triggered. However, in some receiver designs it is desirable to achieve the maximum sensitivity, even if the frequency reported does not meet the required accuracy.

In order to keep the sensitivity measurement consistent, the frequency accuracy should be evaluated to confirm the sensitivity. It is arbitrarily defined that at the sensitivity level, 90% of the frequency reading should be correct. Thus, when the input signal is detected by the receiver, the input signal should be repetitively sent 100 times at this power level. At the same time, the frequency should be recorded. If 90 frequency readings are within a predetermined frequency accuracy, this power is the sensitivity level. If fewer than 90 correct frequency readings are obtained, the power level should be increased by one step and the same procedure repeated until over 90% of the frequency readings are correct.

To find the single-signal dynamic range, keep increasing the input amplitude until additional signals are detected, which are referred to as spurs. In some receivers, when the input is high the receiver may miss the input signal. Sometimes a receiver can take a high input signal without causing any problem. Under this condition, it might be difficult to find the upper limit of the dynamic range because of the limited input power the signal generator can provide. The receiver can be claimed to have a minimum single-signal dynamic range over a certain value. If amplifiers are used to increase the input power level, precautions must be taken to keep the power amplifier from generating undesired spurs. Figure 16.10 shows a typical sensitivity of a receiver.

In this figure, when two signals are recorded the number 2 is printed. If the receiver misses the signal, a 0 is printed. If the frequency reading is outside the desired value, an x is printed. In this figure, the upper limit of the dynamic range is above 0 dBM.

16.11 PULSE AMPLITUDE AND PULSE WIDTH MEASUREMENTS

These measurements are straightforward. In the pulse amplitude measurement, the input frequency and the PW are kept constant. Increase the input power in steps and record the pulse amplitude output at each step. One can either plot the

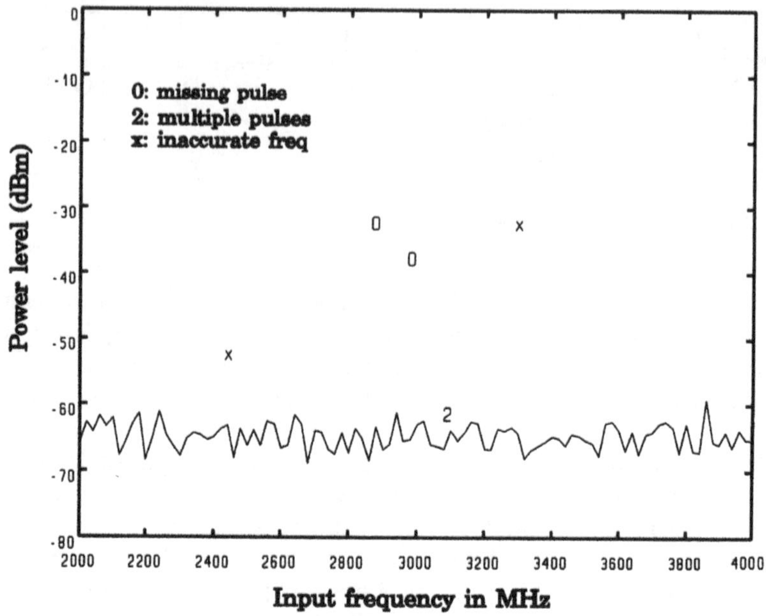

Figure 16.10 Sensitivity as a function of frequency.

error versus the input amplitude or the output versus the input amplitude. Although the former plot may provide more accurate results, the latter may provide a better pictorial display. Figure 16.11 shows the plot of input amplitude versus output amplitude.

In the PW measurement, the frequency and pulse amplitude are kept constant. Increase the PW through either uniform or nonuniform steps from a minimum to a maximum value. The minimum value equals to the shortest PW the receiver can process. The maximum value is the PW when the receiver declares a continuous wave (CW) signal. The PW output at each step is recorded. A plot of input PW versus output PW is shown in Figure 16.12.

In these tests, if the input signal is applied repetitively at the same value for a number of times, the standard deviation can be obtained.

16.12 AOA ACCURACY TEST

In this test, if the receiver uses amplitude comparison to obtain the AOA information, the input can be applied through several RF connectors. The amplitude to each input is controlled separately to simulate the input antenna pattern. If the input antenna pattern cannot be faithfully simulated, the AOA measured can only provide a qualitative result.

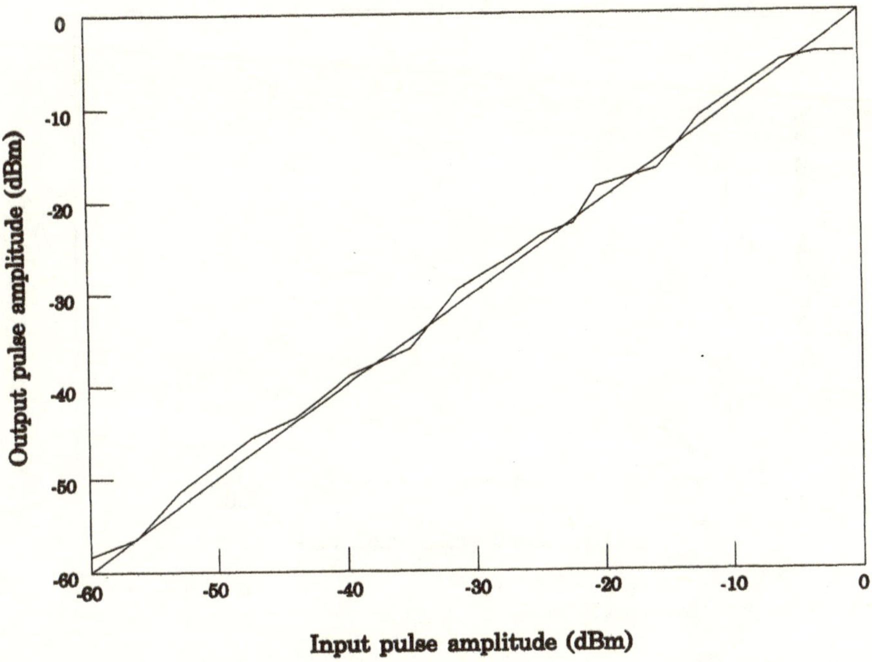

Figure 16.11 Output versus input pulse amplitude plot.

If the receiver is a phase interferometric system, the AOA test can be carried out in an anechoic chamber. One input source is required and its position is fixed. The input frequency pulse amplitude and PW are also all fixed. The receiving antenna is placed on a turntable. Rotating the turntable in steps can change the input AOA. Record the output AOA at each step. Figure 16.13 shows the result of an AOA test.

A standard deviation test can be carried out to evaluate the consistency of the receiver performance.

16.13 TOA TEST

In this test, the frequency, pulse amplitude, and PW are fixed. The input signal must be a pulse train with a very stable pulse repetition frequency (PRF), also referred to as pulse repetition interval (PRI). Each pulse received has a TOA associated with it. This TOA is referenced to an internal clock; thus, it is difficult to compare measured TOA against input TOA directly. Usually, the measured difference TOA (or PRI) is compared with the input PRI. The difference TOA is defined as

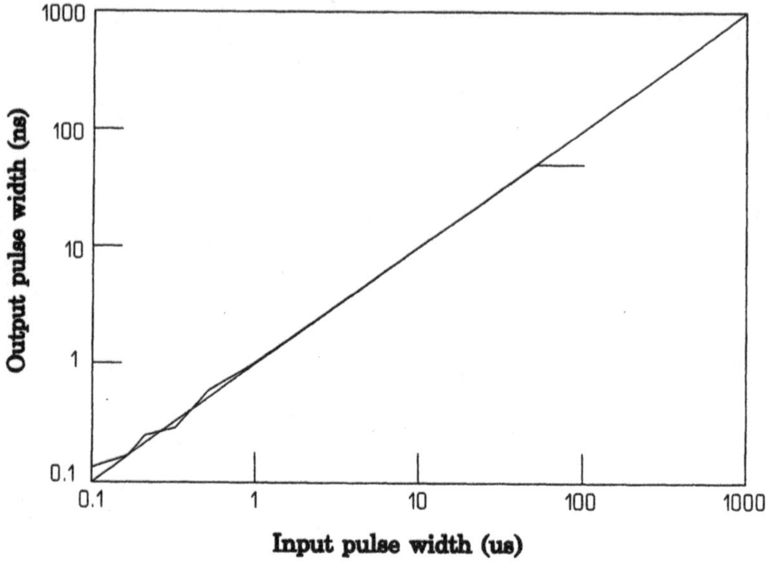

Figure 16.12 Output versus input PW plot.

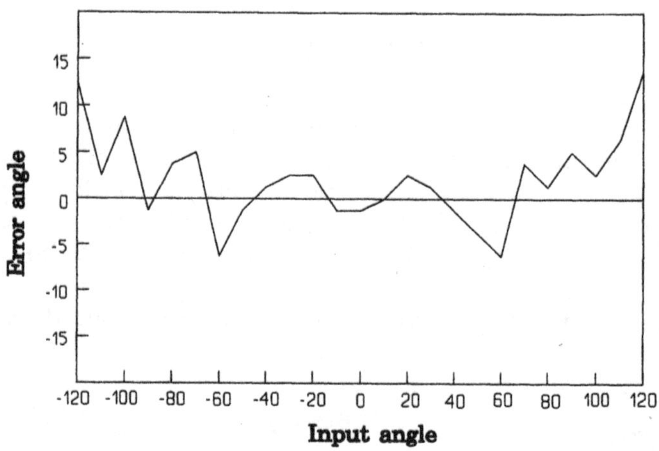

Figure 16.13 Output versus input AOA plot.

$$\Delta TOA_i = TOA_i - TOA_{i-1} = PRI \tag{16.4}$$

which is obtained from two consecutive TOAs. In this test, the result will be presented as the error PRI ΔTOA_i. The error PRI is defined as

$$\Delta PRI = PRI_m - PRI_i \tag{16.5}$$

where PRI_m and PRI_i are the measured and input PRI, respectively. Usually the TOA generated by a receiver does not have much of a problem.

16.14 SHADOW TIME, THROUGHPUT RATE, AND LATENCY TESTS

The shadow time is the time required for a receiver to accept a second pulse after the first pulse. In this section, the shadow time is defined from the end of the first pulse to the leading edge of the second pulse. Sometimes the shadow time is PW-dependent. The explanation to this effect is as follows. The frequency and AOA measurements often start at the leading edge of the pulse. It usually takes more time to measure frequency and AOA than to measure the pulse amplitude, PW, and TOA. But the PW measurement cannot start until the end of the pulse is reached. If the PW is short, the PW data may be ready before the frequency and AOA measurements; thus, it requires relatively more time to accept a second pulse. If the PW is long, the frequency and AOA measurements will be ready before the end of the pulse. Under this condition, as soon as the pulse ends, it takes a relatively short time to accept a second pulse.

The above discussion can be concluded as follows. For short pulse, the shadow time may be longer than that of a long pulse. Thus, the shadow time is PW-dependent. However, if the PW is increased beyond a certain value, the shadow time becomes a constant. Therefore, the shadow time should be measured at the minimum anticipated PW as well as at the longer PW. Thus the shadow time may have several values.

Another quantity related to the shadow time is the throughput rate. Throughput rate can be defined as the maximum pulse density a receiver can process. The measurement procedure is similar to the shadow time measurement. The PW should be at the minimum value, and the input frequency and pulse amplitude are kept at fixed values. If the receiver is designed to process four simultaneous signals, four synchronized signal generators are required. If the receiver can process two signals, two synchronized generators are needed.

Let us use the two-signal receiver as an example. Starting at a low PRF with two simultaneous signals of minimum PW, the receiver will be able to process all the input signals. Keep increasing the PRF until the receiver can process only half of the input signals, which indicates the receiver processes two simultaneous signals and misses the next two. The maximum number of pulses a receiver can process without missing pulses is the throughput rate. The corresponding shadow time can be found from the trailing edge of the first pulse to the leading edge of the second one as it is defined. For example, if the receiver under consideration can process a minimum PW of 200 ns and the highest PRF without missing pulses is 2 MHz, the throughput rate is 4 MPs/s by taking into consideration that the receiver can process two simultaneous signals.

Latency time is defined as the time delay from the input pulse arriving at the receiver to the time the digital output word is generated. This time can be measured from a scope. It has a time from tens of nanoseconds to a few microseconds.

16.15 TWO-SIGNAL FREQUENCY RESOLUTION TEST

For the two-signal tests, in order to keep the data in a controllable manner the two signals are limited to the same PW and the leading edges of the two pulses are time-coincident. If one desires to put a certain delay between the two pulses or use two pulses with different PWs, the test conditions should be carefully specified. The presentation of the output data must be carefully planned.

These tests are to determine the capability of the receiver to separate two signals close in frequency. Set the two signals with the same amplitude and close to the maximum power the receiver can process. The PW should be set long enough so that the spectrum spreading caused by the short PW will not affect the measurement. Keep one input signal near the center of the input bandwidth at a fixed frequency. Set the second signal at a frequency far from the first one and make sure that the receiver can measure both signals correctly. Move the frequency of the second signal toward the first one until the receiver misses one of the input signals or reads one of the input frequencies incorrectly. The minimum frequency separation at which the receiver can process both signals is the frequency resolution.

16.16 TWO-SIGNAL SPURIOUS FREE DYNAMIC RANGE TEST

In this test, the frequencies of the two input signals are separated at a constant value. Both frequencies will be changed by the same value; therefore, the difference frequency between the two signals is kept constant. The minimum frequency separation must be greater than the desired two-signal frequency resolution. Both signals are kept at the same amplitude.

To start the test, both signals are set slightly below the receiver sensitivity level. Increase both signal amplitudes in steps. If both signals are properly received, the power is marked as the lower limit of the dynamic range and the corresponding frequency is the average frequency of the signals. Increase the amplitude of both signals until more than two signals are reported. In order to make sure the extra signal is produced by the third-order intermodulation, one of the extra signals must have a frequency of

$$f = 2f_1 - f_2 \qquad \text{or} \qquad f = 2f_2 - f_1 \tag{16.6}$$

where f_1 and f_2 are the two input frequencies. If the extra signal does not match this condition, it might be caused by some other spurious response. The amplitude of the signals keeps increasing until the extra signal fulfills the above condition. This power level is marked as the upper limit of the dynamic range.

During the test, if the receiver reports more or less than two signals (including no output), the number of signals is recorded. If the frequency reading of one or both signals is reported erroneously, an x is recorded. This test can be repeated for different frequency separations. Figure 16.14 shows a typical result. The frequency scale represents the average frequency of the two input signals.

16.17 INSTANTANEOUS DYNAMIC RANGE TEST

This test is to find the capability of the receiver to simultaneously receive a strong and a weak signal. One of the input signals, the first one, is set at a fixed frequency (say, at the center of the input band), and close to the upper limit of the single-signal dynamic range. The second signal starts at the lower frequency limit of the receiver with an amplitude less than the sensitivity level. Increase the amplitude of the second signal until the receiver reports both signals correctly and this is the lower limit of instantaneous dynamic range at this frequency. Increase the frequency of the second signal and repeat the above procedure to obtain another lower limit. The second signal should cover the entire frequency range of the receiver. The minimum frequency between the two signals should be equal to or slightly greater than the two-signal frequency resolution. Figure 16.15 shows the result of a typical instantaneous dynamic range test.

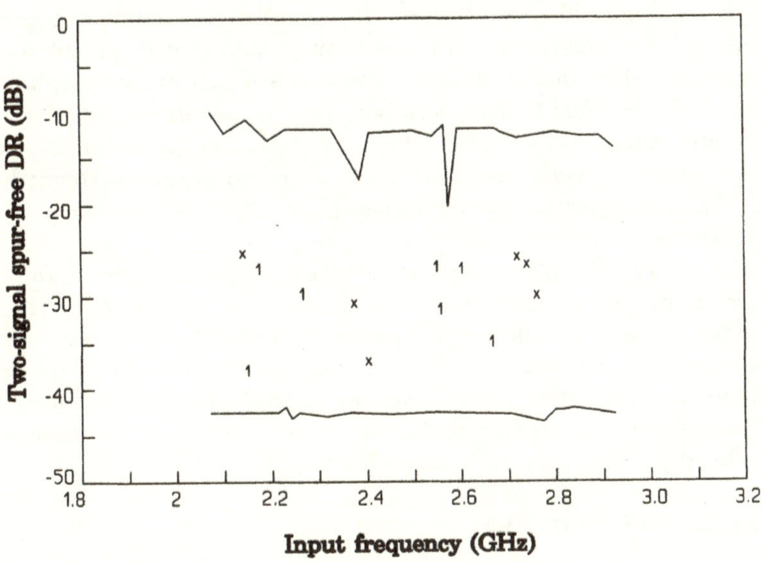

Figure 16.14 Two-signal spur-free dynamic range.

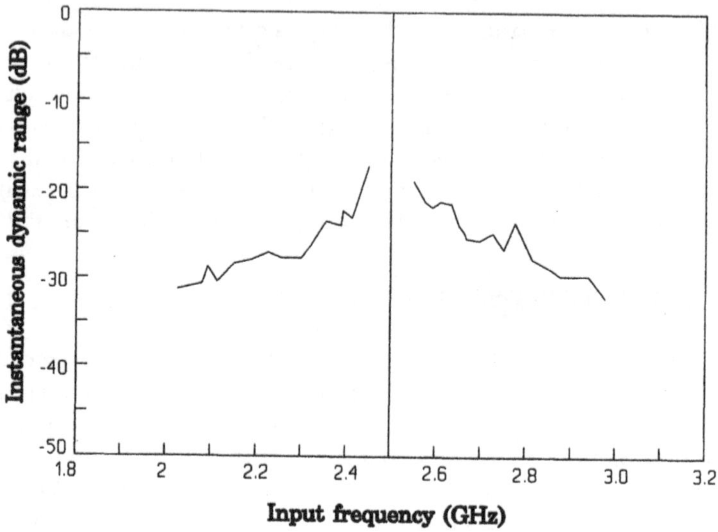

Figure 16.15 Instantaneous dynamic range.

16.18 ANECHOIC CHAMBER TEST

If one cannot feed the input of an EW receiver through an RF connector (i.e., the input of the receiver is an antenna array), all the tests mentioned in Sections 16.7 though 16.17 should be performed in the anechoic chamber to generate the performance of the receiver. However, it is suggested that if possible one should avoid testing a receiver in a chamber because it is much more complicated to set up in an anechoic chamber than in the laboratory. In addition, an anechoic chamber is not generally available. Even if the input of a receiver has antenna array as in a phase interferometric AOA system, one of the input antennas might be disconnected and replaced with an RF connector. If possible, this approach can be used to measure all the parameters except the AOA.

In an anechoic chamber, the distortion of the phase front of the incoming wave should be properly analyzed and measured. The amount of reflection in an anechoic chamber as a function of frequency should also be known. The receiver should be placed at the far field of the transmitting antenna if possible. The wavefront curvature affecting the receiver test should be taken into account. Thus, the AOA test carried out in an anechoic chamber should produce more reliable data than through antenna simulation.

16.19 AOA RESOLUTION TEST

The purpose of this test is to determine capability of the receiver to separate two signals with different AOAs. This test should be carried out in an anechoic chamber.

If two signals arrive at the receiver with the same frequency but different AOAs, some receivers can differentiate them (i.e., AOA is obtained from FFT in the space domain). Some receivers cannot process them (i.e., AOA obtained from a phase interferometric system). Since a small number of antennas can be installed in an airborne system, it is anticipated that most of the AOA measurement in EW receivers will be phase interferometric system. As discussed in Chapter 13, in some AOA measurements the signals are separated by their input frequencies first. If two signals have the same frequency, they cannot be separated and the phase of the interferometric cannot be used to measure the AOA.

The capability of separating two signals of the same frequency by AOA is the true AOA resolution. The true AOA resolution should be frequency-independent. Since most of the EW receivers do not have this capability, the AOA resolution considered here will be frequency-dependent.

In order to simplify the measurement procedure, the two input signals should be separated widely in frequency such that they can be separated by the receiver under test. The two input signals will have the same pulse amplitude and PW. The receiving antenna is kept stationary. One of the input sources is placed at the center of the receiving antenna beam. The second source is placed at one end of the beam and moved toward the other end of the beam as shown in Figure 16.16. At each step the AOA of both signals is recorded. If the receiver is properly designed, the receiver should correctly encode both AOAs across the entire angle range, even if the two signals have the same AOA. The results can be represented by two curves.

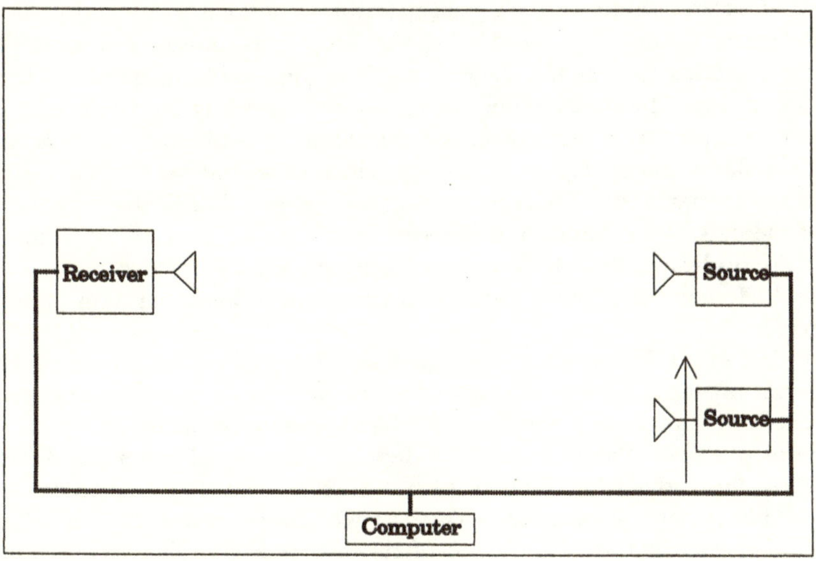

Figure 16.16 Two-signal AOA measurement setup.

Each curve represents the input AOA versus the output AOA of each source. The frequency difference between the two input signals should be decreased to repeat the same test.

16.20 SIMULATOR TEST

The purpose of a simulator test is to find out how a receiver performs under a dense signal environment. A simulator simulates an electronic environment that an EW receiver is anticipated to operate in. A simulator, in general, can produce a very dense environment containing many radars. It usually can simulate the antenna pattern of the radar transmitter, such as the scan rate of a radar. A sophisticated simulator has many parallel output ports. These ports can be connected to the parallel input ports of the receiver under test to simulate the input antenna amplitude pattern. Thus, if the receiver has an amplitude comparison AOA system, it can be tested through the simulator. It might be difficult to simulate the wavefront of many input signals to test a phase interferometric AOA system with the present technology, but it is possible to simulate the wavefront of one signal.

A simulator should be able to generate hundreds of radar beams, but it is impractical to have hundreds of signal generators to simulate all the beams. Even if one can afford to buy many signal generators, it is difficult to sum them all up into one output without creating significant power loss. Because of this difficulty, a simulator often contains a few signal generators and each one is time-shared to simulate many beams. This kind of design may miss pulses. For example, if a simulator contains four signal generators, it is impossible to generate five time-coincident pulses. The signal generator has a certain settling time before producing the correct frequency, which makes the time-sharing problem more severe.

A simulator usually can generate two kinds of scenarios: static and dynamic. In the static scenario, the EW receiver is at a fixed position in the battlefield. The types of radars stay the same, but they can be turned on and off and their beams can scan periodically. In a dynamic environment, the EW receiver is moving in a battlefield. Hence, the types of radars can appear and disappear along the path of the receiver.

Testing of an EW receiver through a simulator faces one major problem. If the receiver misses a pulse or reports an extra pulse, one does not know whether it is a receiver problem or a simulator problem. Two approaches are often used to resolve this problem. The first one is to dedicate one signal generator to a certain radar; thus, this radar is guaranteed to not miss a pulse. The second approach is to use a narrowband superheterodyne receiver to receive a certain signal and compare its result against the output of the EW receiver. The second approach can be used to check both missing pulses and spurious responses generated by the EW receiver.

The simulator test can only produce a qualitative result because it is difficult to report the result in a quantitative manner. However, if the receiver is operated with a real-time signal processor, quantitative results can be obtained from a system operation performance point of view. If the overall system including an EW receiver and an EW processor does not perform satisfactorily, the simulator test can be used to test the EW receiver to determine whether the problem is caused by the receiver or the processor. Under this condition, the output of the EW receiver must be recorded at high speed on a pulse-by-pulse basis.

16.21 FIELD TEST

It may appear that the previous tests can cover all the possible performance of a receiver. Past experience suggests this is not necessarily true. As mentioned before, it is impossible to test an EW receiver under all signal conditions. A field test is used to test the receiver under real radar signals. The receiver must be connected to an antenna or antenna array and placed in the field. The field test can be divided into two groups. One can be considered as a controlled test and the other one as an uncontrolled test.

In the controlled field test, one can order certain radars in the field to start and stop transmitting. Thus, this test is somewhat similar to an anechoic test. In order to assess the performance of the receiver, a superheterodyne receiver is needed to confirm the received signal. For example, if the receiver reports several signals at certain frequencies with a certain PRF, the PRF can be generated from the TOA measurements. The superheterodyne receiver can be tuned to these frequencies. Each signal should be checked separately to make sure whether the frequency and PRF reported by the EW receiver are correct.

The uncontrolled field test is to take the EW receiver to some area where anticipated radar signals might be received. One would have no prior knowledge when and what types of radars will be under operation. It is very difficult to perform this test. If a receiver reports certain data, it is difficult to confirm their fidelity, especially when the radar pulses are emitted for a short time duration. Under this condition, it is even difficult to use a superheterodyne receiver to check the result because there might not be enough time to tune the superheterodyne receiver to the desired frequency. Sometimes two different EW receivers can be used to check the results against each other. If two different results are reported, however, it is difficult to determine which receiver reported the correct one.

One useful bit of information obtained from past field test will be briefly mentioned here. A radar beam scanning past an EW receiver always produces the condition that the signal strength is crossing the sensitivity level. If the receiver threshold is not designed with a hysteresis loop, the receiver may report multiple pulses every time the radar signal crosses the threshold.

Let me conclude this book with another funny situation that occurred during a field test. After tuning the receiver through some frequency ranges to search for a signal, finally a signal was received. However, this signal was later confirmed as the frequency generated by the local oscillator of another receiver.

REFERENCES

[1] Hahn, P., and Jeruchim, M. "Developments in the Theory and Application of Importance Sampling," *IEEE Trans. Communication*, Vol. COM-35, July 1987, pp. 706–716.

[2] Jeruchim, M. C., Hahn, P. M., Smyntek, K. P., and Ray, R. T. "An Experimental Investigation of Conventional and Efficient Importance Sampling," *IEEE Trans. Communication*, Vol. COM-37, June 1989, pp. 578–587.

[3] Jeruchim, M. C., Balaban, P., and Shanmugan, K. S. *Simulation of Communication Systems*, New York NY: Plenum Publishing, 1992.

[4] Bahr, R. K., and Bucklew, J. A. "Quick Simulation of Detector Error Probabilities in the Presence of Memory and Nonlinearity," *IEEE Trans. Communication*, Vol. COM-41, Nov. 1993, pp. 1610–1617.

[5] Tsui, J. B. Y. "Microwave Receivers With Electronic Warfare Applications," New York, NY: John Wiley & Sons, 1986.

[6] Xia, W. Wright State University, Dayton, OH, Private communication.

APPENDIX 16.A

```
% file name: falseala.m
% Find probability of false alarm under normal operating conditions( signal
% is not present).

clear

fprintf(' n The Test of False Alarm Tfa under normal conditions:  n');
fprintf(' --------------------------------------------------- n');
M=input('  the noise source expressed in power ratio M = ? in dB   ');
A=input('  attenuator A = ? in dB  ');
dcac=input('  dc-coupled (1)  or  ac-coupled (0) ?   ');
NF=input('  noise figure of the receiver NF = ? in dB   ');
Br=input('  RF bandwidth Br in Hz = ?   ');
Bv=input(' video bandwidth Bv in Hz = ?   ');
Tfa_i=input('  with noise generator: Tfa = ? in sec.  ');

r=Br/Bv;
pfa_i=1/(Tfa_i*Br);
F=A+NF;
R=1+10^((M-F)/10);
N=R;

if dcac==0,
k1=1;
else
```

```
k1=N;
end

k2=(N^2)/sqrt(1+(r^2)/2);
k3=4*(N^3)/(2+3*(r^2)/4);
k4=(k1^2)/k2;

vt=10;
vt1=0;
vt2=0;
for ii=1:20,
fprintf(' n%g n',ii);
k5=(vt-k1)/sqrt(k2);
b1=k3*((k5^2)-1)*exp(-(k5^2)/2)/sqrt(2*pi)/6/sqrt(k2^3);
b=b1+quad8('xp',k5,(vt*3)/sqrt(k2))/sqrt(2*pi);
if b<=pfa_i,
vt2=vt;
vt=(vt+vt1)/2;
else
vt1=vt;
vt=(vt+vt2)/2;
end
fprintf('vt=%g n',vt);
end

kk1=1;
kk2=1/sqrt(1+(r^2)/2);
kk3=4/(2+3*(r^2)/4);
kk5=(vt-kk1)/sqrt(kk2);

bb1=kk3*((kk5^2)-1)*exp(-(kk5^2)/2)/sqrt(2*pi)/6/sqrt(kk2^3);
pfa=bb1+quad8('xp',kk5,(vt*3)/sqrt(kk2))/sqrt(2*pi);
Tfa=1/(pfa*Br);
fprintf(' n Under Normal Conditions in sec : Tfa=%g n n',Tfa);

end
```

APPENDIX 16.B

```
% filename xp.m
function y=xp(x)
y=exp(-(x.^2)/2);
end
```

Glossary

ADC	analog-to-digital converter
AIC	Akaike information criterion
AM	amplitude modulation
AOA	angle of arrival
AR	autoregression
ARMA	autoregressive moving average
BPSK	biphase shift keying
CAT	criterion autoregression transfer
CW	continuous wave
DC	direct current
DFT	discrete Fourier transform
DMA	direct memory addressing
ECCM	electronic counter-countermeasures
ECM	electronic countermeasure
ELINT	electronic intelligence
EM	electromagnetic
EOB	electronic order battle
ESM	electronic support measures
ESPRIT	estimation of signal parameters via rotational invariance
EW	electronic warfare
FFT	fast Fourier transform

FIR	finite impulse response
FM	frequency modulation
FPE	final prediction error
IDFT	inverse discrete Fourier transform
IF	intermediate frequency
IFM	instantaneous frequency measurement
IIR	infinite impulse response
LMS	least mean square
LSB	least significant bit
MA	moving average
MDL	minimum description length
MEM	maximum entropy method
MSB	most significant bit
MUSIC	multiple signal classification
PC	personal computer
PDW	pulse descriptor word
PRF	pulse repetition frequency
PRI	pulse repetition interval
PSK	phase shift keying
RF	radio frequency
SNDR	signal to noise plus spur distortion ratio
STFT	short time Fourier transform
TOA	time of arrival

About the Author

James Tsui is an electronics engineer with the Air Force Research Laboratory at Wright-Patterson Air Force Base, Ohio. He earned his Ph.D. in electrical engineering from the University of Illinois. He is a fellow of the Air Force Research Laboratory and the IEEE. He is the author of *Microwave Receivers and Related Components* (National Technical Information Services, and Peninsula Publishing Co., 1983), *Microwave Receivers with Electronic Warfare Applications* (John Wiley, 1986), *Digital Microwave Receivers: Theory and Concepts* (Artech House, 1989), and *Fundamentals of Global Positioning System Receivers: A Software Approach* (John Wiley, 2000).

Index

Acquisition time, 162
Adaptive spectrum estimation, 491–97
  illustrated, 497
  noise sensitivity, 496
  See also High-resolution spectrum estimation
Advanced Research Projects Agency (ARPA), 2
Akaike information criterion (AIC), 468
Aliasing effect, 83
  band overlapping caused by, 437
  overlapping, 84–87
All-pole model, 451
Amplifier/ADC interface, 232–34
  amplitude, 232
  dynamic range, 234
  illustrated, 221
  maximum allowable ADC power, 233
  noise figure, 231–32
  third-order intercept point, 233, 234
  third-order intermodulation, 232
Amplifier chain
  amplifier characteristics, 225–30
  in analog receiver, 222
  calculated performance, 236
  computer program and results, 235–37
  defined, 225
  design example, 237–38
  designing, 226
  experimental setup, 238–44
  gain, 225, 226
  noise figure, 225, 226, 235
  performance, 220
  as single amplifier, 219
  third-order intercept point, 226
Amplifiers
  amplifier chain, 225–30
  characteristics, 227
  in front of ADC, 219
  gains, 219
  input power, 225
  limiting, 405
  noise figures, 219
  output noise, 231
  output power, 225
  third-order intercept points, 219
  See also Attenuators
Amplitude comparison system, 510
  accuracy, 509
  defined, 509
Amplitudes
  of delay line, 511
  envelope, 337, 338
  FFT outputs, 413
  I and *Q* channels output, 256
  imbalance, 258, 259
  of *n*-th harmonic, 175
  pulse, 13, 28, 33, 553–55
  simultaneous signals, 335–36
  sine wave, 187
  spur, 176–79
  spur, analysis, 172–76
Analog filters, 428–29, 434–39
  followed by monobit receivers, 434–39
  followed by phase comparators, 428–29
  See also Filters
Analog *I* and *Q* downconverters, 259–62
  advantage, 259
  with balanced outputs, 262

Analog *I* and *Q* downconverters (*continued*)
drawbacks, 259
illustrated, 260
Analog receivers
comparison sensitivity, 222
component selection, 220
input frequency, 249
RF chain, 222
two blocks of, 545
types of, 220
See also Digital receivers
Analog-to-digital converters (ADCs), xvii, 25, 155–212, 230–31, 250
3-bit, 173, 174, 179, 180, 181
4-bit, 156
8-bit, FFT output of, 199
amplifiers in front of, 219
apparent maximum/minimum signals to, 163–66
coarse, 156
defined, 78
dynamic range, 155, 166
dynamic range vs. bits, 182
FFT chip, 408–9
flash converter, 155
folding system, 155
high-speed outputs, 3
histogram of, 190
ideal, 163
ideal, performance, 165
ideal, quantization noise, 166–67
impact of, 155
improvements in, 4
input bandwidth, 162–63
input vs. output, 165
maximum input power, 230
maximum sampling frequency, 209
monobit receiver, 403–8
noise effects, 179–82
nonideal, 219
nonlinear effect, 240
output noise, 231
outputs, with input frequency between bins, 181
outputs, many levels missing, 180
outputs, no levels missing, 179
outputs, time/frequency domains of, 176, 177, 178
output signal, 172
in parallel, 163
parameters, 155
performance, 2, 155, 162–63, 209
quantization error, 165
quantization noise, 230
requirements on, 209–12
sample-and-hold circuit, 160–62
sampling rate, 287
sigma-delta, 155, 157–60
SNR vs. frequency of, 187
speed, 3
spurious responses, 168, 169–72
technology, 2
Tektronix TKAD20C, 237, 239
test through FFT operation, 189–209
test through histogram, 186–88
test through sine curve fitting, 188–89
through folding technique, 155–57
transfer characteristic function, 172
transfer function, 164
two-stage, 156
types of, 155
Anechoic chamber tests, 541, 560
defined, 541
setup, 546–47
setup illustration, 549
See also Receiver tests
Angle of arrival (AOA), 13
accuracy, 17, 30
accuracy test, 554–55
ambiguity, 529
data collection, 510
data resolution, 28
estimation, 529
information, 17–18, 509
narrowband system, 17
parameter, 17
of received pulses, 31
resolution, 29, 523, 528
resolution test, 560–62
simultaneous signal measurement, 18
See also AOA measurements; AOA systems
Angle quantization cells, 333
Angle resolution
number of ADC bits vs., 332
quantization levels and, 332–33
Antenna arrays
circular, 514–17
linear, 512–14
phase, 517–19
problem, 535
with weighting function, 525
Antennas
diameter, 530
one receiver shared among, 537

spacing, minimum, 529–31
spiral, 530
AOA measurements, 509–37, 557
amplitude comparison, 509, 510
analog, 536
calibration table, 510
Chinese remainder theorem application to, 532–35
digital, 536–37
Doppler frequency, 510
frequency sorting followed by, 527–29
hardware considerations, 536–37
importance of, 13, 509
methods, 17, 25, 509
phase comparison, 509, 510
two-signal setup, 561
through zero crossing, 519–20
See also Angle of arrival (AOA)
AOA systems
phase interferometric, 560, 561
phase measurement in, 520–21
queuing, 511
two-antenna element, 517
wideband phase measurement, 527
See also Angle of arrival (AOA)
Application specific integrated circuit (ASIC) technology, 418
Attenuators
amplifier connection results, 229
cascading amplifiers and, 228
characteristics, 227
placement, 229
Autocorrelation, 130–36
autovariance vs., 132
biased, 133
defined, 130, 453
example, 131
FFT application to spectrum estimation, 134–36
input data manipulation method, 461–62
lag of, 130, 477
lags, 465
matrix, 458
physical interpretation of, 132
"sample," 130–31, 453
spectrum estimation, 132–34
of time series, 465
Autoregressive (AR) model, 450–51
autocorrelation method, 461–62
backward predication, 462–65
Burg method, 465–67
covariance method, 459–61
defined, 451
demonstration, 456–57
equivalent circuit, 452
input data manipulation, 457–62
Levinson-Durbin algorithm and, 455–57
modified covariance method, 462–65
moving average (ARMA), 450
order selection, 467–70
plane result, 452
power spectrum from, 456
process results, 458–60
Yule-Walker equation and, 452–55
See also High-resolution spectrum estimation
Autovariance
autocorrelation vs., 132
defined, 131
"sample," 132
Averaged periodogram, 105–7
Average linear prediction error, 464, 465

Backward prediction, 462–65
coefficient, 462
defined, 462
error, 463
See also Linear prediction
Band overlapping, 437
Bandwidth, 1
ADC input, 162–63
digital filters, 431–33, 440
equivalent video, 291
filter, 430
folding, 434
input, 380, 399
instantaneous, 35, 249
maximum unambiguous input, 235
minimum filter, 285
monobit receiver, 438
optimum, 12
output, 372
processing, 168, 235
processing circuit, 428
resolution, 235
RF, 209, 222
sampling, 143
system noise, 15
Baseband receivers, frequency selection, 250
Bias frequency, 550
Biphase shift keying (BPSK), 11
Blackman filter, 178
Blackman-Tukey method, 134
Boltzmann's constant, 15

Bragg cell receivers, 22–23
defined, 22
disadvantages, 23
encoder design, 23
input RF signal, 22
interferometric approach, 23
optical portion, 23
BT coefficients, 360
Burg method, 465–67
defined, 465–66
high spectrum peaks, 466
program, 501
result, 467
spectrum generated by, 467–71
window function, 466

Calibration table, 510
Causality, 268
Channelization, 363–95
approach illustration, 389
digital, 363, 422
goal, 380
in hardware for wideband digital receiver, 440
with polyphase filter, 388–90
processing methods after, 421–47
Channelized approach, 422–23, 434
Channelized receivers, 21–22
amplifiers, 21
amplitude comparison scheme, 21
center frequency, 21
design, 22
IFM, 30, 428
monobit, 435, 438
Characteristic function, 293, 294–95
Chinese remainder theorem, 531–36
for antenna array problem, 535
application to AOA measurements, 532–35
defined, 531
examples, 531–32
practical considerations, 535–36
problems, 535–36
Chirp rate, 14
Circular antenna array
azimuth/elevation information, 517
illustrated, 515
outputs from, 514–17
results, 517
See also Antenna arrays
Coherent digitizing error, 170, 171, 172
Coherent Doppler radar measurement, 349–52
defined, 350
methods, 349
phase error, 351
pulse train, 351
requirements, 350
Comb function, 51–54, 65
defined, 51
Fourier series, 54
Fourier transform, 51, 54, 74–75
Communication receivers
EW receivers vs., 1, 9–10
intercept receivers vs., 9–10
Complementary metal oxide semiconductor (CMOS) logic, 407, 418
Compressive (microscan) receivers, 23–25
defined, 23
front end, 24
illustrated, 24
parametric encoder, 24
used for AOA measurement, 29–30
Continuous wave (CW) signals, 11
Controlled field test, 563, 564
Convolution, 55–58
duality of, 58, 63
examples, 59–71
FFT and, 365–66
graphic display of, 56
of impulse function, 55
linear, 114, 115, 116
mathematical representation of, 132
periodic, 114, 117
physical interpretation of, 132
summary, 73
in time domain, 113
Cooley-Tukey FFT algorithm diagram, 100
Correction method, 279
Cosine function, 74
Cosine wave, 174
Cosine window, 91–92
defined, 91
frequency domain, 96
time domain, 95
Covariance method, 459–61
modified, 462–65
R matrix, 461
T matrix, 459–60
See also Input data manipulation
Cramer-Rao bound
defined, 15
EW applications and, 17
illustrated, 16
Criterion autoregression transfer (CAT), 468
Critical sampling rate, 440

Data points grouping, 286
Data resolution
  AOA, 28
  frequency, 27, 334, 335, 550
  pulse amplitude, 28
  pulse width, 28
  TOA, 28, 366
Data sample
  false alarm time for, 287–88
  probability of false alarm, 287–88
  threshold setting for, 288–89
Decimated FFT, 384
Decimation, 146–52, 374–77
  applications to EW receivers, 148–50
  defined, 146, 374
  DFT through, 146–48
  effects on DFT, 377–78
  example of, 376
  in frequency domain, 380–84
  by *M*, 374
  process, 377
  simplified method, 150–52
Deinterleaving, 31–33
Delta functions, 141
Demultiplexer, 403–8
Derivative, 49
  step function, 71
  summary, 73
Detection
  circuit, 284
  frequency domain, 285–86, 308–11
  multiple-sample, 291–92, 299–302
  N sample, 306–8
  probability of, 289–90, 292, 296, 299, 301, 303
  single-sample, 289–90, 297–99
  time domain, 286–87
Digital EW processors, 9
Digital filters
  bandwidth, 431–33, 440
  defined, 431
  followed by FFT, 446–47
  followed by monobit receivers, 440, 443–46
  followed by monobit receivers and phase comparators, 446
  followed by phase comparators, 431–34
  generation, 443
  monobit receiver vs., 431
Digital mixing/filtering, 274
  ambiguity problem, 331–32
  I and *Q* channels, 331
  with modified filters, 275
  phase discontinuity, 332
Digital phase measurement, 330–32
  IFM receiver approach vs., 330
  illustrated, 331
Digital receivers, 25–26
  comparison sensitivity, 222
  dynamic range, 352
  EW, 25–26, 30–31
  function illustration, 26
  narrowband, 26
  three blocks of, 542
  time frequency and space-AOA transforms, 545
  See also Analog receivers
Digital signal processing
  performance of, 3
  as problem solver, 1–2
Digital-to-analog converter (DAC), 156
Digitized points, 195–98
Digitizing speed, 126
Direct memory addressing (DMA) mode, 546
Discontinuity, reducing, 88
Discrete Fourier transform (DFT), 77–107
  advantages over FFT, 101–4
  analytic approach to, 81–82
  computation intensive, 93
  computing, 93–94
  decimation effect on, 377–78
  defined, 77
  graphical description, 78–81
  graphic illustration, 80
  initial data accumulation, 102
  input signal bandwidth, 86
  interpolation effect on, 377–78
  inverse, 82
  Kernel function of, 399
  length limitation, 83
  limited frequency bandwidth, 83–84
  monobit receiver, 398
  number of operations for computing, 101
  output spectra, 85
  overlapping aliasing effect, 84–87
  properties, 83–87
  sliding, 102–4
  through decimation, 146–48
  unmatched time interval, 84
  use of, 84
  See also Fourier transform
Discrete Hilbert transform, 265–71
  data points, 265
  examples, 268–71
  See also Hilbert transform

Doppler frequency measurement, 510
Doppler pulse train
  frequency resolution, 352
  illustrated, 351
Duality, 45–47
  of convolution, 58, 63
  defined, 45
  examples, 46–47
  summary, 71
  See also Fourier transform
Dynamic range
  ADC, 166
  ADC impact on, 155
  amplifier/ADC interface, 234
  defined, 27
  desire for, 220
  digital receivers, 352
  experimental amplifier chain setup, 240–44
  instantaneous, 28, 200, 415, 438
  instantaneous, test, 559–60
  lower limit of, 219
  single signal, 27, 552–54
  single-signal spur-free, 201–2
  spurious free, 28, 225, 558, 559
  test, 240–44
  third-order intermodulation, 200
  two-signal spur-free, 225
  two-signal spurious free test, 558
  two-tone spur-free, 200, 208
  types of, 13, 27

Eigendecomposition
  defined, 477
  ESPRIT method, 485, 486
  minimum norm method, 487
Eigenvalues, 477–79
  application of, 477–78
  minimum, 478, 479
  MUSIC method, 479
  of *R*, 478
Eigenvectors, 477–79
  application of, 477–78
  MUSIC method, 479
  of *R*, 478
Electronic counter-countermeasures (ECCM), 8
Electronic countermeasures (ECM), 8
Electronic order battle (EOB), 12
Electronic support measure (ESM), 7
Electronic warfare (EW) receivers. *See* EW receivers
Emitter coupled logic (ECL), 407
Envelope
  amplitude, 337, 338
  detector, 294
  time-dependent, 336
  time vs., 337
Error frequency, 348
  maximum variation of, 349
  variation of, 348
ESPRIT method, 482–86
  defined, 482
  eigendecomposition, 485, 486
  input frequencies, 486
  matrices definition, 485–86
  program, 502–4
  steps, 482–86
  See also High-resolution spectrum estimation
Estimation of signal parameters via rotational invariance techniques. *See* ESPRIT method
Even function, 44–45, 71
EW processors, 31–35
  deinterleaving, 31–33
  functions, 31
  PRI generation, 34
  radar identification, 34
  revisiting, 34–35
  tracking, 34
EW receivers, xvii
  advantages in digital EW receiver, 285–87
  analog, overview of, 19–20
  AOA resolution, 28, 29
  characteristics and performance of, 26–29
  communications receiver vs., 1, 9–10
  conventional, 10
  decimation method applications to, 148–50
  design goals, 35–36
  detection approach, 284–85
  detection circuit, 284
  development obstacles, 4
  development trends, 29–31
  digital, 25–26, 30–31
  dynamic range, 13, 27
  false alarm rate, 27
  frequency data resolution, 27
  frequency domain, 285–86
  frequency measurement accuracy, 27
  frequency measurement precision, 27
  frequency resolution, 28
  incoming information, 10
  input RF signal, 284
  instantaneous dynamic range, 28

intermediate frequency (IF), 284
introduction to, 7
latency time, 28
noise level, 180
optimum bandwidth, 12
outputs, 18–19
parameters measured by, 13
performance goals, 35
performance standards, 539–40
pulse amplitude data resolution, 28
pulse width data resolution, 28
queuing receiver, 29
requirements of, 12–13
search stimulation, 539
sensitivity, 13, 27
shadow time, 28
signal environment, 10–12
single signal, 27–28
spurious free dynamic range, 28
theoretical problem solutions, 29
throughput rate, 28
TOA data resolution, 28
two simultaneous signals, 28–29
EW systems, 7–9
ECM, 8
functions, 9
input signal response, 11
intercept, 8
uses, 7
Experimental amplifier chain setup, 238–44
defined, 238–39
dynamic range, 240–44
gain, 241
illustrated, 239
noise figure management, 239–40
output spectrum, 242–44
third-order intercept point, 241
third-order intermodulation product, 241, 242
See also Amplifier chain

False alarm probability, 287–88, 291, 300
calculation, 311, 317
determining, 296
in frequency domain, 311–12
one data sample, 287–88
representation, 291
False alarms
based on multiple data points, 290
as binomial distribution problem, 292
measurement setup, 553
number of, 301
rate, 27
test, 552
threshold, 412
time, 287–88
Fast Fourier transform (FFT), 77, 93–101, 363
32-point, 112, 383, 387, 389, 440, 443
64-point, 333
128-point, 111
256-point, 379, 380, 400, 408
512-point, 178
16384-point, 209
ADC test through, 189–209
application to spectrum estimation, 134–36
butterflies, 99
butterflies illustration, 99
convolution operations and, 365–66
Cooley-Tukey diagram, 100
decimated, 384
derivation, 94
DFT advantages over, 101–4
DFT calculation with, 94
effect with interpolated data, 377
frame time, 416
frequency components, of Hanning window, 123
frequency components, of rectangular window, 122
frequency components, of sine wave, 128
frequency determination by, 126–28
highest output of, 313
imaginary part of, 129, 130
of interpolated outputs, 378–79
inverse, 100
length, 106, 177, 191, 192
N-points, 308, 383
number of operations for computing, 101
numerical frequency bins, 413
on filled window, 368, 370
on one chip, 397
on partially filled window (frequency domain), 371
on partially filled window (time domain), 369
operation example, 95
overlapping consecutive outputs, 285
overlapping input data and, 366–72
power spectrum, 104
processing, 334
real part of, 129, 130
of rectangular window, 373
in time domain, 528
See also FFT chips; FFT outputs

FFT chips, 3, 25
  basic design, 408–9
  high-speed, 30
  input data to, 408
  input signal processing, 415
  layout illustration, 409
  outputs, 408
FFT operators
  design, 128
  parallel, 146
  real input computed by, 128–30
FFT outputs, 104, 366, 397
  of 8-bit ADC, 199
  ambiguity, 139
  amplitudes, 413
  complexity, 410, 441
  crossing threshold, 410, 413
  data rate from, 372–74
  frequency bins, 177, 191
  of input signal, 457
  monobit receivers at, 440
  overlapping consecutive, 285
  peaks, 449
  peaks of correct frequencies, 410
  See also Fast Fourier transform (FFT)
Field programmable gate array (FPGA), 418
Field test, 563–64
  controlled, 563, 564
  defined, 563
  uncontrolled, 563, 564
  See also Receiver tests
Filter banks, 364–65
  analog, followed by monobit receivers, 435
  building, 364
  designed, 395
  design methodology, 378–80
  digital, 363, 364
  followed by phase comparators, 429
  illustrated, 364
  shape, 433
  two signal conditions, 424
Filters
  analog, 428–29, 434–39
  bandwidth, 430
  Blackman, 178
  design, 390–95
  FIR, 266, 364, 390
  IIR, 364, 451
  impulse function of, 386
  with limiting skirt, 425
  lowpass, 158, 160, 273
  modified structure, 444
  with narrower skirt, 427
  output, 384–87
  oversampling, 442, 445
  polyphase, 378, 388–90
  steady state, 441
  transient effect, 422
  in wideband channelized receiver, 423
  with wider skirt, 426
Filter shapes, 423–28
  illustrated, 425–27
  selection, 423–28
Final prediction error (FPE), 468
Finite impulse response (FIR) filter, 266, 364, 390
Flash converter, 155
Folding technique ADC, 155–57
  illustrated, 157
  input vs. output, 156
  two-stage approach vs., 156
  See also Analog-to-digital converters (ADCs)
Forward prediction
  coefficients, 463
  defined, 462
  error, 463
  See also Linear prediction
Fourier series, 50–51, 173
  comb function, 54
  defined, 51
  exponential, 353
  forms, 50–51
  introduction, 39
Fourier transform, 39–41
  comb function, 51, 54, 74–75
  complex plane, 60
  concept, 39
  continuous, 83
  cosine function, 74
  cosine wave, 45, 59, 60
  derivative, 49
  discrete (DFT), 77–107
  duality, 45–47
  even and odd functions, 44–45
  examples, 40, 59–71
  frequency domain, 45, 60
  frequency shift, 49
  Gaussian function, 66, 75
  generalized cosine window function, 64–65, 75
  impulse function, 43, 74
  integral, 49–50
  inverse, 40, 42, 57, 58, 527
  isosceles triangle, 61, 75

linearity, 44, 59
properties, 43–50
quadrature phase shift, 76
rectangular function, 41, 74
related operations, 111–52
RF pulse train, 64, 69, 76
scaling, 47–48
sine function, 45, 46, 59, 74, 264
over space domain, 521–24
summary, 71–76
time domain, 45, 524
time shift, 48–49
two-dimensional, 524–27
windowed cosine function, 61, 75
Frequency
accuracy, 550, 553
accuracy test, 549–51
bias, 550
channelization, 363–95, 421–47
determination by FFT, 126–28
error, 340, 549, 550
filter cutoff, 273
information, 13–17
information, obtaining, 310
instantaneous, 254, 336, 337
local oscillator, 253, 272
measured, 549
periodic variation, 339
phase vs., 141, 142
precision test, 551
sampling, 142, 169, 266, 272
sensitivity as function of, 553
shift, 49, 73
sorting, 527–29
spurious, 252
test signal, 278
Frequency conversion, 250–53
achieving, 250–51
I and *Q* channel, 253–55
purpose, 250
upconversion, 251
Frequency domain
ADC outputs, 176, 177, 178
ADC outputs with input frequency between bins, 181
ADC outputs with many levels missing, 180
ADC outputs with no levels missing, 179
cosine window, 96
decimation in, 380–84
Gaussian window, 94
general cosine window, 65
Hamming window, 98, 270
multiplication in, 113
noise in, 316
noise variance, 317
probability density in, 312
probability of detection in, 312, 314–17
probability of false alarm in, 311–12
rectangular function, 121
rectangular window, 90
response of Parks-McClellan window, 387
S/N in, 315, 316
time alone in, 309
window performance and, 88
See also Time domain
Frequency domain detection, 285–86
approach to, 309–11
comments on, 318–19
examples on, 317–18
input signal conditions in, 312–14
introduction to, 408–9
sensitivity and, 318–19
See also Detection
Frequency downconverters, 249–79
defined, 251
I and *Q*, 255
introduction, 249–50
See also Frequency conversion
Frequency encoder, 410–11
design, 410–11
design functions, 410
functional block, 411
functions, 410
Frequency measurement
accuracy, 27, 350
coherent Doppler radar, 349–52
illustrated, 550
rms values of, 551
of two signals, 338–40
Frequency modulated (FM) signals, 329
Frequency resolution, 14, 15
cell, 377, 551
data, 27, 334, 335, 550
defined, 28
generated from short delay, 335
uniform, 365

Gain, 219
amplifier chain, 225, 226
correction coefficients for, 277
experimental amplifier chain setup, 241
highest, 230
noise figure vs., 240
required RF chain, 408

Gaussian function, 63, 66, 75
Gaussian window, 90–91
  defined, 90–91
  frequency domain, 94
  response of, 93–94
  sidelobes and, 91
  time domain, 93
Generalized cosine window function, 64–65
  Fourier transform, 75
  frequency domain, 65
  time domain, 64
Global Positioning System (GPS) receiver, 397

Half-cycle sine wave, 187–88
Hamming window, 63, 269
  FFT output of, 372
  FIR filter with, 268
  frequency domain, 98
  in frequency domain, 270
  generalized, 92–93
  response of, 97–98
  sidelobes, 93
  time domain, 97, 375
Hanning window, 93
  frequency components from FFT of, 123
  peak position estimation for, 123–24
  in time domain, 123
Hard limiter effect, 401–2
High-resolution spectrum estimation, 449–97
  adaptive, 491–97
  autoregressive (AR) method, 450–51
  backward prediction, 462–65
  Burg method, 465–67
  covariance method, 459–61, 462–65
  ESPRIT method, 482–86
  input data, 450
  input data manipulation, 457–62
  introduction to, 449–50
  least squares Prony's method, 475–77
  Levinson-Durbin recursive algorithm, 455–57
  methods, 449–50
  minimum norm method, 487–89
  minimum norm method with DFT, 489–91
  MUSIC method, 479–82
  order of operation determination, 450
  order selection, 467–70
  outputs, 450
  prediction model, 450
  Prony's method, 470–75
  quality improvement, 458
  visual displays, 450
  Yule-Walker equation, 452–55
Hilbert transform, 262–65
  defined, 262
  direct form of filter for, 269
  discrete, 265–71
  filter input/output, 270
  imaginary part, 267
  result, 263
  in time domain, 262–63
Histogram
  of ADC, 190
  of sine wave, 189

I and *Q* channels, 129, 145
  amplitude error, 276
  analog downconverters, 259–62
  balanced output, 256
  conversion, 253–55
  defined, 249
  digital approach to, 262
  digitizing, 331
  graphical representation of outputs, 258
  imbalance in, 250, 255–59
  for instantaneous frequency determination, 254
  narrowband, 271
  outputs, 256, 330, 428
  outputs, complex form, 277–78
  outputs, from filters, 276, 276–79
  phase error, 276
  relative phase between, 261
  sub-Nyquist sampling with, 146
  wideband, 271–74
  wideband, filter design hardware considerations for, 274–76
IF receivers, 249
Importance sampling, 552
Impulse function, 42–43
  amplitudes, 49
  convolution of, 55
  defined, 42
  of filter, 386
  Fourier transform, 43, 74
  properties, 73
  representation, 42
  time-shifted, 43
Impulse response, 55
Infinite impulse response (IIR) filter, 364, 451
Input data manipulation, 457–62
  autocorrelation method, 461–62
  covariance method, 459–61

Input data rate, 392, 394
Input sampling rate, 388, 389
Instantaneous dynamic range, 415
  free, 28
  high, 438
  illustrated, 560
  test, 559–60
  See also Dynamic range
Instantaneous frequency
  defined, 336
  I-Q channels for determining, 254
  maximum value, 337
  measurement, 336, 337
  minimum value, 338
  negative, 338
  time vs., 339
Instantaneous frequency measurement (IFM) receivers, 16, 20–21, 250, 329, 422
  channelized, 30, 428
  defined, 20
  digital phase measurement vs., 330
  illustrated, 20
  limiting amplifiers, 405
  narrowband, 428
  performance, 542
  phase relation, 140
  phase vs. frequency, 140
  simultaneous signal processing and, 19, 20
  sub-Nyquist sampling vs., 137
Integral, 49–50, 73
Interface, 219–20
Interleaving, 32–33
Interpolation, 374–77
  defined, 374
  effects on DFT, 377–78
  example of, 376
  process, 377
  representation, 376
  zero padding, 378
Inverse DFT (IDFT), 82
Inverse FFT, 100
Inverse Fourier transform, 40, 42, 57, 527
  finding, 58
  for quadrature phase shift, 70
Isosceles triangle, Fourier transform, 61, 75

Jitter effect
  distribution, 184
  illustrated, 185, 186
  measurement, 183
  sampling window, 182–86
Joint density function, 290

Kernel function, 398, 443
  changing, 417
  of DFT, 399
  digitized into 1 bit, 399, 400
  increasing number of bits in, 417
  values, 398
  values, digitizing, 419
  values, moving, 418
  values, spacing, 399
Kronecker delta, 478

Laboratory tests
  importance, 540
  preliminary considerations, 542–44
  results presentation, 544
  setup, 545–46
  setup illustration, 546
  See also Receiver tests
Latency
  test, 557
  time, 28, 558
Leakage effect
  explanation, 88
  sine wave, 85
Least mean square (LMS) approach, 454–55
Least squares Prony's method, 475–77
  performance, 477
  results, 476
  See also High-resolution spectrum estimation
Levinson-Durbin recursive algorithm, 455–57
  defined, 456
  results, 456
  using, 461
L'Hospital's rule, 53
Limiting amplifiers, 405
Linear antenna array, 521
  data output from, 526
  digital data from, 512–14
  four-element, 521
  illustrated, 513
  placement, 512
  qth element, 513–14
  results, 517
  with uniform spacing, 522
Linear convolution, 114, 115, 116
  of each section, 118
  illustrated, 115, 116
  output, 117
  overall, 118
  with overlap-add, 118–19
  See also Convolution

Linearity
  defined, 44
  summary, 71
  use example, 59–60
  See also Fourier transform
Linear prediction
  average error, 464, 465
  backward, 462–65
  Burg method, 465–67
  forward, 462, 463
  model, 451
Local oscillator frequency, 253, 272, 337
Logarithmic form equations, 220
Look-through time, 12
L-out-of-*N* approach, 291, 292, 296
  sampled data crossing threshold, 308
  S/N, 318
  summation method vs., 307
Lowpass filters, 158, 160, 273
Low probability of intercept (LPI) radars, 11

MATLAB program, 384, 393, 477
Maximum entropy method (MEM). *See* Burg method
Microscan receivers, 311
Minimum antenna spacing, 529–31
Minimum description length (MDL), 468
Minimum norm method, 487–89
  defined, 487
  eigendecomposition, 487
  frequency response, 489
  procedure, 487–88
  program, 504–5
  result, 489
  See also High-resolution spectrum estimation
Minimum norm method with DFT, 489–91
  defined, 489–90
  frequency response, 492
  program, 505–6
  signal matrix, 490–91
  signal subspace matrix, 491
  steps, 490–91
  See also High-resolution spectrum estimation
Mixer
  circuit, 251
  spur chart, 252
M/M' meaning, 234
Modified Bessel function, 290, 296, 307
Modified covariance method, 462–65
  data use, 465
  defined, 464
  noise sensitivity, 465
  See also Covariance method
Monobit receivers, 397–419, 422
  ADC, 403–8
  analog filters followed by, 434–39
  bandwidth, 438
  channelized, 435, 438
  chip layout, 418–19
  components, 403
  deficiencies, 417
  deficiency, 403
  demultiplexer, 403–8
  design criteria, 399–403
  digital filter bank followed by phase comparators and, 446
  digital filters followed by, 440, 443–46
  digital filters vs., 431
  discrete Fourier transform, 398
  experimental, 407
  followed by phase comparators, 429–32
  frequency encoder, 410–11
  idea, 398–99
  illustrated portions, 404
  input frequency range, 415
  introduction to, 397–98
  narrowband, 434, 440
  original concept, 398
  output rate, 431
  outputs, 414, 430
  oversampling filters followed by, 445
  possible improvements, 417–18
  preliminary performance of, 413–17
  RF chain, 403–8
  summing outputs from, 439
  testing, 413–14
  threshold selection, 412–13
  timing analysis, 419
Most significant bits (MSBs), 156
Moving average (MA) model, 451
Multiple-sample detection, 291–92
  based on summation, 296
  detection based on, 291
  detection scheme for, 292
  example, 299–302
  See also Detection
MUltiple SIgnal Classification. *See* MUSIC method
MUSIC method, 479–82
  calculation, 481
  defined, 479
  eigenvalues, 479

eigenvectors, 479
idea behind, 480
program, 501–2
result, 481–85
steps, 480
See also High-resolution spectrum estimation

Narrowband receivers, 422–23
digital, 26
IFM, 428
monobit, 434, 440
Noise
in ADC, 179–82
floor, 241
in frequency domain, 316
probability density of, 288
sensitivity and, 181
spur reduction and, 180, 181
total, 185
Noise dithering, 168
defined, 168
signal detected with, 169
Noise figures, 219, 223–25
of amplifier/ADC combination, 231–32
amplifier chain, 225, 226, 235
defined, 223
gain vs., 240
lowest, 230
management, 239–40
overall, 231–32
Noise levels
EW receiver, 180
notation, 221
Notations, 220–22
list of, 221–22
noise levels, 221
quantities, 221
N sample detection, 306–8
Nyquist sampling
criterion, 25, 399, 434
frequency, 399
rate, 86, 139, 271
Nyquist theorem, 530

Odd function, 44–45, 71
One and two-signal tests, 544
Order selection, 467–70
empirical approach, 469
methods, 468
results, 468
Organization, this book, 4–5

Output filter
shape from decimated FFT, 384
widening with weighting function, 384–87
Output frequency bins, 440
Output noise, 231, 232
Output sampling rate, 372, 443
changing, 387–88
equals input sampling rate, 388
increasing, 431
increasing, by two, 440–43
output frequency bin number and, 440
See also Sampling rates
Overlap-add method, 118–19
Overlapping input data, 366–73
Oversampling filters
by two arrangement, 442
followed by monobit receivers, 445

Parametric encoder, 24
Parks-McClellan window, 384
frequency domain, 387
response of, 386–87
time domain, 386
Parseval's theorem, 58–59
Peak position estimation, 119–26
for Hanning window, 123–24
for rectangular window, 119–22
through iteration, 124–26
Percentage error, 351
Periodic convolution, 114, 117
avoiding, 117
illustrated, 117
See also Convolution
Periodic equation, 338
Periodic function, 50
Periodogram, 104–7
averaged, 105–7
of noise only input, 106, 107
square roots of, 105
Phase
antenna array, two-element, 517–19
comparison system, 428–29, 509, 510
correction coefficients for, 277
of delay line, 511–12
difference, 141, 142, 334
difference angles, 335
discontinuity, 332
frequency vs., 141, 142
between *I* and *Q* channels, 261
instantaneous, 336
Phase comparators
analog filters followed by, 428–29

Phase comparators (*continued*)
 digital filter bank followed by monobit receivers and, 446
 digital filters followed by, 431–34
 filter bank illustration, 429
 monobit receiver followed by, 429–31
 phase at edge of, 431
Phase measurement, 329, 330–32, 349
 in AOA systems with multiple antennas, 520–21
 application to coherent Doppler radar, 349–52
 digital, 330–32
 FFT results comparison, 333–34
 from first/last data points, 333–34
 for obtaining fine frequency, 334, 335
 scheme application, 334–35
 wideband AOA system, 527
Phase relation, 137
 IFM receiver, 140
 in sub-Nyquist sampling, 140–43
Phase shift keying (PSK) modulation, 329
Phase shifts, 142
 negative, 141
 positive, 141
Polyphase coded signals, 11
Polyphase filters, 378
 channelization through, 388–90
 individual, illustrated, 391
 operation of, 390
 time resolution by, 390
Power spectrum, 133
Prediction error power, 456
Prediction model, 450
Preliminary tests, 547–49
Probability density
 of envelope, 288
 in frequency domain, 312
 of noise, 288
Probability density function, 293, 312
 combined, 293
 of sum of random variables, 293
 of sum samples with square law detector, 294–96
Probability of detection
 90% criterion, 301, 318
 determining, 296, 303
 in frequency domain, 312, 314–17
 from multiple samples, 292
 for signal conditions, 315
 for single-sample detection, 289–90
 S/N vs., 299–300
 See also Detection
Probability of false alarm, 287–88, 291, 300
 calculation, 311, 317
 determining, 296
 in frequency domain, 311–12
 one data sample, 287–88
 representation, 291
 See also False alarms
Prony's method, 470–75
 defined, 470–71
 deriving, 470
 least squares, 475–77
 matrix form of equation, 474
 origination, 452
 results, 471–72
 S/N and, 475
 steps, 475
 through *z* transform, 475
 See also High-resolution spectrum estimation
Pulse amplitude (PA), 13, 554–55
 data resolution, 28
 measurement, 33
Pulse descriptor words (PDWs), 8, 10, 285
 passing, 18
 TOA readings in, 18
 typical format, 18
Pulse Doppler radar, 70
Pulse parameters, 14
Pulse repetition frequency (PRF), 54, 555
Pulse repetition interval (PRI), 11, 555
 agile capability, 31
 generation, 34
 of pulse train, 65
Pulse width (PW), 13
 data resolution, 28
 measurement, 553–54
 measurement accuracy, 33
 minimum, 14
 multipath problem and, 422

Quadrature phase shift, 70
 defined, 70
 Fourier transform, 76
 illustrated, 72
Quantization error, 171
Quantization levels, 300
 angle resolution and, 332–33
 probability of noise crossing, 300
 as threshold, 302–3
Quantization noise, 160
 as ADC parameter, 230
 ideal ADC, 166–67

Quantization steps, 179–80
Queuing concept, 510–12
  in AOA system, 511
  defined, 510
Queuing receiver, 29

Radar identification, 34
Radio frequency (RF)
  amplifiers, 4
  bandwidth, 222
  components, 19
  front end, 20
  input bandwidth, 209
  input signal, 2, 10, 22
  pulse train, 69, 76
  See also RF chain
Rayleigh distribution, 298
Real zero crossings, generating, 354–56
Receiver tests, 539–64
  anechoic chamber, 541, 546–47, 561
  AOA accuracy, 554–55
  AOA resolution, 560–62
  false alarm, 552
  field, 563–64
  introduction to, 539–40
  laboratory, 540–41, 542–44, 545–46
  one and two-signal, 544
  preliminary, 547–49
  simulator, 541–42, 544–45, 562–63
  single-signal, 543, 548
  software simulation, 544–45
  TOA, 555–57
  two-signal, 543, 548
  two-signal frequency resolution, 558
  two-signal spurious free dynamic range, 558, 559
  types of, 540–42
Rectangular function, 40, 41, 74, 81
Rectangular window, 89–90
  FFT of, 373
  frequency components from FFT of, 122
  frequency domain, 90
  frequency response after shift, 91
  frequency response near peak, 92
  peak position estimation for, 119–22
  response of, 89–92
  time domain, 89
  use, 89
  See also Window functions
Resolution bandwidth, 235
Revisiting, 34–35
RF chain
  arrangement, 405
  design in wideband receiver, 403
  frequency plan of, 406
  monobit receiver, 403–8
  photograph, 406
  required gain of, 408
  wideband, 415
  See also Radio frequency (RF)
Rician distribution, 298

Sample-and-hold circuit, 160–61
  acquisition time, 162
  aperture time, 162
  aperture window, 160
  illustrated, 161
  sample mode, 161
  sampling window, 182
  time domain response, 162
Sampling
  bandwidth, 143
  frequency, 142, 169, 266, 272, 393
  importance, 552
  Nyquist, 25, 139, 271, 399, 434
  period, 266
  periodic property of, 266
  sub-Nyquist, 137–46
Sampling rates
  ADC, 287
  critical, 440
  input, 388, 389
  Nyquist, 86, 139, 271
  output, 372, 387–88
  sub-Nyquist sampling and, 145
Sampling window
  jittering effect, 182–86
  sample-and-hold circuit, 182
Scaling, 47–48, 71
Sensitivity, 13
  analog vs. digital receivers comparison, 222
  comparison at threshold levels, 306
  defined, 27
  frequency domain detection and, 318–19
  as function of frequency, 554
  measurement at one frequency value, 553
  single-signal dynamic range and, 552–54
  tangential, 222
  wideband system receiver, 220
Shadow time, 557–58
  defined, 28, 557
  measurement, 558
  PW dependence, 557–58
  short pulse, 557

Sidelobes
Gaussian window and, 91
Hamming window and, 93
spatial Fourier transform, 523
as spurs, 172
two-dimensional Fourier transform, 527
Sigma-delta ADC, 155, 157–60
defined, 157–58
functional diagram, 158
lowpass digital filter, 158, 160
noise spectrum density, 161
output, 159, 160
parts, 158
sigma-delta modulator, 158, 159
uses, 158
See also Analog-to-digital converters (ADCs)
Signal digitization, 77–78
Signal processing, data length, 35–36
Signal-to-noise and distortion (SINAD) ratio, 200
effective number of bits obtained from, 209
effective number of bits vs. input frequency, 211
expressed in decibels, 209
Signal-to-noise ratio (S/N), 167, 185, 186
8-bit ADC, 200
effective number of bits obtained from, 209
effective number of bits vs. input frequency, 210
expressed in decibels, 209
in frequency domain, 315, 316
frequency vs., for ADC, 187
input frequency range vs., 203–6
for *L*-out-of-*N* method, 318
L value vs., 307
obtaining, 230–31
probability of detection vs., 299–300
in time domain, 315, 316, 318
Sign function, 70
Simplified decimation method, 150–52
advantages, 151
defined, 150–51
disadvantages, 151–52
illustrated, 151
See also Decimation
Simulator tests, 541–42, 544–45, 562–63
defined, 541
EW receiver, 563
purpose, 562
qualitative result, 563
scenarios, 563
See also Receiver tests
Simultaneous signals
amplitude, 335–36
analysis, 335–38
detection scheme, 335–36
monobit receiver performance on, 416
Sinc function, 40–41
Sine curve fitting, 188–89
Sine function, Fourier transform, 45, 46, 59, 74, 264
Sine wave
amplitude, 187
frequency components from FFT of, 128
half-cycle, 187–88
histogram of, 189
leakage effect, 85
one bit quantizer and, 167
pure, generation, 186–87
sampled at five points, 273
Single-frequency zero crossing method, 340–43
experimental results, 347–49
ill condition, 343–44
simplified, 344–47
See also Zero crossings
Single-sample detection, 289–90
example, 297–99
probability of detection for, 289–90
See also Detection
Single-signal test, 543, 548, 549–51
frequency accuracy, 549–51
frequency precision, 551
spur-free dynamic range, 201–2
See also Receiver tests
Sliding DFT, 102–4
defined, 103
generalization, 104
illustrated, 103
real-time, 104
See also Discrete Fourier transform
Space domain, 524
discrete data in, 527
Fourier transform over, 521–24
Spatial Fourier transform, 521–24
outputs, 524–25
peaks, 523
sidelobes, 523
Spectral splitting, 467
Spiral antennas, 530
Spur chart, 252
Spurious free dynamic range
defined, 28
test, 559, 560
two-signal, 225, 558, 559
See also Dynamic range

Spurious responses, 168, 169–72
not identified as probable harmonics, 200
representation, 173
Spurs
amplitudes, 176–79
amplitudes analysis, 172–76
caused by quantization error, 171
defined, 168
FFT length and, 171
generation, 172
higher-order, 253
highest level vs. number of bits, 175
noise and, 180, 181
sidelobes as, 172
Square law detector, 294–96
mathematical analysis, 294
probability density function from, 295
Step function, 70, 71
Sub-Nyquist sampling, 137–46
defined, 137
digitized output processing, 138
"holes," 145
with *I* and *Q* channels, 146
IFM receiver vs., 137
illustrated, 138
input band vs. output band, 139
input signal, 138
phase relation in, 137, 140–43
phase vs. frequency, 141, 142
problems and potential solutions, 143–46
with two sampling rates, 145
Summation method, 306–8
defined, 306–7
L-out-of-*N* method comparison, 307
Summing outputs, 439

Tektronix TKAD20C, 237, 239
Third-order intercept points, 219, 223–25
accuracy, 225
amplifier/ADC interface, 233, 234
amplifier chain, 226
calculation, 226
defined, 223
experimental amplifier chain setup, 241, 242
highest, 230
illustrated, 224
obtaining, 223
Third-order intermodulation, 200, 207
defined, 200, 223
dynamic range, 200
illustrated, 207
products, 223, 224
Three-element phase interferometric system, 533
Threshold, 317
false alarm, 412
first quantization level as, 302
level selection, 302–4
from numerical integration equation, 306–7
preliminary, 412
second quantization level as, 302, 304
selection, 412–13
selection optimization, 304–6
sensitivity comparison, 306
Throughput rate
defined, 28, 557
test, 557
Time domain
ADC outputs, 176, 177, 178
ADC outputs with input frequency between bins, 181
ADC outputs with many levels missing, 180
ADC outputs with no levels missing, 179
convolution in, 113
cosine window, 95
data string in, 367
detection, 286–87
FFT in, 528
Fourier transform in, 524
Gaussian window, 93
general cosine window, 64
Hamming window, 97, 375
Hanning window, 123
Hilbert transform in, 262–63
noise alone in, 309
plot of two signals at same amplitude, 241
rectangular function, 121
rectangular window, 89
response of Parks-McClellan window, 386
S/N, 315, 316, 318
transient effect and, 88–89
See also Frequency domain
Time of arrival (TOA), 13, 390
condition reported in reverse order, 18–19
data resolution, 28, 366
difference, 31
readings in PDWs, 18
test, 555–57
See also Angle of arrival (AOA)
Time shift, 48–49
defined, 48
example, 48–49
summary, 73
See also Fourier transform

Toeplitz matrix, 455–56
Tracking, 34
Transfer function, 450, 451
    ADC, 164
    Hilbert, 266
Triangular window, 134
Two-dimensional Fourier transform, 524–27
    with continuous/discrete data, 527
    discrete, 527
    of input data, 524
    sidelobes, 527
    See also Fourier transform
Two-element phase array antenna, 517–19
    illustrated, 517
    zero crossing, 520
Two-signal tests, 543, 548
    frequency resolution, 558
    spurious free dynamic range, 558, 559
    See also Receiver tests
Two-threshold arrangement, 413, 414
Two-tone spur-free dynamic range, 200, 208
    defined, 200
    illustrated, 208

Uncontrolled field test, 563, 564

Weighting function, 384–87
White Gaussian noise, 477
Wideband RF delay lines, 511
Window functions, 87–93, 141
    256-point, 388
    Burg method, 466
    cosine window, 91–92
    Gaussian window, 90–91
    Hamming window, 92–93
    rectangular window, 89–90
Windowing, 268
Wright Laboratory (WL), 2–3

Yule-Walker equations, 452–55
    defined, 454
    from LMS approach, 454–55
    matrix form, 454
    obtaining, 454–55, 462

Zero crossings
    advantages, 352
    AOA measurement through, 519–20
    coefficient calculation, 356–59
    data point before, 343
    error frequency calculated from, 347–50
    for general frequency determination, 352
    limitation, 341
    measurement accuracy, 354
    of multiple signals, 355–56
    nth, 348
    real, 354–56
    single-frequency measurement from, 340–43
    spectrum analysis definition, 353–54
    spectrum analyzer, 359–60
    straight-line approximation, worst error, 345–46
    straight-line approximation illustration, 345
    three points selected to calculate, 342
    time, 353, 357
    time calculation, 344
    total, 355
    true time, 345
Zero overlapping approach, 373–74
Zero padding, 111–13, 119, 367
    defined, 111
    effect of, 113, 114
    interpolation, 378
    peak frequency component selection and, 113
    scheme, 90

Recent Titles in the Artech House Radar Library

David K. Barton, Series Editor

Advanced Techniques for Digital Receivers, Phillip E. Pace

Airborne Pulsed Doppler Radar, Second Edition, Guy V. Morris and Linda Harkness, editors

Bayesian Multiple Target Tracking, Lawrence D. Stone, Carl A. Barlow, and Thomas L. Corwin

CRISP: Complex Radar Image and Signal Processing, Software and User's Manual, August W. Rihaczek, et al.

Design and Analysis of Modern Tracking Systems, Samuel Blackman and Robert Popoli

Digital Techniques for Wideband Receivers, Second Edition, James Tsui

Electronic Intelligence: The Analysis of Radar Signals, Second Edition, Richard G. Wiley

Electronic Warfare in the Information Age, D. Curtis Schleher

EW 101: A First Course in Electronic Warfare, David Adamy

Fundamentals of Electronic Warfare, Sergei A. Vakin, Lev N. Shustov, and Robert H. Dunwell

High-Resolution Radar, Second Edition, Donald R. Wehner

Introduction to Electronic Warfare, D. Curtis Schleher

Microwave Radar: Imaging and Advanced Concepts, Roger J. Sullivan

Millimeter-Wave and Infrared Multisensor Design and Signal Processing, Lawrence A. Klein

Modern Radar System Analysis, David K. Barton

Multitarget-Multisensor Tracking: Applications and Advances Volume III, Yaakov Bar-Shalom and William Dale Blair, editors

Principles of High-Resolution Radar, August W. Rihaczek

Radar Cross Section, Second Edition, Eugene F. Knott, et al.

Radar Evaluation Handbook, David K. Barton, et al.

Radar Meteorology, Henri Sauvageot

Radar Reflectivity of Land and Sea, Third Edition, Maurice W. Long

Radar Resolution and Complex-Image Analysis, August W. Rihaczek and Stephen J. Hershkowitz

Radar Signal Processing and Adaptive Systems, Ramon Nitzberg

Radar System Performance Modeling, G. Richard Curry

Radar Technology Encyclopedia, David K. Barton and Sergey A. Leonov, editors

Range-Doppler Radar Imaging and Motion Compensation, Jae Sok Son, et al.

Theory and Practice of Radar Target Identification, August W. Rihaczek and Stephen J. Hershkowitz

For further information on these and other Artech House titles, including previously considered out-of-print books now available through our In-Print-Forever® (IPF®) program, contact:

Artech House
685 Canton Street
Norwood, MA 02062
Phone: 781-769-9750
Fax: 781-769-6334
e-mail: artech@artechhouse.com

Artech House
46 Gillingham Street
London SW1V 1AH UK
Phone: +44 (0)20 7596-8750
Fax: +44 (0)20 7630-0166
e-mail: artech-uk@artechhouse.com

www.ingramcontent.com/pod-product-compliance
Lightning Source LLC
LaVergne TN
LVHW101450200926
848483LV00001B/106
* 9 7 8 1 5 8 0 5 3 2 9 9 0 *